# CONDENSED INSTRUCTIONS ON THE USE OF VOLUME 5*

EXAMPLE

Obtain references on the specific heat of a mixture of helium and nitrogen in the liquid state.

## STEP 1

a. **Select** volume containing material of interest.

b. Scan the *Table of Contents;* note also that **instructions** on use are given at the beginning of each part.

a. Mixtures are indexed in *Volume 5* (Mixtures and Solutions).



## STEP 2

a. Locate the **names of the components** of the mixture in Part A, the *Materials Directory* (see example). Note the **properties** listed.

b. Record the seven-digit **substance numbers** for the components.

a. In Class 512 (Components of Nonaqueous Binary Mixtures), on pages A7 and A10, locate the following entries:

| Substance Name | Property | Number |
|---|---|---|
| Helium | aefl | 512-0061 |
| Nitrogen | aefl | 512-0106 |

The property of interest e (specific heat) is listed for both components.

b. The seven-digit number for helium is 512-0061. For nitrogen the number is 512-0106.

## STEP 3

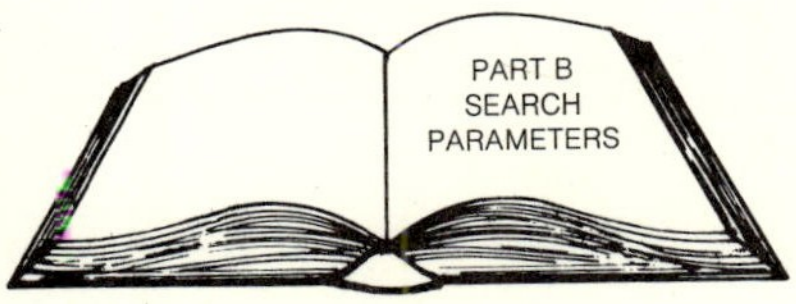


a. Select **property chapter** of interest, and locate the seven-digit **substance numbers** (obtained from Part A, the *Materials Directory*).

b. Use the five **search parameters** (physical state, subject, language, temperature, and year) to locate and record **TPRC numbers** of interest.

a. Locate property chapter Specific Heat (Chapter 3, page B7). The seven-digit number 512-0061 is listed on page B7, column 2; 512-0106 is listed on page B7, column 3.

b. For the parameter of interest (liquid state) locate the following entry:

| Substance Number | Phys. State | Subject | Language | Temperature | Year | TPRC Number |
|---|---|---|---|---|---|---|
| 512-0061 | L | C | E | L | 1971 | 74103 |

Check to see if the **same** TPRC number is also listed for the seven-digit number 512-0106:

| | | | | | | |
|---|---|---|---|---|---|---|
| 512-0106 | L | C | E | L | 1971 | 74103 |

## STEP 4

a. Locate **TPRC numbers** of interest (recorded from Part B, the *Search Parameters*) to obtain complete **bibliographic citation** information of documents for the material and properties of interest.

a. Page C43, column 2, has the following citation for TPRC number 74103:

74103 THE CALCULAT... FUGACITIES FROM VOLUMETRIC AND
PHASE EQUILI...
GIBBS-DUHE...
BARIEAU, R. E...
IND. ENG. CHI...
10 (3), 428-...



## NOTE

a. Locate author's name in the *Author Index* to obtain TPRC numbers of all documents written by this author and reported in this volume.

a. To obtain TPRC numbers ... lished by author R. E. Barieau, turn to page D1, column 3, and locate the following entry:

BARIEAU R E 74103

The complete bibliographic citation for the document is given on page C43, column 2.

*For additional details on the use of this volume, see the introductory remarks for Parts A, B, C, and D, scan the *Table of Contents*, and also note the *Condensed Materials Index* on the opposite page.

# thermophysical properties research literature retrieval guide

## supplement II
1971-1977

### volume 5
mixtures and solutions

---

Editors

J. KOOLHAAS GERRITSEN

V. RAMDAS

T. M. PUTNAM

# thermophysical properties research literature retrieval guide

## supplement II

1971-1977

A Comprehensive Compilation of Scientific and Technical Literature by the Center for Information and Numerical Data Analysis and Synthesis (CINDAS), Purdue University

*Volume 1. Elements and Inorganic Compounds*

*Volume 2. Organic Compounds and Polymeric Materials*

*Volume 3. Alloys, Intermetallic Compounds, and Cermets*

*Volume 4. Oxide Mixtures and Minerals*

*Volume 5. Mixtures and Solutions*

*Volume 6. Coatings, Systems, and Composites*

See inside front and back covers for CONDENSED MATERIALS INDEX to the SIX-VOLUME RETRIEVAL GUIDE SUPPLEMENT II.

New literature on thermophysical properties is being constantly accumulated at CINDAS. Contact CINDAS and use its interim updating search services for the most current scientific information

# thermophysical properties research literature retrieval guide

## supplement II
## 1971-1977

## volume 5
## mixtures and solutions

---

Editors

J. KOOLHAAS GERRITSEN

V. RAMDAS

T. M. PUTNAM

PART A. MATERIALS DIRECTORY
PART B. SEARCH PARAMETERS
PART C. BIBLIOGRAPHY
PART D. AUTHOR INDEX

IFI/PLENUM · NEW YORK · WASHINGTON · LONDON

Library of Congress Cataloging in Publication Data

Purdue University, Lafayette, Ind. Center for Information and Numerical Data Analysis and Synthesis (CINDAS).

Thermophysical properties research literature retrieval guide, Supplement II, 1971-1977.

Includes indexes.

CONTENTS: v. 1. Elements and inorganic compounds.—v. 2. Organic compounds and polymeric materials.—v. 3. Alloys, intermetallic compounds, and cermets. [etc.]

1. Materials—Thermal properties—Bibliography. I. Gerritsen, Jacqueline Koolhaas, 1912- II. Ramdas, Vasanthalakshmi. III. Putnam, Thomas Milton. IV. Title.

Z5853.M38P82 1967, suppl. 2 [TA418.52] 016.6201'121 79-16324

ISBN 0-306-67210-3





IFI/Plenum Data Company
A Division of Plenum Publishing Corporation
227 West 17th Street, New York, N.Y. 10011



Printed in the United States of America

# CONTENTS OF VOLUME 5

# PREFACE

The phenomenal growth of science and technology has brought about a universal appreciation of the fact that present limitations in many technical developments are often a direct result of the paucity of knowledge on material properties. Engineering developments in the years ahead will be closely linked to the research that is done today to contribute to a better understanding of the properties of matter, of which thermophysical properties constitute a major segment.

While research on the properties of materials continues, adequate steps are not being taken to ensure that this invaluable body of information is coordinated, synthesized, organized, and disseminated to the ultimate user, namely, the individual scientist and engineer.

It is generally agreed that the present level of research support on thermophysical properties of matter falls short of existing needs and anticipated future demands; but what is even more disturbing is the fact that engineering groups across the nation are using no more than a fraction of the information already available, either because it is in a form not directly useful to them or, often, because its existence is not generally known. As a result, such information remains buried in the world's scientific literature. The repercussions of this latter condition are indeed serious since it leads to unintentional duplication of research effort with the resultant waste of time and scientific manpower.

In conjunction with its research activities, CINDAS screens the world's literature and collects published information on a wide range of materials in the field of thermophysics. This information concerns data, theoretical estimation methods, and experimental measurement techniques. Technical papers come from journals, abstracting services, reports, doctoral dissertations, masters theses, and many other sources. Full evaluation and analysis of the collected raw data are needed before publications on recommended values can be prepared. Such an effort is obviously time consuming and expensive, and therefore this critical evaluation is currently performed at a rather modest funding level. The end result is that much of the available world literature is not being processed and distilled.

As a complementary effort to its Data Tables Series, CINDAS published in 1967 a work entitled "Thermophysical Properties Research Literature Retrieval Guide." This three-book work reported 33,700 references on seven thermophysical property groups and about 45,000 materials. This Basic Edition systematically covered the world's unclassified literature published essentially between 1920 and mid-1964, in many instances going much earlier. Then, in 1973, Supplement I to the Basic Edition was published, with 26,000 additional references on sixteen thermophysical properties of 20,000 materials, covering the years mid-1964 to 1970.

The present work, referred to as Supplement II to the Basic Edition, reports over 15,000 new references on fourteen thermophysical properties of 12,674 substances. It essentially covers the years 1971 to 1977, going beyond these dates in a few instances. In the materials listing an additional 8000 synonyms and trade names have been incorporated to assist the user in identifying the material or substance of interest.

Supplement II follows the same format of presentation as Supplement I and the Basic Edition. Its structure is identical to that of Supplement I, viz., a series of six Retrieval Guides for the various materials classes.

It is sincerely hoped that the Basic Edition (1967), Supplement I (1973), and Supplement II (1979) of the "Thermophysical Properties Research Literature Retrieval Guide" will constitute a permanent and valuable contribution to science and technology as well as to scientific documentation. These volumes, and those to follow, should prove to be an invaluable source of information to every scientist and engineer, with a scope of knowledge humanly impossible to master for any one individual or any group of individuals. Perhaps even more important, it is hoped that a wealth of information, heretofore unknown, will have been made available to many, including the specialist. The CINDAS staff is most anxious to receive comments, suggestions, and criticism from all users of these volumes. All communications will be gratefully appreciated. Specific information concerning CINDAS' operations, services, publications, and research activities can be obtained by communicating with the Director.

The preparation of these volumes was made possible through the collective financial support received from a large number of governmental, industrial, and nonprofit research organizations. Their interest and support are gratefully acknowledged. I wish to specifically acknowledge the major support received from the Department of Defense in connection with the operation of the Thermophysical and Electronic Properties Information Analysis Center (TEPIAC), which has made this work possible.

In closing, I wish to acknowledge the individual and collective accomplishments of CINDAS' Scientific Documentation Division: Mrs. C. R. Rodriguez and Mr. G. Kvakovszky, Technical Coding; Mrs. M. R. Troyer, Documentalist; Mrs. F. L. Frist, Mrs. C. Greene, and Mrs. G. M. Halstead, Computer Operations; Mr. C. C. Hsiao, Computer Programmer; Mrs. S. J. Creamer and Mrs. J. A. Brittingham, Library. Special thanks to Mr. J. F. Chaney for helpful suggestions and assistance. Special thanks are also extended to the staff of R. R. Donnelley and Sons, Chicago, Illinois, for their cooperation and the computer-assisted phototypesetting of this new Retrieval Guide.

Y. S. Touloukian

Director, Center for Information and Numerical Data Analysis and Synthesis
Distinguished Atkins Professor of Engineering

# Part A
# MATERIALS DIRECTORY

PART A

# USE OF MATERIALS DIRECTORY

## I. ORGANIZATION AND GENERAL CONSIDERATIONS

The organization of thermophysical properties information at CINDAS is by substance/material. A *Condensed Materials Index* to the six-volume Retrieval Guide Supplement is located on the inside front and back covers of this volume.

In order to index the world literature the classification system must be systematic, and yet flexible. Therefore, CINDAS has adopted a classification scheme which arranges materials into logical groups that have closely related chemical composition. However, certain materials do not lend themselves to a purely chemical classification and a more logical method is to classify them instead into compatible groups by their physical form and/or by their use and application. These materials are normally of engineering interest and consist primarily of what is referred to as systems and composites.

Within this system of classification by substance name, there exists a class of pertinent information for which no specific substance name would be appropriate, e.g., "Theory of the Thermal Conductivity of Gases" or "New Technique for the Viscosity Measurement of Liquids" or "Emissivity of a Black-Body Cavity." Literature covering this class is not reported in this series but is available at CINDAS. A special computer search and retrieval can be made upon request.

The index to materials in this volume, arranged alphabetically, is divided into four groups:

Components of Nonaqueous Binary Mixtures—Class 512
Components of Nonaqueous Multiple Mixtures—Class 514
Components of Aqueous Binary Mixtures—Class 515
Components of Aqueous Multiple Mixtures—Class 517

Synonyms, trade names, and cross-references have been incorporated in the *Materials Directory*. A cross-reference is preceded by the words "See Also" and a synonym or trade name is followed by the word "See." An example on the use of this volume is given on the inside of the front and back covers.

The user of this index should not be concerned with the structure of the seven-digit number associated with each material. These numbers uniquely identify a material, and their structure is of significance only to CINDAS.

## II. DEFINITIONS AND LIMITATIONS USED IN MATERIAL CLASSIFICATION

For the effective use of the material index of this volume certain definitions and limitations of terms accepted by CINDAS should be understood. These are listed below:

1. *Impurities*—For the purposes of classification, CINDAS defines the "impurity" limits as follows:

   *Elements.* Total of impurities must be <0.5 percent and individual impurity <0.2 percent by weight. A metallic element having impurities equal to or in excess of these limits is considered an alloy whereas a nonmetallic element is considered a mixture.

   In CINDAS' classification scheme, arsenic, boron, selenium, silicon, and tellurium are considered metals.

   *Compounds.* Total of impurities must be <5.0 percent and individual impurity <2.0 percent by weight. A compound with impurities equal to or in excess of these limits is considered a mixture.

2. *Doping*—Doped materials are indicated only by the physical state letter D in Part B of this directory. The dopant is not specified.
3. *Isotopes*—Isotopes are listed under the corresponding elements as a single entry. Exceptions are the isotopes of hydrogen and helium, which are designated separately as deuterium, tritium, helium-3, and helium-4.
4. *Names of Chemical Compounds*—CINDAS has based the naming of compounds on the cumulative indices issued by the Chemical Abstracts Service.
5. Mixtures and solutions which do not belong in more specific groups are found here. (Mixtures of metallic oxides are located in Volume 4; mixtures involving metals are in Volume 3; aggregates and systems are in Volume 6.)

## III. EXCLUSIONS TO MATERIAL AND PROPERTY COVERAGE

While CINDAS attempts to cover the world research literature on all matter for the fourteen thermophysical properties it monitors, for reasons of scientific and technical rationale and practical expediency it has become necessary to put a number of constraints on certain classes of materials in the coverage of specific properties.

### 1. GENERAL CONSTRAINTS

a. Nonoriginal papers, patents, promotional literature, and unidentified product catalogs are excluded. However, review articles, major handbooks, and data compilations are included.

b. Data reported under unsteady-state and nonequilibrium conditions are excluded.

c. Data reported in arbitrary units or relative ratio without specification of reference used are excluded.

d. Data on all polymeric mixtures and solutions of undefined character are excluded.

### 2. SPECIFIC CONTRAINTS RELATIVE TO CERTAIN PROPERTIES

| | *Property* | *Constraint* |
|---|---|---|
| a. | Viscosity | Data for slags and scales, ceramics, glasses, polymers, oils, and suspensions are excluded. |
| b. | Emittance and reflectance | Data outside the wavelength range of 10 to $2 \times 10^5$ $cm^{-1}$ or 0.5 to 1000 $\mu m$ or $1.24 \times 10^{-3}$ to 25 eV are excluded. |
| c. | Absorptance, transmittance, and absorptance-to-emittance ratio | Data outside the same wavelength range as above and data for liquids, gases, organic compounds, and inorganic complexes are included. |
| d. | Thermal linear and volumetric expansion | Data for all liquids, gases, and irreversible processes are excluded. |

## IV. USEFUL REFERENCES FOR MATERIALS IDENTIFICATION

The latest editions of the following selected references have been used by CINDAS in the classification and identification of materials:

1. *Chemical Abstracts—Subject Index and Formula Index* (annual), Chemical Abstracts Service, American Chemical Society, Columbus, Ohio.
2. *Chemical Synonyms and Trade Names: A Dictionary and Commercial Handbook,* W. Gardner and E. I. Cooke, CRC Press, Cleveland, Ohio.
3. *The Condensed Chemical Dictionary,* G. G. Hawley, Van Nostrand Reinhold Co., New York, New York.
4. *Handbook of Material Trade Names,* O. T. Zimmerman and I. Lavine, Industrial Research Service, Inc., Dover, N.H.
5. *The Ring Index,* A. M. Patterson, L. T. Capell, and D. F. Walker, Chemical Abstracts Service, American Chemical Society, Columbus, Ohio.
6. *Synthetic Organic Chemical Manufacturers Association Handbook: Commercial Organic Chemical Names,* Chemical Abstracts Service, American Chemical Society, Columbus, Ohio.
7. *The Merck Index of Chemicals and Drugs,* Merck and Co., Inc., Rahway, N.J.
8. *Webster's New International Dictionary, Unabridged,* G. & C. Merriam Co., Springfield, Mass.
9. *Glossary of Geology,* M. Gary, R. McAfee, Jr., and C. L. Wolf, American Geological Institute, Falls Church, Virginia.

# VOLUME 5. MIXTURES AND SOLUTIONS

## COMPONENTS OF NONAQUEOUS BINARY MIXTURES CLASS 512

| Substance Name | Property | Number |
|---|---|---|
| Acanthite | | |
| see Silver Sulfide | en | 512-0050 |
| Acetic Acid | aef | 512-0189 |
| Acetic Acid Dimethylamide | | |
| see N, N- Dimethylacetamide | f | 512-0207 |
| Acetic Anhydride | af | 512-0880 |
| Acetic Ester | | |
| see Ethyl Acetate | f | 512-0086 |
| Acetic Ether | | |
| see Ethyl Acetate | f | 512-0086 |
| Acetomethylamide | | |
| see N- Methylacetamide | f | 512-0154 |
| Acetone | aef | 512-0136 |
| Acetonitrile | aef | 512-0108 |
| Acetophenone | f | 512-0217 |
| Acetylbenzene | | |
| see Acetophenone | f | 512-0217 |
| Acrylic Acid | a | 512-0909 |
| Acrylonitrile | e | 512-1074 |
| Adeps | | |
| see Fat, Pig | a | 512-0341 |
| Adiponitrile | f | 512-0218 |
| A D P | | |
| see Ammonium Phosphate $NH_4H_2PO_4$ | f | 512-0647 |
| Aerosil Powder | | |
| see Silicon Oxide $SiO_2$ | aj | 512-0683 |
| Agate | | |
| see Silicon Oxide $SiO_2$ | aj | 512-0683 |
| Alcan C X-30 | | |
| see Aluminum Oxide | afn | 512-0433 |
| Alcohol | | |
| see Ethanol | ef | 512-0042 |
| Alcohol C-7 | | |
| see 1- Heptanol | a | 512-0319 |
| Alkane | af | 512-0911 |
| Allyl Isosulfocyanate | | |
| see Allyl Isothiocyanate | f | 512-0160 |
| Allyl Isothiocyanate | f | 512-0160 |
| Allyl Mustard Oil | | |
| see Allyl Isothiocyanate | f | 512-0160 |
| Allyl Rhodanide | | |
| see Allyl Thiocyanate | f | 512-0161 |
| Allyl Sulfocyanate | | |
| see Allyl Thiocyanate | f | 512-0161 |
| Allyl Thiocyanate | f | 512-0161 |
| Alon - Cabot | | |
| see Aluminum Oxide | afn | 512-0433 |
| Alon C | | |
| see Silicon Oxide $SiO_2$ | aj | 512-0683 |
| Alumina | | |
| see Aluminum Oxide | afn | 512-0433 |
| Aluminum Fluoride | j | 512-0435 |
| Aluminum Hydroxide | j | 512-0875 |
| Aluminum Lithium Bromide | f | 512-0399 |
| Aluminum Oxide | afn | 512-0433 |
| Aluminum Potassium Bromide | f | 512-0401 |
| Aluminum Sodium Bromide | f | 512-0400 |
| Aluminum Sodium Fluoride | | |
| see Sodium Hexafluoroaluminate | f | 512-0380 |
| Aluminum Sulfide | j | 512-0964 |
| Alundum | | |
| see Aluminum Oxide | afn | 512-0433 |
| Amchlor | | |
| see Ammonium Chloride | f | 512-0600 |
| American Ashes | | |
| see Potassium Carbonate | f | 512-0065 |
| Amersil | | |
| see Silicon Oxide $SiO_2$ | aj | 512-0683 |
| Aminobenzene | | |
| see Aniline | aef | 512-0084 |
| p- Aminochlorobenzene | | |
| see p- Chloroaniline | f | 512-0335 |
| 1- Amino-2,5-dimethylbenzene | | |
| see 2,5- Xylidine | f | 512-0360 |
| 1- Amino-2,6-dimethylbenzene | | |
| see 2,6- Xylidine | f | 512-0359 |
| 1- Amino-3,4-dimethylbenzene | | |
| see 3,4- Xylidine | f | 512-0361 |
| 2- Aminoethanol | f | 512-0305 |
| β- Aminoethyl Alcohol | | |
| see 2- Aminoethanol | f | 512-0305 |
| Aminomethane | | |
| see Methylamine | af | 512-0110 |
| 1- Amino-2-nitrobenzene | | |
| see o- Nitroaniline | f | 512-0272 |
| p- Aminotoluene | | |
| see p- Toluidine | f | 512-0271 |
| Ammonia | aefj | 512-0011 |
| Ammonia, Trideuterated | afj | 512-0192 |
| Ammonium Acid Phosphate | | |
| see Ammonium Phosphate $NH_4H_2PO_4$ | f | 512-0647 |
| Ammonium Bromide | f | 512-0500 |
| Ammonium Chloride | f | 512-0600 |
| Ammonium Iodide | f | 512-0145 |
| Ammonium Muriate | | |
| see Ammonium Chloride | f | 512-0600 |
| Ammonium Nitrate | f | 512-0155 |
| Ammonium Perchlorate | a | 512-0709 |
| Ammonium Phosphate $NH_4H_2PO_4$ | f | 512-0647 |
| Ammonium Phosphate $(NH_4)_2HPO_4$ | f | 512-0766 |
| Ammonium Pyrophosphate | f | 512-0889 |
| Ammonium Sulfate | f | 512-0157 |
| Amyl Acetate | | |
| see Pentyl Acetate | f | 512-0416 |
| Amyl Acetic Ester | | |
| see Pentyl Acetate | f | 512-0416 |
| Amyl Alcohol | | |
| see 1- Pentanol | af | 512-0045 |
| pri- Amyl Alcohol | | |
| see 1- Pentanol | af | 512-0045 |
| Amyl Carbinol | | |
| see 1- Hexanol | aef | 512-0046 |
| Amyldimethylmethane | | |
| see 2- Methylheptane | ae | 512-0142 |
| β- Amylene | | |
| see cis-2- Pentene | e | 512-0410 |
| Anesthesia Ether | | |
| see Ethyl Ether | aef | 512-0134 |
| Aniline | aef | 512-0084 |
| Aniline N, N-dimethyl | | |
| see N, N- Dimethylaniline | f | 512-0385 |
| m- Anisaldehyde | f | 512-0349 |
| o- Anisaldehyde | f | 512-0348 |
| p- Anisaldehyde | f | 512-0350 |
| m- Anisic Aldehyde | | |
| see m- Anisaldehyde | f | 512-0349 |
| o- Anisic Aldehyde | | |
| see o- Anisaldehyde | f | 512-0348 |
| p- Anisic Aldehyde | | |
| see p- Anisaldehyde | f | 512-0350 |
| Anone | | |
| see Cyclohexanone | f | 512-0122 |
| Anthracene | e | 512-1071 |
| Antimony Oxide | j | 512-1004 |
| Antimony Rubidium Sulfide | a | 512-1025 |

**Property:** **a.** Thermal conductivity; **b.** Accommodation coefficient; **c.** Thermal contact resistance; **d.** Thermal diffusivity; **e.** Specific heat; **f.** Viscosity; **g.** Emittance; **h.** Reflectance; **i.** Absorptance; **j.** Transmittance; **k.** $\alpha_s/\epsilon_t$ ratio; **l.** Prandtl number; **n.** Thermal linear expansion coefficient; **o.** Thermal volumetric expansion coefficient

| Substance Name | Property | Number |
|---|---|---|
| Antimony Sulfide | aj | 512-1005 |
| A 0 600 | | |
| see Carbon | adefghn | 512-0174 |
| Aqua Fortis | | |
| see Nitric Acid | adef | 512-0311 |
| Aragonite | | |
| see Calcium Carbonate | f | 512-0156 |
| Arcton 3 | | |
| see Chlorotrifluoromethane | al | 512-0329 |
| Arcton 4 | | |
| see Chlorodifluoromethane | ae | 512-0009 |
| Arcton 6 | | |
| see Dichlorodifluoromethane | aefl | 512-0008 |
| Argentite | | |
| see Silver Sulfide | en | 512-0050 |
| Argon | aefjl | 512-0012 |
| Armco Iron | | |
| see Iron | aef | 512-0176 |
| Arsenic Sulfide | n | 512-0418 |
| Arsenic Sulfide | j | 512-0960 |
| Arsenic Sulfide, Nonstoichiometric | j | 512-0962 |
| Associated Substance | a | 512-0409 |
| Astroquartz | | |
| see Silicon Oxide Si $O_2$ | aj | 512-0683 |
| Atomite | | |
| see Calcium Carbonate | f | 512-0156 |
| Aubepine | | |
| see p- Anisaldehyde | f | 512-0350 |
| Aurora Yellow | | |
| see Cadmium Sulfide | h | 512-0422 |
| 3 β- Chlorocholest-5-ene | f | 512-0366 |
| Banana Oil | | |
| see Pentyl Acetate | f | 512-0416 |
| Barite | | |
| see Barium Sulfate | adej | 512-0139 |
| Barium Chloride | f | 512-0089 |
| Barium Fluoride | hjn | 512-0240 |
| Barium Nitrate | f | 512-0453 |
| Barium Oxide | ag | 512-0778 |
| Barium Sulfate | adej | 512-0139 |
| Barium Tetracyanonickelate | h | 512-0704 |
| Barium Tetracyanopalladate | h | 512-0703 |
| Barium Tetracyanoplatinate | h | 512-0702 |
| Barytes | | |
| see Barium Sulfate | adej | 512-0139 |
| Battery Acid | | |
| see Sulfuric Acid | f | 512-0644 |
| 1- Benzazine | | |
| see Quinoline | f | 512-0206 |
| 2- Benzazine | | |
| see Isoquinoline | f | 512-0346 |
| Benzene | adef | 512-0006 |
| Benzene, Hexadeuterated | f | 512-0164 |
| 1,2- Benzenediamine | | |
| see o- Phenylenediamine | f | 512-0339 |
| 1,4- Benzene Dibromide | | |
| see p- Dibromobenzene | f | 512-0334 |
| 1,2- Benzenediol | | |
| see Pyrocatechol | f | 512-0332 |
| 1,3- Benzenediol | | |
| see Resorcinol | f | 512-0333 |
| Benzo[B]pyridine | | |
| see Quinoline | f | 512-0206 |
| Benzo[C]pyridine | | |
| see Isoquinoline | f | 512-0346 |
| Benzoic Acid | f | 512-0724 |
| Benzol | | |
| see Benzene | adef | 512-0006 |
| Benzole | | |
| see Benzene | adef | 512-0006 |
| Benzophenone | f | 512-0123 |
| Benzoylbenzene | | |
| see Benzophenone | f | 512-0123 |

| Substance Name | Property | Number |
|---|---|---|
| Beryllium Fluoride | ejn | 512-0454 |
| Beryllium Oxide | an | 512-1009 |
| Bicarburetted Hydrogen | | |
| see Ethylene | aef | 512-0002 |
| Bichloroacetic Acid | | |
| see Dichloroacetic Acid | f | 512-0406 |
| Bicyclo[4,4,0]decane | | |
| see Decahydronaphthalene | f | 512-0245 |
| Bihexyl | | |
| see Dodecane | e | 512-0115 |
| Bimethyl | | |
| see Ethane | af | 512-0019 |
| Biphenyl | a | 512-0315 |
| Bischofite | | |
| see Magnesium Chloride | f | 512-0387 |
| 4,4'- Bis(Hexyloxy)azoxybenzene | h | 512-0203 |
| 4,4'- Bis(2- Methylbutoxy)azoxybenzene | h | 512-0204 |
| Black Lead | | |
| see Graphite | aegin | 512-0262 |
| Blanc Fixe | | |
| see Barium Sulfate | adej | 512-0139 |
| Boralloy | | |
| see Boron Nitride | a | 512-0432 |
| Borazon | | |
| see Boron Nitride | a | 512-0432 |
| Boron | aen | 512-0287 |
| Boron Carbide Nitride | an | 512-0441 |
| Boron Fluoride | f | 512-0212 |
| Boron Fluoride - Methyl Ether Complex | f | 512-0383 |
| Boron Fluoride - Trimethyl Borate Complex | f | 512-0384 |
| Boron Nitride | a | 512-0432 |
| Brimstone | | |
| see Sulfur | ae | 512-0177 |
| Bromobenzene | adf | 512-0195 |
| Bromobenzol | | |
| see Bromobenzene | adf | 512-0195 |
| 1- Bromo-4-chlorobenzene | f | 512-0337 |
| p- Bromochlorobenzene | | |
| see 1- Bromo-4-chlorobenzene | f | 512-0337 |
| Bromoform | f | 512-0284 |
| Bromomethane | a | 512-0017 |
| m- Bromotoluene | f | 512-0424 |
| Bromyride | | |
| see Silver Bromide | e | 512-0024 |
| Butane | af | 512-0022 |
| Butanoic Acid | | |
| see Butyric Acid | af | 512-0394 |
| 1- Butanol | af | 512-0044 |
| 2- Butanol | af | 512-0214 |
| 2- Butanone | f | 512-0085 |
| 1- Butene | a | 512-0231 |
| trans-2- Butene | a | 512-0232 |
| 2- Butine | | |
| see 2- Butyne | f | 512-0163 |
| 1- Butoxybutane | | |
| see Butyl Ether | af | 512-0072 |
| Butter Of Tin | | |
| see Tin Chloride | ef | 512-0403 |
| Butter Of Zinc | | |
| see Zinc Chloride | ef | 512-0054 |
| Butyl Acetate | f | 512-0199 |
| Butyl Alcohol | | |
| see 1- Butanol | af | 512-0044 |
| sec- Butyl Alcohol | | |
| see 2- Butanol | af | 512-0214 |
| sym- Butyl Alcohol | | |
| see 2- Butanol | af | 512-0214 |
| tert- Butyl Alcohol | | |
| see 2- Methyl-2-propanol | ef | 512-0113 |
| Butylamine | f | 512-0733 |
| Butyl Carbinol | | |
| see 1- Pentanol | af | 512-0045 |
| Butyl Chloride | | |
| see 1- Chlorobutane | f | 512-0075 |

**Property: a.** Thermal conductivity; **b.** Accommodation coefficient; **c.** Thermal contact resistance; **d.** Thermal diffusivity; **e.** Specific heat; **f.** Viscosity; **g.** Emittance; **h.** Reflectance; **i.** Absorptance; **j.** Transmittance; **k.** $\alpha_s/\epsilon_t$ ratio; **l.** Prandtl number; **n.** Thermal linear expansion coefficient; **o.** Thermal volumetric expansion coefficient

**Property:** **a.** Thermal conductivity; **b.** Accommodation coefficient; **c.** Thermal contact resistance; **d.** Thermal diffusivity; **e.** Specific heat; **f.** Viscosity; **g.** Emittance; **h.** Reflectance; **i.** Absorptance; **j.** Transmittance; **k.** $\alpha_s/\epsilon_t$ ratio; **l.** Prandtl number; **n.** Thermal linear expansion coefficient; **o.** Thermal volumetric expansion coefficient

**Property: a.** Thermal conductivity; **b.** Accommodation coefficient; **c.** Thermal contact resistance; **d.** Thermal diffusivity; **e.** Specific heat; **f.** Viscosity; **g.** Emittance; **h.** Reflectance; **i.** Absorptance; **j.** Transmittance; **k.** $\alpha_s/\epsilon_t$ ratio; **l.** Prandtl number; **n.** Thermal linear expansion coefficient; **o.** Thermal volumetric expansion coefficient

| Substance Name | Property | Number |
|---|---|---|
| N, N- Dimethylaniline | f | 512-0385 |
| o- Dimethylaniline | | |
| see 2,6- Xylidine | f | 512-0359 |
| p- Dimethylaniline | | |
| see 3,4- Xylidine | f | 512-0361 |
| 1,2- Dimethylbenzene | | |
| see o- Xylene | f | 512-0236 |
| 1,3- Dimethylbenzene | | |
| see m- Xylene | f | 512-0201 |
| 1,4- Dimethylbenzene | | |
| see p- Xylene | ef | 512-0126 |
| Dimethyl Carbinol | | |
| see 2- Propanol | af | 512-0141 |
| trans-1,4- Dimethylcyclohexane | e | 512-0324 |
| Dimethyl Ether | | |
| see Methyl Ether | f | 512-0068 |
| sym- Dimethylethylene | | |
| see trans-2- Butene | a | 512-0232 |
| N, N- Dimethylformamide | f | 512-0041 |
| 1,1- Dimethylhydrazine | ae | 512-0735 |
| Dimethyl Ketone | | |
| see Acetone | aef | 512-0136 |
| Dimethylmethane | | |
| see Propane | aef | 512-0021 |
| 2,5- Dimethylnitrobenzene | | |
| see 2- Nitro-p-xylene | f | 512-0364 |
| 2,6- Dimethylnitrobenzene | | |
| see 2- Nitro-m-xylene | f | 512-0363 |
| 3,4- Dimethylnitrobenzene | | |
| see 4- Nitro-o-xylene | f | 512-0362 |
| Dimethylnitrosamine | | |
| see N- Nitrosodimethylamine | h | 512-0438 |
| Dimethylphenylene | | |
| see Xylene | f | 512-0402 |
| Dimethylpropane | af | 512-0228 |
| Dimethylpropylmethane | | |
| see 2- Methylpentane | f | 512-0168 |
| 2,5- Dimethylpyrazine | f | 512-0283 |
| Dimethyl Sulfite | | |
| see Methyl Methanesulfonate | f | 512-0197 |
| Dimethyl Sulfoxide | | |
| see Methyl Sulfoxide | ef | 512-0098 |
| Dioxane | fh | 512-0237 |
| 1,4- Dioxane | ef | 512-0127 |
| p- Dioxane | | |
| see 1,4- Dioxane | ef | 512-0127 |
| Dioxolone-2 | | |
| see Ethylene Carbonate | f | 512-0153 |
| Dioxyethylene | | |
| see 1,4- Dioxane | ef | 512-0127 |
| Dioxyethylene Glycol | | |
| see Diethylene Glycol | f | 512-0537 |
| Diphenyl | | |
| see Biphenyl | a | 512-0315 |
| Diphenyl Ketone | | |
| see Benzophenone | f | 512-0123 |
| Dipping Acid | | |
| see Sulfuric Acid | f | 512-0644 |
| Dipropyl Ether | | |
| see Propyl Ether | f | 512-0417 |
| D M A C | | |
| see N, N- Dimethylacetamide | f | 512-0207 |
| D M F | | |
| see N, N- Dimethylformamide | f | 512-0041 |
| D M S O | | |
| see Methyl Sulfoxide | ef | 512-0098 |
| Dodecahydrosqualene | | |
| see 2,6,10,15,19,23- Hexamethyltetracosane | f | 512-0321 |
| Dodecane | e | 512-0115 |
| Dodecyl Chloride | | |
| see 1- Chlorododecane | f | 512-0074 |
| Driosan | e | 512-0034 |
| Dry Ice | | |
| see Carbon Oxide C $O_2$ | aefgl | 512-0193 |
| Dutch Liquid | | |
| see 1,2- Dichloroethane | ae | 512-0358 |
| Dutch Oil | | |
| see 1,2- Dichloroethane | ae | 512-0358 |
| Dynaquartz | | |
| see Silicon Oxide Si $O_2$ | aj | 512-0683 |
| Dynasil | | |
| see Silicon Oxide Si $O_2$ | aj | 512-0683 |
| Dysprosium Ethyl Sulfate | e | 512-0267 |
| E D B | | |
| see 1,2- Dibromoethane | e | 512-0007 |
| Elayl | | |
| see Ethylene | aef | 512-0002 |
| Electrolyte | af | 512-0290 |
| E L F-0 | | |
| see Carbon | adefghn | 512-0174 |
| Emery | | |
| see Aluminum Oxide | afn | 512-0433 |
| 1,3,5- Estratrien-3-ol-17-one | | |
| see Estrone | h | 512-0353 |
| Estrone | h | 512-0353 |
| Ethane | af | 512-0019 |
| 1,2- Ethanediamine | | |
| see Ethylenediamine | e | 512-0190 |
| 1,2- Ethanediol Bis-(2-hydroxyethyl Ether) | | |
| see Triethylene Glycol | a | 512-0140 |
| Ethanenitrile | | |
| see Acetonitrile | aef | 512-0108 |
| Ethanoic Acid | | |
| see Acetic Acid | aef | 512-0189 |
| Ethanol | ef | 512-0042 |
| Ethanolamine | | |
| see 2- Aminoethanol | f | 512-0305 |
| Ethene | | |
| see Ethylene | aef | 512-0002 |
| Ether | | |
| see Ethyl Ether | aef | 512-0134 |
| Ethiops Mineral | | |
| see Mercury Sulfide | h | 512-0423 |
| Ethoxyethane | | |
| see Ethyl Ether | aef | 512-0134 |
| Ethyl Acetate | f | 512-0086 |
| Ethylacetic Acid | | |
| see Butyric Acid | af | 512-0394 |
| Ethyl Alcohol | | |
| see Ethanol | ef | 512-0042 |
| Ethylbenzene | af | 512-0405 |
| Ethyl Carbinol | | |
| see 1- Propanol | ef | 512-0043 |
| Ethyl Chloride | | |
| see Chloroethane | ae | 512-0016 |
| Ethylene | aef | 512-0002 |
| Ethylene Bromide | | |
| see 1,2- Dibromoethane | e | 512-0007 |
| Ethylene Carbonate | f | 512-0153 |
| Ethylene Chloride | | |
| see 1,2- Dichloroethane | ae | 512-0358 |
| Ethylenediamine | e | 512-0190 |
| Ethylene Dibromide | | |
| see 1,2- Dibromoethane | e | 512-0007 |
| Ethylene Dichloride | | |
| see 1,2- Dichloroethane | ae | 512-0358 |
| 2,2'- Ethylenedioxydiethanol | | |
| see Triethylene Glycol | a | 512-0140 |
| Ethylene Glycol | f | 512-0063 |
| Ethyl Ethanoate | | |
| see Ethyl Acetate | f | 512-0086 |
| Ethyl Ether | aef | 512-0134 |
| Ethylethylene | | |
| see 1- Butene | a | 512-0231 |
| Ethyl Hydroxide | | |
| see Ethanol | ef | 512-0042 |

**Property:** **a.** Thermal conductivity; **b.** Accommodation coefficient; **c.** Thermal contact resistance; **d.** Thermal diffusivity; **e.** Specific heat; **f.** Viscosity; **g.** Emittance; **h.** Reflectance; **i.** Absorptance; **j.** Transmittance; **k.** $\alpha_s/\epsilon_t$ ratio; **l.** Prandtl number; **n.** Thermal linear expansion coefficient; **o.** Thermal volumetric expansion coefficient

| Substance Name | Property | Number |
|---|---|---|
| Ethylidene Fluoride | | |
| see 1,1- Difluoroethane | efl | 512-0365 |
| Ethyl Methyl Carbinol | | |
| see 2- Butanol | af | 512-0214 |
| Ethyl Methyl Ketone | | |
| see 2- Butanone | f | 512-0085 |
| Ethylolamine | | |
| see 2- Aminoethanol | f | 512-0305 |
| Ethyl Oxide | | |
| see Ethyl Ether | aef | 512-0134 |
| Ethyl Peroxide | a | 512-0726 |
| Europium Sulfide Eu S | j | 512-0961 |
| Europium Sulfide Eu $S_2$ | j | 512-0959 |
| Explosive Oil | | |
| see Nitroglycerin | f | 512-0062 |
| False Topaz | | |
| see Silicon Oxide Si $O_2$ | aj | 512-0683 |
| Fat, Pig | a | 512-0341 |
| Fermentation Alcohol | | |
| see Ethanol | ef | 512-0042 |
| Ferro- Vac- E | | |
| see Iron | aef | 512-0176 |
| Ferrum | | |
| see Iron | aef | 512-0176 |
| Flint | | |
| see Silicon Oxide Si $O_2$ | aj | 512-0683 |
| Flowers Of Sulfur | | |
| see Sulfur | ae | 512-0177 |
| Fluorine | aef | 512-0293 |
| Fluorine, Monatomic | af | 512-0294 |
| Fluorine Oxide | | |
| see Oxygen Fluoride | ej | 512-0729 |
| Fluorite | | |
| see Calcium Fluoride | fhjn | 512-0241 |
| Fluoroethylene | a | 512-0774 |
| Fluoroform | ael | 512-0105 |
| Fluoromethane | a | 512-0107 |
| Fluorosulfuric Acid | | |
| see Sulfuryl Fluoride | a | 512-0421 |
| Follicular Hormone | | |
| see Estrone | h | 512-0353 |
| Folliculin | | |
| see Estrone | h | 512-0353 |
| Formamide | af | 512-0103 |
| Formic Acid | a | 512-0719 |
| N- Formyldimethylamine | | |
| see N, N- Dimethylformamide | f | 512-0041 |
| Formyl Tribromide | | |
| see Bromoform | f | 512-0284 |
| Freon 10 | | |
| see Carbon Tetrachloride | aefi | 512-0117 |
| Freon 12 | | |
| see Dichlorodifluoromethane | aefl | 512-0008 |
| Freon 13 | | |
| see Chlorotrifluoromethane | al | 512-0329 |
| Freon 20 | | |
| see Chloroform | aef | 512-0135 |
| Freon 22 | | |
| see Chlorodifluoromethane | ae | 512-0009 |
| Freon 23 | | |
| see Fluoroform | ael | 512-0105 |
| Freon 30 | | |
| see Dichloromethane | af | 512-0235 |
| Freon 31 | | |
| see Chlorofluoromethane | a | 512-0328 |
| Freon 32 | | |
| see Difluoromethane | l | 512-1082 |
| Freon 40 | | |
| see Chloromethane | af | 512-0015 |
| Freon 41 | | |
| see Fluoromethane | a | 512-0107 |
| Freon 50 | | |
| see Methane | aefl | 512-0001 |
| Freon 115 | | |
| see Chloropentafluoroethane | ael | 512-0291 |
| Freon 152a | | |
| see 1,1- Difluoroethane | efl | 512-0365 |
| Freon 160 | | |
| see Chloroethane | ae | 512-0016 |
| Freon 290 | | |
| see Propane | aef | 512-0021 |
| Freon 846 | | |
| see Sulfur Fluoride | aef | 512-0205 |
| Fused Quartz | | |
| see Silicon Oxide Si $O_2$ | aj | 512-0683 |
| G 74 | | |
| see Nitrogen | aefl | 512-0106 |
| Gallium Chloride | f | 512-0119 |
| Gallium Phosphide | a | 512-0369 |
| Gallium Sulfide | j | 512-0967 |
| Ganister | | |
| see Silicon Oxide Si $O_2$ | aj | 512-0683 |
| Gas, Inert | af | 512-0109 |
| Genetron 12 | | |
| see Dichlorodifluoromethane | aefl | 512-0008 |
| Genetron 13 | | |
| see Chlorotrifluoromethane | al | 512-0329 |
| Genetron 22 | | |
| see Chlorodifluoromethane | ae | 512-0009 |
| Genetron 31 | | |
| see Chlorofluoromethane | a | 512-0328 |
| Germane | j | 512-0976 |
| Germane, Tetradeuterated | j | 512-0977 |
| Germanic Acid | | |
| see Germanium Oxide | j | 512-0434 |
| Germanium Oxide | j | 512-0434 |
| Glacial Acetic Acid | | |
| see Acetic Acid | aef | 512-0189 |
| Glasrock | | |
| see Silicon Oxide Si $O_2$ | aj | 512-0683 |
| Globar | | |
| see Silicon Carbide | aehjn | 512-0175 |
| Glonoin Oil | | |
| see Nitroglycerin | f | 512-0062 |
| Glycerin | | |
| see Glycerol | aefh | 512-0032 |
| Glycerol | aefh | 512-0032 |
| Glycerol Glass | | |
| see Glycerol | aefh | 512-0032 |
| Glycerol Trinitrate | | |
| see Nitroglycerin | f | 512-0062 |
| Glyceryl Nitrate | | |
| see Nitroglycerin | f | 512-0062 |
| Glycol Bis(Hydroxyethyl) Ether | | |
| see Triethylene Glycol | a | 512-0140 |
| Glycol Carbonate | | |
| see Ethylene Carbonate | f | 512-0153 |
| Glycol Dibromide | | |
| see 1,2- Dibromoethane | e | 512-0007 |
| Glycol Dichloride | | |
| see 1,2- Dichloroethane | ae | 512-0358 |
| Glycol Ethylene Ether | | |
| see 1,4- Dioxane | ef | 512-0127 |
| Glycyl Alcohol | | |
| see Glycerol | aefh | 512-0032 |
| Grahams Salt | | |
| see Sodium Metaphosphate | f | 512-0088 |
| Grain Alcohol | | |
| see Ethanol | ef | 512-0042 |
| Graphite | aegin | 512-0262 |
| Greenland Spar | | |
| see Sodium Hexafluoroaluminate | f | 512-0380 |
| Greenockite | | |
| see Cadmium Sulfide | h | 512-0422 |

**Property: a.** Thermal conductivity; **b.** Accommodation coefficient; **c.** Thermal contact resistance; **d.** Thermal diffusivity; **e.** Specific heat; **f.** Viscosity; **g.** Emittance; **h.** Reflectance; **i.** Absorptance; **j.** Transmittance; **k.** $\alpha_s/\epsilon_t$ ratio; **l.** Prandtl number; **n.** Thermal linear expansion coefficient; **o.** Thermal volumetric expansion coefficient

| Substance Name | Property | Number |
|---|---|---|
| Hafnium | n | 512-1028 |
| Hafnium Boride | adn | 512-1043 |
| Hafnium Carbide | an | 512-0092 |
| Hafnium Oxide | a | 512-0810 |
| Halite | | |
| see Sodium Chloride | adefhjno | 512-0030 |
| Heavy Hydrogen | | |
| see Deuterium | aefj | 512-0069 |
| Heavy Spar | | |
| see Barium Sulfate | adej | 512-0139 |
| Heavy Water | | |
| see Water, Dideuterated | f | 512-0213 |
| Helium | aefl | 512-0061 |
| Helium 3 | aef | 512-0055 |
| Helium 4 | aef | 512-0056 |
| Helium 3, Adsorbed on Argon, Adsorbed on Copper | e | 512-0059 |
| Helium 4, Adsorbed on Argon, Adsorbed on Copper | e | 512-0060 |
| Helium 3, Adsorbed on Copper | ef | 512-0128 |
| Helium 4, Adsorbed on Copper | e | 512-0132 |
| Helium 3, Adsorbed on Vycor Glass | e | 512-0169 |
| Helium 4, Adsorbed on Vycor Glass | e | 512-0170 |
| Heptane | ef | 512-0216 |
| 1-Heptanol | a | 512-0319 |
| Heptyl Alcohol | | |
| see 1-Heptanol | a | 512-0319 |
| Herasil | | |
| see Silicon Oxide Si $O_2$ | aj | 512-0683 |
| Hexadecane | e | 512-0116 |
| 1-Hexadecanol | h | 512-0436 |
| Hexadecyl Chloride | | |
| see 1-Chlorohexadecane | f | 512-0076 |
| Hexadeuteriobenzene | | |
| see Benzene, Hexadeuterated | f | 512-0164 |
| 2,4-Hexadiyne | f | 512-0165 |
| Hexafluorobenzene | aef | 512-0269 |
| Hexahydrobenzene | | |
| see Cyclohexane | ef | 512-0039 |
| Hexahydromesitylene | | |
| see 1-trans-3,5-Trimethylcyclohexane | e | 512-0325 |
| Hexahydrophenol | | |
| see Cyclohexanol | f | 512-0239 |
| Hexahydrotoluene | | |
| see Methylcyclohexane | ef | 512-0166 |
| Hexahydro-m-xylene | | |
| see trans-1,4-Dimethylcyclohexane | e | 512-0324 |
| Hexamethylene | | |
| see Cyclohexane | ef | 512-0039 |
| 2,6,10,15,19,23-Hexamethyltetracosane | f | 512-0321 |
| Hexane | efh | 512-0114 |
| Hexanedinitrile | | |
| see Adiponitrile | f | 512-0218 |
| 1-Hexanol | aef | 512-0046 |
| Hexone | | |
| see 4-Methyl-2-pentanone | f | 512-0289 |
| Hexyl Alcohol | | |
| see 1-Hexanol | aef | 512-0046 |
| Hexyl Chloride | | |
| see 1-Chlorohexane | f | 512-0077 |
| H M G 50 | | |
| see Graphite | aegin | 512-0262 |
| H N | | |
| see Hydrazine Nitrate | af | 512-0708 |
| Homosil | | |
| see Silicon Oxide Si $O_2$ | aj | 512-0683 |
| Hornstone | | |
| see Silicon Oxide Si $O_2$ | aj | 512-0683 |
| Hydrazine | aef | 512-0370 |
| Hydrazine, Anhydrous | | |
| see Hydrazine | aef | 512-0370 |
| Hydrazine, Tetradeuterated | af | 512-0882 |
| Hydrazine Base | | |
| see Hydrazine | aef | 512-0370 |
| Hydrazine Nitrate | af | 512-0708 |
| Hydrobromic Acid | a | 512-0137 |
| Hydrobromide | | |
| see Hydrobromic Acid | a | 512-0137 |
| Hydrocarbon | aef | 512-0356 |
| Hydrochloric Acid | afhi | 512-0194 |
| Hydrochloric Acid, Monodeuterated | af | 512-0699 |
| Hydrochloride | | |
| see Hydrochloric Acid | afhi | 512-0194 |
| Hydrofluoric Acid | af | 512-0296 |
| Hydrofluoric Acid, Monodeuterated | af | 512-0297 |
| Hydrogen | aefjl | 512-0003 |
| Hydrogen, Monatomic | af | 512-0202 |
| Hydrogen, Monodeuterated | af | 512-0104 |
| Hydrogen Bromide | | |
| see Hydrobromic Acid | a | 512-0137 |
| Hydrogen Chloride | | |
| see Hydrochloric Acid | afhi | 512-0194 |
| Hydrogen Fluoride | | |
| see Hydrofluoric Acid | af | 512-0296 |
| Hydrogen Fluoride, Monodeuterated | | |
| see Hydrofluoric Acid, Monodeuterated | af | 512-0297 |
| Hydrogen Sulfate | | |
| see Sulfuric Acid | f | 512-0644 |
| Hydrogen Sulfide | a | 512-0013 |
| Hydrosulfuric Acid | | |
| see Hydrogen Sulfide | a | 512-0013 |
| Hydroxybenzene | | |
| see Phenol | f | 512-0208 |
| 3-Hydroxyestra-1,3,5 (10)-trien-17-one | | |
| see Estrone | h | 512-0353 |
| β-Hydroxyethylamine | | |
| see 2-Aminoethanol | f | 512-0305 |
| 3-Hydroxyphenol | | |
| see Resorcinol | f | 512-0333 |
| 2-Hydroxy-1,2,3-propanetricarboxylic Acid | | |
| see Citric Acid | e | 512-0033 |
| m-Hydroxytoluene | | |
| see m-Cresol | f | 512-0371 |
| p-Hydroxytoluene | | |
| see p-Cresol | f | 512-0372 |
| β-Hydroxytricarballylic Acid | | |
| see Citric Acid | e | 512-0033 |
| Hypnone | | |
| see Acetophenone | f | 512-0217 |
| Hytrol O | | |
| see Cyclohexanone | f | 512-0122 |
| Ice Spar | | |
| see Sodium Hexafluoroaluminate | f | 512-0380 |
| Ice Stone | | |
| see Sodium Hexafluoroaluminate | f | 512-0380 |
| Indium Phosphide | a | 512-0368 |
| Indium Sulfide | h | 512-0449 |
| Infrasil | | |
| see Silicon Oxide Si $O_2$ | aj | 512-0683 |
| Iodyride | | |
| see Silver Iodide | e | 512-0124 |
| I P A | | |
| see 2-Propanol | af | 512-0141 |
| Ipsen | | |
| see Aluminum Oxide | afn | 512-0433 |
| I R F N A | | |
| see Nitric Acid | adef | 512-0311 |
| Iron | aef | 512-0176 |
| Iron Sulfide | f | 512-0693 |
| Irtran 2 | | |
| see Zinc Sulfide | adeh | 512-0138 |
| Irtran 3 | | |
| see Calcium Fluoride | fhjn | 512-0241 |
| Irtran 5 | | |
| see Magnesium Oxide | an | 512-0431 |
| Isoamyl Alcohol | | |
| see 3-Methyl-1-butanol | f | 512-0200 |
| pri-Isoamyl Alcohol | | |
| see 3-Methyl-1-butanol | f | 512-0200 |

**Property:** **a.** Thermal conductivity; **b.** Accommodation coefficient; **c.** Thermal contact resistance; **d.** Thermal diffusivity; **e.** Specific heat; **f.** Viscosity; **g.** Emittance; **h.** Reflectance; **i.** Absorptance; **j.** Transmittance; **k.** $\alpha_s/\epsilon_t$ ratio; **l.** Prandtl number; **n.** Thermal linear expansion coefficient; **o.** Thermal volumetric expansion coefficient

**Property:** **a.** Thermal conductivity; **b.** Accommodation coefficient; **c.** Thermal contact resistance; **d.** Thermal diffusivity; **e.** Specific heat; **f.** Viscosity; **g.** Emittance; **h.** Reflectance; **i.** Absorptance; **j.** Transmittance; **k.** $a_s/\epsilon_t$ ratio; **l.** Prandtl number; **n.** Thermal linear expansion coefficient; **o.** Thermal volumetric expansion coefficient

**Property:** **a.** Thermal conductivity; **b.** Accommodation coefficient; **c.** Thermal contact resistance; **d.** Thermal diffusivity; **e.** Specific heat; **f.** Viscosity; **g.** Emittance; **h.** Reflectance; **i.** Absorptance; **j.** Transmittance; **k.** $\alpha_s/\epsilon_t$ ratio; **l.** Prandtl number; **n.** Thermal linear expansion coefficient; **o.** Thermal volumetric expansion coefficient

**Property:** **a.** Thermal conductivity; **b.** Accommodation coefficient; **c.** Thermal contact resistance; **d.** Thermal diffusivity; **e.** Specific heat; **f.** Viscosity; **g.** Emittance; **h.** Reflectance; **i.** Absorptance; **j.** Transmittance; **k.** $\alpha_s/\epsilon_t$ ratio; **l.** Prandtl number; **n.** Thermal linear expansion coefficient; **o.** Thermal volumetric expansion coefficient

**Property: a.** Thermal conductivity; **b.** Accommodation coefficient; **c.** Thermal contact resistance; **d.** Thermal diffusivity; **e.** Specific heat; **f.** Viscosity; **g.** Emittance; **h.** Reflectance; **i.** Absorptance; **j.** Transmittance; **k.** $\alpha_s/\epsilon_t$ ratio; **l.** Prandtl number; **n.** Thermal linear expansion coefficient; **o.** Thermal volumetric expansion coefficient

| Substance Name | Property | Number |
|---|---|---|
| Rubidium Chloride | adej | 512-0603 |
| Rubidium Hydrogen Perchlorylamide | j | 512-0253 |
| Rubidium Hydrogen Perchlorylamide, Monodeuterated | j | 512-0252 |
| Rubidium Nitrate | af | 512-0180 |
| Sal Ammoniac | | |
| see Ammonium Chloride | f | 512-0600 |
| Salicylaldehyde Methyl Ether | | |
| see o- Anisaldehyde | f | 512-0348 |
| Salmiac | | |
| see Ammonium Chloride | f | 512-0600 |
| Sal Soda | | |
| see Sodium Carbonate $Na_2 C O_3$ | f | 512-0067 |
| Salt | | |
| see Sodium Chloride | adefhjno | 512-0030 |
| Salt | af | 512-0130 |
| Salt, Electrolytic | | |
| see Electrolyte | af | 512-0290 |
| Salt Of Tartar | | |
| see Potassium Carbonate | f | 512-0065 |
| Saltpeter | | |
| see Potassium Nitrate | adf | 512-0057 |
| Santocel | | |
| see Silicon Oxide $Si O_2$ | aj | 512-0683 |
| Scacchite | | |
| see Manganese Chloride | e | 512-0611 |
| Scheelite | j | 512-0246 |
| Sea Salt | | |
| see Sodium Chloride | adefhjno | 512-0030 |
| Sedoneural | | |
| see Sodium Bromide | fj | 512-0144 |
| Seleninyl Chloride | f | 512-0902 |
| Selenium | a | 512-0459 |
| Shell Sol A | | |
| see Hydrocarbon | aef | 512-0356 |
| Silane | j | 512-0083 |
| Silane, Tetradeuterated | j | 512-0082 |
| Silica | | |
| see Silicon Oxide $Si O_2$ | aj | 512-0683 |
| Silica Aerogel | | |
| see Silicon Oxide $Si O_2$ | aj | 512-0683 |
| Silica Gel | | |
| see Silicon Oxide $Si O_2$ | aj | 512-0683 |
| Silicane | | |
| see Silane | j | 512-0083 |
| Silicon | an | 512-0445 |
| Silicon Carbide | aehjn | 512-0175 |
| Silicon Chloride | e | 512-1045 |
| Silicon Hydride | | |
| see Silane | j | 512-0083 |
| Silicon Nitride | an | 512-0430 |
| Silicon Oxide $Si O_2$ | aj | 512-0683 |
| Silicon Tetradeuteride | | |
| see Silane, Tetradeuterated | j | 512-0082 |
| Silit | | |
| see Silicon Carbide | aehjn | 512-0175 |
| Silver | a | 512-0794 |
| Silver Bromide | e | 512-0024 |
| Silver Chloride | e | 512-0049 |
| Silver Iodide | e | 512-0124 |
| Silver Nitrate | f | 512-0310 |
| Silver Sulfide | en | 512-0050 |
| Silver Thiocyanate | f | 512-0374 |
| Soda | | |
| see Sodium Carbonate $Na_2 C O_3$ | f | 512-0067 |
| Soda, Calcined | | |
| see Sodium Carbonate $Na_2 C O_3$ | f | 512-0067 |
| Soda, Carbonating | | |
| see Sodium Carbonate $Na_2 C O_3$ | f | 512-0067 |
| Soda, Monohydrate | | |
| see Sodium Carbonate $Na_2 C O_3$ | f | 512-0067 |
| Soda Ash | | |
| see Sodium Carbonate $Na_2 C O_3$ | f | 512-0067 |
| Sodaniter | | |
| see Sodium Nitrate | adf | 512-0182 |
| Sodium | j | 512-0730 |
| Sodium Aluminum Fluoride | | |
| see Sodium Hexafluoroaluminate | f | 512-0380 |
| Sodium Bromide | fj | 512-0144 |
| Sodium Carbonate $Na_2 C O_3$ | f | 512-0067 |
| Sodium Chloride | adefhjno | 512-0030 |
| Sodium Fluoaluminate | | |
| see Sodium Hexafluoroaluminate | f | 512-0380 |
| Sodium Fluoride | efj | 512-0280 |
| Sodium Hexafluoroaluminate | f | 512-0380 |
| Sodium Hydrate | | |
| see Sodium Hydroxide | e | 512-0643 |
| Sodium Hydroxide | e | 512-0643 |
| Sodium Hyperchlorate | | |
| see Sodium Perchlorate | ef | 512-0221 |
| Sodium Iodide | aefn | 512-0099 |
| Sodium Isothiocyanate | f | 512-0219 |
| Sodium Metaphosphate | f | 512-0088 |
| Sodium Nitrate | adf | 512-0182 |
| Sodium Perchlorate | ef | 512-0221 |
| Sodium Pyrophosphate | f | 512-0087 |
| Sodium Tetrafluoroborate | f | 512-0804 |
| Sodium Tetraphenylborate | f | 512-0220 |
| Sodium Trifluoromethanesulfonate | f | 512-0224 |
| Solution | | |
| see Mixture (General) | adef | 512-0064 |
| Spectrosil (Numbered Series) | | |
| see Silicon Oxide $Si O_2$ | aj | 512-0683 |
| Sphalerite | | |
| see Zinc Sulfide | adeh | 512-0138 |
| Spinacane | | |
| see 2,6,10,15,19,23- Hexamethyltetracosane | f | 512-0321 |
| Squalane | | |
| see 2,6,10,15,19,23- Hexamethyltetracosane | f | 512-0321 |
| Stannic Chloride | | |
| see Tin Chloride | ef | 512-0403 |
| Starch | a | 512-0710 |
| Starch, Corn | a | 512-0390 |
| Starch, Maize | | |
| see Starch, Corn | a | 512-0390 |
| Strontium Fluoride | hj | 512-0257 |
| Strontium Nitrate | f | 512-0452 |
| Sulfamide | f | 512-1044 |
| Sulfolane | | |
| see Tetrahydrothiophene 1,1-dioxide | f | 512-0462 |
| Sulfonyldiamide | | |
| see Sulfamide | f | 512-1044 |
| Sulfonyl Fluoride | | |
| see Sulfuryl Fluoride | a | 512-0421 |
| Sulfur | ae | 512-0177 |
| Sulfuretted Hydrogen | | |
| see Hydrogen Sulfide | a | 512-0013 |
| Sulfur Fluoride | aef | 512-0205 |
| Sulfuric Acid | f | 512-0644 |
| Sulfuric Ether | | |
| see Ethyl Ether | aef | 512-0134 |
| Sulfuric Fluoride | | |
| see Sulfuryl Fluoride | a | 512-0421 |
| Sulfuric Oxyfluoride | | |
| see Sulfuryl Fluoride | a | 512-0421 |
| Sulfurous Acid Anhydride | | |
| see Sulfur Oxide | af | 512-0014 |
| Sulfur Oxide | af | 512-0014 |
| Sulfuryl Amide | | |
| see Sulfamide | f | 512-1044 |
| Sulfuryl Fluoride | a | 512-0421 |
| Sulpholane | | |
| see Tetrahydrothiophene 1,1-dioxide | f | 512-0462 |
| Sylvite | | |
| see Potassium Chloride | adefjno | 512-0048 |

**Property:** **a.** Thermal conductivity; **b.** Accommodation coefficient; **c.** Thermal contact resistance; **d.** Thermal diffusivity; **e.** Specific heat; **f.** Viscosity; **g.** Emittance; **h.** Reflectance; **i.** Absorptance; **j.** Transmittance; **k.** $a_s/\epsilon_t$ ratio; **l.** Prandtl number; **n.** Thermal linear expansion coefficient; **o.** Thermal volumetric expansion coefficient

**Property: a.** Thermal conductivity; **b.** Accommodation coefficient; **c.** Thermal contact resistance; **d.** Thermal diffusivity; **e.** Specific heat; **f.** Viscosity; **g.** Emittance; **h.** Reflectance; **i.** Absorptance; **j.** Transmittance; **k.** $\alpha_s/\epsilon_t$ ratio; **l.** Prandtl number; **n.** Thermal linear expansion coefficient; **o.** Thermal volumetric expansion coefficient

| Substance Name | Property | Number |
|---|---|---|
| Vinegar Naphtha | | |
| see Ethyl Acetate | f | 512-0086 |
| Vitreosil (Numbered Series) | | |
| see Silicon Oxide $SiO_2$ | aj | 512-0683 |
| Vitreous Silica | | |
| see Silicon Oxide $SiO_2$ | aj | 512-0683 |
| Vulcasil | | |
| see Silicon Oxide $SiO_2$ | aj | 512-0683 |
| Washing Soda | | |
| see Sodium Carbonate $Na_2CO_3$ | f | 512-0067 |
| Water, Dideuterated | f | 512-0213 |
| White Caustic | | |
| see Sodium Hydroxide | e | 512-0643 |
| White Copperas | | |
| see Zinc Sulfate | a | 512-0340 |
| White Vitriol | | |
| see Zinc Sulfate | a | 512-0340 |
| Whiting | | |
| see Calcium Carbonate | f | 512-0156 |
| Wood Alcohol | | |
| see Methanol | aef | 512-0028 |
| Wood Ether | | |
| see Methyl Ether | f | 512-0068 |
| Xenon | aefil | 512-0037 |
| Xylene | f | 512-0402 |
| m-Xylene | f | 512-0201 |
| o-Xylene | f | 512-0236 |
| p-Xylene | ef | 512-0126 |
| 2,5-Xylidine | f | 512-0360 |
| 2,6-Xylidine | f | 512-0359 |
| 3,4-Xylidine | f | 512-0361 |
| Xylol | | |
| see Xylene | f | 512-0402 |
| Yttrium Ethyl Sulfate | e | 512-0268 |
| Zinc Blende | | |
| see Zinc Sulfide | adeh | 512-0138 |
| Zinc Chloride | ef | 512-0054 |
| Zinc Compound | e | 512-0191 |
| Zinc Iodide | f | 512-0102 |
| Zinc Sulfate | a | 512-0340 |
| Zinc Sulfide | adeh | 512-0138 |
| Zirconium | an | 512-0740 |
| Zirconium Boride | adegn | 512-1048 |
| Zirconium Carbide | aegn | 512-0091 |
| Zirconium Oxide | a | 512-0808 |
| Zirconium Silicate $Zr_2SiO_4$ | an | 512-0756 |

## COMPONENTS OF NONAQUEOUS MULTIPLE MIXTURES CLASS 514

| Substance Name | Property | Number |
|---|---|---|
| Acanthite | | |
| see Silver Sulfide | n | 514-0050 |
| Acetic Acid | e | 514-0189 |
| Acetone | f | 514-0136 |
| A D P | | |
| see Ammonium Phosphate $NH_4H_2PO_4$ | f | 514-0647 |
| Air | aefghl | 514-0020 |
| Air, Dissociating | aefl | 514-0386 |
| Alcohol | | |
| see Ethanol | f | 514-0042 |
| Allyl Isosulfocyanate | | |
| see Allyl Isothiocyanate | f | 514-0160 |
| Allyl Isothiocyanate | f | 514-0160 |
| Allyl Mustard Oil | | |
| see Allyl Isothiocyanate | f | 514-0160 |
| Allyl Rhodanide | | |
| see Allyl Thiocyanate | f | 514-0161 |
| Allyl Sulfocyanate | | |
| see Allyl Thiocyanate | f | 514-0161 |
| Allyl Thiocyanate | f | 514-0161 |
| Aluminum | an | 514-0440 |
| Aluminum Nitride | an | 514-0437 |
| Aluminum Sodium Fluoride | | |
| see Sodium Hexafluoroaluminate | f | 514-0380 |
| Aminomethane | | |
| see Methylamine | a | 514-0110 |
| Ammonia | af | 514-0011 |
| Ammonium Acid Phosphate | | |
| see Ammonium Phosphate $NH_4H_2PO_4$ | f | 514-0647 |
| Ammonium Iodide | f | 514-0145 |
| Ammonium Nitrate | f | 514-0155 |
| Ammonium Phosphate $NH_4H_2PO_4$ | f | 514-0647 |
| Ammonium Phosphate $(NH_4)_2HPO_4$ | f | 514-0766 |
| Ammonium Pyrophosphate | f | 514-0889 |
| Ammonium Sulfate | f | 514-0157 |
| Amyl Carbinol | | |
| see 1-Hexanol | f | 514-0046 |
| A O 600 | | |
| see Carbon | adegn | 514-0174 |
| Aragonite | | |
| see Calcium Carbonate | f | 514-0156 |
| Arcton 6 | | |
| see Dichlorodifluoromethane | f | 514-0008 |
| Argentite | | |
| see Silver Sulfide | n | 514-0050 |
| Argon | afi | 514-0012 |
| Arsenic Sulfide | j | 514-0418 |
| Atmosphere, Jupiter | h | 514-1001 |
| Atmosphere, Mercury | h | 514-1002 |
| Atmosphere, Planet | afhl | 514-0129 |
| Atmosphere, Saturn Satellite | h | 514-0753 |
| Atmosphere, Sun | a | 514-0427 |
| Atmosphere, Venus | h | 514-0309 |
| Atmosphere, Venus, Synthetic | h | 514-0249 |
| Atomite | | |
| see Calcium Carbonate | f | 514-0156 |
| Barium Chloride | f | 514-0089 |
| Benzene | aef | 514-0006 |
| Benzine (German, Dutch) | | |
| see Gasoline | f | 514-0475 |
| Benzol | | |
| see Benzene | aef | 514-0006 |
| Benzole | | |
| see Benzene | aef | 514-0006 |
| Beryllium Fluoride | f | 514-0454 |
| Bimethyl | | |
| see Ethane | e | 514-0019 |
| Biphenyl | f | 514-0315 |
| 1,2-Biphenylbenzene | | |
| see o-Terphenyl | f | 514-0312 |
| 1,3-Biphenylbenzene | | |
| see m-Terphenyl | f | 514-0313 |
| 1,4-Biphenylbenzene | | |
| see p-Terphenyl | f | 514-0314 |
| Bischofite | | |
| see Magnesium Chloride | f | 514-0387 |
| Bis-(2,2-dimethylheptyl) | | |
| 1,10-decanedicarboxylate | | |
| see Bis-(2,2-dimethylheptyl) Dodecanedioate | f | 514-0921 |
| Bis-(2,2-dimethylheptyl) Dodecanedioate | f | 514-0921 |
| Bis-(2,2-dimethylhexyl) | | |
| 1,10-decanedicarboxylate | | |
| see Bis-(2,2-dimethylhexyl) Dodecanedioate | f | 514-0920 |
| Bis-(2,2-dimethylhexyl) Dodecanedioate | f | 514-0920 |
| Bis-(2,2-dimethylnonyl) | | |
| 1,10-decanedicarboxylate | | |
| see Bis-(2,2-dimethylnonyl) Dodecanedioate | f | 514-0923 |
| Bis-(2,2-dimethylnonyl) Dodecanedioate | f | 514-0923 |
| Bis-(2,2-dimethyloctyl) | | |
| 1,10-decanedicarboxylate | | |
| see Bis-(2,2-dimethyloctyl) Dodecanedioate | f | 514-0922 |
| Bis-(2,2-dimethyloctyl) Dodecanedioate | f | 514-0922 |

**Property:** **a.** Thermal conductivity; **b.** Accommodation coefficient; **c.** Thermal contact resistance; **d.** Thermal diffusivity; **e.** Specific heat; **f.** Viscosity; **g.** Emittance; **h.** Reflectance; **i.** Absorptance; **j.** Transmittance; **k.** $\alpha_s/\epsilon_t$ ratio; **l.** Prandtl number; **n.** Thermal linear expansion coefficient; **o.** Thermal volumetric expansion coefficient

| Substance Name | Property | Number |
|---|---|---|
| Bis-(2,2-dimethylpentyl) | | |
| 1,10-decanedicarboxylate | | |
| see Bis-(2,2-dimethylpentyl) Dodecanedioate | f | 514-0919 |
| Bis-(2,2-dimethylpentyl) Dodecanedioate | f | 514-0919 |
| Black Lead | | |
| see Graphite | aegn | 514-0262 |
| Boralloy | | |
| see Boron Nitride | an | 514-0432 |
| Borazon | | |
| see Boron Nitride | an | 514-0432 |
| Boric Acid | j | 514-0412 |
| Boric Acid, Dideuterated | j | 514-0414 |
| Boric Acid, Monodeuterated | j | 514-0415 |
| Boric Acid, Trideuterated | j | 514-0413 |
| Boron | an | 514-0287 |
| Boron Carbide | an | 514-0439 |
| Boron Nitride | an | 514-0432 |
| Brimstone | | |
| see Sulfur | j | 514-0177 |
| Bromobenzene | a | 514-0195 |
| Bromobenzol | | |
| see Bromobenzene | a | 514-0195 |
| 1-Bromooctane | f | 514-0867 |
| Bromyride | | |
| see Silver Bromide | n | 514-0024 |
| 1-Butanol | f | 514-0044 |
| 2-Butanone | f | 514-0085 |
| Butyl Acetate | f | 514-0199 |
| Butyl Alcohol | | |
| see 1-Butanol | f | 514-0044 |
| tert-Butyl Alcohol | | |
| see 2-Methyl-2-propanol | f | 514-0113 |
| Butyl Ethanoate | | |
| see Butyl Acetate | f | 514-0199 |
| Butyric Alcohol | | |
| see 1-Butanol | f | 514-0044 |
| Cadmium Iodide | f | 514-0097 |
| Calcium Carbonate | f | 514-0156 |
| Calcium Chromate | e | 514-0827 |
| Calcspar | | |
| see Calcium Carbonate | f | 514-0156 |
| Carbamide | | |
| see Urea | f | 514-0507 |
| Carbinol | | |
| see Methanol | ef | 514-0028 |
| Carbolac | | |
| see Carbon | adegn | 514-0174 |
| Carbon | adegn | 514-0174 |
| Carbon Black | | |
| see Carbon | adegn | 514-0174 |
| Carbonic Acid | | |
| see Carbon Oxide $CO_2$ | aef | 514-0193 |
| Carbonic Anhydride | | |
| see Carbon Oxide $CO_2$ | aef | 514-0193 |
| Carbon Oxide $CO_2$ | aef | 514-0193 |
| Carbon Sulfide | f | 514-0162 |
| Carbon Tetrachloride | af | 514-0117 |
| Carbon Tetrafluoride | e | 514-0451 |
| Celanese G Y-70 | | |
| see Graphite | aegn | 514-0262 |
| Cesium Chloride | n | 514-0031 |
| Cesium Iodide | f | 514-0101 |
| Chalk | | |
| see Calcium Carbonate | f | 514-0156 |
| Channel Black | | |
| see Carbon | adegn | 514-0174 |
| Chlorine | i | 514-0388 |
| Chlorine Fluoride Cl F | i | 514-0461 |
| Chlorobenzene | a | 514-0196 |
| Chloroform | f | 514-0135 |
| Chloromethane | f | 514-0015 |
| Chromosphere | | |
| see Plasma, Sun Atmosphere | a | 514-0687 |

| Substance Name | Property | Number |
|---|---|---|
| C K-4 | | |
| see Carbon | adegn | 514-0174 |
| Common Salt | | |
| see Sodium Chloride | f | 514-0030 |
| Cornstarch | | |
| see Starch, Corn | a | 514-0390 |
| Courtaulds H M S | | |
| see Graphite | aegn | 514-0262 |
| Courtaulds H T-S | | |
| see Graphite | aegn | 514-0262 |
| Cryolite | | |
| see Sodium Hexafluoroaluminate | f | 514-0380 |
| Cumene | a | 514-1012 |
| Cyclohexane | ef | 514-0039 |
| Cyclohexanol | f | 514-0239 |
| Cyclooctane | f | 514-0718 |
| Cyclotol | | |
| see Explosive | dn | 514-1051 |
| Decane | e | 514-0238 |
| Diamine | | |
| see Hydrazine | ae | 514-0370 |
| Diazotizing Salts | | |
| see Sodium Nitrite | ah | 514-0667 |
| Dichlorodifluoromethane | f | 514-0008 |
| Didecyl Adipate | f | 514-0792 |
| Diethylene Glycol Caproate | | |
| see 2-(2-Hydroxyethoxy)ethyl Caproate | f | 514-0785 |
| Diethylene Glycol Caprylate | | |
| see 2-(2-Hydroxyethoxy)ethyl Caprylate | f | 514-0787 |
| Diethylene Glycol Heptanoate | | |
| see 2-(2-Hydroxyethoxy)ethyl Heptanoate | f | 514-0786 |
| Diethylene Glycol Nonanoate | | |
| see 2-(2-Hydroxyethoxy)ethyl Nonanoate | f | 514-0788 |
| Diethylene Glycol Valerate | | |
| see 2-(2-Hydroxyethoxy)ethyl Valerate | f | 514-0784 |
| Difluorodichloromethane | | |
| see Dichlorodifluoromethane | f | 514-0008 |
| cis-1,2-Difluoroethylene | a | 514-0772 |
| trans-1,2-Difluoroethylene | a | 514-0773 |
| Dimethyl | | |
| see Ethane | e | 514-0019 |
| Dimethylamine | a | 514-0701 |
| Dimethyl Ether | | |
| see Methyl Ether | f | 514-0068 |
| N, N-Dimethylformamide | f | 514-0041 |
| 1,1-Dimethylhydrazine | e | 514-0735 |
| Dimethyl Ketone | | |
| see Acetone | f | 514-0136 |
| Dimethyl Sulfoxide | | |
| see Methyl Sulfoxide | f | 514-0098 |
| Dinonyl Adipate | f | 514-0790 |
| Dioctyl Adipate | f | 514-0789 |
| Dioctyl Azelate | f | 514-0791 |
| Dioctyl Sebacate | f | 514-0793 |
| Diphenyl | | |
| see Biphenyl | f | 514-0315 |
| 1,2-Diphenylbenzene | | |
| see o-Terphenyl | f | 514-0312 |
| 1,3-Diphenylbenzene | | |
| see m-Terphenyl | f | 514-0313 |
| 1,4-Diphenylbenzene | | |
| see p-Terphenyl | f | 514-0314 |
| D M F | | |
| see N, N-Dimethylformamide | f | 514-0041 |
| D M S O | | |
| see Methyl Sulfoxide | f | 514-0098 |
| Dry Ice | | |
| see Carbon Oxide $CO_2$ | aef | 514-0193 |
| Dust | ghj | 514-0308 |
| E L F-0 | | |
| see Carbon | adegn | 514-0174 |
| Ethane | e | 514-0019 |

**Property:** **a.** Thermal conductivity; **b.** Accommodation coefficient; **c.** Thermal contact resistance; **d.** Thermal diffusivity; **e.** Specific heat; **f.** Viscosity; **g.** Emittance; **h.** Reflectance; **i.** Absorptance; **j.** Transmittance; **k.** $\alpha_s/\epsilon_t$ ratio; **l.** Prandtl number; **n.** Thermal linear expansion coefficient; **o.** Thermal volumetric expansion coefficient

| Substance Name | Property | Number |
|---|---|---|
| Ethanoic Acid | | |
| see Acetic Acid | e | 514-0189 |
| Ethanol | f | 514-0042 |
| Ethyl Alcohol | | |
| see Ethanol | f | 514-0042 |
| Ethylbenzene | a | 514-0405 |
| Ethyl Carbinol | | |
| see 1-Propanol | f | 514-0043 |
| Ethyl Hydroxide | | |
| see Ethanol | f | 514-0042 |
| Ethyl Methyl Ketone | | |
| see 2-Butanone | f | 514-0085 |
| Europium Sulfide Eu $S_2$ | j | 514-0959 |
| Explosive | dn | 514-1051 |
| Explosive Lx-04 | | |
| see Explosive | dn | 514-1051 |
| Fermentation Alcohol | | |
| see Ethanol | f | 514-0042 |
| Fertilizer | h | 514-0698 |
| Flowers Of Sulfur | | |
| see Sulfur | j | 514-0177 |
| Fluorine | ei | 514-0293 |
| Fluoroform | e | 514-0105 |
| Formamide | f | 514-0103 |
| N-Formyldimethylamine | | |
| see N, N-Dimethylformamide | f | 514-0041 |
| Freon 10 | | |
| see Carbon Tetrachloride | af | 514-0117 |
| Freon 12 | | |
| see Dichlorodifluoromethane | f | 514-0008 |
| Freon 20 | | |
| see Chloroform | f | 514-0135 |
| Freon 23 | | |
| see Fluoroform | e | 514-0105 |
| Freon 40 | | |
| see Chloromethane | f | 514-0015 |
| Freon 50 | | |
| see Methane | aef | 514-0001 |
| Freon 729 | | |
| see Air | aefghl | 514-0020 |
| G 74 | | |
| see Nitrogen | aefi | 514-0106 |
| Gas, Lean Natural, Synthetic | e | 514-0707 |
| Gasoline | f | 514-0475 |
| Genetron 12 | | |
| see Dichlorodifluoromethane | f | 514-0008 |
| Germanium Sulfide Ge S | n | 514-0420 |
| Germanium Sulfide Ge $S_2$ | n | 514-0419 |
| Glacial Acetic Acid | | |
| see Acetic Acid | e | 514-0189 |
| Globar | | |
| see Silicon Carbide | adegn | 514-0175 |
| Grain Alcohol | | |
| see Ethanol | f | 514-0042 |
| Graphite | aegn | 514-0262 |
| Greenland Spar | | |
| see Sodium Hexafluoroaluminate | f | 514-0380 |
| Hafnium Boride | adn | 514-1043 |
| Halite | | |
| see Sodium Chloride | f | 514-0030 |
| H B R | f | 514-0316 |
| Helium | f | 514-0061 |
| Hendecane | | |
| see Undecane | e | 514-0318 |
| Heptane | e | 514-0216 |
| Hexadecane | f | 514-0116 |
| Hexahydrobenzene | | |
| see Cyclohexane | ef | 514-0039 |
| Hexahydrophenol | | |
| see Cyclohexanol | f | 514-0239 |

| Substance Name | Property | Number |
|---|---|---|
| Hexamethylene | | |
| see Cyclohexane | ef | 514-0039 |
| Hexane | f | 514-0114 |
| 1-Hexanol | f | 514-0046 |
| Hexyl Alcohol | | |
| see 1-Hexanol | f | 514-0046 |
| H M G 50 | | |
| see Graphite | aegn | 514-0262 |
| H N | | |
| see Hydrazine Nitrate | ae | 514-0708 |
| Hydrazine | ae | 514-0370 |
| Hydrazine, Anhydrous | | |
| see Hydrazine | ae | 514-0370 |
| Hydrazine Base | | |
| see Hydrazine | ae | 514-0370 |
| Hydrazine Nitrate | ae | 514-0708 |
| Hydrocarbon | ef | 514-0356 |
| Hydrogen | af | 514-0003 |
| Hydrogen Sulfide | e | 514-0013 |
| Hydrosulfuric Acid | | |
| see Hydrogen Sulfide | e | 514-0013 |
| 2-(2-Hydroxyethoxy)ethyl Caproate | f | 514-0785 |
| 2-(2-Hydroxyethoxy)ethyl Caprylate | f | 514-0787 |
| 2-(2-Hydroxyethoxy)ethyl Heptanoate | f | 514-0786 |
| 2-(2-Hydroxyethoxy)ethyl Nonanoate | f | 514-0788 |
| 2-(2-Hydroxyethoxy)ethyl Valerate | f | 514-0784 |
| Ice Spar | | |
| see Sodium Hexafluoroaluminate | f | 514-0380 |
| Ice Stone | | |
| see Sodium Hexafluoroaluminate | f | 514-0380 |
| Isoamyl Alcohol | | |
| see 3-Methyl-1-butanol | f | 514-0200 |
| pri-Isoamyl Alcohol | | |
| see 3-Methyl-1-butanol | f | 514-0200 |
| Isotron 12 | | |
| see Dichlorodifluoromethane | f | 514-0008 |
| Kaltron 23 | | |
| see Fluoroform | e | 514-0105 |
| Kerosene | e | 514-1031 |
| Kerosine | | |
| see Kerosene | e | 514-1031 |
| Ketopropane | | |
| see Acetone | f | 514-0136 |
| Kodak Metal Etch Resist | f | 514-0425 |
| Kodak Metal Etch Resist Thinner | f | 514-0426 |
| Kosmos 40 | | |
| see Carbon | adegn | 514-0174 |
| Kryolith | | |
| see Sodium Hexafluoroaluminate | f | 514-0380 |
| Krypton | af | 514-0036 |
| Lampblack | | |
| see Carbon | adegn | 514-0174 |
| Lithium Chloride | e | 514-0029 |
| Lithium Fluoride | fj | 514-0281 |
| L X-04 | | |
| see Explosive | dn | 514-1051 |
| Magnesium Chloride | f | 514-0387 |
| Marsh Gas | | |
| see Methane | aef | 514-0001 |
| Mascagnite | | |
| see Ammonium Sulfate | f | 514-0157 |
| Mesitylene | f | 514-0282 |
| Methanamide | | |
| see Formamide | f | 514-0103 |
| Methane | aef | 514-0001 |
| Methanecarboxylic Acid | | |
| see Acetic Acid | e | 514-0189 |
| Methanol | ef | 514-0028 |
| Methoxymethane | | |
| see Methyl Ether | f | 514-0068 |

**Property: a.** Thermal conductivity; **b.** Accommodation coefficient; **c.** Thermal contact resistance; **d.** Thermal diffusivity; **e.** Specific heat; **f.** Viscosity; **g.** Emittance; **h.** Reflectance; **i.** Absorptance; **j.** Transmittance; **k.** $a_s/\epsilon_t$ ratio; **l.** Prandtl number; **n.** Thermal linear expansion coefficient; **o.** Thermal volumetric expansion coefficient

| *Substance Name* | *Property* | *Number* |
|---|---|---|
| Methyl Alcohol | | |
| see Methanol | ef | 514-0028 |
| Methylamine | a | 514-0110 |
| Methylbenzene | | |
| see Toluene | aef | 514-0047 |
| 2-Methylbutane | f | 514-0227 |
| 3-Methyl-1-butanol | f | 514-0200 |
| Methyl Carbinol | | |
| see Ethanol | f | 514-0042 |
| Methyl Chloride | | |
| see Chloromethane | f | 514-0015 |
| Methyl Ether | f | 514-0068 |
| Methyl Ethyl Ketone | | |
| see 2-Butanone | f | 514-0085 |
| Methylhydrazine | ae | 514-0736 |
| Methylmethane | | |
| see Ethane | e | 514-0019 |
| Methyl Oxide | | |
| see Methyl Ether | f | 514-0068 |
| 2-Methyl-2-propanol | f | 514-0113 |
| Methylsulfinylmethane | | |
| see Methyl Sulfoxide | f | 514-0098 |
| Methyl Sulfoxide | f | 514-0098 |
| Mixture (General) | aef | 514-0064 |
| Mogul (Lettered Series) | | |
| see Carbon | adegn | 514-0174 |
| Monatomic Substance | a | 514-0112 |
| Morganite | | |
| see Graphite | aegn | 514-0262 |
| Motor Spirits | | |
| see Gasoline | f | 514-0475 |
| Mustard Oil, Artificial | | |
| see Allyl Isothiocyanate | f | 514-0160 |
| Naphthalene 1,2,3,4-Tetrahydride | | |
| see 1,2,3,4-Tetrahydronaphthalene | e | 514-0244 |
| 1-Naphthylamine | f | 514-0159 |
| α-Naphthylamine | | |
| see 1-Naphthylamine | f | 514-0159 |
| Neon | af | 514-0018 |
| Nickel | a | 514-0179 |
| Niter | | |
| see Potassium Nitrate | ah | 514-0057 |
| Nitrobenzene | f | 514-0392 |
| Nitrobenzol | | |
| see Nitrobenzene | f | 514-0392 |
| Nitrogen | aefi | 514-0106 |
| Nitrogen Oxide N O | aef | 514-0391 |
| Nitrogen Oxide $N O_2$ | aef | 514-0051 |
| Nitrogen Oxide $N_2 O_3$ | e | 514-1027 |
| Nitrogen Oxide $N_2 O_4$ | aef | 514-0052 |
| Nonane | e | 514-0317 |
| Norway Saltpeter | | |
| see Ammonium Nitrate | f | 514-0155 |
| Octane | e | 514-0131 |
| Oil Of Mirbane | | |
| see Nitrobenzene | f | 514-0392 |
| Orpiment | | |
| see Arsenic Sulfide | j | 514-0418 |
| Oxygen | aef | 514-0152 |
| Pentaerithrytol Caproate | f | 514-0780 |
| Pentaerithrytol Caprylate | f | 514-0782 |
| Pentaerithrytol Heptanoate | f | 514-0781 |
| Pentaerithrytol Nonanoate | f | 514-0783 |
| Pentaerithrytol Valerate | f | 514-0779 |
| Pentane | e | 514-0151 |
| Perchloromethane | | |
| see Carbon Tetrachloride | af | 514-0117 |
| Petrol (British) | | |
| see Gasoline | f | 514-0475 |
| Phene | | |
| see Benzene | aef | 514-0006 |
| Phenylbenzene | | |
| see Biphenyl | f | 514-0315 |
| m-Phenylbiphenyl | | |
| see m-Terphenyl | f | 514-0313 |
| o-Phenylbiphenyl | | |
| see o-Terphenyl | f | 514-0312 |
| p-Phenylbiphenyl | | |
| see p-Terphenyl | f | 514-0314 |
| Phenyl Bromide | | |
| see Bromobenzene | a | 514-0195 |
| Phenyl Chloride | | |
| see Chlorobenzene | a | 514-0196 |
| Phenylethane | | |
| see Ethylbenzene | a | 514-0405 |
| Phenylmethane | | |
| see Toluene | aef | 514-0047 |
| Phosphoric Acid | f | 514-0121 |
| Phosphoric Anhydride | | |
| see Phosphorus Oxide | f | 514-0288 |
| Phosphorus Oxide | f | 514-0288 |
| Plasma, Air | afl | 514-0171 |
| Plasma, Sun Atmosphere | a | 514-0687 |
| Plasma, Sun Atmosphere, Partially Ionized | a | 514-0811 |
| Polyatomic Substance | a | 514-0881 |
| Potassamide | | |
| see Potassium Amide | f | 514-0186 |
| Potassium Amide | f | 514-0186 |
| Potassium Chloride | efjn | 514-0048 |
| Potassium Hexafluorozirconate | f | 514-0071 |
| Potassium Iodide K I | f | 514-0100 |
| Potassium Muriate | | |
| see Potassium Chloride | efjn | 514-0048 |
| Potassium Nitrate | ah | 514-0057 |
| Potassium Sulfate | f | 514-0447 |
| Potassium Zirconium Fluoride | | |
| see Potassium Hexafluorozirconate | f | 514-0071 |
| 1-Propanol | f | 514-0043 |
| 2-Propanone | | |
| see Acetone | f | 514-0136 |
| 2-Propenyl Isothiocyanate | | |
| see Allyl Isothiocyanate | f | 514-0160 |
| Propyl Alcohol | | |
| see 1-Propanol | f | 514-0043 |
| Propyl Carbinol | | |
| see 1-Butanol | f | 514-0044 |
| Protonosphere | | |
| see Air | aefghl | 514-0020 |
| Pyridine | e | 514-0184 |
| Pyroacetic Ether | | |
| see Acetone | f | 514-0136 |
| R 10 | | |
| see Carbon Tetrachloride | af | 514-0117 |
| R 12 | | |
| see Dichlorodifluoromethane | f | 514-0008 |
| R 20 | | |
| see Chloroform | f | 514-0135 |
| R 23 | | |
| see Fluoroform | e | 514-0105 |
| R 40 | | |
| see Chloromethane | f | 514-0015 |
| R 50 | | |
| see Methane | aef | 514-0001 |
| R 729 | | |
| see Air | aefghl | 514-0020 |
| Reacting Substance | f | 514-0476 |
| Salt | | |
| see Sodium Chloride | f | 514-0030 |
| Saltpeter | | |
| see Potassium Nitrate | ah | 514-0057 |
| Sea Salt | | |
| see Sodium Chloride | f | 514-0030 |

**Property: a.** Thermal conductivity; **b.** Accommodation coefficient; **c.** Thermal contact resistance; **d.** Thermal diffusivity; **e.** Specific heat; **f.** Viscosity; **g.** Emittance; **h.** Reflectance; **i.** Absorptance; **j.** Transmittance; **k.** $\alpha_s/\epsilon_t$ ratio; **l.** Prandtl number; **n.** Thermal linear expansion coefficient; **o.** Thermal volumetric expansion coefficient

## COMPONENTS OF AQUEOUS BINARY MIXTURES CLASS 515



**Property:** **a.** Thermal conductivity; **b.** Accommodation coefficient; **c.** Thermal contact resistance; **d.** Thermal diffusivity; **e.** Specific heat; **f.** Viscosity; **g.** Emittance; **h.** Reflectance; **i.** Absorptance; **j.** Transmittance; **k.** $\alpha_s/\epsilon_t$ ratio; **l.** Prandtl number; **n.** Thermal linear expansion coefficient; **o.** Thermal volumetric expansion coefficient

| Substance Name | Property | Number |
|---|---|---|
| Aminobenzene | | |
| see Aniline | e | 515-0084 |
| 4-Aminobutanoic Acid | | |
| see 4-Aminobutyric Acid | f | 515-0534 |
| 4-Aminobutyric Acid | f | 515-0534 |
| ε-Aminocaproic Acid | | |
| see 6-Aminohexanoic Acid | f | 515-0555 |
| Aminoethanoic Acid | | |
| see Glycine | f | 515-0519 |
| 2-Aminoethanol | f | 515-0305 |
| β-Aminoethyl Alcohol | | |
| see 2-Aminoethanol | f | 515-0305 |
| Aminoform | | |
| see Hexamethylenetetramine | f | 515-0551 |
| 6-Aminohexanoic Acid | f | 515-0555 |
| 3-Aminopropionic Acid | | |
| see β-Alanine | f | 515-0525 |
| β-Aminopropionic Acid | | |
| see β-Alanine | f | 515-0525 |
| Ammoform | | |
| see Hexamethylenetetramine | f | 515-0551 |
| Ammonia | aef | 515-0011 |
| Ammonioformaldehyde | | |
| see Hexamethylenetetramine | f | 515-0551 |
| Ammonium Acetate | e | 515-0830 |
| Ammonium Acid Phosphate | | |
| see Ammonium Phosphate $NH_4H_2PO_4$ | efh | 515-0647 |
| Ammonium Bicarbonate | | |
| see Ammonium Carbonate $NH_4HCO_3$ | aef | 515-0509 |
| Ammonium Bromide | f | 515-0500 |
| Ammonium Carbonate $NH_4HCO_3$ | aef | 515-0509 |
| Ammonium Chloride | aef | 515-0600 |
| Ammonium Muriate | | |
| see Ammonium Chloride | aef | 515-0600 |
| Ammonium Nitrate | f | 515-0155 |
| Ammonium Nitrite | f | 515-0646 |
| Ammonium Perchlorate | f | 515-0709 |
| Ammonium Phosphate $NH_4H_2PO_4$ | efh | 515-0647 |
| Ammonium Sulfate | f | 515-0157 |
| Anhydrone | | |
| see Magnesium Perchlorate | ef | 515-0610 |
| Aniline | e | 515-0084 |
| Aniline Chloride | | |
| see Aniline Hydrochloride | f | 515-0547 |
| Aniline Hydrochloride | f | 515-0547 |
| Aniline Salt | | |
| see Aniline Hydrochloride | f | 515-0547 |
| Anilinium Chloride | | |
| see Aniline Hydrochloride | f | 515-0547 |
| Anone | | |
| see Cyclohexanone | f | 515-0122 |
| Aqua Fortis | | |
| see Nitric Acid | aefl | 515-0311 |
| 3β-Chlorocholest-5-ene | h | 515-0366 |
| Bakers Sugar | | |
| see Dextrose | f | 515-0554 |
| Baking Soda | | |
| see Sodium Carbonate $NaHCO_3$ | aefh | 515-0506 |
| Barium Bromide | af | 515-0715 |
| Barium Chloride | aef | 515-0089 |
| Barium Iodide | af | 515-0716 |
| Barium Nitrate | f | 515-0453 |
| Barium Nitrite | f | 515-0495 |
| Barium Perchlorate | ef | 515-0494 |
| Barium Sodium (Ethylenedinitrilo)tetraacetate | f | 515-0574 |
| Basic Zirconium Chloride | | |
| see Dichlorooxozirconium | f | 515-0614 |
| Battery Acid | | |
| see Sulfuric Acid | aefh | 515-0644 |
| Beet Sugar | | |
| see Sucrose | def | 515-0578 |
| Benzene | f | 515-0006 |

| Substance Name | Property | Number |
|---|---|---|
| Benzine (German, Dutch) | | |
| see Gasoline | f | 515-0475 |
| Benzol | | |
| see Benzene | f | 515-0006 |
| Benzole | | |
| see Benzene | f | 515-0006 |
| Beryllium Chloride | e | 515-0496 |
| Beryllium Nitrate | f | 515-0906 |
| Beryllium Perchlorate | f | 515-0497 |
| Bischofite | | |
| see Magnesium Chloride | aefh | 515-0387 |
| cis-Bis(Glycinato)copper Complex | f | 515-0917 |
| Black Lead | | |
| see Graphite | a | 515-0262 |
| Blue Copperas | | |
| see Copper Sulfate | af | 515-0636 |
| Blue Stone | | |
| see Copper Sulfate | af | 515-0636 |
| Blue Vitriol | | |
| see Copper Sulfate | af | 515-0636 |
| Boric Acid | ef | 515-0412 |
| Butanamide | | |
| see Butyramide | f | 515-0533 |
| 1,4-Butanedicarboxylic Acid | | |
| see Adipic Acid | f | 515-0550 |
| 1-Butanol | f | 515-0044 |
| 2-Butanol | f | 515-0214 |
| 2-Butanone | f | 515-0085 |
| 2-Butoxyethanol | e | 515-0981 |
| 1-Butoxy-2-methoxyethane | e | 515-0987 |
| Butter Of Zinc | | |
| see Zinc Chloride | af | 515-0054 |
| Butyl Alcohol | | |
| see 1-Butanol | f | 515-0044 |
| sec-Butyl Alcohol | | |
| see 2-Butanol | f | 515-0214 |
| sym-Butyl Alcohol | | |
| see 2-Butanol | f | 515-0214 |
| tert-Butyl Alcohol | | |
| see 2-Methyl-2-propanol | f | 515-0113 |
| Butylamine | f | 515-0733 |
| tert-Butylamine | f | 515-1019 |
| Butylammonium Chloride | f | 515-0824 |
| Butyl Dicresyl Phosphate | | |
| see Butyl Ditolyl Phosphate | f | 515-0587 |
| Butyl Ditolyl Phosphate | f | 515-0587 |
| Butyl Phosphate | | |
| see Tributyl Phosphate | f | 515-0580 |
| Butylurea | f | 515-0541 |
| Butyramide | f | 515-0533 |
| Butyric Alcohol | | |
| see 1-Butanol | f | 515-0044 |
| Butyric Amide | | |
| see Butyramide | f | 515-0533 |
| Cadmium Chloride | f | 515-0396 |
| Cadmium Hyperchlorate | | |
| see Cadmium Perchlorate | ef | 515-0597 |
| Cadmium Nitrite | f | 515-0598 |
| Cadmium Perchlorate | ef | 515-0597 |
| Calcium Bromide | f | 515-0503 |
| Calcium Chloride | aef | 515-0460 |
| Calcium Chromate | f | 515-0827 |
| Calcium Hyperchlorate | | |
| see Calcium Perchlorate | ef | 515-0595 |
| Calcium Iodide | f | 515-0958 |
| Calcium Nitrate | f | 515-0058 |
| Calcium Nitrite | f | 515-0596 |
| Calcium Perchlorate | ef | 515-0595 |
| Calcium Sodium (Ethylenedinitrilo)tetraacetate | f | 515-0575 |
| Cane Sugar | | |
| see Sucrose | def | 515-0578 |
| Capramol | | |
| see 6-Aminohexanoic Acid | f | 515-0555 |

**Property:** **a.** Thermal conductivity; **b.** Accommodation coefficient; **c.** Thermal contact resistance; **d.** Thermal diffusivity; **e.** Specific heat; **f.** Viscosity; **g.** Emittance; **h.** Reflectance; **i.** Absorptance; **j.** Transmittance; **k.** $\alpha_s/\epsilon_t$ ratio; **l.** Prandtl number; **n.** Thermal linear expansion coefficient; **o.** Thermal volumetric expansion coefficient

**Property:** **a.** Thermal conductivity; **b.** Accommodation coefficient; **c.** Thermal contact resistance; **d.** Thermal diffusivity; **e.** Specific heat; **f.** Viscosity; **g.** Emittance; **h.** Reflectance; **i.** Absorptance; **j.** Transmittance; **k.** $\alpha_s/\epsilon_t$ ratio; **l.** Prandtl number; **n.** Thermal linear expansion coefficient; **o.** Thermal volumetric expansion coefficient

| Substance Name | Property | Number |
|---|---|---|
| Diethylene Dioxide | | |
| see 1,4- Dioxane | e | 515-0127 |
| Diethylene Ether | | |
| see 1,4- Dioxane | e | 515-0127 |
| Diethylene Glycol | af | 515-0537 |
| Diethylene Oxide | | |
| see 1,4- Dioxane | e | 515-0127 |
| Diethylmethylamine | | |
| see N- Methyldiethylamine | e | 515-0841 |
| Diethylmethylammonium Chloride | | |
| see N- Methyldiethylammonium Chloride | e | 515-0839 |
| 1,1- Diethylpiperidinium Iodide | f | 515-0573 |
| 1,1- Diethylpyrrolidinium Iodide | f | 515-0569 |
| 1,1- Diethylurea | f | 515-0542 |
| 1,3- Diethylurea | f | 515-0543 |
| Dihydroxyethyl Ether | | |
| see Diethylene Glycol | af | 515-0537 |
| 1,2- Dimethoxyethane | e | 515-0983 |
| N, N- Dimethylacetamide | f | 515-0207 |
| Dimethylacetic Acid | | |
| see Isobutyric Acid | f | 515-0532 |
| Dimethylamine | e | 515-0701 |
| Dimethylammonium Chloride | e | 515-0834 |
| 1,1'- Dimethyl-4,4'-bipyridinium Dichloride | ef | 515-0577 |
| Dimethyl Carbinol | | |
| see 2- Propanol | ef | 515-0141 |
| N, N- Dimethylformamide | af | 515-0041 |
| Dimethyl Ketone | | |
| see Acetone | f | 515-0136 |
| Dimethylphenylene | | |
| see Xylene | h | 515-0402 |
| 1,1- Dimethylpiperidinium Iodide | f | 515-0566 |
| 2,6- Dimethylpyridine | | |
| see 2,6- Lutidine | f | 515-0562 |
| 1,1- Dimethylpyrrolidinium Iodide | f | 515-0556 |
| Dimethyl Sulfoxide | | |
| see Methyl Sulfoxide | f | 515-0098 |
| 1,1- Dimethylurea | e | 515-0912 |
| 1,1- Dimethylurea, Octadeuterated | e | 515-0914 |
| 1,3- Dimethylurea | ef | 515-0526 |
| 1,3- Dimethylurea, Octadeuterated | e | 515-0915 |
| N, N'- Dimethylurea | | |
| see 1,3- Dimethylurea | ef | 515-0526 |
| sym- Dimethylurea | | |
| see 1,3- Dimethylurea | ef | 515-0526 |
| Dinitratodioxouranium | f | 515-0673 |
| Dioxane | ef | 515-0237 |
| 1,4- Dioxane | e | 515-0127 |
| p- Dioxane | | |
| see 1,4- Dioxane | e | 515-0127 |
| Dioxodiperchloratouranium | f | 515-0617 |
| 1,3- Dioxolane | e | 515-0849 |
| Dioxosulfatouranium | f | 515-0684 |
| Dioxyethylene | | |
| see 1,4- Dioxane | e | 515-0127 |
| Dioxyethylene Glycol | | |
| see Diethylene Glycol | af | 515-0537 |
| Dipping Acid | | |
| see Sulfuric Acid | aefh | 515-0644 |
| 1,2- Dipropoxyethane | e | 515-0988 |
| Dipropylamine | e | 515-0846 |
| Dipropylammonium Chloride | e | 515-0836 |
| D M A C | | |
| see N, N- Dimethylacetamide | f | 515-0207 |
| D M F | | |
| see N, N- Dimethylformamide | af | 515-0041 |
| D M S O | | |
| see Methyl Sulfoxide | f | 515-0098 |
| Dodecyl Sodium Sulfate | f | 515-0579 |
| Dry Ice | | |
| see Carbon Oxide C $O_2$ | f | 515-0193 |
| Dysprosium Nitrate | e | 515-0638 |
| Dysprosium Perchlorate | e | 515-0620 |

| Substance Name | Property | Number |
|---|---|---|
| E A C A Kabi | | |
| see 6- Aminohexanoic Acid | f | 515-0555 |
| E D T A - Calcium Chloride | | |
| see (Ethylenedinitrilo)tetraacetic Acid - Calcium Chloride Complex | f | 515-0800 |
| E D T A - Cerium Chloride | | |
| see (Ethylenedinitrilo)tetraacetic Acid - Cerium Chloride Complex | f | 515-0801 |
| Electrolyte | aef | 515-0290 |
| Emery | | |
| see Aluminum Oxide | ef | 515-0433 |
| Emulsion | f | 515-0734 |
| Entobacterin | ade | 515-0472 |
| Epsamon | | |
| see 6- Aminohexanoic Acid | f | 515-0555 |
| Epsikapron | | |
| see 6- Aminohexanoic Acid | f | 515-0555 |
| Epsilcapramin | | |
| see 6- Aminohexanoic Acid | f | 515-0555 |
| Epsom Salts | | |
| see Manganese Sulfate | f | 515-0665 |
| Erbium Chloride | e | 515-0872 |
| Erbium Perchlorate | e | 515-0621 |
| Etambro | | |
| see Tetraethylammonium Bromide | f | 515-0571 |
| Etamon Chloride | | |
| see Tetraethylammonium Chloride | f | 515-0572 |
| Ethanamide | | |
| see Acetamide | f | 515-0518 |
| 1,2- Ethanediol Bis-(2-hydroxyethyl Ether) | | |
| see Triethylene Glycol | af | 515-0140 |
| Ethanoic Acid | | |
| see Acetic Acid | ef | 515-0189 |
| Ethanol | adefl | 515-0042 |
| Ethanolamine | | |
| see 2- Aminoethanol | f | 515-0305 |
| 2- Ethoxyethanol | e | 515-0979 |
| 1- Ethoxy-2-methoxyethane | e | 515-0984 |
| Ethyl Acetate | f | 515-0086 |
| Ethyl Alcohol | | |
| see Ethanol | adefl | 515-0042 |
| Ethylammonium Chloride | f | 515-0822 |
| Ethyl Carbinol | | |
| see 1- Propanol | ef | 515-0043 |
| (Ethylenedinitrilo)tetraacetic Acid - Calcium Chloride Complex | f | 515-0800 |
| (Ethylenedinitrilo)tetraacetic Acid - Cerium Chloride Complex | f | 515-0801 |
| 2,2'- Ethylenedioxydiethanol | | |
| see Triethylene Glycol | af | 515-0140 |
| Ethylene Glycol | af | 515-0063 |
| Ethyl Ethanoate | | |
| see Ethyl Acetate | f | 515-0086 |
| Ethyl Hydroxide | | |
| see Ethanol | adefl | 515-0042 |
| Ethylmethylamine | | |
| see N- Methylethylamine | e | 515-0844 |
| Ethyl Methyl Carbinol | | |
| see 2- Butanol | f | 515-0214 |
| Ethyl Methyl Ketone | | |
| see 2- Butanone | f | 515-0085 |
| 1- Ethyl-1-methylpiperidinium Iodide | f | 515-0570 |
| 1- Ethyl-1-methylpyrrolidinium Iodide | f | 515-0567 |
| Ethylolamine | | |
| see 2- Aminoethanol | f | 515-0305 |
| N- Ethylpropylamine | e | 515-0845 |
| 1- Ethylpyridinium Iodide | f | 515-0564 |
| Ethyltrimethylammonium Iodide | f | 515-0546 |
| Ethylurea | f | 515-0527 |
| E T P Copper | | |
| see Copper | a | 515-0634 |
| Etylon | | |
| see Tetraethylammonium Bromide | f | 515-0571 |

**Property: a.** Thermal conductivity; **b.** Accommodation coefficient; **c.** Thermal contact resistance; **d.** Thermal diffusivity; **e.** Specific heat; **f.** Viscosity; **g.** Emittance; **h.** Reflectance; **i.** Absorptance; **j.** Transmittance; **k.** $\alpha_s/\epsilon_t$ ratio; **l.** Prandtl number; **n.** Thermal linear expansion coefficient; **o.** Thermal volumetric expansion coefficient

| Substance Name | Property | Number |
|---|---|---|
| Europium Chloride | f | 515-0926 |
| Fermentation Alcohol | | |
| see Ethanol | adefl | 515-0042 |
| Fertilizer | f | 515-0698 |
| Formamide | af | 515-0103 |
| Formamine | | |
| see Hexamethylenetetramine | f | 515-0551 |
| Formic Acid | f | 515-0719 |
| Formin | | |
| see Hexamethylenetetramine | f | 515-0551 |
| N- Formyldimethylamine | | |
| see N, N- Dimethylformamide | af | 515-0041 |
| Francium Bromide | e | 515-0499 |
| Francium Chloride | e | 515-0599 |
| Francium Fluoride | e | 515-0639 |
| Francium Iodide | e | 515-0641 |
| Freon 10 | | |
| see Carbon Tetrachloride | f | 515-0117 |
| Fructose | f | 515-0553 |
| Fruit Sugar | | |
| see Fructose | f | 515-0553 |
| G 74 | | |
| see Nitrogen | a | 515-0106 |
| Gadolinium Chloride | f | 515-0927 |
| Gallium Chloride | f | 515-0119 |
| Gallium Perchlorate | e | 515-0862 |
| Gasoline | f | 515-0475 |
| Glacial Acetic Acid | | |
| see Acetic Acid | ef | 515-0189 |
| Glaubers Salt | | |
| see Sodium Sulfate | aef | 515-0681 |
| D- Glucitol | f | 515-0557 |
| Glucose | f | 515-0552 |
| see also Dextrose | | 515-0554 |
| D- Glucose | | |
| see Dextrose | f | 515-0554 |
| ($\alpha$- D- Glucosido)-$\beta$- D-fructofuranoside | | |
| see Sucrose | def | 515-0578 |
| Glycerin | | |
| see Glycerol | aefl | 515-0032 |
| Glycerol | aefl | 515-0032 |
| Glycerol Glass | | |
| see Glycerol | aefl | 515-0032 |
| Glycine | f | 515-0519 |
| Glycocoll | | |
| see Glycine | f | 515-0519 |
| Glycol Bis(Hydroxyethyl) Ether | | |
| see Triethylene Glycol | af | 515-0140 |
| Glycol Ethylene Ether | | |
| see 1,4- Dioxane | e | 515-0127 |
| Glycyl Alcohol | | |
| see Glycerol | aefl | 515-0032 |
| (Glycylglycinato)copper Complex | f | 515-0918 |
| N- Glycylglycine | f | 515-0530 |
| Grain Alcohol | | |
| see Ethanol | adefl | 515-0042 |
| Grape Sugar | | |
| see Dextrose | f | 515-0554 |
| Graphite | a | 515-0262 |
| Hafnium Sulfate | f | 515-0895 |
| Halite | | |
| see Sodium Chloride | aefh | 515-0030 |
| Heavy Water | | |
| see Water, Dideuterated | aef | 515-0213 |
| Helium | e | 515-0061 |
| Hemocaprol | | |
| see 6- Aminohexanoic Acid | f | 515-0555 |
| Hempa | | |
| see Hexamethylphosphoric Triamide | f | 515-0559 |
| Hepin | | |
| see 6- Aminohexanoic Acid | f | 515-0555 |

| Substance Name | Property | Number |
|---|---|---|
| Heptamethyleneimine | | |
| see Octahydroazocine | e | 515-0852 |
| 1- Heptanol | f | 515-0319 |
| Heptyl Alcohol | | |
| see 1- Heptanol | f | 515-0319 |
| Hexadecyltrimethylammonium | f | 515-0759 |
| Hexafluoroacetone | | |
| see Hexafluoro-2-propanone | f | 515-0521 |
| Hexafluoro-2-propanone | f | 515-0521 |
| Hexahydric Alcohol | | |
| see D- Glucitol | f | 515-0557 |
| Hexahydroazepine | e | 515-0851 |
| Hexahydroazepine Hydrochloride | e | 515-0850 |
| Hexamethylenamine | | |
| see Hexamethylenetetramine | f | 515-0551 |
| Hexamethyleneimine | | |
| see Hexahydroazepine | e | 515-0851 |
| Hexamethyleneimine Hydrochloride | | |
| see Hexahydroazepine Hydrochloride | e | 515-0850 |
| Hexamethylenetetramine | f | 515-0551 |
| Hexamethylphosphoric Triamide | f | 515-0559 |
| Hexamine | | |
| see Hexamethylenetetramine | f | 515-0551 |
| Hexanedioic Acid | | |
| see Adipic Acid | f | 515-0550 |
| 1,2,3,4,5,6- Hexanehexol | | |
| see D- Glucitol | f | 515-0557 |
| H M G 50 | | |
| see Graphite | a | 515-0262 |
| Holmium Nitrate | e | 515-0648 |
| Holmium Perchlorate | e | 515-0622 |
| H S | | |
| see Hydroxylamine Sulfate | f | 515-0876 |
| Hydrochloric Acid | efh | 515-0194 |
| Hydrochloride | | |
| see Hydrochloric Acid | efh | 515-0194 |
| Hydrocyanite | | |
| see Copper Sulfate | af | 515-0636 |
| Hydrogen | aef | 515-0003 |
| Hydrogen (+ 1) Ion | e | 515-0955 |
| Hydrogen Chloride | | |
| see Hydrochloric Acid | efh | 515-0194 |
| Hydrogen Sulfate | | |
| see Sulfuric Acid | aefh | 515-0644 |
| Hydrogen Sulfide | ef | 515-0013 |
| Hydrosulfuric Acid | | |
| see Hydrogen Sulfide | ef | 515-0013 |
| Hydroxide (-1) Ion O H | e | 515-1083 |
| Hydroxybenzene | | |
| see Phenol | e | 515-0208 |
| $\beta$- Hydroxyethylamine | | |
| see 2- Aminoethanol | f | 515-0305 |
| Hydroxylamine Sulfate | f | 515-0876 |
| Hydroxylammonium Sulfate | | |
| see Hydroxylamine Sulfate | f | 515-0876 |
| 2- Hydroxy-1,2,3-propanetricarboxylic Acid | | |
| see Citric Acid | f | 515-0033 |
| o- Hydroxytoluene | | |
| see o- Cresol | f | 515-0561 |
| $\beta$- Hydroxytricarballylic Acid | | |
| see Citric Acid | f | 515-0033 |
| Hytrol O | | |
| see Cyclohexanone | f | 515-0122 |
| Indium Perchlorate | e | 515-0864 |
| Ion, Monatomic | e | 515-0934 |
| I P A | | |
| see 2- Propanol | ef | 515-0141 |
| Ipsen | | |
| see Aluminum Oxide | ef | 515-0433 |
| Ipsilon | | |
| see 6- Aminohexanoic Acid | f | 515-0555 |
| I R F N A | | |
| see Nitric Acid | aefl | 515-0311 |

**Property: a.** Thermal conductivity; **b.** Accommodation coefficient; **c.** Thermal contact resistance; **d.** Thermal diffusivity; **e.** Specific heat; **f.** Viscosity; **g.** Emittance; **h.** Reflectance; **i.** Absorptance; **j.** Transmittance; **k.** $\alpha_s/\epsilon_t$ ratio; **l.** Prandtl number; **n.** Thermal linear expansion coefficient; **o.** Thermal volumetric expansion coefficient

**Property: a.** Thermal conductivity; **b.** Accommodation coefficient; **c.** Thermal contact resistance; **d.** Thermal diffusivity; **e.** Specific heat; **f.** Viscosity; **g.** Emittance; **h.** Reflectance; **i.** Absorptance; **j.** Transmittance; **k.** $\alpha_s/\epsilon_t$ ratio; **l.** Prandtl number; **n.** Thermal linear expansion coefficient; **o.** Thermal volumetric expansion coefficient

| Substance Name | Property | Number |
|---|---|---|
| Morenosite | | |
| see Nickel Sulfate | f | 515-0682 |
| Morganite | | |
| see Graphite | a | 515-0262 |
| Motor Spirits | | |
| see Gasoline | f | 515-0475 |
| Mucilage, Rye | f | 515-0477 |
| Nadone | | |
| see Cyclohexanone | f | 515-0122 |
| Neodymium Chloride | e | 515-0628 |
| Nickel Chloride | f | 515-0747 |
| Nickel Nitrate | f | 515-0670 |
| Nickel Perchlorate | ef | 515-0613 |
| Nickel Salts, Single | | |
| see Nickel Sulfate | f | 515-0682 |
| Nickel Sulfate | f | 515-0682 |
| Niter | | |
| see Potassium Nitrate | ef | 515-0057 |
| Nitric Acid | aefl | 515-0311 |
| Nitrilotriacetic Acid | f | 515-0549 |
| Nitrobarite | | |
| see Barium Nitrate | f | 515-0453 |
| Nitrocalcite | | |
| see Calcium Nitrate | f | 515-0058 |
| 1- Nitroethylsodium | f | 515-0516 |
| Nitrogen | a | 515-0106 |
| Nonelectrolyte | f | 515-0478 |
| Norge Niter | | |
| see Calcium Nitrate | f | 515-0058 |
| Norton R A 5058 | | |
| see Aluminum Oxide | ef | 515-0433 |
| Norway Saltpeter | | |
| see Ammonium Nitrate | f | 515-0155 |
| Norwegian Saltpeter | | |
| see Calcium Nitrate | f | 515-0058 |
| N T A | | |
| see Nitrilotriacetic Acid | f | 515-0549 |
| Octahydroazocine | e | 515-0852 |
| O F H C Copper | | |
| see Copper | a | 515-0634 |
| Oil Of Vitriol | | |
| see Sulfuric Acid | aefh | 515-0644 |
| 2,2'- Oxydiethanol | | |
| see Diethylene Glycol | af | 515-0537 |
| Oxygen | aef | 515-0152 |
| o- Oxytoluene | | |
| see o- Cresol | f | 515-0561 |
| Pearl Ash | | |
| see Potassium Carbonate | aef | 515-0065 |
| Pentamethylene Oxide | | |
| see Tetrahydropyran | e | 515-0211 |
| Pentanoic Acid | | |
| see Valeric Acid | f | 515-0540 |
| Pentylammonium Chloride | f | 515-0825 |
| Perchloric Acid | e | 515-0408 |
| Perchloromethane | | |
| see Carbon Tetrachloride | f | 515-0117 |
| Perhydroazepine | | |
| see Hexahydroazepine | e | 515-0851 |
| Perhydroazocine | | |
| see Octahydroazocine | e | 515-0852 |
| Petrol (British) | | |
| see Gasoline | f | 515-0475 |
| Phene | | |
| see Benzene | f | 515-0006 |
| Phenol | e | 515-0208 |
| Phenylamine | | |
| see Aniline | e | 515-0084 |
| p- Phenylene Tetrabutyl Diphosphate | f | 515-0591 |
| Phenylic Acid | | |
| see Phenol | e | 515-0208 |

| Substance Name | Property | Number |
|---|---|---|
| Phosphoric Acid | aef | 515-0121 |
| Phosphoric Anhydride | | |
| see Phosphorus Oxide | ef | 515-0288 |
| Phosphorus Oxide | ef | 515-0288 |
| 2- Picoline | f | 515-0344 |
| α- Picoline | | |
| see 2- Picoline | f | 515-0344 |
| Pimelic Ketone | | |
| see Cyclohexanone | f | 515-0122 |
| Piperidic Acid | | |
| see 4- Aminobutyric Acid | f | 515-0534 |
| Piperidine | e | 515-0847 |
| Piperidinium Chloride | e | 515-0838 |
| Potassa | | |
| see Potassium Hydroxide | efh | 515-0642 |
| Potassium Acetate | f | 515-0854 |
| Potassium Acid Carbonate | | |
| see Potassium Carbonate $K H C O_3$ | efh | 515-0504 |
| Potassium Benzoate | aef | 515-0560 |
| Potassium Bicarbonate | | |
| see Potassium Carbonate $K H C O_3$ | efh | 515-0504 |
| Potassium Bromide | aefh | 515-0259 |
| Potassium Carbonate | aef | 515-0065 |
| Potassium Carbonate $K H C O_3$ | efh | 515-0504 |
| Potassium Caseinate | f | 515-0479 |
| Potassium Chloride | aefh | 515-0048 |
| Potassium Fluoride | efh | 515-0070 |
| Potassium Hydrate | | |
| see Potassium Hydroxide | efh | 515-0642 |
| Potassium Hydroxide | efh | 515-0642 |
| Potassium Iodide K I | efh | 515-0100 |
| Potassium Iodide $K I_3$ | f | 515-0652 |
| Potassium Molybdate | | |
| see Molybdenum Potassium Oxide | f | 515-0654 |
| Potassium Muriate | | |
| see Potassium Chloride | aefh | 515-0048 |
| Potassium Nitrate | ef | 515-0057 |
| Potassium Nitrite | f | 515-0653 |
| Potassium Oxide | f | 515-0655 |
| Potassium Phosphate $K H_2 P O_4$ | ef | 515-0898 |
| Potassium Phthalate | f | 515-0771 |
| Potassium Pyrophosphate | f | 515-0657 |
| Potassium Rhodanate | | |
| see Potassium Thiocyanate | f | 515-0510 |
| Potassium Rhodanide | | |
| see Potassium Thiocyanate | f | 515-0510 |
| Potassium Silicate | f | 515-0656 |
| Potassium Sulfate | afh | 515-0447 |
| Potassium Sulfocyanate | | |
| see Potassium Thiocyanate | f | 515-0510 |
| Potassium Sulfocyanide | | |
| see Potassium Thiocyanate | f | 515-0510 |
| Potassium Thiocyanate | f | 515-0510 |
| Praseodymium Chloride | ef | 515-0871 |
| Praseodymium Nitrate | e | 515-0674 |
| Praseodymium Perchlorate | e | 515-0623 |
| Propanamide | | |
| see Propionamide | f | 515-0524 |
| 1,2,3- Propanetriol | | |
| see Glycerol | aefl | 515-0032 |
| Propanoic Acid | | |
| see Propionic Acid | f | 515-0393 |
| 1- Propanol | ef | 515-0043 |
| 2- Propanol | ef | 515-0141 |
| 2- Propanone | | |
| see Acetone | f | 515-0136 |
| Propionamide | f | 515-0524 |
| Propionic Acid | f | 515-0393 |
| Propionic Acid Amide | | |
| see Propionamide | f | 515-0524 |
| 2- Propoxyethanol | e | 515-0980 |
| Propyl Alcohol | | |
| see 1- Propanol | ef | 515-0043 |

**Property: a.** Thermal conductivity; **b.** Accommodation coefficient; **c.** Thermal contact resistance; **d.** Thermal diffusivity; **e.** Specific heat; **f.** Viscosity; **g.** Emittance; **h.** Reflectance; **i.** Absorptance; **j.** Transmittance; **k.** $\alpha_s/\epsilon_t$ ratio; **l.** Prandtl number; **n.** Thermal linear expansion coefficient; **o.** Thermal volumetric expansion coefficient

**Property:** **a.** Thermal conductivity; **b.** Accommodation coefficient; **c.** Thermal contact resistance; **d.** Thermal diffusivity; **e.** Specific heat; **f.** Viscosity; **g.** Emittance; **h.** Reflectance; **i.** Absorptance; **j.** Transmittance; **k.** $\alpha_s/\epsilon_t$ ratio; **l.** Prandtl number; **n.** Thermal linear expansion coefficient; **o.** Thermal volumetric expansion coefficient

**Property:** **a.** Thermal conductivity; **b.** Accommodation coefficient; **c.** Thermal contact resistance; **d.** Thermal diffusivity; **e.** Specific heat; **f.** Viscosity; **g.** Emittance; **h.** Reflectance; **i.** Absorptance; **j.** Transmittance; **k.** $\alpha_s/\epsilon_t$ ratio; **l.** Prandtl number; **n.** Thermal linear expansion coefficient; **o.** Thermal volumetric expansion coefficient

| *Substance Name* | *Property* | *Number* |
|---|---|---|
| Uranium Sulfate | | |
| see Dioxosulfatouranium | f | 515-0684 |
| Uranyl Chloride | | |
| see Dichlorodioxouranium | f | 515-0615 |
| Uranyl Nitrate | | |
| see Dinitratodioxouranium | f | 515-0673 |
| Uranyl Perchlorate | | |
| see Dioxodiperchloratouranium | f | 515-0617 |
| Uranyl Sulfate | | |
| see Dioxosulfatouranium | f | 515-0684 |
| Urea | aef | 515-0507 |
| Urea, Tetradeuterated | e | 515-0913 |
| Uritone | | |
| see Hexamethylenetetramine | f | 515-0551 |
| Urotropin | | |
| see Hexamethylenetetramine | f | 515-0551 |
| Valerianic Acid | | |
| see Valeric Acid | f | 515-0540 |
| Valeric Acid | f | 515-0540 |
| Vinegar Acid | | |
| see Acetic Acid | ef | 515-0189 |
| Vinegar Naphtha | | |
| see Ethyl Acetate | f | 515-0086 |
| Washing Soda | | |
| see Sodium Carbonate $Na_2 C O_3$ | aef | 515-0067 |
| Water, Dideuterated | aef | 515-0213 |
| Water, Sea, Artificial | ef | 515-0493 |
| Waterglass | | |
| see Sodium Silicate | f | 515-0680 |
| White Caustic | | |
| see Sodium Hydroxide | efhj | 515-0643 |
| Wood Alcohol | | |
| see Methanol | ef | 515-0028 |
| Xylene | h | 515-0402 |
| Xylite | | |
| see Xylitol | f | 515-0545 |
| Xylitol | f | 515-0545 |
| Xylol | | |
| see Xylene | h | 515-0402 |
| Yeast | ade | 515-0869 |
| Yellow Salt | | |
| see Dinitratodioxouranium | f | 515-0673 |
| Ytterbium Chloride | f | 515-0929 |
| Ytterbium Nitrate | e | 515-0678 |
| Ytterbium Perchlorate | e | 515-0627 |
| Yttrium Chloride | e | 515-0873 |
| Yttrium Perchlorate | e | 515-0863 |
| Zeolite, Synthetic | e | 515-0795 |
| Zinc Chloride | af | 515-0054 |
| Zinc Compound | e | 515-0191 |
| Zinc Nitrate | f | 515-0672 |
| Zinc Perchlorate | ef | 515-0616 |
| Zirconium Oxychloride | | |
| see Dichlorooxozirconium | f | 515-0614 |
| Zirconium Sulfate | f | 515-0686 |
| Zirconyl Chloride | | |
| see Dichlorooxozirconium | f | 515-0614 |

## COMPONENTS OF AQUEOUS MULTIPLE MIXTURES CLASS 517

| *Substance Name* | *Property* | *Number* |
|---|---|---|
| Acetaldehyde | af | 517-0517 |
| Acetic Acid | f | 517-0189 |
| Acetic Aldehyde | | |
| see Acetaldehyde | af | 517-0517 |
| Acetic Ester | | |
| see Ethyl Acetate | a | 517-0086 |
| Acetic Ether | | |
| see Ethyl Acetate | a | 517-0086 |
| Acetone | ef | 517-0136 |
| Adipic Acid | f | 517-0550 |
| A D P | | |
| see Ammonium Phosphate $N H_4 H_2 P O_4$ | f | 517-0647 |
| Aerosil Powder | | |
| see Silicon Oxide $Si O_2$ | f | 517-0683 |
| Agar | a | 517-0464 |
| Agar-agar | | |
| see Agar | a | 517-0464 |
| Agate | | |
| see Silicon Oxide $Si O_2$ | f | 517-0683 |
| Air | f | 517-0020 |
| Alcan C X-30 | | |
| see Aluminum Oxide | ef | 517-0433 |
| Alcohol | | |
| see Ethanol | af | 517-0042 |
| Alcohol C-7 | | |
| see 1- Heptanol | f | 517-0319 |
| Aldehyde | | |
| see Acetaldehyde | af | 517-0517 |
| Alon - Cabot | | |
| see Aluminum Oxide | ef | 517-0433 |
| Alon C | | |
| see Silicon Oxide $Si O_2$ | f | 517-0683 |
| Alumina | | |
| see Aluminum Oxide | ef | 517-0433 |
| Aluminum Nitrate | f | 517-0765 |
| Aluminum Oxide | ef | 517-0433 |
| Aluminum Sulfate | f | 517-0770 |
| Alundum | | |
| see Aluminum Oxide | ef | 517-0433 |
| Amchlor | | |
| see Ammonium Chloride | aef | 517-0600 |
| American Ashes | | |
| see Potassium Carbonate | ef | 517-0065 |
| Amersil | | |
| see Silicon Oxide $Si O_2$ | f | 517-0683 |
| Aminoacetic Acid | | |
| see Glycine | f | 517-0519 |
| Aminoethanoic Acid | | |
| see Glycine | f | 517-0519 |
| Aminoform | | |
| see Hexamethylenetetramine | f | 517-0551 |
| Ammoform | | |
| see Hexamethylenetetramine | f | 517-0551 |
| Ammonia | aef | 517-0011 |
| Ammonioformaldehyde | | |
| see Hexamethylenetetramine | f | 517-0551 |
| Ammonium Acetate | f | 517-0830 |
| Ammonium Acid Phosphate | | |
| see Ammonium Phosphate $N H_4 H_2 P O_4$ | f | 517-0647 |
| Ammonium Bicarbonate | | |
| see Ammonium Carbonate $N H_4 H C O_3$ | ae | 517-0509 |
| Ammonium Bromide | f | 517-0500 |
| Ammonium Carbonate $N H_4 H C O_3$ | ae | 517-0509 |
| Ammonium Chloride | aef | 517-0600 |
| Ammonium Muriate | | |
| see Ammonium Chloride | aef | 517-0600 |
| Ammonium Nitrate | f | 517-0155 |
| Ammonium Nitrite | f | 517-0646 |
| Ammonium Perchlorate | f | 517-0709 |
| Ammonium Phosphate $N H_4 H_2 P O_4$ | f | 517-0647 |
| Ammonium Phosphate $(N H_4)_2 H P O_4$ | f | 517-0766 |
| Ammonium Phosphate $(N H_4)_3 P O_4$ | f | 517-0879 |
| Ammonium Pyrophosphate | f | 517-0889 |
| Ammonium Sulfate | f | 517-0157 |
| Amyl Alcohol | | |
| see 1- Pentanol | a | 517-0045 |
| pri- Amyl Alcohol | | |
| see 1- Pentanol | a | 517-0045 |
| A N B-3254 (Aerojet) | | |
| see Combustion Products Of Propellant A N B 3254 | l | 517-1053 |
| Anesthesia Ether | | |
| see Ethyl Ether | a | 517-0134 |

**Property:** **a.** Thermal conductivity; **b.** Accommodation coefficient; **c.** Thermal contact resistance; **d.** Thermal diffusivity; **e.** Specific heat; **f.** Viscosity; **g.** Emittance; **h.** Reflectance; **i.** Absorptance; **j.** Transmittance; **k.** $\alpha_s/\epsilon_t$ ratio; **l.** Prandtl number; **n.** Thermal linear expansion coefficient; **o.** Thermal volumetric expansion coefficient

| Substance Name | Property | Number |
|---|---|---|
| Anhydrone | | |
| see Magnesium Perchlorate | f | 517-0610 |
| Aniline Chloride | | |
| see Aniline Hydrochloride | f | 517-0547 |
| Aniline Hydrochloride | f | 517-0547 |
| Aniline Salt | | |
| see Aniline Hydrochloride | f | 517-0547 |
| Anilinium Chloride | | |
| see Aniline Hydrochloride | f | 517-0547 |
| Aqua Fortis | | |
| see Nitric Acid | af | 517-0311 |
| Astroquartz | | |
| see Silicon Oxide Si $O_2$ | f | 517-0683 |
| Bakers Sugar | | |
| see Dextrose | f | 517-0554 |
| Baking Soda | | |
| see Sodium Carbonate Na H C $O_3$ | ef | 517-0506 |
| Barium (+2) Ion | e | 517-0946 |
| Barium Chloride | e | 517-0089 |
| Barium Nitrate | f | 517-0453 |
| Basic Zirconium Chloride | | |
| see Dichlorooxozirconium | f | 517-0614 |
| Battery Acid | | |
| see Sulfuric Acid | f | 517-0644 |
| Beet Sugar | | |
| see Sucrose | f | 517-0578 |
| Bengalese Gelatin | | |
| see Agar | a | 517-0464 |
| Bengalese Isinglass | | |
| see Agar | a | 517-0464 |
| Benzene | f | 517-0006 |
| Benzol | | |
| see Benzene | f | 517-0006 |
| Benzole | | |
| see Benzene | f | 517-0006 |
| Beryllium Chloride | e | 517-0496 |
| Beryllium Nitrate | f | 517-0906 |
| Bihexyl | | |
| see Dodecane | f | 517-0115 |
| Bischofite | | |
| see Magnesium Chloride | ef | 517-0387 |
| Bis-(2-ethylhexyl)phosphate | f | 517-0993 |
| Blood, Human | f | 517-0750 |
| Blood, Plasma, Human | f | 517-0751 |
| Blood, Red Corpuscle, Human | f | 517-0807 |
| Blue Copperas | | |
| see Copper Sulfate | af | 517-0636 |
| Blue Stone | | |
| see Copper Sulfate | af | 517-0636 |
| Blue Vitriol | | |
| see Copper Sulfate | af | 517-0636 |
| Boric Acid | f | 517-0412 |
| Bromine (-1) Ion | e | 517-0937 |
| Butaldehyde | | |
| see Butyraldehyde | af | 517-0531 |
| Butalyde | | |
| see Butyraldehyde | af | 517-0531 |
| Butanal | | |
| see Butyraldehyde | af | 517-0531 |
| 1,4- Butanedicarboxylic Acid | | |
| see Adipic Acid | f | 517-0550 |
| 1- Butanol | af | 517-0044 |
| 2- Butenal | | |
| see Crotonaldehyde | af | 517-0528 |
| Butter Of Zinc | | |
| see Zinc Chloride | f | 517-0054 |
| Butyl Alcohol | | |
| see 1- Butanol | af | 517-0044 |
| tert- Butyl Alcohol | | |
| see 2- Methyl-2-propanol | e | 517-0113 |
| Butyl Aldehyde | | |
| see Butyraldehyde | af | 517-0531 |
| Butyl Carbinol | | |
| see 1- Pentanol | a | 517-0045 |
| Butyl Phosphate | | |
| see Tributyl Phosphate | f | 517-0580 |
| Butyraldehyde | af | 517-0531 |
| Butyric Alcohol | | |
| see 1- Butanol | af | 517-0044 |
| Butyric Aldehyde | | |
| see Butyraldehyde | af | 517-0531 |
| Cadmium Chloride | f | 517-0396 |
| Calcium (+2) Ion | e | 517-0944 |
| Calcium Chloride | aef | 517-0460 |
| Calcium Chromate | f | 517-0827 |
| Calcium Hyperchlorate | | |
| see Calcium Perchlorate | f | 517-0595 |
| Calcium Iodide | f | 517-0958 |
| Calcium Nitrate | f | 517-0058 |
| Calcium Perchlorate | f | 517-0595 |
| Calcium Phosphate | f | 517-0877 |
| Cane Sugar | | |
| see Sucrose | f | 517-0578 |
| Carbamide | | |
| see Urea | af | 517-0507 |
| Carbinol | | |
| see Methanol | af | 517-0028 |
| Carbonic Acid | | |
| see Carbon Oxide C $O_2$ | aefgl | 517-0193 |
| Carbonic Anhydride | | |
| see Carbon Oxide C $O_2$ | aefgl | 517-0193 |
| Carbon Oxide C O | ael | 517-0158 |
| Carbon Oxide C $O_2$ | aefgl | 517-0193 |
| Carbon Tetrachloride | f | 517-0117 |
| Carboxylic Acid | f | 517-0992 |
| Caustic Potash | | |
| see Potassium Hydroxide | f | 517-0642 |
| Caustic Soda | | |
| see Sodium Hydroxide | f | 517-0643 |
| Cesium (+1) Ion | e | 517-0940 |
| Cesium Bromide | f | 517-0498 |
| Cesium Chloride | f | 517-0031 |
| Cesium Iodide | f | 517-0101 |
| Cesium Nitrate | f | 517-0181 |
| Cesium Nitrite | f | 517-0633 |
| Cesium Sulfate | f | 517-0448 |
| Cetrimide | | |
| see Hexadecyltrimethylammonium | f | 517-0759 |
| Ceylon Gelatin | | |
| see Agar | a | 517-0464 |
| Ceylon Isinglass | | |
| see Agar | a | 517-0464 |
| Chalcanthite | | |
| see Copper Sulfate | af | 517-0636 |
| Chinese Gelatin | | |
| see Agar | a | 517-0464 |
| Chinese Isinglass | | |
| see Agar | a | 517-0464 |
| Chlorine (-1) Ion | e | 517-0936 |
| Chlorobenzene | f | 517-0196 |
| Chloroform | f | 517-0135 |
| Chromium Chloride | f | 517-0732 |
| Chromium Lithium Oxide | | |
| see Lithium Chromate | f | 517-0630 |
| Chromium Magnesium Oxide | | |
| see Magnesium Chromate | f | 517-0631 |
| Cloud | gh | 517-0466 |
| Cloud, Ice | | |
| see Cloud | gh | 517-0466 |
| Cobalt (+2) Ion | e | 517-0947 |
| Cobalt Chloride | f | 517-0604 |
| Combustion Products Of (Boron + Fuel Oil, Jet) | e | 517-1018 |
| Combustion Products Of (Chlorine Fluoride (Cl $F_5$) + Hydrazine) | e | 517-1036 |
| Combustion Products Of Coal | aefg | 517-0890 |

**Property:** **a.** Thermal conductivity; **b.** Accommodation coefficient; **c.** Thermal contact resistance; **d.** Thermal diffusivity; **e.** Specific heat; **f.** Viscosity; **g.** Emittance; **h.** Reflectance; **i.** Absorptance; **j.** Transmittance; **k.** $\alpha_s/\epsilon_t$ ratio; **l.** Prandtl number; **n.** Thermal linear expansion coefficient; **o.** Thermal volumetric expansion coefficient

**Property:** **a.** Thermal conductivity; **b.** Accommodation coefficient; **c.** Thermal contact resistance; **d.** Thermal diffusivity; **e.** Specific heat; **f.** Viscosity; **g.** Emittance; **h.** Reflectance; **i.** Absorptance; **j.** Transmittance; **k.** $\alpha_s/\epsilon_t$ ratio; **l.** Prandtl number; **n.** Thermal linear expansion coefficient; **o.** Thermal volumetric expansion coefficient

| Substance Name | Property | Number |
|---|---|---|
| Fermentation Alcohol | | |
| see Ethanol | af | 517-0042 |
| Flame | g | 517-0833 |
| Flame, (Acetylene + Air) | a | 517-0754 |
| Flame, (Acetylene + Helium + Oxygen) | a | 517-0755 |
| Flame, (Carbon Oxide $CO_2$ + Soot + Water) | g | 517-0998 |
| Flame, (Carbon Oxide $CO_2$ + Water) | g | 517-0997 |
| Flame, Luminous | g | 517-0473 |
| Flame, Nonluminous | g | 517-0996 |
| Flame (Carbon + Natural Gas) | g | 517-1080 |
| Flint | | |
| see Silicon Oxide $SiO_2$ | f | 517-0683 |
| Fluorine $_{-1}$ Ion | e | 517-0941 |
| Fog | | |
| see Cloud | gh | 517-0466 |
| Formamide | af | 517-0103 |
| Formamine | | |
| see Hexamethylenetetramine | f | 517-0551 |
| Formin | | |
| see Hexamethylenetetramine | f | 517-0551 |
| N-Formyldimethylamine | | |
| see N, N-Dimethylformamide | f | 517-0041 |
| Freon 10 | | |
| see Carbon Tetrachloride | f | 517-0117 |
| Freon 20 | | |
| see Chloroform | f | 517-0135 |
| Freon 50 | | |
| see Methane | aefl | 517-0001 |
| Freon 729 | | |
| see Air | f | 517-0020 |
| Fructose | f | 517-0553 |
| Fruit Sugar | | |
| see Fructose | f | 517-0553 |
| Fused Quartz | | |
| see Silicon Oxide $SiO_2$ | f | 517-0683 |
| G 74 | | |
| see Nitrogen | g | 517-0106 |
| Gadolinium (+3) Ion | e | 517-0956 |
| Ganister | | |
| see Silicon Oxide $SiO_2$ | f | 517-0683 |
| Gas, Exhaust, Rocket | g | 517-0474 |
| Gelose | | |
| see Agar | a | 517-0464 |
| Glacial Acetic Acid | | |
| see Acetic Acid | f | 517-0189 |
| Glasrock | | |
| see Silicon Oxide $SiO_2$ | f | 517-0683 |
| Glaubers Salt | | |
| see Sodium Sulfate | ef | 517-0681 |
| Glucose | f | 517-0552 |
| see also Dextrose | | 517-0554 |
| D-Glucose | | |
| see Dextrose | f | 517-0554 |
| (α-D-Glucosido)-β-D-fructofuranoside | | |
| see Sucrose | f | 517-0578 |
| Glycerin | | |
| see Glycerol | f | 517-0032 |
| Glycerol | f | 517-0032 |
| Glycerol Glass | | |
| see Glycerol | f | 517-0032 |
| Glycine | f | 517-0519 |
| Glycocoll | | |
| see Glycine | f | 517-0519 |
| Glycol Bis(Hydroxyethyl) Ether | | |
| see Triethylene Glycol | af | 517-0140 |
| Glycyl Alcohol | | |
| see Glycerol | f | 517-0032 |
| Grain Alcohol | | |
| see Ethanol | af | 517-0042 |
| Grape Sugar | | |
| see Dextrose | f | 517-0554 |

| Substance Name | Property | Number |
|---|---|---|
| Halite | | |
| see Sodium Chloride | aef | 517-0030 |
| Heavy Water | | |
| see Water, Dideuterated | f | 517-0213 |
| Heptane | f | 517-0216 |
| 1-Heptanol | f | 517-0319 |
| Heptyl Alcohol | | |
| see 1-Heptanol | f | 517-0319 |
| Herasil | | |
| see Silicon Oxide $SiO_2$ | f | 517-0683 |
| Hexadecyltrimethylammonium | f | 517-0759 |
| Hexahydrobenzene | | |
| see Cyclohexane | f | 517-0039 |
| Hexamethylenamine | | |
| see Hexamethylenetetramine | f | 517-0551 |
| Hexamethylene | | |
| see Cyclohexane | f | 517-0039 |
| Hexamethylenetetramine | f | 517-0551 |
| Hexamine | | |
| see Hexamethylenetetramine | f | 517-0551 |
| Hexanedioic Acid | | |
| see Adipic Acid | f | 517-0550 |
| Homosil | | |
| see Silicon Oxide $SiO_2$ | f | 517-0683 |
| Hornstone | | |
| see Silicon Oxide $SiO_2$ | f | 517-0683 |
| Hydrochloric Acid | f | 517-0194 |
| Hydrochloride | | |
| see Hydrochloric Acid | f | 517-0194 |
| Hydrocyanite | | |
| see Copper Sulfate | af | 517-0636 |
| Hydrogen | aefl | 517-0003 |
| Hydrogen (+1) Ion | e | 517-0955 |
| Hydrogen Chloride | | |
| see Hydrochloric Acid | f | 517-0194 |
| Hydrogen Sulfate | | |
| see Sulfuric Acid | f | 517-0644 |
| o-Hydroxytoluene | | |
| see o-Cresol | f | 517-0561 |
| Infrasil | | |
| see Silicon Oxide $SiO_2$ | f | 517-0683 |
| Iodine (-1) Ion | e | 517-0942 |
| Ion, Monatomic | e | 517-0934 |
| I P A | | |
| see 2-Propanol | aef | 517-0141 |
| Ipsen | | |
| see Aluminum Oxide | ef | 517-0433 |
| I R F N A | | |
| see Nitric Acid | af | 517-0311 |
| Iron Chloride $FeCl_2$ | f | 517-0608 |
| Iron Oxide | f | 517-0907 |
| Isoamyl Acetate | | |
| see Isopentyl Acetate | af | 517-0565 |
| Isoamyl Alcohol | | |
| see 3-Methyl-1-butanol | af | 517-0200 |
| pri-Isoamyl Alcohol | | |
| see 3-Methyl-1-butanol | af | 517-0200 |
| Isobutanol | | |
| see 2-Methyl-1-propanol | af | 517-0260 |
| Isobutyl Alcohol | | |
| see 2-Methyl-1-propanol | af | 517-0260 |
| Isopentyl Acetate | af | 517-0565 |
| Isopentyl Ethanoate | | |
| see Isopentyl Acetate | af | 517-0565 |
| Isopropanol | | |
| see 2-Propanol | aef | 517-0141 |
| Isopropyl Alcohol | | |
| see 2-Propanol | aef | 517-0141 |
| Isopropylcarbinol | | |
| see 2-Methyl-1-propanol | af | 517-0260 |

**Property: a.** Thermal conductivity; **b.** Accommodation coefficient; **c.** Thermal contact resistance; **d.** Thermal diffusivity; **e.** Specific heat; **f.** Viscosity; **g.** Emittance; **h.** Reflectance; **i.** Absorptance; **j.** Transmittance; **k.** $\alpha_s/\epsilon_t$ ratio; **l.** Prandtl number; **n.** Thermal linear expansion coefficient; **o.** Thermal volumetric expansion coefficient

**Property:** **a.** Thermal conductivity; **b.** Accommodation coefficient; **c.** Thermal contact resistance; **d.** Thermal diffusivity; **e.** Specific heat; **f.** Viscosity; **g.** Emittance; **h.** Reflectance; **i.** Absorptance; **j.** Transmittance; **k.** $a_s/\epsilon_t$ ratio; **l.** Prandtl number; **n.** Thermal linear expansion coefficient; **o.** Thermal volumetric expansion coefficient

**Property: a.** Thermal conductivity; **b.** Accommodation coefficient; **c.** Thermal contact resistance; **d.** Thermal diffusivity; **e.** Specific heat; **f.** Viscosity; **g.** Emittance; **h.** Reflectance; **i.** Absorptance; **j.** Transmittance; **k.** $\alpha_s/\epsilon_t$ ratio; **l.** Prandtl number; **n.** Thermal linear expansion coefficient; **o.** Thermal volumetric expansion coefficient

| Substance Name | Property | Number |
|---|---|---|
| Sodaniter | | |
| see Sodium Nitrate | ef | 517-0182 |
| Sodium (+1) Ion | e | 517-0938 |
| Sodium Acetate | f | 517-0721 |
| Sodium Acid Carbonate | | |
| see Sodium Carbonate $NaHCO_3$ | ef | 517-0506 |
| Sodium Bicarbonate | | |
| see Sodium Carbonate $NaHCO_3$ | ef | 517-0506 |
| Sodium Bromate | f | 517-0501 |
| Sodium Bromide | f | 517-0144 |
| Sodium Carbonate $NaHCO_3$ | ef | 517-0506 |
| Sodium Carbonate $Na_2CO_3$ | ef | 517-0067 |
| Sodium Chlorate | f | 517-0601 |
| Sodium Chloride | aef | 517-0030 |
| Sodium Citrate | f | 517-0760 |
| Sodium Dodecyl Sulfate | | |
| see Dodecyl Sodium Sulfate | f | 517-0579 |
| Sodium Fluoride | f | 517-0280 |
| Sodium Hydrate | | |
| see Sodium Hydroxide | f | 517-0643 |
| Sodium Hydroxide | f | 517-0643 |
| Sodium Hyperchlorate | | |
| see Sodium Perchlorate | f | 517-0221 |
| Sodium Iodide | ef | 517-0099 |
| Sodium Molybdate | | |
| see Molybdenum Sodium Oxide | f | 517-0666 |
| Sodium Nitrate | ef | 517-0182 |
| Sodium Nitrite | f | 517-0667 |
| Sodium Octadecanoate | | |
| see Sodium Stearate | f | 517-0588 |
| Sodium Octadecyl Sulfate | | |
| see Octadecyl Sodium Sulfate | f | 517-0589 |
| Sodium Oxide | ef | 517-0679 |
| Sodium Perchlorate | f | 517-0221 |
| Sodium Phosphate $NaH_2PO_4$ | f | 517-0762 |
| Sodium Phosphate $Na_2HPO_4$ | f | 517-0761 |
| Sodium Phosphate $Na_3PO_4$ | f | 517-0764 |
| Sodium Pyrophosphate | f | 517-0087 |
| Sodium Rhodanate | | |
| see Sodium Thiocyanate | f | 517-0512 |
| Sodium Rhodanide | | |
| see Sodium Thiocyanate | f | 517-0512 |
| Sodium Stearate | f | 517-0588 |
| Sodium Sulfate | ef | 517-0681 |
| Sodium Sulfocyanate | | |
| see Sodium Thiocyanate | f | 517-0512 |
| Sodium Sulfocyanide | | |
| see Sodium Thiocyanate | f | 517-0512 |
| Sodium Tartrate | f | 517-0763 |
| Sodium Tetradecyl Sulfate | f | 517-0583 |
| Sodium Tetraphenylborate | e | 517-0220 |
| Sodium Thiocyanate | f | 517-0512 |
| Spectrosil (Numbered Series) | | |
| see Silicon Oxide $SiO_2$ | f | 517-0683 |
| Starch Syrup | | |
| see Glucose | f | 517-0552 |
| Strontium (+2) Ion | e | 517-0945 |
| Strontium Chloride | e | 517-0619 |
| Strontium Hydrate | | |
| see Strontium Hydroxide | f | 517-0645 |
| Strontium Hydroxide | f | 517-0645 |
| Strontium Nitrate | f | 517-0452 |
| Sucrose | f | 517-0578 |
| Sugar | | |
| see Sucrose | f | 517-0578 |
| Sulfuric Acid | f | 517-0644 |
| Sulfuric Ether | | |
| see Ethyl Ether | a | 517-0134 |
| Sylvite | | |
| see Potassium Chloride | ef | 517-0048 |
| Sympatektoman | | |
| see Tetraethylammonium Bromide | f | 517-0571 |

| Substance Name | Property | Number |
|---|---|---|
| T B P | | |
| see Tributyl Phosphate | f | 517-0580 |
| T E A B | | |
| see Tetraethylammonium Bromide | f | 517-0571 |
| T E A C | | |
| see Tetraethylammonium Chloride | f | 517-0572 |
| T E A Chloride | | |
| see Tetraethylammonium Chloride | f | 517-0572 |
| T E G | | |
| see Triethylene Glycol | af | 517-0140 |
| Tempone | | |
| see 2,2,6,6- Tetramethyl-4-oxopiperidinooxy Radical | f | 517-0713 |
| Tetrabutylammonium Bromide | ef | 517-0118 |
| Tetrabutylammonium Chloride | f | 517-0585 |
| Tetrabutylammonium Iodide | f | 517-0147 |
| Tetrabutylammonium Nitrate | f | 517-0586 |
| Tetrachloromethane | | |
| see Carbon Tetrachloride | f | 517-0117 |
| Tetradecyl Sodium Sulfate | | |
| see Sodium Tetradecyl Sulfate | f | 517-0583 |
| Tetraethylammonium Bromide | f | 517-0571 |
| Tetraethylammonium Chloride | f | 517-0572 |
| Tetramethylammonium Bromide | f | 517-0357 |
| Tetramethylammonium Chloride | f | 517-0538 |
| 2,2,6,6- Tetramethyl-4-oxopiperidinooxy Radical | f | 517-0713 |
| 2,2,6,6- Tetramethylpiperidone- N-oxyl Radical | | |
| see 2,2,6,6- Tetramethyl-4-oxopiperidinooxy Radical | f | 517-0713 |
| Tetranium | | |
| see Tetraethylammonium Bromide | f | 517-0571 |
| Tetraphenylboronsodium | | |
| see Sodium Tetraphenylborate | e | 517-0220 |
| Tetrapropylammonium Bromide | f | 517-0581 |
| Tetrapropylammonium Chloride | f | 517-0582 |
| Thallium Perchlorate | e | 517-0866 |
| Thenardite | | |
| see Sodium Sulfate | ef | 517-0681 |
| Thermonatrite | | |
| see Sodium Carbonate $Na_2CO_3$ | ef | 517-0067 |
| Thiocarbamide | | |
| see Thiourea | f | 517-0508 |
| Thiourea | f | 517-0508 |
| Titanium Chloride | f | 517-0629 |
| Titanium Sulfate | f | 517-0685 |
| p- Toluidine Hydrochloride | f | 517-0563 |
| Trialkylamine | f | 517-0994 |
| Tributyl Phosphate | f | 517-0580 |
| Trichloromethane | | |
| see Chloroform | f | 517-0135 |
| Triethylene Glycol | af | 517-0140 |
| Trioxyethylene Glycol | | |
| see Triethylene Glycol | af | 517-0140 |
| Triphosphoric Acid | f | 517-0968 |
| Tris(2,2'-dipyridine)copper Dinitrate | f | 517-0593 |
| T S P P | | |
| see Sodium Pyrophosphate | f | 517-0087 |
| Ultrasil | | |
| see Silicon Oxide $SiO_2$ | f | 517-0683 |
| Uranium Perchlorate | h | 517-0602 |
| Urea | af | 517-0507 |
| Uritone | | |
| see Hexamethylenetetramine | f | 517-0551 |
| Urotropin | | |
| see Hexamethylenetetramine | f | 517-0551 |
| Vinegar Acid | | |
| see Acetic Acid | f | 517-0189 |
| Vinegar Naphtha | | |
| see Ethyl Acetate | a | 517-0086 |
| Vitreosil (Numbered Series) | | |
| see Silicon Oxide $SiO_2$ | f | 517-0683 |
| Vitreous Silica | | |
| see Silicon Oxide $SiO_2$ | f | 517-0683 |

**Property: a.** Thermal conductivity; **b.** Accommodation coefficient; **c.** Thermal contact resistance; **d.** Thermal diffusivity; **e.** Specific heat; **f.** Viscosity; **g.** Emittance; **h.** Reflectance; **i.** Absorptance; **j.** Transmittance; **k.** $\alpha_s/\epsilon_t$ ratio; **l.** Prandtl number; **n.** Thermal linear expansion coefficient; **o.** Thermal volumetric expansion coefficient

**Property: a.** Thermal conductivity; **b.** Accommodation coefficient; **c.** Thermal contact resistance; **d.** Thermal diffusivity; **e.** Specific heat; **f.** Viscosity; **g.** Emittance; **h.** Reflectance; **i.** Absorptance; **j.** Transmittance; **k.** $\alpha_s/\epsilon_t$ ratio; **l.** Prandtl number; **n.** Thermal linear expansion coefficient; **o.** Thermal volumetric expansion coefficient

# Part B
# SEARCH PARAMETERS

PART B

# USE OF SEARCH PARAMETERS

This part is arranged in 11 chapters, each representing a specific property. Within each chapter, the seven-digit substance numbers are listed in numerical order followed by physical state within each substance, subject, the language, the temperature range, and the year. The last column gives the TPRC serial number pertaining to the document retrieved. The code designations for the search parameters need not be memorized since they are interpreted at the bottom of each page of this section.

The substance numbers link this portion of the volume with the *Materials Directory* (Part A), and the TPRC numbers provide the link to the *Bibliography* (Part C). The procedure for the retrieval of references in literature searches is as follows:

1. Locate the desired substances (the individual components in a mixture) in Part A, the *Materials Directory*. Check if the desired property is listed, then record the seven-digit substance numbers.
2. In Part B (this section) locate the property chapter of interest and the seven-digit substance number of the *first* component.
3. Use as many of the five search parameters (physical state, subject, language, temperature, and year) as desired to locate and record TPRC numbers.
4. Locate the seven-digit substance number of the *second* component( and for each component of the mixture) in the same chapter and follow step 3.
5. If the *same* TPRC number is listed for each component's seven-digit number, then it is likely that there is a mixture of these components.
6. Locate the complete bibliographic citations for the recorded TPRC numbers in Part C, the *Bibliography*.

An example is given below.

*Example:* Obtain references on the viscosity of a mixture of acetone + methanol + carbon disulfide.

1. The components of the mixture, Acetone, Methanol, and Carbon Disulfide (Carbon Sulfide) are listed in Class 514 (Components of Nonaqueous Multiple Mixture), on page A14, column 1, page A16, column 2, and page A15, column 1. Viscosity, Property *f*, is listed for the materials. The seven-digit numbers are, respectively, 514-0136, 514-0028, and 514-0162.
2. In this section locate Chapter 4 (Viscosity), page B11. Locate the substance number of the *first* component (Acetone) 514-0136, on page B15, column 2.
3. For the substance number 514-0136, the following entries are listed:

| Substance Number | Phys. State | Subject | Language | Temperature | Year | TPRC Number |
|---|---|---|---|---|---|---|
| 514-0136 | L | D | R | N | 1971 | 63380 |
| 0136 | L | D | E | N | 1971 | 63381 |
| 0136 | L | D | E | N | 1973 | 75791 |

(The code letters of the parameters are interpreted below.*)

*Phys. State: D, Doped; E, Expanded; F, Fibrous or Whisker; G, Gas; L, Liquid; M, Multiphase; P, Powder; S. Solid.
Subject: C, Calculated Data + Theory; D, Data; E, Experiment; G, General (Data + Theory + Experiment); M, Measured Data + Experiment; S, Survey (Review, Compendium, etc.); T, Theory.
Language: C, Czech; D, Dutch; E, English; F, French; G, German; I, Italian; J, Japanese; O, Other Languages; P, Polish; R, Russian; S, Spanish.
Temperature: F, Full Range (Low + Normal + High); L, Low (0 to 75 K) + Overlap into Normal; N, Normal (75 to 1273 K); H, High (above 1273 K) + Overlap into Normal; U, Undefined.

4. Locate the substance number of the *second* component (Methanol) 514-0028, on page B15, column 1. The entries listed are:

| Substance Number | Phys. State | Sub-ject | Lan-guage | Temper-ature | Year | TPRC Number |
|---|---|---|---|---|---|---|
| 514-0028 | L | C | E | N | 1971 | 63598 |
| 0028 | L | D | R | N | 1972 | 70041 |
| 0028 | L | D | E | N | 1973 | 75791 |

Locate the substance number of the *third* component (Carbon Sulfide) 514-0162 on page B15, column 2. The entry listed is:

| Substance Number | Phys. State | Sub-ject | Lan-guage | Temper-ature | Year | TPRC Number |
|---|---|---|---|---|---|---|
| 514-0162 | L | D | E | N | 1973 | 75791 |

5. TPRC number 75791 is the only one common for the three components. It is very likely that this document contains the viscosity of the mixture of the three components.

6. Obtain complete bibliographic citation for TPRC number 75791 in Part C (the *Bibliography*).

# Chapter 1 Thermal Conductivity

| Substance Number | Phys. State | Sub-ject | Lan-guage | Temper-ature | Year | TPRC Number |
|---|---|---|---|---|---|---|
| 512-0001 | G | G | G | N | 1965 | 36223 |
| 0001 | G | C | G | N | 1968 | 50037 |
| 0001 | G | D | R | N | 1971 | 66597 |
| 0001 | G | D | G | N | 1973 | 73931 |
| 0001 | G | D | E | N | 1977 | 93626 |
| 0001 | G | T | E | N | 1977 | 93626 |
| 0001 | L | D | E | N | 1977 | 93626 |
| 0001 | L | T | E | N | 1977 | 93626 |
| 0002 | G | C | R | N | 1969 | 64255 |
| 0002 | G | D | R | N | 1971 | 67115 |
| 0002 | G | D | E | N | 1974 | 73190 |
| 0002 | G | C | E | N | 1974 | 74757 |
| 0003 | G | C | E | H | 1972 | 38084 |
| 0003 | G | C | R | N | 1965 | 43291 |
| 0003 | G | C | R | N | 1967 | 49226 |
| 0003 | G | D | E | N | 1971 | 61674 |
| 0003 | G | D | E | N | 1971 | 62470 |
| 0003 | G | C | E | N | 1971 | 64142 |
| 0003 | G | C | E | N | 1967 | 64189 |
| 0003 | G | D | R | N | 1971 | 66597 |
| 0003 | G | D | R | N | 1971 | 67115 |
| 0003 | G | D | G | N | 1973 | 73931 |
| 0003 | G | C | E | H | 1973 | 74430 |
| 0006 | L | D | R | N | 1973 | 72941 |
| 0006 | L | M | E | N | 1974 | 74791 |
| 0006 | L | D | E | N | 1974 | 74854 |
| 0006 | L | D | R | N | 1974 | 76115 |
| 0006 | L | D | E | N | 1974 | 76691 |
| 0008 | G | C | E | N | 1972 | 65975 |
| 0008 | L | G | G | N | 1966 | 44432 |
| 0008 | L | M | G | N | 1972 | 69283 |
| 0009 | G | C | E | N | 1972 | 65975 |
| 0009 | L | G | G | N | 1966 | 44432 |
| 0009 | L | M | E | N | 1967 | 63003 |
| 0009 | L | M | G | N | 1972 | 69283 |
| 0010 | L | G | G | N | 1966 | 44432 |
| 0010 | M | D | O | N | 1966 | 49792 |
| 0011 | G | C | E | H | 1972 | 38084 |
| 0011 | G | C | R | N | 1965 | 43291 |
| 0011 | G | G | G | N | 1968 | 49902 |
| 0011 | G | C | R | N | 1968 | 49920 |
| 0011 | G | G | E | N | 1971 | 54274 |
| 0011 | G | C | R | N | 1970 | 59019 |
| 0011 | G | D | E | U | 1971 | 64762 |
| 0011 | G | C | E | N | 1968 | 65122 |
| 0011 | G | C | E | N | 1970 | 71943 |
| 0011 | G | C | E | H | 1973 | 74430 |
| 0011 | G | D | R | N | 1973 | 75277 |
| 0012 | G | G | G | N | 1965 | 36223 |
| 0012 | G | D | F | N | 1971 | 43208 |
| 0012 | G | G | G | N | 1968 | 49902 |
| 0012 | G | C | G | N | 1968 | 50037 |
| 0012 | G | G | E | N | 1969 | 50258 |
| 0012 | G | D | E | N | 1970 | 58534 |
| 0012 | G | C | R | N | 1970 | 59019 |
| 0012 | G | C | E | N | 1971 | 61334 |
| 0012 | G | M | E | N | 1971 | 62078 |
| 0012 | G | D | E | N | 1971 | 62470 |
| 0012 | G | C | E | H | 1971 | 63651 |
| 0012 | G | C | E | N | 1971 | 64142 |
| 0012 | G | C | E | N | 1971 | 65187 |
| 0012 | G | D | E | N | 1971 | 66322 |
| 0012 | G | D | E | N | 1971 | 66354 |
| 0012 | G | D | E | N | 1972 | 67157 |
| 0012 | G | D | E | N | 1972 | 67158 |
| 0012 | G | M | R | N | 1972 | 67680 |
| 0012 | G | C | E | N | 1972 | 71386 |
| 0012 | G | C | E | U | 1973 | 71594 |
| 0012 | G | C | E | N | 1970 | 71943 |
| 0012 | G | D | E | H | 1973 | 72974 |
| 0012 | G | D | E | N | 1974 | 73190 |
| 0012 | G | D | G | N | 1973 | 73931 |
| 0012 | G | C | E | H | 1973 | 74430 |
| 0012 | G | C | E | N | 1974 | 74462 |
| 0012 | G | C | R | H | 1973 | 74799 |
| 0012 | G | M | R | N | 1974 | 74918 |
| 0012 | G | M | E | N | 1974 | 75005 |
| 0013 | G | G | G | N | 1968 | 49902 |
| 0014 | G | G | G | N | 1968 | 49902 |
| 0014 | G | G | E | N | 1970 | 65884 |
| 0015 | G | G | G | N | 1968 | 49902 |
| 0015 | G | C | G | N | 1968 | 50037 |
| 0015 | G | C | E | N | 1971 | 65187 |
| 0016 | G | G | G | N | 1968 | 49902 |
| 0016 | G | C | G | N | 1968 | 50037 |
| 0017 | G | G | G | N | 1968 | 49902 |
| 0017 | G | D | E | N | 1972 | 66548 |
| 0018 | G | G | G | N | 1968 | 49902 |
| 0018 | G | D | E | N | 1970 | 58534 |
| 0018 | G | C | R | N | 1970 | 59019 |
| 0018 | G | C | E | N | 1971 | 61334 |
| 0018 | G | M | E | N | 1971 | 62078 |
| 0018 | G | D | E | N | 1971 | 62470 |
| 0018 | G | C | E | H | 1971 | 63651 |
| 0018 | G | D | E | N | 1972 | 66090 |
| 512-0018 | G | D | E | N | 1971 | 66322 |
| 0018 | G | D | E | N | 1971 | 66354 |
| 0018 | G | C | E | N | 1972 | 66533 |
| 0018 | G | C | E | N | 1972 | 71386 |
| 0018 | G | C | E | N | 1970 | 71943 |
| 0018 | G | C | E | N | 1974 | 74462 |
| 0018 | G | C | R | H | 1973 | 74799 |
| 0019 | G | G | G | N | 1965 | 36223 |
| 0019 | G | C | G | N | 1968 | 50037 |
| 0019 | G | D | E | N | 1977 | 93626 |
| 0019 | G | T | E | N | 1977 | 93626 |
| 0019 | L | D | E | N | 1977 | 93626 |
| 0019 | L | T | E | N | 1977 | 93626 |
| 0021 | G | G | G | N | 1965 | 36223 |
| 0021 | G | D | E | N | 1977 | 93626 |
| 0021 | G | T | E | N | 1977 | 93626 |
| 0022 | G | G | G | N | 1965 | 36223 |
| 0022 | G | D | E | N | 1972 | 67157 |
| 0022 | G | D | E | N | 1977 | 93626 |
| 0022 | G | T | E | N | 1977 | 93626 |
| 0022 | L | D | E | N | 1977 | 93626 |
| 0022 | L | T | E | N | 1977 | 93626 |
| 0023 | G | G | G | N | 1965 | 36223 |
| 0023 | G | D | R | N | 1973 | 72955 |
| 0026 | G | M | G | H | 1965 | 34150 |
| 0026 | G | M | E | H | 1967 | 47654 |
| 0026 | G | G | G | H | 1968 | 54290 |
| 0026 | G | T | R | H | 1970 | 62620 |
| 0026 | G | T | E | H | 1971 | 62621 |
| 0026 | G | C | E | H | 1970 | 68072 |
| 0026 | G | C | E | H | 1972 | 68719 |
| 0026 | G | D | E | H | 1972 | 68848 |
| 0026 | G | T | R | H | 1973 | 71356 |
| 0026 | G | T | R | H | 1973 | 72250 |
| 0026 | G | T | E | H | 1973 | 72251 |
| 0026 | G | T | E | H | 1973 | 73481 |
| 0026 | G | D | E | H | 1977 | 89755 |
| 0026 | G | T | E | H | 1977 | 89755 |
| 0026 | G | D | E | H | 1976 | 91157 |
| 0026 | G | T | E | H | 1976 | 91157 |
| 0028 | G | C | R | N | 1970 | 59019 |
| 0028 | G | C | E | N | 1970 | 71943 |
| 0028 | L | C | R | N | 1968 | 50756 |
| 0028 | L | C | E | N | 1968 | 65336 |
| 0028 | L | M | E | N | 1974 | 74791 |
| 0028 | L | C | E | N | 1974 | 74854 |
| 0028 | L | D | E | N | 1974 | 76691 |
| 0030 | L | D | R | N | 1974 | 76115 |
| 0030 | L | D | E | N | 1974 | 77220 |
| 0030 | S | M | R | N | 1965 | 42140 |
| 0032 | L | C | R | N | 1968 | 50756 |
| 0032 | L | C | E | N | 1968 | 65336 |
| 0036 | G | G | E | N | 1969 | 50258 |
| 0036 | G | C | R | N | 1970 | 59019 |
| 0036 | G | C | E | N | 1971 | 61334 |
| 0036 | G | D | E | N | 1971 | 62470 |
| 0036 | G | C | E | H | 1971 | 63651 |
| 0036 | G | D | E | N | 1972 | 66090 |
| 0036 | G | C | E | N | 1972 | 71386 |
| 0036 | G | C | E | U | 1973 | 71594 |
| 0036 | G | C | E | N | 1970 | 71943 |
| 0036 | G | C | R | H | 1973 | 74799 |
| 0037 | G | C | R | N | 1970 | 59019 |
| 0037 | G | C | E | N | 1971 | 61334 |
| 0037 | G | D | E | N | 1971 | 62470 |
| 0037 | G | C | E | H | 1971 | 63651 |
| 0037 | G | C | E | N | 1971 | 65187 |
| 0037 | G | C | E | U | 1972 | 65366 |
| 0037 | G | C | E | N | 1972 | 71386 |
| 0037 | G | C | E | N | 1970 | 71943 |
| 0037 | G | C | R | H | 1973 | 74799 |
| 0038 | G | C | E | H | 1970 | 37602 |
| 0038 | G | T | E | H | 1969 | 55181 |
| 0038 | G | T | R | H | 1970 | 62368 |
| 0038 | G | T | E | H | 1971 | 62369 |
| 0038 | G | S | E | H | 1971 | 63222 |
| 0038 | G | G | E | H | 1971 | 63637 |
| 0038 | G | C | E | H | 1971 | 63638 |
| 0038 | G | T | E | H | 1970 | 65541 |
| 0038 | G | T | R | H | 1970 | 65632 |
| 0038 | G | T | E | H | 1970 | 65633 |
| 0038 | G | C | E | U | 1971 | 66242 |
| 0038 | G | C | E | H | 1969 | 66267 |
| 0038 | G | T | E | U | 1972 | 66642 |
| 0038 | G | T | R | H | 1972 | 67514 |
| 0038 | G | T | E | H | 1971 | 68305 |
| 0038 | G | T | R | H | 1972 | 68453 |
| 0038 | G | C | E | H | 1972 | 68719 |
| 0038 | G | T | E | H | 1973 | 69252 |
| 0038 | G | T | E | H | 1972 | 69670 |
| 0038 | G | T | E | H | 1972 | 69683 |
| 0038 | G | T | E | H | 1972 | 69807 |
| 0038 | G | T | E | H | 1972 | 69808 |
| 0038 | G | T | E | H | 1973 | 70289 |
| 0038 | G | C | E | H | 1972 | 70292 |
| 512-0038 | G | T | E | H | 1973 | 73157 |
| 0038 | G | S | E | H | 1973 | 73279 |
| 0038 | G | T | E | H | 1972 | 73327 |
| 0038 | G | C | R | H | 1972 | 73463 |
| 0038 | G | C | E | H | 1972 | 73464 |
| 0038 | G | C | F | H | 1973 | 73933 |
| 0038 | G | T | R | H | 1974 | 76793 |
| 0038 | G | T | E | H | 1974 | 76794 |
| 0044 | L | C | R | N | 1968 | 50756 |
| 0044 | L | C | E | N | 1968 | 65336 |
| 0044 | L | M | E | N | 1974 | 74791 |
| 0044 | L | C | E | N | 1974 | 74854 |
| 0044 | L | D | E | N | 1974 | 76691 |
| 0045 | L | M | E | N | 1974 | 74791 |
| 0046 | L | M | E | N | 1974 | 74791 |
| 0046 | L | C | E | N | 1974 | 74854 |
| 0047 | L | D | R | N | 1973 | 72941 |
| 0047 | L | M | E | N | 1974 | 74791 |
| 0047 | L | C | E | N | 1974 | 74854 |
| 0047 | L | D | E | N | 1974 | 76691 |
| 0048 | L | D | R | N | 1973 | 72533 |
| 0048 | L | D | E | N | 1974 | 77220 |
| 0048 | S | D | E | N | 1972 | 68858 |
| 0048 | S | D | E | N | 1974 | 75755 |
| 0051 | G | G | E | N | 1971 | 65727 |
| 0051 | G | S | R | N | 1971 | 66284 |
| 0051 | G | S | E | N | 1971 | 67539 |
| 0052 | G | G | E | N | 1971 | 65727 |
| 0052 | G | S | R | N | 1971 | 66284 |
| 0052 | G | S | E | N | 1971 | 67539 |
| 0052 | L | M | R | N | 1970 | 63894 |
| 0052 | S | M | R | N | 1970 | 63894 |
| 0053 | G | T | E | H | 1971 | 65758 |
| 0055 | G | T | E | U | 1972 | 68415 |
| 0055 | L | S | E | L | 1964 | 38454 |
| 0055 | L | S | J | L | 1968 | 54356 |
| 0055 | L | T | E | U | 1972 | 68415 |
| 0055 | L | D | R | L | 1973 | 73290 |
| 0055 | L | C | E | L | 1974 | 74745 |
| 0055 | L | G | E | L | 1974 | 76831 |
| 0055 | S | S | E | L | 1964 | 38454 |
| 0055 | S | D | E | L | 1972 | 66112 |
| 0056 | G | T | E | U | 1972 | 68415 |
| 0056 | L | S | E | L | 1964 | 38454 |
| 0056 | L | S | J | L | 1968 | 54356 |
| 0056 | L | T | E | U | 1972 | 68415 |
| 0056 | L | D | R | L | 1973 | 73290 |
| 0056 | L | C | E | L | 1974 | 74745 |
| 0056 | L | G | E | L | 1974 | 76831 |
| 0056 | S | S | E | L | 1964 | 38454 |
| 0056 | S | D | E | L | 1972 | 66112 |
| 0057 | L | G | E | N | 1969 | 65879 |
| 0057 | L | M | R | N | 1973 | 75169 |
| 0061 | G | C | E | H | 1972 | 38084 |
| 0061 | G | D | F | N | 1971 | 43208 |
| 0061 | G | C | R | N | 1967 | 49226 |
| 0061 | G | G | E | N | 1969 | 50258 |
| 0061 | G | D | E | N | 1970 | 58534 |
| 0061 | G | C | R | N | 1970 | 59019 |
| 0061 | G | C | E | N | 1971 | 61334 |
| 0061 | G | D | E | N | 1971 | 61674 |
| 0061 | G | M | E | N | 1971 | 62078 |
| 0061 | G | D | E | N | 1971 | 62816 |
| 0061 | G | C | E | H | 1971 | 63651 |
| 0061 | G | C | E | N | 1971 | 64142 |
| 0061 | G | C | E | N | 1967 | 64189 |
| 0061 | G | M | E | N | 1970 | 64299 |
| 0061 | G | C | E | U | 1972 | 65366 |
| 0061 | G | D | E | N | 1971 | 66322 |
| 0061 | G | D | E | N | 1971 | 66354 |
| 0061 | G | C | E | N | 1972 | 66533 |
| 0061 | G | M | R | N | 1972 | 67680 |
| 0061 | G | C | E | U | 1972 | 69764 |
| 0061 | G | C | E | N | 1972 | 71386 |
| 0061 | G | C | E | U | 1973 | 71594 |
| 0061 | G | C | E | N | 1970 | 71943 |
| 0061 | G | C | E | H | 1973 | 74430 |
| 0061 | G | C | E | N | 1974 | 74462 |
| 0061 | G | C | R | H | 1973 | 74799 |
| 0061 | G | M | E | N | 1974 | 75005 |
| 0061 | G | D | E | N | 1974 | 76378 |
| 0064 | D | S | E | L | 1970 | 61232 |
| 0064 | D | S | E | N | 1970 | 61232 |
| 0064 | G | S | E | N | 1963 | 35381 |
| 0064 | G | T | R | H | 1962 | 37009 |
| 0064 | G | E | R | U | 1965 | 38921 |
| 0064 | G | T | R | U | 1969 | 55757 |
| 0064 | G | S | R | L | 1969 | 55851 |
| 0064 | G | T | G | U | 1969 | 56735 |
| 0064 | G | E | R | U | 1968 | 57538 |
| 0064 | G | T | R | U | 1968 | 57538 |
| 0064 | G | T | R | U | 1970 | 60734 |
| 0064 | G | T | E | U | 1971 | 61331 |
| 0064 | G | T | E | U | 1971 | 62463 |
| 0064 | G | E | E | U | 1971 | 63422 |

**Phys. State:** **D.** Doped; **E.** Expanded; **F.** Fibrous or Whisker; **G.** Gas; **L.** Liquid; **M.** Multiphase; **P.** Powder; **S.** Solid
**Subject:** **C.** Calculated Data + Theory; **D.** Data; **E.** Experiment; **G.** General (Data + Theory + Experiment); **M.** Measured Data + Experiment; **S.** Survey (Review, Compendium, etc.); **T.** Theory
**Language:** **C.** Czech; **D.** Dutch; **E.** English; **F.** French; **G.** German; **I.** Italian; **J.** Japanese; **O.** Other Languages; **P.** Polish; **R.** Russian; **S.** Spanish
**Temperature:** **F.** Full Range (Low + Normal + High); **L.** Low (0 to 75K) + Overlap into Normal; **N.** Normal (75 to 1273K); **H.** High (above 1273K) + Overlap into Normal; **U.** Undefined

| Substance Number | Phys. State | Subject | Language | Temperature | Year | TPRC Number |
|---|---|---|---|---|---|---|
| 512-0064 | G | T | R | N | 1970 | 63875 |
| 0064 | G | E | R | L | 1970 | 63893 |
| 0064 | G | E | E | U | 1971 | 64047 |
| 0064 | G | T | E | U | 1971 | 64047 |
| 0064 | G | E | E | U | 1971 | 64761 |
| 0064 | G | T | E | U | 1962 | 64901 |
| 0064 | G | C | F | U | 1971 | 65623 |
| 0064 | G | C | E | N | 1972 | 66545 |
| 0064 | G | T | R | U | 1972 | 67090 |
| 0064 | G | T | E | N | 1972 | 67501 |
| 0064 | G | E | R | N | 1972 | 67680 |
| 0064 | G | T | E | U | 1969 | 69502 |
| 0064 | G | S | E | L | 1969 | 69503 |
| 0064 | G | T | R | N | 1972 | 70272 |
| 0064 | G | T | E | U | 1970 | 72172 |
| 0064 | G | S | E | N | 1973 | 73278 |
| 0064 | G | T | R | U | 1971 | 73559 |
| 0064 | G | T | E | U | 1972 | 73560 |
| 0064 | G | C | R | H | 1973 | 74799 |
| 0064 | G | T | E | U | 1974 | 74827 |
| 0064 | G | T | E | U | 1974 | 74952 |
| 0064 | G | E | E | N | 1974 | 75005 |
| 0064 | G | T | E | U | 1974 | 76389 |
| 0064 | G | S | R | H | 1971 | 77228 |
| 0064 | G | S | E | H | 1971 | 77229 |
| 0064 | L | T | R | U | 1970 | 60734 |
| 0064 | L | T | E | U | 1971 | 62463 |
| 0064 | L | T | E | N | 1972 | 67501 |
| 0064 | L | E | C | U | 1972 | 69474 |
| 0064 | L | S | E | N | 1972 | 69620 |
| 0064 | L | S | E | N | 1971 | 69767 |
| 0064 | L | T | E | U | 1970 | 72172 |
| 0064 | L | T | E | U | 1974 | 74827 |
| 0064 | L | T | E | U | 1974 | 74952 |
| 0064 | M | T | R | H | 1962 | 37009 |
| 0064 | S | S | E | L | 1970 | 61232 |
| 0064 | S | S | E | N | 1970 | 61232 |
| 0064 | S | T | E | U | 1967 | 61760 |
| 0069 | G | C | E | H | 1972 | 38084 |
| 0069 | G | D | E | N | 1971 | 62470 |
| 0069 | G | C | E | N | 1971 | 64142 |
| 0069 | G | M | E | N | 1970 | 64299 |
| 0069 | G | T | E | L | 1972 | 66544 |
| 0069 | G | C | E | U | 1972 | 69764 |
| 0069 | G | C | E | H | 1973 | 74430 |
| 0072 | L | C | E | N | 1974 | 74854 |
| 0084 | L | M | E | N | 1974 | 74791 |
| 0084 | L | C | E | N | 1974 | 74854 |
| 0090 | S | M | G | N | 1964 | 36882 |
| 0091 | D | D | E | N | 1973 | 73823 |
| 0091 | S | M | G | N | 1964 | 36882 |
| 0091 | S | D | E | N | 1973 | 73823 |
| 0092 | S | M | G | N | 1964 | 36882 |
| 0092 | S | D | E | N | 1973 | 73823 |
| 0093 | S | M | G | N | 1964 | 36882 |
| 0094 | S | M | G | N | 1964 | 36882 |
| 0094 | S | D | E | H | 1962 | 77281 |
| 0095 | G | C | G | H | 1965 | 47645 |
| 0095 | G | C | R | H | 1967 | 49693 |
| 0095 | G | M | E | H | 1969 | 61076 |
| 0095 | G | G | E | H | 1971 | 63637 |
| 0095 | G | D | R | H | 1971 | 64641 |
| 0095 | G | D | E | H | 1971 | 64642 |
| 0095 | G | C | E | H | 1970 | 64975 |
| 0095 | G | C | E | H | 1971 | 66217 |
| 0095 | G | G | R | H | 1971 | 66262 |
| 0095 | G | D | E | H | 1972 | 68848 |
| 0095 | G | D | E | H | 1972 | 68908 |
| 0095 | G | C | E | H | 1973 | 71182 |
| 0095 | G | D | E | H | 1976 | 91157 |
| 0095 | G | T | E | H | 1976 | 91157 |
| 0096 | G | C | G | H | 1965 | 47645 |
| 0096 | G | C | E | H | 1967 | 47653 |
| 0096 | G | T | E | H | 1967 | 47655 |
| 0096 | G | G | G | H | 1970 | 60128 |
| 0096 | G | G | O | H | 1969 | 60796 |
| 0096 | G | D | E | H | 1969 | 61077 |
| 0096 | G | G | E | H | 1971 | 63637 |
| 0096 | G | C | E | H | 1970 | 64975 |
| 0096 | G | D | E | H | 1972 | 67279 |
| 0096 | G | C | F | H | 1972 | 67745 |
| 0096 | G | D | E | H | 1972 | 68848 |
| 0096 | G | D | E | H | 1977 | 89755 |
| 0096 | G | T | E | H | 1977 | 89755 |
| 0099 | S | D | R | N | 1971 | 65414 |
| 0099 | S | D | E | N | 1971 | 65415 |
| 0103 | L | C | R | N | 1968 | 50756 |
| 0103 | L | C | E | N | 1968 | 65336 |
| 0104 | G | C | E | N | 1971 | 64142 |
| 0104 | G | T | E | L | 1972 | 66544 |
| 0104 | G | C | E | N | 1974 | 74462 |
| 0105 | G | G | E | N | 1971 | 54274 |
| 0105 | G | M | E | N | 1971 | 64761 |
| 0105 | G | C | E | N | 1972 | 65975 |
| 0106 | G | C | R | N | 1965 | 43291 |
| 0106 | G | C | R | N | 1968 | 49920 |
| 0106 | G | G | E | N | 1971 | 54274 |
| 0106 | G | D | E | N | 1970 | 58534 |
| 0106 | G | C | R | N | 1970 | 59019 |
| 0106 | G | M | E | N | 1971 | 62078 |
| 0106 | G | D | E | N | 1971 | 62470 |

| Substance Number | Phys. State | Subject | Language | Temperature | Year | TPRC Number |
|---|---|---|---|---|---|---|
| 512-0106 | G | D | E | N | 1971 | 62816 |
| 0106 | G | C | R | N | 1969 | 64255 |
| 0106 | G | M | E | N | 1971 | 64761 |
| 0106 | G | D | E | U | 1971 | 64762 |
| 0106 | G | C | E | N | 1968 | 65122 |
| 0106 | G | D | E | N | 1971 | 66322 |
| 0106 | G | D | E | N | 1971 | 66354 |
| 0106 | G | D | R | N | 1971 | 66597 |
| 0106 | G | D | R | N | 1971 | 67115 |
| 0106 | G | D | R | H | 1972 | 70621 |
| 0106 | G | D | E | H | 1972 | 70622 |
| 0106 | G | C | E | N | 1970 | 71943 |
| 0106 | G | D | R | N | 1973 | 72955 |
| 0106 | G | D | E | H | 1973 | 72974 |
| 0106 | G | C | E | H | 1973 | 74430 |
| 0106 | G | C | E | N | 1974 | 74462 |
| 0106 | G | D | R | N | 1973 | 75277 |
| 0106 | G | D | E | N | 1977 | 93626 |
| 0106 | G | T | E | N | 1977 | 93626 |
| 0106 | L | D | E | N | 1977 | 93626 |
| 0106 | L | T | E | N | 1977 | 93626 |
| 0107 | G | G | E | N | 1971 | 54274 |
| 0107 | G | M | E | N | 1971 | 64761 |
| 0108 | G | G | E | N | 1971 | 54274 |
| 0108 | G | M | E | N | 1971 | 64761 |
| 0109 | G | T | E | L | 1972 | 65367 |
| 0110 | G | G | E | N | 1972 | 71404 |
| 0112 | G | S | E | U | 1972 | 38291 |
| 0112 | G | S | E | H | 1971 | 62464 |
| 0112 | G | S | R | U | 1970 | 63876 |
| 0117 | L | M | E | N | 1974 | 74791 |
| 0117 | L | C | E | N | 1974 | 74854 |
| 0117 | L | D | E | N | 1974 | 76691 |
| 0117 | M | M | E | N | 1969 | 36336 |
| 0120 | S | D | R | N | 1971 | 65414 |
| 0120 | S | D | E | N | 1971 | 65415 |
| 0130 | L | S | E | N | 1971 | 61580 |
| 0131 | L | C | R | N | 1968 | 50756 |
| 0131 | L | C | E | N | 1968 | 65336 |
| 0133 | G | T | R | N | 1968 | 49920 |
| 0133 | G | T | E | N | 1968 | 65122 |
| 0134 | G | C | R | N | 1968 | 49920 |
| 0134 | G | D | E | N | 1970 | 58534 |
| 0134 | G | C | E | N | 1968 | 65122 |
| 0134 | G | D | E | N | 1972 | 66548 |
| 0134 | L | M | E | N | 1974 | 74791 |
| 0134 | L | C | E | N | 1974 | 74854 |
| 0134 | L | D | E | N | 1974 | 76691 |
| 0135 | G | C | R | N | 1968 | 49920 |
| 0135 | G | C | E | N | 1968 | 65122 |
| 0135 | L | M | E | N | 1974 | 74791 |
| 0136 | G | C | R | N | 1968 | 49920 |
| 0136 | G | C | E | N | 1968 | 65122 |
| 0136 | L | M | E | N | 1974 | 74791 |
| 0136 | L | C | E | N | 1974 | 74854 |
| 0136 | L | D | E | N | 1974 | 76691 |
| 0137 | G | C | R | N | 1968 | 49920 |
| 0137 | G | C | E | N | 1968 | 65122 |
| 0138 | P | M | R | U | 1968 | 49504 |
| 0138 | P | M | E | U | 1968 | 65123 |
| 0139 | P | M | R | U | 1968 | 49504 |
| 0139 | P | M | E | U | 1968 | 65123 |
| 0140 | L | C | R | N | 1968 | 50756 |
| 0140 | L | C | E | N | 1968 | 65336 |
| 0141 | L | C | R | N | 1968 | 50756 |
| 0141 | L | C | E | N | 1968 | 65336 |
| 0141 | L | M | E | N | 1974 | 74791 |
| 0142 | L | C | R | N | 1968 | 50756 |
| 0142 | L | C | E | N | 1968 | 65336 |
| 0151 | G | D | E | N | 1972 | 67157 |
| 0152 | G | C | R | N | 1965 | 43291 |
| 0152 | G | C | R | N | 1970 | 59019 |
| 0152 | G | D | E | N | 1971 | 62470 |
| 0152 | G | C | E | N | 1970 | 71943 |
| 0152 | G | D | E | N | 1974 | 76378 |
| 0158 | G | C | R | N | 1965 | 43291 |
| 0158 | G | D | R | N | 1971 | 66597 |
| 0172 | G | G | E | H | 1971 | 63637 |
| 0172 | G | D | R | H | 1971 | 63941 |
| 0172 | G | D | E | H | 1971 | 63942 |
| 0174 | G | T | E | H | 1973 | 73019 |
| 0174 | P | M | E | U | 1973 | 77930 |
| 0174 | S | D | J | H | 1964 | 43236 |
| 0174 | S | D | E | N | 1972 | 68837 |
| 0174 | S | D | E | N | 1970 | 77563 |
| 0174 | S | M | E | H | 1969 | 77573 |
| 0175 | P | M | E | U | 1973 | 77930 |
| 0175 | S | D | J | H | 1964 | 43236 |
| 0175 | S | D | R | H | 1971 | 67751 |
| 0175 | S | D | E | N | 1972 | 68819 |
| 0175 | S | D | E | N | 1973 | 71359 |
| 0175 | S | M | E | H | 1974 | 77928 |
| 0176 | M | C | E | N | 1970 | 46359 |
| 0177 | S | M | R | N | 1963 | 36064 |
| 0180 | L | G | E | N | 1969 | 65879 |
| 0181 | L | G | E | N | 1969 | 65879 |
| 0182 | L | G | E | N | 1969 | 65879 |
| 0182 | L | M | R | N | 1973 | 75169 |
| 0183 | L | G | E | N | 1969 | 65879 |
| 0188 | G | G | E | N | 1972 | 71404 |
| 0189 | L | D | R | N | 1973 | 75840 |

| Substance Number | Phys. State | Subject | Language | Temperature | Year | TPRC Number |
|---|---|---|---|---|---|---|
| 512-0189 | L | D | E | N | 1973 | 75841 |
| 0192 | G | C | E | H | 1973 | 74430 |
| 0193 | G | C | R | N | 1965 | 43291 |
| 0193 | G | D | E | N | 1971 | 62816 |
| 0193 | G | C | R | N | 1969 | 64255 |
| 0193 | G | D | R | H | 1972 | 70621 |
| 0193 | G | D | E | H | 1972 | 70622 |
| 0193 | G | D | R | N | 1973 | 72955 |
| 0193 | G | M | R | N | 1974 | 74918 |
| 0193 | G | D | E | N | 1977 | 93626 |
| 0193 | G | T | E | N | 1977 | 93626 |
| 0193 | L | D | E | N | 1977 | 93626 |
| 0193 | L | T | E | N | 1977 | 93626 |
| 0194 | G | C | E | N | 1971 | 69329 |
| 0194 | G | C | E | H | 1973 | 74430 |
| 0195 | L | D | R | N | 1973 | 72941 |
| 0202 | G | C | E | H | 1972 | 38084 |
| 0202 | G | C | E | H | 1973 | 74430 |
| 0205 | G | C | E | U | 1972 | 65366 |
| 0205 | G | G | E | N | 1970 | 65884 |
| 0210 | G | T | R | U | 1968 | 53433 |
| 0210 | G | T | E | U | 1968 | 67624 |
| 0214 | L | M | E | N | 1974 | 74791 |
| 0222 | G | C | E | H | 1970 | 64975 |
| 0222 | G | C | E | H | 1970 | 68072 |
| 0222 | G | E | R | H | 1972 | 68630 |
| 0222 | G | E | E | H | 1973 | 69671 |
| 0225 | G | C | E | H | 1970 | 68072 |
| 0226 | G | D | E | N | 1972 | 67157 |
| 0227 | G | D | E | N | 1972 | 67157 |
| 0228 | G | D | E | N | 1972 | 67157 |
| 0229 | G | D | E | N | 1972 | 67158 |
| 0230 | G | D | E | N | 1972 | 67158 |
| 0231 | G | D | E | N | 1972 | 67158 |
| 0232 | G | D | E | N | 1972 | 67158 |
| 0233 | S | D | E | H | 1973 | 70463 |
| 0235 | L | M | E | N | 1974 | 74791 |
| 0235 | L | D | E | N | 1974 | 74854 |
| 0235 | L | D | E | N | 1974 | 76691 |
| 0243 | G | D | E | H | 1967 | 47658 |
| 0243 | S | G | E | H | 1967 | 48010 |
| 0256 | G | T | E | H | 1972 | 70286 |
| 0259 | S | D | E | H | 1972 | 68858 |
| 0259 | S | D | E | N | 1974 | 75755 |
| 0260 | L | M | E | N | 1974 | 74791 |
| 0262 | S | D | E | N | 1965 | 45152 |
| 0262 | S | D | R | H | 1970 | 61943 |
| 0262 | S | D | R | N | 1970 | 63212 |
| 0262 | S | D | E | N | 1963 | 70328 |
| 0262 | S | D | R | N | 1972 | 71641 |
| 0262 | S | D | R | H | 1970 | 74076 |
| 0262 | S | D | E | H | 1970 | 74077 |
| 0262 | S | G | R | H | 1974 | 75073 |
| 0262 | S | G | E | H | 1974 | 75074 |
| 0262 | S | D | E | H | 1962 | 77281 |
| 0262 | S | M | E | H | 1974 | 77928 |
| 0269 | L | M | E | N | 1974 | 74791 |
| 0270 | G | D | E | N | 1971 | 66322 |
| 0270 | G | D | E | N | 1971 | 66354 |
| 0287 | S | D | R | H | 1970 | 61943 |
| 0287 | S | D | R | N | 1970 | 63212 |
| 0287 | S | D | E | N | 1963 | 70328 |
| 0290 | L | T | E | U | 1973 | 75142 |
| 0291 | L | M | E | N | 1967 | 63003 |
| 0292 | G | C | E | H | 1972 | 38084 |
| 0292 | G | C | E | H | 1973 | 74430 |
| 0293 | G | C | E | H | 1972 | 38084 |
| 0293 | G | C | E | H | 1973 | 74430 |
| 0294 | G | C | E | H | 1972 | 38084 |
| 0294 | G | C | E | H | 1973 | 74430 |
| 0295 | G | C | E | H | 1972 | 38084 |
| 0295 | G | C | E | H | 1973 | 74430 |
| 0296 | G | C | E | H | 1972 | 38084 |
| 0296 | G | C | E | N | 1971 | 69329 |
| 0296 | G | C | E | H | 1973 | 74430 |
| 0297 | G | C | E | H | 1972 | 38084 |
| 0297 | G | C | E | N | 1971 | 69329 |
| 0297 | G | C | E | H | 1973 | 74430 |
| 0311 | L | M | R | N | 1970 | 63894 |
| 0311 | S | M | R | N | 1970 | 63894 |
| 0315 | L | D | E | N | 1973 | 71002 |
| 0319 | L | M | E | N | 1974 | 74791 |
| 0328 | G | C | E | N | 1972 | 65975 |
| 0329 | G | C | E | N | 1972 | 65975 |
| 0340 | M | C | E | N | 1970 | 46359 |
| 0340 | M | C | E | U | 1972 | 66009 |
| 0341 | M | C | E | N | 1970 | 46359 |
| 0341 | M | C | E | U | 1972 | 66009 |
| 0342 | M | C | E | N | 1970 | 46359 |
| 0342 | M | C | E | U | 1972 | 66009 |
| 0343 | M | C | E | N | 1970 | 46359 |
| 0343 | M | C | E | U | 1972 | 66009 |
| 0356 | G | E | R | N | 1971 | 66570 |
| 0356 | L | E | R | N | 1971 | 66570 |
| 0358 | L | C | E | N | 1974 | 74854 |
| 0368 | S | D | R | N | 1963 | 44787 |
| 0369 | S | D | R | N | 1963 | 44787 |
| 0370 | G | C | E | H | 1973 | 74430 |
| 0370 | L | M | G | N | 1967 | 47301 |
| 0370 | L | D | E | N | 1966 | 70125 |
| 0370 | L | D | E | N | 1966 | 70128 |

**Phys. State:** **D.** Doped; **E.** Expanded; **F.** Fibrous or Whisker; **G.** Gas; **L.** Liquid; **M.** Multiphase; **P.** Powder; **S.** Solid
**Subject:** **C.** Calculated Data + Theory; **D.** Data; **E.** Experiment; **G.** General (Data + Theory + Experiment); **M.** Measured Data + Experiment; **S.** Survey (Review, Compendium, etc.); **T.** Theory
**Language:** **C.** Czech; **D.** Dutch; **E.** English; **F.** French; **G.** German; **I.** Italian; **J.** Japanese; **O.** Other Languages; **P.** Polish; **R.** Russian; **S.** Spanish
**Temperature:** **F.** Full Range (Low + Normal + High); **L.** Low (0 to 75K) + Overlap into Normal; **N.** Normal (75 to 1273K); **H.** High (above 1273K) + Overlap into Normal; **U.** Undefined

| Substance Number | Phys. State | Subject | Language | Temperature | Year | TPRC Number |
|---|---|---|---|---|---|---|
| 512-0370 | L | D | E | N | 1966 | 70129 |
| 0370 | L | G | E | N | 1968 | 76272 |
| 0390 | M | M | E | N | 1969 | 36336 |
| 0392 | L | M | E | N | 1974 | 74791 |
| 0392 | L | C | E | N | 1974 | 74854 |
| 0392 | L | D | E | N | 1974 | 76691 |
| 0393 | L | D | R | N | 1973 | 75840 |
| 0393 | L | D | E | N | 1973 | 75841 |
| 0394 | L | D | R | N | 1973 | 75840 |
| 0394 | L | D | E | N | 1973 | 75841 |
| 0395 | G | T | E | H | 1972 | 67737 |
| 0395 | G | S | E | H | 1973 | 73279 |
| 0397 | L | D | R | N | 1973 | 72533 |
| 0405 | L | D | R | N | 1973 | 72941 |
| 0409 | L | C | E | N | 1972 | 66037 |
| 0421 | G | G | E | N | 1970 | 65884 |
| 0428 | S | G | E | N | 1972 | 71401 |
| 0429 | G | C | E | H | 1970 | 64975 |
| 0429 | G | D | E | H | 1972 | 68908 |
| 0429 | G | D | R | H | 1973 | 73337 |
| 0429 | G | D | E | H | 1973 | 73338 |
| 0430 | S | D | R | H | 1971 | 67751 |
| 0430 | S | D | E | N | 1972 | 68819 |
| 0430 | S | D | E | N | 1973 | 71359 |
| 0431 | S | D | R | H | 1971 | 67751 |
| 0432 | P | M | E | U | 1973 | 77930 |
| 0432 | S | D | R | H | 1971 | 67751 |
| 0433 | M | D | O | N | 1966 | 49792 |
| 0433 | P | M | E | U | 1973 | 77930 |
| 0433 | S | D | R | H | 1971 | 67751 |
| 0441 | S | D | G | H | 1972 | 70392 |
| 0442 | S | D | G | H | 1972 | 70392 |
| 0444 | L | D | E | N | 1971 | 70169 |
| 0445 | P | M | E | U | 1973 | 77930 |
| 0445 | S | D | R | H | 1970 | 61943 |
| 0445 | S | D | E | N | 1972 | 68837 |
| 0445 | S | D | R | H | 1970 | 74076 |
| 0445 | S | D | E | H | 1970 | 74077 |
| 0450 | L | D | E | N | 1973 | 71002 |
| 0455 | S | G | E | N | 1972 | 71401 |
| 0456 | L | M | G | N | 1972 | 69283 |
| 0457 | L | M | G | N | 1972 | 69283 |
| 0459 | S | M | R | N | 1963 | 36064 |
| 0460 | S | M | R | N | 1965 | 42140 |
| 0463 | S | D | E | N | 1965 | 45152 |
| 0476 | G | T | R | N | 1968 | 51001 |
| 0476 | G | T | R | U | 1969 | 56446 |
| 0476 | G | T | E | U | 1970 | 64427 |
| 0476 | G | G | E | N | 1971 | 65727 |
| 0476 | G | T | E | N | 1968 | 65913 |
| 0603 | S | D | E | N | 1974 | 75755 |
| 0683 | M | D | O | N | 1966 | 49792 |
| 0699 | G | C | E | N | 1971 | 69329 |
| 0699 | G | C | E | H | 1973 | 74430 |
| 0701 | G | G | E | N | 1972 | 71404 |
| 0708 | L | M | G | N | 1967 | 47301 |
| 0709 | L | M | R | N | 1971 | 68226 |
| 0710 | L | M | R | N | 1971 | 68226 |
| 0712 | L | S | E | N | 1971 | 64329 |
| 0719 | L | D | R | N | 1973 | 75840 |
| 0719 | L | D | E | N | 1973 | 75841 |
| 0726 | G | D | E | N | 1970 | 58534 |
| 0727 | G | D | E | H | 1972 | 68908 |
| 0735 | L | D | E | N | 1966 | 70128 |
| 0735 | L | D | E | N | 1966 | 70129 |
| 0736 | L | D | E | N | 1966 | 70125 |
| 0736 | L | D | E | N | 1967 | 72364 |
| 0736 | L | G | E | N | 1968 | 76272 |
| 0739 | G | T | E | H | 1973 | 73019 |
| 0740 | S | D | R | H | 1970 | 61943 |
| 0756 | S | D | R | H | 1970 | 74076 |
| 0756 | S | D | E | H | 1970 | 74077 |
| 0772 | G | D | E | N | 1974 | 73190 |
| 0773 | G | D | E | N | 1974 | 73190 |
| 0774 | G | D | E | N | 1974 | 73190 |
| 0774 | G | C | E | N | 1974 | 74757 |
| 0775 | G | D | E | N | 1974 | 73190 |
| 0775 | G | C | E | N | 1974 | 74757 |
| 0776 | G | D | E | N | 1974 | 73190 |
| 0776 | G | C | E | N | 1974 | 74757 |
| 0777 | G | D | E | N | 1974 | 73190 |
| 0777 | G | C | E | N | 1974 | 74757 |
| 0778 | S | G | R | H | 1974 | 75073 |
| 0778 | S | G | E | H | 1974 | 75074 |
| 0794 | S | D | R | N | 1972 | 71641 |
| 0808 | D | D | E | N | 1973 | 73823 |
| 0808 | S | D | E | N | 1973 | 73823 |
| 0809 | S | D | E | N | 1973 | 73823 |
| 0810 | S | D | E | N | 1973 | 73823 |
| 0880 | L | M | E | N | 1974 | 74791 |
| 0881 | G | T | E | N | 1968 | 54378 |
| 0882 | G | C | E | H | 1973 | 74430 |
| 0883 | G | C | E | H | 1973 | 74430 |
| 0884 | G | C | E | H | 1973 | 74430 |
| 0909 | L | D | R | N | 1973 | 75840 |
| 0909 | L | D | E | N | 1973 | 75841 |
| 0911 | L | T | E | N | 1972 | 67762 |
| 1003 | L | D | E | N | 1974 | 74854 |
| 1003 | L | D | E | N | 1974 | 76691 |
| 1005 | S | D | R | N | 1974 | 74924 |
| 1005 | S | D | E | N | 1974 | 77018 |

| Substance Number | Phys. State | Subject | Language | Temperature | Year | TPRC Number |
|---|---|---|---|---|---|---|
| 512-1007 | G | C | E | N | 1974 | 74757 |
| 1008 | L | M | E | N | 1974 | 74791 |
| 1009 | P | M | E | U | 1973 | 77930 |
| 1025 | S | D | R | N | 1974 | 74924 |
| 1025 | S | D | E | N | 1974 | 77018 |
| 1029 | G | D | E | H | 1975 | 77118 |
| 1041 | L | D | R | N | 1974 | 76115 |
| 1041 | L | D | E | N | 1974 | 77220 |
| 1042 | S | D | E | N | 1970 | 77563 |
| 1043 | S | D | E | N | 1970 | 77563 |
| 1043 | S | M | E | H | 1969 | 77573 |
| 1048 | S | M | E | H | 1969 | 77573 |
| 514-0001 | G | D | G | N | 1973 | 73931 |
| 0003 | G | D | G | N | 1973 | 73931 |
| 0006 | L | D | R | N | 1973 | 72941 |
| 0011 | G | G | E | N | 1972 | 71404 |
| 0012 | G | C | E | N | 1973 | 71809 |
| 0012 | G | D | E | N | 1974 | 73190 |
| 0012 | G | D | G | N | 1973 | 73931 |
| 0012 | G | C | E | N | 1974 | 74390 |
| 0018 | G | C | E | N | 1974 | 74390 |
| 0020 | G | M | R | N | 1967 | 46165 |
| 0020 | G | D | E | N | 1967 | 47005 |
| 0020 | G | D | E | N | 1968 | 50730 |
| 0020 | G | C | R | N | 1970 | 59019 |
| 0020 | G | S | E | F | 1971 | 62671 |
| 0020 | G | D | E | H | 1971 | 63204 |
| 0020 | G | D | R | N | 1971 | 63756 |
| 0020 | G | M | E | N | 1967 | 64176 |
| 0020 | G | C | R | N | 1969 | 64255 |
| 0020 | G | G | E | N | 1971 | 65228 |
| 0020 | G | C | E | N | 1971 | 65266 |
| 0020 | G | M | R | N | 1970 | 65443 |
| 0020 | G | G | E | N | 1971 | 65727 |
| 0020 | G | D | E | N | 1971 | 67154 |
| 0020 | G | M | R | N | 1972 | 67681 |
| 0020 | G | C | R | N | 1970 | 67725 |
| 0020 | G | D | E | N | 1972 | 68595 |
| 0020 | G | D | E | N | 1972 | 69324 |
| 0020 | G | M | E | N | 1972 | 69601 |
| 0020 | G | S | E | N | 1971 | 71643 |
| 0020 | G | C | E | N | 1970 | 71943 |
| 0020 | G | C | R | H | 1973 | 72523 |
| 0020 | G | T | E | H | 1973 | 73150 |
| 0020 | G | S | E | U | 1954 | 73367 |
| 0020 | G | D | E | N | 1946 | 74332 |
| 0020 | G | M | E | N | 1946 | 74333 |
| 0020 | G | M | G | N | 1896 | 74338 |
| 0020 | G | C | E | H | 1974 | 74346 |
| 0020 | G | M | E | N | 1974 | 75006 |
| 0020 | G | C | E | H | 1974 | 76376 |
| 0020 | G | D | E | N | 1975 | 76427 |
| 0020 | L | D | R | N | 1971 | 63756 |
| 0020 | L | D | E | N | 1971 | 67154 |
| 0036 | G | C | E | N | 1974 | 74390 |
| 0047 | L | D | R | N | 1973 | 72941 |
| 0051 | G | S | R | N | 1971 | 66284 |
| 0051 | G | S | E | N | 1971 | 67539 |
| 0051 | G | G | R | N | 1972 | 71033 |
| 0051 | G | C | E | N | 1973 | 71809 |
| 0051 | G | M | E | N | 1971 | 73916 |
| 0051 | S | M | E | N | 1971 | 73916 |
| 0052 | G | S | R | N | 1971 | 66284 |
| 0052 | G | S | E | N | 1971 | 67539 |
| 0052 | G | G | R | N | 1972 | 71033 |
| 0052 | G | C | E | N | 1973 | 71809 |
| 0057 | L | M | E | N | 1973 | 69692 |
| 0057 | S | M | E | N | 1973 | 69692 |
| 0064 | G | T | R | U | 1969 | 55757 |
| 0064 | G | E | R | U | 1968 | 57538 |
| 0064 | G | T | R | U | 1968 | 57538 |
| 0064 | G | E | E | U | 1971 | 64047 |
| 0064 | G | T | E | U | 1971 | 64047 |
| 0064 | G | E | E | U | 1971 | 64761 |
| 0064 | G | T | E | U | 1969 | 69502 |
| 0064 | G | S | E | N | 1973 | 73278 |
| 0064 | G | T | G | H | 1972 | 73814 |
| 0064 | S | T | E | U | 1967 | 61760 |
| 0106 | G | C | E | N | 1973 | 71809 |
| 0106 | G | D | G | N | 1973 | 73931 |
| 0110 | G | G | E | N | 1972 | 71404 |
| 0112 | G | S | E | U | 1972 | 38291 |
| 0112 | G | S | R | U | 1970 | 63876 |
| 0117 | M | M | E | N | 1969 | 36336 |
| 0129 | G | T | R | H | 1971 | 66074 |
| 0129 | G | T | E | H | 1971 | 66075 |
| 0129 | G | C | E | H | 1970 | 68072 |
| 0129 | G | D | R | H | 1972 | 70621 |
| 0129 | G | D | E | H | 1972 | 70622 |
| 0152 | G | S | R | N | 1971 | 66284 |
| 0152 | G | S | E | N | 1971 | 67539 |
| 0152 | G | G | R | N | 1972 | 71033 |
| 0152 | G | M | E | N | 1971 | 73916 |
| 0152 | S | M | E | N | 1971 | 73916 |
| 0171 | G | C | E | H | 1967 | 47652 |
| 0171 | G | G | E | H | 1971 | 63637 |
| 0171 | G | D | R | H | 1971 | 63941 |
| 0171 | G | D | E | H | 1971 | 63942 |
| 0171 | G | D | E | H | 1971 | 63981 |
| 0174 | S | D | E | N | 1970 | 77561 |

| Substance Number | Phys. State | Subject | Language | Temperature | Year | TPRC Number |
|---|---|---|---|---|---|---|
| 514-0174 | S | D | E | N | 1970 | 77563 |
| 0174 | S | M | E | H | 1969 | 77573 |
| 0175 | S | D | E | N | 1970 | 77561 |
| 0175 | S | D | E | N | 1970 | 77563 |
| 0175 | S | M | E | H | 1969 | 77573 |
| 0179 | S | D | R | N | 1972 | 71641 |
| 0182 | L | M | E | N | 1973 | 69692 |
| 0182 | S | M | E | N | 1973 | 69692 |
| 0188 | G | G | E | N | 1972 | 71404 |
| 0193 | G | C | E | N | 1973 | 71809 |
| 0195 | L | D | R | N | 1973 | 72941 |
| 0196 | L | D | R | N | 1973 | 72941 |
| 0262 | S | D | R | H | 1970 | 61943 |
| 0262 | S | D | R | N | 1972 | 71641 |
| 0262 | S | D | R | N | 1970 | 74072 |
| 0262 | S | D | E | N | 1970 | 74073 |
| 0262 | S | D | R | H | 1970 | 74076 |
| 0262 | S | D | E | H | 1970 | 74077 |
| 0287 | S | D | R | H | 1970 | 61943 |
| 0287 | S | D | R | N | 1970 | 74072 |
| 0287 | S | D | E | N | 1970 | 74073 |
| 0370 | L | G | E | N | 1968 | 76272 |
| 0386 | G | C | R | N | 1969 | 55396 |
| 0386 | G | T | R | U | 1970 | 73706 |
| 0386 | G | T | E | U | 1970 | 73707 |
| 0386 | G | C | G | H | 1972 | 73814 |
| 0390 | M | M | E | N | 1969 | 36336 |
| 0391 | G | S | R | N | 1971 | 66284 |
| 0391 | G | S | E | N | 1971 | 67539 |
| 0391 | G | G | R | N | 1972 | 71033 |
| 0391 | G | M | E | N | 1971 | 73916 |
| 0391 | S | M | E | N | 1971 | 73916 |
| 0405 | L | D | R | N | 1973 | 72941 |
| 0405 | M | M | E | N | 1969 | 36336 |
| 0427 | G | T | E | H | 1973 | 73328 |
| 0432 | S | D | G | H | 1972 | 70392 |
| 0437 | S | D | G | H | 1972 | 70392 |
| 0439 | S | D | G | H | 1972 | 70392 |
| 0440 | S | D | G | H | 1972 | 70392 |
| 0445 | S | D | R | H | 1970 | 61943 |
| 0445 | S | D | R | N | 1970 | 74072 |
| 0445 | S | D | E | N | 1970 | 74073 |
| 0445 | S | D | R | H | 1970 | 74076 |
| 0445 | S | D | E | H | 1970 | 74077 |
| 0445 | S | D | E | N | 1970 | 77561 |
| 0667 | L | M | E | N | 1973 | 69692 |
| 0667 | S | M | E | N | 1973 | 69692 |
| 0687 | G | S | E | U | 1971 | 68890 |
| 0687 | G | C | E | H | 1972 | 70292 |
| 0687 | G | T | E | H | 1972 | 73829 |
| 0690 | G | T | E | U | 1973 | 70290 |
| 0701 | G | G | E | N | 1972 | 71404 |
| 0708 | L | G | E | N | 1968 | 76272 |
| 0736 | L | G | E | N | 1968 | 76272 |
| 0740 | S | D | R | H | 1970 | 61943 |
| 0740 | S | D | R | N | 1970 | 74072 |
| 0740 | S | D | E | N | 1970 | 74073 |
| 0740 | S | D | R | H | 1970 | 74076 |
| 0740 | S | D | E | H | 1970 | 74077 |
| 0741 | S | D | R | H | 1970 | 61943 |
| 0741 | S | D | R | N | 1970 | 74072 |
| 0741 | S | D | E | N | 1970 | 74073 |
| 0772 | G | D | E | N | 1974 | 73190 |
| 0773 | G | D | E | N | 1974 | 73190 |
| 0794 | S | D | R | N | 1972 | 71641 |
| 0811 | G | T | E | H | 1972 | 73829 |
| 0881 | G | T | E | N | 1968 | 54378 |
| 1012 | L | D | R | N | 1973 | 72941 |
| 1040 | M | T | R | N | 1973 | 75167 |
| 1042 | S | D | E | N | 1970 | 77561 |
| 1042 | S | D | E | N | 1970 | 77563 |
| 1043 | S | D | E | N | 1970 | 77561 |
| 1043 | S | D | E | N | 1970 | 77563 |
| 1043 | S | M | E | H | 1969 | 77573 |
| 1048 | S | M | E | H | 1969 | 77573 |
| 515-0003 | G | C | R | N | 1971 | 65016 |
| 0003 | G | C | E | N | 1973 | 70014 |
| 0010 | L | D | R | N | 1972 | 69566 |
| 0011 | L | M | R | N | 1972 | 69615 |
| 0011 | L | M | E | N | 1972 | 70815 |
| 0029 | L | M | E | N | 1971 | 64090 |
| 0029 | L | G | G | N | 1971 | 66506 |
| 0030 | L | M | R | N | 1970 | 59465 |
| 0030 | L | M | E | N | 1971 | 64090 |
| 0030 | L | M | E | N | 1972 | 66197 |
| 0030 | L | G | G | N | 1971 | 66506 |
| 0030 | L | M | R | N | 1972 | 69615 |
| 0030 | L | M | E | N | 1972 | 70815 |
| 0030 | L | C | R | N | 1973 | 71888 |
| 0030 | L | C | E | N | 1973 | 72558 |
| 0030 | L | S | E | N | 1960 | 94260 |
| 0031 | L | G | G | N | 1971 | 66506 |
| 0032 | L | C | R | N | 1968 | 50756 |
| 0032 | L | D | R | U | 1968 | 50758 |
| 0032 | L | C | E | N | 1968 | 65336 |
| 0032 | L | D | E | U | 1968 | 65912 |
| 0032 | L | D | R | N | 1970 | 75459 |
| 0032 | L | D | E | N | 1970 | 75460 |
| 0041 | L | C | R | N | 1968 | 50756 |
| 0041 | L | C | E | N | 1968 | 65336 |

**Phys. State:** **D.** Doped; **E.** Expanded; **F.** Fibrous or Whisker; **G.** Gas; **L.** Liquid; **M.** Multiphase; **P.** Powder; **S.** Solid
**Subject:** **C.** Calculated Data + Theory; **D.** Data; **E.** Experiment; **G.** General (Data + Theory + Experiment); **M.** Measured Data + Experiment; **S.** Survey (Review, Compendium, etc.); **T.** Theory
**Language:** **C.** Czech; **D.** Dutch; **E.** English; **F.** French; **G.** German; **I.** Italian; **J.** Japanese; **O.** Other Languages; **P.** Polish; **R.** Russian; **S.** Spanish
**Temperature:** **F.** Full Range (Low + Normal + High); **L.** Low (0 to 75K) + Overlap into Normal; **N.** Normal (75 to 1273K); **H.** High (above 1273K) + Overlap into Normal; **U.** Undefined

| Substance Number | Phys. State | Subject | Language | Temperature | Year | TPRC Number |
|---|---|---|---|---|---|---|
| 515-0042 | G | M | R | N | 1971 | 66290 |
| 0042 | L | D | R | N | 1971 | 64236 |
| 0042 | L | D | E | N | 1971 | 69511 |
| 0042 | L | D | R | N | 1970 | 75459 |
| 0042 | L | D | E | N | 1970 | 75460 |
| 0048 | L | M | E | N | 1971 | 64090 |
| 0048 | L | G | G | N | 1971 | 66506 |
| 0048 | L | D | R | N | 1973 | 72165 |
| 0048 | L | D | E | N | 1973 | 73281 |
| 0048 | L | C | R | N | 1973 | 75177 |
| 0054 | L | G | G | N | 1971 | 66506 |
| 0063 | L | C | R | N | 1968 | 50756 |
| 0063 | L | C | E | N | 1968 | 65336 |
| 0064 | L | E | C | U | 1972 | 69474 |
| 0065 | L | D | R | N | 1973 | 72165 |
| 0065 | L | D | E | N | 1973 | 73281 |
| 0067 | L | M | R | N | 1972 | 69615 |
| 0067 | L | M | E | N | 1972 | 70815 |
| 0067 | L | C | R | N | 1973 | 71888 |
| 0067 | L | D | R | N | 1973 | 72165 |
| 0067 | L | C | E | N | 1973 | 72558 |
| 0067 | L | D | E | N | 1973 | 73281 |
| 0089 | L | D | R | N | 1972 | 69713 |
| 0089 | L | C | R | N | 1973 | 75177 |
| 0099 | L | M | E | N | 1971 | 64090 |
| 0099 | L | G | G | N | 1971 | 66506 |
| 0101 | L | C | R | N | 1973 | 75177 |
| 0103 | L | C | R | N | 1968 | 50756 |
| 0103 | L | C | E | N | 1968 | 65336 |
| 0106 | G | C | R | N | 1970 | 59019 |
| 0106 | G | C | E | N | 1970 | 71943 |
| 0121 | L | D | E | N | 1971 | 61177 |
| 0121 | L | M | R | N | 1973 | 72161 |
| 0140 | L | C | R | N | 1968 | 50756 |
| 0140 | L | D | R | N | 1971 | 63531 |
| 0140 | L | D | E | N | 1971 | 63532 |
| 0140 | L | C | E | N | 1968 | 65336 |
| 0144 | L | G | G | N | 1971 | 66506 |
| 0152 | G | C | R | N | 1968 | 49920 |
| 0152 | G | C | R | N | 1971 | 65016 |
| 0152 | G | C | E | N | 1968 | 65122 |
| 0152 | G | C | E | N | 1973 | 70014 |
| 0184 | L | C | R | N | 1968 | 50756 |
| 0184 | L | C | E | U | 1973 | 63558 |
| 0184 | L | C | E | N | 1968 | 65336 |
| 0184 | L | C | R | U | 1973 | 72087 |
| 0213 | G | C | R | N | 1968 | 49920 |
| 0213 | G | C | E | N | 1968 | 65122 |
| 0259 | L | G | E | N | 1971 | 63392 |
| 0262 | M | C | E | N | 1970 | 46359 |
| 0280 | L | G | G | N | 1971 | 66506 |
| 0290 | L | T | E | U | 1973 | 75142 |
| 0311 | L | D | R | N | 1963 | 37245 |
| 0311 | L | G | R | N | 1971 | 73718 |
| 0311 | L | G | E | N | 1971 | 73719 |
| 0387 | L | M | R | N | 1972 | 69615 |
| 0387 | L | M | E | N | 1972 | 70815 |
| 0440 | M | C | E | N | 1970 | 46359 |
| 0447 | L | D | R | N | 1973 | 72165 |
| 0447 | L | D | E | N | 1973 | 73281 |
| 0460 | L | G | G | N | 1971 | 66506 |
| 0460 | L | M | R | N | 1972 | 69615 |
| 0460 | L | M | E | N | 1972 | 70815 |
| 0460 | L | C | R | N | 1973 | 71888 |
| 0460 | L | C | E | N | 1973 | 72558 |
| 0460 | L | C | R | N | 1973 | 75177 |
| 0465 | M | D | R | N | 1964 | 43252 |
| 0465 | M | D | R | N | 1973 | 75636 |
| 0465 | M | D | E | N | 1973 | 75637 |
| 0472 | L | D | E | N | 1972 | 66024 |
| 0472 | M | D | R | N | 1971 | 73714 |
| 0472 | M | D | E | N | 1971 | 73715 |
| 0482 | M | D | E | U | 1972 | 60119 |
| 0482 | M | D | E | N | 1972 | 65118 |
| 0482 | M | D | E | U | 1972 | 66013 |
| 0482 | M | D | E | U | 1972 | 66641 |
| 0498 | L | G | G | N | 1971 | 66506 |
| 0506 | L | M | R | N | 1972 | 69615 |
| 0506 | L | M | E | N | 1972 | 70815 |
| 0507 | L | D | R | N | 1974 | 75670 |
| 0509 | L | M | R | N | 1972 | 69615 |
| 0509 | L | M | E | N | 1972 | 70815 |
| 0537 | L | C | R | N | 1968 | 50756 |
| 0537 | L | D | R | N | 1971 | 63531 |
| 0537 | L | D | E | N | 1971 | 63532 |
| 0537 | L | C | E | N | 1968 | 65336 |
| 0560 | L | D | C | N | 1970 | 62412 |
| 0600 | L | M | R | N | 1972 | 69615 |
| 0600 | L | M | E | N | 1972 | 70815 |
| 0603 | L | G | G | N | 1971 | 66506 |
| 0634 | M | C | E | N | 1970 | 46359 |
| 0636 | L | D | R | N | 1974 | 75670 |
| 0644 | L | G | R | N | 1971 | 73718 |
| 0644 | L | G | E | N | 1971 | 73719 |
| 0681 | L | M | R | N | 1972 | 69615 |
| 0681 | L | M | E | N | 1972 | 70815 |
| 0715 | L | D | R | N | 1972 | 69713 |
| 0715 | L | C | R | N | 1973 | 75177 |
| 0716 | L | D | R | N | 1972 | 69713 |
| 0716 | L | C | R | N | 1973 | 75177 |
| 0869 | M | D | R | N | 1971 | 73714 |
| 515-0869 | M | D | E | N | 1971 | 73715 |
| 1026 | L | D | R | N | 1974 | 74869 |
| 517-0001 | G | D | G | N | 1972 | 68888 |
| 0003 | G | D | G | N | 1972 | 68888 |
| 0011 | L | C | R | N | 1973 | 71888 |
| 0011 | L | C | E | N | 1973 | 72558 |
| 0028 | L | C | R | N | 1968 | 50756 |
| 0028 | L | C | E | N | 1968 | 65336 |
| 0030 | L | C | R | N | 1973 | 71888 |
| 0030 | L | C | E | N | 1973 | 72558 |
| 0042 | L | D | R | N | 1971 | 64327 |
| 0042 | L | D | R | N | 1971 | 65429 |
| 0042 | L | D | R | N | 1972 | 67668 |
| 0042 | L | C | R | N | 1972 | 69136 |
| 0043 | L | D | R | N | 1972 | 67668 |
| 0043 | L | C | R | N | 1972 | 69136 |
| 0044 | L | C | R | N | 1968 | 50756 |
| 0044 | L | C | E | N | 1968 | 65336 |
| 0044 | L | D | R | N | 1972 | 67668 |
| 0044 | L | C | R | N | 1972 | 69136 |
| 0045 | L | D | R | N | 1972 | 67668 |
| 0045 | L | C | R | N | 1972 | 69136 |
| 0052 | L | G | R | N | 1971 | 73718 |
| 0052 | L | G | E | N | 1971 | 73719 |
| 0063 | L | C | R | N | 1968 | 50756 |
| 0063 | L | C | E | N | 1968 | 65336 |
| 0086 | L | D | R | N | 1971 | 64327 |
| 0086 | L | C | R | N | 1972 | 69136 |
| 0103 | L | C | R | N | 1968 | 50756 |
| 0103 | L | C | E | N | 1968 | 65336 |
| 0134 | L | C | R | N | 1972 | 69136 |
| 0140 | L | C | R | N | 1968 | 50756 |
| 0140 | L | C | E | N | 1968 | 65336 |
| 0141 | L | C | R | N | 1968 | 50756 |
| 0141 | L | C | E | N | 1968 | 65336 |
| 0141 | L | D | R | N | 1972 | 67668 |
| 0141 | L | C | R | N | 1972 | 69136 |
| 0158 | G | D | G | N | 1972 | 68888 |
| 0193 | G | D | G | N | 1972 | 68888 |
| 0200 | L | D | R | N | 1972 | 67668 |
| 0200 | L | C | R | N | 1972 | 69136 |
| 0260 | L | D | R | N | 1972 | 67668 |
| 0260 | L | C | R | N | 1972 | 69136 |
| 0311 | L | G | R | N | 1971 | 73718 |
| 0311 | L | G | E | N | 1971 | 73719 |
| 0460 | L | C | R | N | 1973 | 71888 |
| 0460 | L | C | E | N | 1973 | 72558 |
| 0464 | M | C | J | N | 1972 | 66367 |
| 0464 | M | C | E | N | 1972 | 68738 |
| 0468 | G | T | R | H | 1968 | 55010 |
| 0468 | G | T | E | H | 1968 | 64428 |
| 0469 | G | C | R | H | 1968 | 55010 |
| 0469 | G | D | R | H | 1970 | 63896 |
| 0469 | G | D | R | H | 1971 | 63955 |
| 0469 | G | D | E | H | 1971 | 63956 |
| 0469 | G | C | E | H | 1968 | 64428 |
| 0492 | L | D | E | N | 1972 | 73833 |
| 0507 | L | D | R | N | 1974 | 75670 |
| 0509 | L | C | R | N | 1973 | 71888 |
| 0509 | L | C | E | N | 1973 | 72558 |
| 0517 | L | D | R | N | 1971 | 65429 |
| 0517 | L | C | R | N | 1972 | 69136 |
| 0522 | L | D | R | N | 1971 | 65429 |
| 0522 | L | C | R | N | 1972 | 69136 |
| 0523 | L | D | R | N | 1971 | 64327 |
| 0523 | L | C | R | N | 1972 | 69136 |
| 0528 | L | D | R | N | 1971 | 65429 |
| 0528 | L | C | R | N | 1972 | 69136 |
| 0531 | L | D | R | N | 1971 | 65429 |
| 0531 | L | C | R | N | 1972 | 69136 |
| 0565 | L | D | R | N | 1971 | 64327 |
| 0565 | L | C | R | N | 1972 | 69136 |
| 0600 | L | C | R | N | 1973 | 71888 |
| 0600 | L | C | E | N | 1973 | 72558 |
| 0634 | M | C | J | N | 1972 | 66367 |
| 0634 | M | C | E | N | 1972 | 68738 |
| 0636 | L | D | R | N | 1974 | 75670 |
| 0754 | G | D | E | H | 1974 | 73920 |
| 0755 | G | D | E | H | 1974 | 73920 |
| 0857 | L | D | R | N | 1973 | 74394 |
| 0890 | G | D | R | H | 1974 | 75751 |
| 0890 | G | D | E | H | 1974 | 75752 |
| 1052 | G | D | E | H | 1964 | 70137 |

**Phys. State:** **D.** Doped; **E.** Expanded; **F.** Fibrous or Whisker; **G.** Gas; **L.** Liquid; **M.** Multiphase; **P.** Powder; **S.** Solid
**Subject:** **C.** Calculated Data + Theory; **D.** Data; **E.** Experiment; **G.** General (Data + Theory + Experiment); **M.** Measured Data + Experiment; **S.** Survey (Review, Compendium, etc.); **T.** Theory
**Language:** **C.** Czech; **D.** Dutch; **E.** English; **F.** French; **G.** German; **I.** Italian; **J.** Japanese; **O.** Other Languages; **P.** Polish; **R.** Russian; **S.** Spanish
**Temperature:** **F.** Full Range (Low + Normal + High); **L.** Low (0 to 75K) + Overlap into Normal; **N.** Normal (75 to 1273K); **H.** High (above 1273K) + Overlap into Normal; **U.** Undefined

# Chapter 2 Thermal Diffusivity

| Substance Number | Phys. State | Sub-ject | Lan-guage | Temper-ature | Year | TPRC Number |
|---|---|---|---|---|---|---|
| 512-0006 | L | D | E | N | 1973 | 72868 |
| 0030 | S | M | R | N | 1965 | 42140 |
| 0038 | G | C | E | H | 1970 | 37602 |
| 0047 | L | D | E | N | 1973 | 72868 |
| 0048 | S | D | E | N | 1974 | 75755 |
| 0052 | L | M | R | N | 1970 | 63894 |
| 0052 | S | M | R | N | 1970 | 63894 |
| 0057 | S | M | R | N | 1963 | 36022 |
| 0064 | G | T | E | N | 1972 | 67501 |
| 0064 | L | T | E | N | 1972 | 67501 |
| 0064 | L | S | E | N | 1972 | 69620 |
| 0064 | L | M | E | N | 1973 | 76524 |
| 0095 | G | M | R | H | 1972 | 70589 |
| 0095 | G | M | E | H | 1973 | 72302 |
| 0138 | P | M | R | U | 1968 | 49504 |
| 0138 | P | M | E | U | 1968 | 65123 |
| 0139 | P | M | R | U | 1968 | 49504 |
| 0139 | P | M | E | U | 1968 | 65123 |
| 0174 | S | D | E | N | 1970 | 77563 |
| 0174 | S | M | E | H | 1969 | 77573 |
| 0182 | S | M | R | N | 1963 | 36022 |
| 0195 | L | D | E | N | 1973 | 72868 |
| 0243 | G | C | E | H | 1971 | 63144 |
| 0259 | S | D | E | N | 1974 | 75755 |
| 0311 | L | M | R | N | 1970 | 63894 |
| 0311 | S | M | R | N | 1970 | 63894 |
| 0446 | S | D | E | H | 1973 | 70463 |
| 0458 | S | M | R | N | 1963 | 36022 |
| 0460 | S | M | R | N | 1965 | 42140 |
| 0603 | S | D | E | N | 1974 | 75755 |
| 1042 | S | D | E | N | 1970 | 77563 |
| 1043 | S | D | E | N | 1970 | 77563 |
| 1043 | S | M | E | H | 1969 | 77573 |
| 1048 | S | M | E | H | 1969 | 77573 |
| 514-0174 | S | D | E | N | 1970 | 77561 |
| 0174 | S | D | E | N | 1970 | 77563 |
| 0174 | S | M | E | H | 1969 | 77573 |
| 0175 | S | D | E | N | 1970 | 77561 |
| 0175 | S | D | E | N | 1970 | 77563 |
| 0175 | S | M | E | H | 1969 | 77573 |
| 0445 | S | D | E | N | 1970 | 77561 |
| 1042 | S | D | E | N | 1970 | 77561 |
| 1042 | S | D | E | N | 1970 | 77563 |
| 1043 | S | D | E | N | 1970 | 77561 |
| 1043 | S | D | E | N | 1970 | 77563 |
| 1043 | S | M | E | H | 1969 | 77573 |
| 1048 | S | M | E | H | 1969 | 77573 |
| 1051 | S | M | E | N | 1973 | 70160 |
| 515-0010 | L | D | R | N | 1972 | 69566 |
| 0042 | L | C | R | N | 1970 | 62456 |
| 0064 | L | M | E | N | 1973 | 76524 |
| 0465 | M | D | R | N | 1964 | 43252 |
| 0472 | L | D | E | N | 1972 | 66024 |
| 0472 | M | D | R | N | 1971 | 73714 |
| 0472 | M | D | E | N | 1971 | 73715 |
| 0578 | L | C | E | N | 1973 | 76224 |
| 0869 | M | D | R | N | 1971 | 73714 |
| 0869 | M | D | E | N | 1971 | 73715 |
| 517-0857 | L | D | R | N | 1973 | 74394 |

**Phys. State:** **D.** Doped; **E.** Expanded; **F.** Fibrous or Whisker; **G.** Gas; **L.** Liquid; **M.** Multiphase; **P.** Powder; **S.** Solid
**Subject:** **C.** Calculated Data + Theory; **D.** Data; **E.** Experiment; **G.** General (Data + Theory + Experiment); **M.** Measured Data + Experiment; **S.** Survey (Review, Compendium, etc.); **T.** Theory
**Language:** **C.** Czech; **D.** Dutch; **E.** English; **F.** French; **G.** German; **I.** Italian; **J.** Japanese; **O.** Other Languages; **P.** Polish; **R.** Russian; **S.** Spanish
**Temperature:** **F.** Full Range (Low + Normal + High); **L.** Low (0 to 75K) + Overlap into Normal; **N.** Normal (75 to 1273K); **H.** High (above 1273K) + Overlap into Normal; **U.** Undefined

# Chapter 3 Specific Heat

| Substance Number | Phys. State | Subject | Language | Temperature | Year | TPRC Number |
|---|---|---|---|---|---|---|
| 512-0001 | G | M | E | N | 1965 | 38895 |
| 0001 | G | G | G | N | 1965 | 43277 |
| 0001 | G | D | E | N | 1971 | 65773 |
| 0001 | G | C | E | N | 1972 | 66556 |
| 0001 | G | S | R | H | 1974 | 78662 |
| 0001 | L | M | E | N | 1965 | 38895 |
| 0001 | L | C | E | N | 1972 | 66556 |
| 0001 | L | M | E | L | 1972 | 68665 |
| 0001 | S | D | R | L | 1971 | 65971 |
| 0001 | S | D | E | L | 1972 | 65972 |
| 0001 | S | M | E | L | 1972 | 68665 |
| 0002 | G | G | G | N | 1965 | 34644 |
| 0002 | G | G | G | N | 1965 | 43277 |
| 0003 | G | G | G | N | 1965 | 34644 |
| 0003 | G | G | G | N | 1965 | 43277 |
| 0003 | G | S | R | H | 1974 | 78662 |
| 0003 | S | C | E | L | 1972 | 65328 |
| 0003 | S | G | E | L | 1971 | 70957 |
| 0003 | S | D | E | L | 1974 | 74376 |
| 0006 | G | D | R | N | 1969 | 60917 |
| 0006 | L | M | G | N | 1965 | 35673 |
| 0006 | L | D | E | N | 1968 | 50991 |
| 0006 | L | D | J | N | 1969 | 58112 |
| 0006 | L | D | E | N | 1970 | 59027 |
| 0006 | L | D | R | N | 1969 | 60917 |
| 0006 | L | D | E | N | 1971 | 62623 |
| 0006 | L | D | E | N | 1971 | 62624 |
| 0006 | L | D | E | N | 1971 | 62625 |
| 0006 | L | D | E | N | 1971 | 64303 |
| 0006 | L | D | E | N | 1972 | 66774 |
| 0006 | L | C | E | U | 1972 | 68536 |
| 0006 | L | D | E | N | 1974 | 75810 |
| 0007 | L | M | G | N | 1965 | 35673 |
| 0008 | G | C | E | N | 1965 | 44126 |
| 0009 | G | C | E | N | 1965 | 44126 |
| 0009 | G | M | E | N | 1973 | 75790 |
| 0010 | S | D | F | N | 1970 | 70692 |
| 0011 | L | C | R | N | 1965 | 45340 |
| 0012 | L | M | E | L | 1972 | 68665 |
| 0012 | S | D | R | L | 1970 | 61432 |
| 0012 | S | D | E | L | 1971 | 61433 |
| 0012 | S | D | R | L | 1971 | 64321 |
| 0012 | S | D | E | L | 1971 | 64322 |
| 0012 | S | G | R | L | 1969 | 65285 |
| 0012 | S | D | E | L | 1971 | 65677 |
| 0012 | S | M | E | L | 1972 | 68665 |
| 0016 | G | M | E | N | 1973 | 75790 |
| 0018 | S | D | E | L | 1974 | 74376 |
| 0021 | G | M | E | N | 1965 | 38895 |
| 0021 | G | C | E | N | 1972 | 66556 |
| 0021 | G | M | E | N | 1973 | 75790 |
| 0021 | L | M | E | N | 1965 | 38895 |
| 0021 | L | C | E | N | 1972 | 66556 |
| 0024 | S | D | G | L | 1964 | 36131 |
| 0024 | S | D | G | N | 1973 | 71983 |
| 0025 | S | D | G | L | 1964 | 36131 |
| 0025 | S | D | G | N | 1973 | 71983 |
| 0026 | G | C | E | H | 1971 | 62685 |
| 0026 | G | C | E | H | 1972 | 66581 |
| 0028 | G | D | R | N | 1969 | 60917 |
| 0028 | L | D | R | N | 1969 | 60917 |
| 0028 | L | D | R | N | 1971 | 63550 |
| 0028 | L | D | R | N | 1971 | 63767 |
| 0028 | L | D | E | N | 1970 | 64281 |
| 0028 | L | D | R | N | 1971 | 64393 |
| 0028 | L | D | E | N | 1971 | 64394 |
| 0028 | L | M | E | N | 1971 | 65111 |
| 0028 | L | D | R | N | 1971 | 65461 |
| 0028 | L | D | E | N | 1971 | 65743 |
| 0028 | L | G | E | N | 1971 | 66156 |
| 0028 | L | D | E | N | 1972 | 66926 |
| 0028 | L | D | E | N | 1971 | 67243 |
| 0028 | L | D | E | N | 1971 | 67244 |
| 0028 | L | M | E | N | 1971 | 67696 |
| 0028 | L | D | E | N | 1972 | 68704 |
| 0028 | L | C | E | N | 1973 | 71794 |
| 0028 | L | D | R | N | 1973 | 75611 |
| 0028 | L | D | E | N | 1973 | 75612 |
| 0029 | L | M | G | N | 1969 | 56949 |
| 0029 | L | M | E | N | 1973 | 69032 |
| 0029 | L | D | E | N | 1972 | 70522 |
| 0029 | L | D | R | N | 1973 | 72492 |
| 0029 | S | M | E | N | 1973 | 69032 |
| 0029 | S | D | E | N | 1972 | 70522 |
| 0030 | S | D | R | N | 1972 | 66602 |
| 0030 | S | D | E | N | 1972 | 69793 |
| 0032 | L | M | G | N | 1969 | 56949 |
| 0032 | L | D | R | N | 1973 | 72492 |
| 0033 | L | M | G | N | 1969 | 56949 |
| 0034 | L | M | G | N | 1969 | 56949 |
| 0036 | S | D | R | L | 1971 | 65971 |
| 0036 | S | D | E | L | 1972 | 65972 |
| 0037 | G | C | E | U | 1972 | 65366 |
| 0037 | S | D | R | L | 1970 | 61432 |
| 0037 | S | D | E | L | 1971 | 61433 |

| Substance Number | Phys. State | Subject | Language | Temperature | Year | TPRC Number |
|---|---|---|---|---|---|---|
| 512-0037 | S | D | R | L | 1971 | 65971 |
| 0037 | S | D | E | L | 1972 | 65972 |
| 0039 | L | D | J | N | 1969 | 58112 |
| 0039 | L | D | E | N | 1970 | 59027 |
| 0039 | L | D | E | N | 1971 | 62622 |
| 0039 | L | D | E | N | 1971 | 62625 |
| 0039 | L | D | R | N | 1971 | 65461 |
| 0039 | L | D | E | N | 1972 | 68704 |
| 0039 | L | G | G | N | 1974 | 78380 |
| 0042 | L | D | J | N | 1969 | 58112 |
| 0042 | L | C | E | U | 1972 | 68536 |
| 0043 | L | D | J | N | 1969 | 58112 |
| 0046 | L | D | E | N | 1973 | 72308 |
| 0047 | L | D | J | N | 1969 | 58112 |
| 0047 | L | D | E | N | 1970 | 59027 |
| 0047 | L | D | E | N | 1974 | 75810 |
| 0048 | L | M | E | N | 1971 | 61756 |
| 0048 | L | M | E | N | 1973 | 69032 |
| 0048 | L | D | E | N | 1972 | 70522 |
| 0048 | L | D | E | N | 1973 | 74822 |
| 0048 | S | D | R | N | 1972 | 66602 |
| 0048 | S | M | E | N | 1973 | 69032 |
| 0048 | S | D | E | N | 1972 | 69793 |
| 0048 | S | D | E | N | 1972 | 70522 |
| 0048 | S | D | E | N | 1974 | 75755 |
| 0049 | L | M | E | N | 1971 | 61756 |
| 0050 | L | M | E | N | 1971 | 61756 |
| 0051 | G | D | E | H | 1974 | 79055 |
| 0052 | G | D | E | H | 1974 | 79055 |
| 0052 | L | D | R | N | 1963 | 37245 |
| 0054 | L | C | E | U | 1972 | 68536 |
| 0055 | G | S | E | L | 1970 | 68510 |
| 0055 | G | C | E | L | 1973 | 69230 |
| 0055 | L | S | E | L | 1964 | 38454 |
| 0055 | L | M | E | L | 1971 | 42261 |
| 0055 | L | S | J | L | 1968 | 54356 |
| 0055 | L | G | E | L | 1971 | 64056 |
| 0055 | L | G | E | L | 1970 | 65564 |
| 0055 | L | T | E | L | 1972 | 66423 |
| 0055 | L | S | E | L | 1970 | 68510 |
| 0055 | L | C | E | L | 1973 | 69230 |
| 0055 | L | T | E | U | 1971 | 71765 |
| 0055 | L | C | E | L | 1974 | 74759 |
| 0055 | S | S | E | L | 1964 | 38454 |
| 0055 | S | C | E | L | 1972 | 66806 |
| 0056 | G | S | E | L | 1970 | 68510 |
| 0056 | G | C | E | L | 1973 | 69230 |
| 0056 | L | S | E | L | 1964 | 38454 |
| 0056 | L | M | E | L | 1971 | 42261 |
| 0056 | L | S | J | L | 1968 | 54356 |
| 0056 | L | G | E | L | 1971 | 64056 |
| 0056 | L | G | E | L | 1970 | 65564 |
| 0056 | L | T | E | L | 1972 | 66423 |
| 0056 | L | S | E | L | 1970 | 68510 |
| 0056 | L | C | E | L | 1973 | 69230 |
| 0056 | L | T | E | U | 1971 | 71765 |
| 0056 | L | C | E | L | 1974 | 74759 |
| 0056 | S | S | E | L | 1964 | 38454 |
| 0056 | S | C | E | L | 1972 | 66806 |
| 0059 | S | T | E | U | 1972 | 65329 |
| 0060 | S | T | E | U | 1972 | 65329 |
| 0061 | G | C | E | U | 1972 | 65366 |
| 0061 | G | G | E | N | 1973 | 73805 |
| 0061 | G | C | E | N | 1971 | 74103 |
| 0061 | G | D | E | N | 1974 | 76378 |
| 0061 | L | C | E | L | 1971 | 74103 |
| 0064 | G | S | E | U | 1969 | 53699 |
| 0064 | G | T | R | U | 1969 | 57535 |
| 0064 | G | C | F | U | 1971 | 65623 |
| 0064 | G | T | E | U | 1972 | 66540 |
| 0064 | G | E | E | N | 1972 | 71388 |
| 0064 | L | S | E | U | 1969 | 53699 |
| 0064 | L | T | R | U | 1969 | 57535 |
| 0064 | L | E | E | N | 1971 | 65173 |
| 0064 | L | E | E | N | 1972 | 65842 |
| 0064 | L | E | E | N | 1971 | 67696 |
| 0064 | L | T | E | N | 1973 | 70788 |
| 0064 | M | T | R | U | 1969 | 57535 |
| 0069 | S | C | E | L | 1972 | 65328 |
| 0069 | S | G | E | L | 1971 | 70957 |
| 0084 | L | C | R | N | 1965 | 45340 |
| 0091 | S | D | R | H | 1973 | 72239 |
| 0091 | S | D | E | H | 1974 | 72240 |
| 0094 | S | D | R | H | 1973 | 72239 |
| 0094 | S | D | E | H | 1974 | 72240 |
| 0095 | G | C | E | H | 1972 | 68474 |
| 0096 | G | C | I | H | 1970 | 58473 |
| 0096 | G | C | E | H | 1971 | 63424 |
| 0098 | L | D | R | N | 1973 | 73063 |
| 0098 | L | D | E | N | 1973 | 74679 |
| 0099 | L | D | R | N | 1971 | 63550 |
| 0099 | L | D | R | N | 1971 | 63767 |
| 0099 | L | D | R | N | 1971 | 64393 |
| 0099 | L | D | E | N | 1971 | 64394 |
| 0099 | L | D | E | N | 1971 | 67243 |

| Substance Number | Phys. State | Subject | Language | Temperature | Year | TPRC Number |
|---|---|---|---|---|---|---|
| 512-0099 | L | D | E | N | 1971 | 67244 |
| 0099 | L | D | R | N | 1973 | 73063 |
| 0099 | L | D | E | N | 1973 | 74679 |
| 0099 | L | D | R | N | 1973 | 75611 |
| 0099 | L | D | E | N | 1973 | 75612 |
| 0101 | L | D | E | N | 1972 | 68736 |
| 0101 | L | D | R | N | 1973 | 73063 |
| 0101 | L | D | E | N | 1973 | 74679 |
| 0101 | S | D | E | N | 1972 | 68736 |
| 0105 | G | G | E | N | 1972 | 71388 |
| 0105 | L | D | E | N | 1972 | 66860 |
| 0106 | G | D | E | N | 1972 | 69256 |
| 0106 | G | C | R | N | 1973 | 71112 |
| 0106 | G | G | E | N | 1972 | 71388 |
| 0106 | G | C | E | N | 1971 | 74103 |
| 0106 | G | C | E | N | 1973 | 74224 |
| 0106 | G | M | R | N | 1974 | 76466 |
| 0106 | G | M | E | N | 1974 | 76467 |
| 0106 | L | C | E | L | 1971 | 74103 |
| 0106 | S | G | R | L | 1969 | 65285 |
| 0106 | S | D | E | L | 1971 | 65677 |
| 0108 | L | D | E | N | 1973 | 69908 |
| 0113 | L | D | E | N | 1971 | 44504 |
| 0114 | L | D | E | N | 1971 | 44504 |
| 0114 | L | C | E | N | 1973 | 71794 |
| 0114 | L | M | S | N | 1974 | 75861 |
| 0115 | L | D | E | N | 1971 | 44504 |
| 0115 | L | D | E | N | 1973 | 72308 |
| 0116 | G | C | E | N | 1972 | 66556 |
| 0116 | L | D | E | N | 1971 | 44504 |
| 0116 | L | D | J | N | 1969 | 58112 |
| 0116 | L | D | E | N | 1971 | 62623 |
| 0116 | L | C | E | N | 1972 | 66556 |
| 0116 | L | D | E | N | 1973 | 72308 |
| 0116 | L | M | S | N | 1974 | 75861 |
| 0117 | L | G | G | N | 1974 | 78380 |
| 0121 | L | D | R | N | 1970 | 63207 |
| 0124 | L | D | E | N | 1972 | 68736 |
| 0124 | S | D | G | N | 1964 | 36129 |
| 0124 | S | C | G | N | 1964 | 36130 |
| 0124 | S | D | E | N | 1972 | 68736 |
| 0125 | S | D | G | N | 1964 | 36129 |
| 0125 | S | C | G | N | 1964 | 36130 |
| 0126 | L | D | E | N | 1970 | 59027 |
| 0126 | L | D | E | N | 1971 | 64303 |
| 0127 | L | D | E | N | 1971 | 64303 |
| 0128 | G | S | E | U | 1969 | 53699 |
| 0128 | S | D | E | L | 1973 | 72899 |
| 0131 | L | D | E | N | 1971 | 62624 |
| 0132 | S | D | E | L | 1973 | 72899 |
| 0134 | L | C | E | U | 1972 | 68536 |
| 0135 | L | C | E | U | 1972 | 68536 |
| 0136 | L | D | E | N | 1971 | 61837 |
| 0136 | L | D | R | N | 1971 | 63550 |
| 0136 | L | D | R | N | 1971 | 63767 |
| 0136 | L | D | R | N | 1971 | 64393 |
| 0136 | L | D | E | N | 1971 | 64394 |
| 0136 | L | D | E | N | 1971 | 67243 |
| 0136 | L | D | E | N | 1971 | 67244 |
| 0136 | L | D | R | N | 1973 | 75611 |
| 0136 | L | D | E | N | 1973 | 75612 |
| 0138 | P | M | R | U | 1968 | 49504 |
| 0138 | P | M | E | U | 1968 | 65123 |
| 0139 | P | M | R | U | 1968 | 49504 |
| 0139 | P | M | E | U | 1968 | 65123 |
| 0142 | L | D | E | N | 1974 | 75810 |
| 0150 | L | D | E | N | 1971 | 62466 |
| 0151 | G | C | E | N | 1972 | 66556 |
| 0151 | L | D | E | N | 1971 | 62466 |
| 0151 | L | D | E | N | 1971 | 62622 |
| 0151 | L | D | E | N | 1971 | 63930 |
| 0151 | L | C | E | N | 1972 | 66556 |
| 0151 | L | D | E | N | 1972 | 66774 |
| 0152 | G | D | E | L | 1973 | 71776 |
| 0152 | G | D | E | N | 1974 | 76378 |
| 0152 | L | D | E | L | 1973 | 71776 |
| 0158 | S | D | R | L | 1971 | 64321 |
| 0158 | S | D | E | L | 1971 | 64322 |
| 0162 | L | D | E | N | 1970 | 64281 |
| 0162 | L | M | E | N | 1971 | 65111 |
| 0162 | L | C | E | U | 1972 | 68536 |
| 0162 | L | D | E | N | 1973 | 69908 |
| 0166 | L | D | J | N | 1969 | 58112 |
| 0166 | L | D | E | N | 1970 | 59027 |
| 0169 | M | D | E | L | 1971 | 62325 |
| 0169 | S | D | E | L | 1970 | 65181 |
| 0170 | M | D | E | L | 1971 | 62325 |
| 0170 | S | D | E | L | 1970 | 65181 |
| 0174 | S | D | E | F | 1971 | 63728 |
| 0174 | S | D | R | H | 1971 | 65303 |
| 0174 | S | D | E | H | 1971 | 65304 |
| 0174 | S | M | E | H | 1969 | 77573 |
| 0175 | S | D | E | F | 1971 | 63728 |
| 0175 | S | D | R | H | 1971 | 65303 |
| 0175 | S | D | E | H | 1971 | 65304 |

**Phys. State:** **D.** Doped; **E.** Expanded; **F.** Fibrous or Whisker; **G.** Gas; **L.** Liquid; **M.** Multiphase; **P.** Powder; **S.** Solid
**Subject:** **C.** Calculated Data + Theory; **D.** Data; **E.** Experiment; **G.** General (Data + Theory + Experiment); **M.** Measured Data + Experiment; **S.** Survey (Review, Compendium, etc.); **T.** Theory
**Language:** **C.** Czech; **D.** Dutch; **E.** English; **F.** French; **G.** German; **I.** Italian; **J.** Japanese; **O.** Other Languages; **P.** Polish; **R.** Russian; **S.** Spanish
**Temperature:** **F.** Full Range (Low + Normal + High); **L.** Low (0 to 75K) + Overlap into Normal; **N.** Normal (75 to 1273K); **H.** High (above 1273K) + Overlap into Normal; **U.** Undefined

| Substance Number | Phys. State | Subject | Language | Temperature | Year | TPRC Number |
|---|---|---|---|---|---|---|
| 512-0176 | L | D | R | H | 1971 | 64929 |
| 0176 | L | D | E | H | 1971 | 64930 |
| 0177 | L | D | R | H | 1971 | 64929 |
| 0177 | L | D | E | H | 1971 | 64930 |
| 0178 | L | D | R | H | 1971 | 64929 |
| 0178 | L | D | E | H | 1971 | 64930 |
| 0179 | L | D | R | H | 1971 | 64929 |
| 0179 | L | D | E | H | 1971 | 64930 |
| 0189 | L | D | E | N | 1971 | 61837 |
| 0190 | L | C | R | N | 1965 | 45340 |
| 0191 | L | C | R | N | 1965 | 45340 |
| 0193 | G | D | E | N | 1971 | 65773 |
| 0193 | G | D | E | N | 1972 | 69256 |
| 0193 | G | C | R | N | 1973 | 71112 |
| 0193 | G | C | E | N | 1973 | 74224 |
| 0193 | G | M | R | N | 1974 | 76466 |
| 0193 | G | M | E | N | 1974 | 76467 |
| 0205 | G | C | E | U | 1972 | 65366 |
| 0216 | L | D | E | N | 1968 | 50991 |
| 0216 | L | D | J | N | 1969 | 58112 |
| 0216 | L | D | E | N | 1971 | 65743 |
| 0216 | L | M | E | N | 1971 | 67696 |
| 0216 | L | D | E | N | 1973 | 72308 |
| 0221 | L | G | E | N | 1971 | 66156 |
| 0222 | G | C | E | H | 1971 | 62685 |
| 0222 | G | C | E | H | 1972 | 66581 |
| 0225 | G | C | E | H | 1971 | 62685 |
| 0225 | G | C | E | H | 1972 | 66581 |
| 0226 | G | M | E | N | 1973 | 75790 |
| 0227 | G | C | E | N | 1972 | 66556 |
| 0227 | L | C | E | N | 1972 | 66556 |
| 0233 | S | D | E | L | 1964 | 37642 |
| 0233 | S | G | F | H | 1972 | 68940 |
| 0258 | L | D | R | N | 1973 | 72492 |
| 0259 | S | D | E | N | 1974 | 75755 |
| 0262 | S | D | R | L | 1970 | 63212 |
| 0267 | S | T | F | L | 1972 | 66729 |
| 0268 | S | T | F | L | 1972 | 66729 |
| 0269 | L | D | E | N | 1970 | 59027 |
| 0280 | L | D | E | N | 1973 | 72365 |
| 0281 | L | D | E | N | 1973 | 72365 |
| 0282 | L | D | E | N | 1970 | 59027 |
| 0287 | S | D | R | L | 1970 | 63212 |
| 0288 | L | D | R | N | 1970 | 63207 |
| 0291 | G | C | E | N | 1965 | 44126 |
| 0293 | G | D | E | L | 1973 | 71776 |
| 0293 | L | D | E | L | 1973 | 71776 |
| 0311 | L | D | R | N | 1963 | 37245 |
| 0320 | S | D | E | L | 1964 | 37642 |
| 0324 | L | D | E | N | 1970 | 59027 |
| 0325 | L | D | E | N | 1970 | 59027 |
| 0326 | L | D | E | N | 1970 | 59027 |
| 0327 | L | D | E | N | 1970 | 59027 |
| 0356 | G | T | R | U | 1966 | 58281 |
| 0356 | L | S | E | N | 1966 | 41432 |
| 0357 | L | G | E | N | 1971 | 66156 |
| 0357 | L | D | E | N | 1972 | 66926 |
| 0358 | L | C | E | U | 1972 | 68536 |
| 0365 | G | C | E | N | 1965 | 44126 |
| 0370 | G | S | R | H | 1974 | 78662 |
| 0370 | L | D | E | N | 1967 | 72364 |
| 0370 | L | G | E | N | 1968 | 76272 |
| 0397 | L | D | E | N | 1973 | 74822 |
| 0403 | L | G | G | N | 1974 | 78380 |
| 0410 | L | D | E | N | 1971 | 63930 |
| 0451 | G | G | E | N | 1972 | 71388 |
| 0451 | L | D | E | N | 1972 | 66860 |
| 0454 | L | D | E | N | 1973 | 72365 |
| 0476 | G | T | R | U | 1970 | 59295 |
| 0603 | S | D | E | N | 1974 | 75755 |
| 0611 | S | D | R | L | 1974 | 74893 |
| 0611 | S | D | E | L | 1974 | 76536 |
| 0643 | L | D | R | N | 1973 | 72492 |
| 0728 | S | G | F | H | 1972 | 68940 |
| 0729 | G | S | R | H | 1974 | 78662 |
| 0735 | G | S | R | H | 1974 | 78662 |
| 0735 | L | D | E | N | 1967 | 72364 |
| 0735 | L | G | E | N | 1968 | 76272 |
| 0736 | L | G | E | N | 1968 | 76272 |
| 0742 | S | D | F | N | 1970 | 70692 |
| 0747 | S | D | R | L | 1974 | 74893 |
| 0747 | S | D | E | L | 1974 | 76536 |
| 0797 | S | M | F | L | 1973 | 71477 |
| 0798 | S | M | F | L | 1973 | 71477 |
| 0806 | G | G | E | N | 1973 | 73805 |
| 0827 | L | D | E | N | 1972 | 70522 |
| 0827 | S | D | E | N | 1972 | 70522 |
| 0884 | G | S | R | H | 1974 | 78662 |
| 1029 | G | D | E | H | 1975 | 77118 |
| 1030 | G | S | R | H | 1974 | 78662 |
| 1045 | L | G | G | N | 1974 | 78380 |
| 1048 | S | M | E | H | 1969 | 77573 |
| 1071 | S | D | E | N | 1970 | 62419 |
| 1072 | S | D | E | N | 1970 | 62419 |
| 1073 | S | D | F | N | 1970 | 70692 |
| 1074 | S | D | F | N | 1970 | 70692 |
| 1075 | S | D | F | N | 1970 | 70692 |
| 1076 | S | D | F | N | 1970 | 70692 |
| 1081 | G | D | E | H | 1977 | 93604 |
| 514-0001 | G | M | E | N | 1972 | 68529 |
| 514-0001 | G | S | R | H | 1974 | 78662 |
| 0006 | G | D | R | N | 1968 | 53702 |
| 0006 | G | D | R | N | 1969 | 60917 |
| 0006 | G | D | E | N | 1972 | 67826 |
| 0006 | G | D | E | N | 1972 | 67827 |
| 0006 | L | D | R | N | 1968 | 53702 |
| 0006 | L | C | R | N | 1967 | 53703 |
| 0006 | L | D | R | N | 1969 | 60917 |
| 0006 | L | D | E | N | 1972 | 67826 |
| 0006 | L | D | E | N | 1972 | 67827 |
| 0013 | G | M | E | N | 1974 | 73526 |
| 0019 | G | M | E | N | 1972 | 68529 |
| 0020 | G | D | G | H | 1965 | 42486 |
| 0020 | G | D | G | F | 1961 | 45651 |
| 0020 | G | S | E | F | 1971 | 62671 |
| 0020 | G | M | E | N | 1967 | 64178 |
| 0020 | G | G | E | N | 1971 | 65228 |
| 0020 | G | C | R | N | 1970 | 67725 |
| 0020 | G | D | E | N | 1972 | 68595 |
| 0020 | G | S | E | N | 1971 | 71643 |
| 0020 | G | C | R | N | 1973 | 72861 |
| 0020 | G | C | E | N | 1973 | 72862 |
| 0020 | G | D | R | N | 1970 | 73874 |
| 0020 | G | D | E | N | 1970 | 73875 |
| 0020 | G | C | R | N | 1973 | 75016 |
| 0020 | G | C | E | N | 1974 | 75017 |
| 0020 | G | D | E | N | 1974 | 76016 |
| 0020 | G | C | E | H | 1974 | 76376 |
| 0020 | G | D | E | N | 1975 | 76427 |
| 0020 | G | D | E | H | 1965 | 78046 |
| 0020 | L | D | G | F | 1961 | 45651 |
| 0028 | G | D | R | N | 1968 | 53702 |
| 0028 | G | D | R | N | 1969 | 60917 |
| 0028 | L | D | R | N | 1968 | 53702 |
| 0028 | L | D | R | N | 1969 | 60917 |
| 0029 | L | D | E | N | 1972 | 70522 |
| 0029 | S | D | E | N | 1972 | 70522 |
| 0039 | G | D | E | N | 1972 | 67826 |
| 0039 | L | D | E | N | 1972 | 67826 |
| 0047 | G | D | R | N | 1968 | 53702 |
| 0047 | G | D | R | N | 1969 | 60917 |
| 0047 | L | D | R | N | 1968 | 53702 |
| 0047 | L | C | R | N | 1967 | 53703 |
| 0047 | L | D | R | N | 1969 | 60917 |
| 0048 | L | D | E | N | 1972 | 70522 |
| 0048 | S | D | E | N | 1972 | 70522 |
| 0051 | G | M | R | N | 1966 | 44589 |
| 0051 | G | D | R | N | 1972 | 71034 |
| 0051 | G | D | E | H | 1974 | 79055 |
| 0052 | G | M | R | N | 1966 | 44589 |
| 0052 | G | D | R | N | 1972 | 71034 |
| 0052 | G | D | E | H | 1974 | 79055 |
| 0064 | L | C | R | N | 1967 | 53703 |
| 0091 | S | D | E | H | 1977 | 93601 |
| 0091 | S | D | E | N | 1977 | 93601 |
| 0091 | S | E | E | H | 1977 | 93601 |
| 0091 | S | E | E | N | 1977 | 93601 |
| 0105 | G | G | E | N | 1972 | 71388 |
| 0106 | G | M | E | N | 1972 | 68529 |
| 0106 | G | G | E | N | 1972 | 71388 |
| 0106 | G | M | E | N | 1974 | 73526 |
| 0131 | G | D | E | N | 1972 | 67827 |
| 0131 | L | M | P | N | 1965 | 35709 |
| 0131 | L | D | E | N | 1972 | 67827 |
| 0151 | G | D | E | N | 1972 | 67826 |
| 0151 | L | D | E | N | 1972 | 67826 |
| 0152 | G | M | R | N | 1966 | 44589 |
| 0152 | G | D | R | N | 1972 | 71034 |
| 0152 | G | S | R | H | 1974 | 78662 |
| 0174 | S | M | E | H | 1969 | 77573 |
| 0175 | S | M | E | H | 1969 | 77573 |
| 0184 | L | M | P | N | 1965 | 35709 |
| 0189 | L | M | P | N | 1965 | 35709 |
| 0193 | G | M | E | N | 1974 | 73526 |
| 0216 | L | M | P | N | 1965 | 35709 |
| 0233 | S | D | E | H | 1977 | 93601 |
| 0233 | S | D | E | N | 1977 | 93601 |
| 0233 | S | E | E | H | 1977 | 93601 |
| 0233 | S | E | E | N | 1977 | 93601 |
| 0238 | L | M | P | N | 1965 | 35709 |
| 0244 | G | D | E | N | 1972 | 67827 |
| 0244 | L | D | E | N | 1972 | 67827 |
| 0262 | S | D | E | H | 1977 | 93601 |
| 0262 | S | D | E | N | 1977 | 93601 |
| 0293 | G | S | R | H | 1974 | 78662 |
| 0317 | L | M | P | N | 1965 | 35709 |
| 0318 | L | M | P | N | 1965 | 35709 |
| 0356 | L | T | R | U | 1973 | 76185 |
| 0370 | L | G | E | N | 1968 | 76272 |
| 0370 | L | M | E | N | 1964 | 78042 |
| 0386 | G | C | R | N | 1969 | 55396 |
| 0386 | G | T | R | U | 1970 | 73706 |
| 0386 | G | T | E | U | 1970 | 73707 |
| 0391 | G | M | R | N | 1966 | 44589 |
| 0391 | G | D | R | N | 1972 | 71034 |
| 0451 | G | G | E | N | 1972 | 71388 |
| 0707 | G | M | E | N | 1971 | 70136 |
| 0708 | L | G | E | N | 1968 | 76272 |
| 0708 | L | M | E | N | 1964 | 78042 |
| 0735 | G | S | R | H | 1974 | 78662 |
| 0736 | L | G | E | N | 1968 | 76272 |
| 514-0736 | L | M | E | N | 1964 | 78042 |
| 0827 | L | D | E | N | 1972 | 70522 |
| 0827 | S | D | E | N | 1972 | 70522 |
| 0931 | G | D | E | H | 1974 | 79055 |
| 1027 | G | D | E | H | 1974 | 79055 |
| 1031 | G | S | R | H | 1974 | 78662 |
| 1040 | M | T | R | N | 1973 | 75167 |
| 1048 | S | M | E | H | 1969 | 77573 |
| 515-0003 | G | T | R | N | 1971 | 65016 |
| 0005 | L | D | E | N | 1973 | 73415 |
| 0010 | L | D | R | N | 1972 | 69566 |
| 0011 | L | M | R | N | 1971 | 64245 |
| 0011 | L | C | E | U | 1972 | 68536 |
| 0011 | L | M | E | N | 1971 | 70814 |
| 0013 | G | D | E | N | 1977 | 93616 |
| 0013 | L | D | E | N | 1977 | 93616 |
| 0013 | M | E | E | N | 1977 | 93615 |
| 0013 | M | D | E | N | 1977 | 93616 |
| 0028 | L | C | E | U | 1972 | 68536 |
| 0029 | L | C | R | N | 1970 | 60855 |
| 0029 | L | C | R | U | 1971 | 65821 |
| 0029 | L | C | E | U | 1971 | 67354 |
| 0029 | L | D | E | N | 1974 | 76260 |
| 0030 | L | C | R | N | 1970 | 60855 |
| 0030 | L | M | R | N | 1971 | 63260 |
| 0030 | L | M | E | N | 1971 | 65173 |
| 0030 | L | C | R | U | 1971 | 65821 |
| 0030 | L | M | E | N | 1971 | 66485 |
| 0030 | L | D | R | N | 1972 | 66602 |
| 0030 | L | C | E | U | 1971 | 67354 |
| 0030 | L | C | E | U | 1972 | 68536 |
| 0030 | L | M | E | N | 1972 | 69342 |
| 0030 | L | D | E | N | 1972 | 69793 |
| 0030 | L | D | E | N | 1973 | 70917 |
| 0030 | L | D | E | N | 1973 | 71555 |
| 0030 | L | D | E | N | 1973 | 72373 |
| 0030 | L | D | E | N | 1973 | 75574 |
| 0030 | L | C | R | N | 1973 | 75656 |
| 0030 | L | C | E | N | 1973 | 75657 |
| 0030 | L | D | E | N | 1974 | 76260 |
| 0030 | L | S | E | N | 1960 | 91174 |
| 0030 | L | S | E | N | 1960 | 94260 |
| 0030 | S | S | E | N | 1960 | 91174 |
| 0031 | L | C | R | N | 1970 | 60855 |
| 0031 | L | D | E | N | 1974 | 76260 |
| 0032 | L | D | R | N | 1970 | 75459 |
| 0032 | L | D | E | N | 1970 | 75460 |
| 0042 | L | D | R | N | 1970 | 75459 |
| 0042 | L | D | E | N | 1970 | 75460 |
| 0043 | L | C | E | U | 1972 | 68536 |
| 0048 | L | C | R | N | 1970 | 60855 |
| 0048 | L | M | R | N | 1971 | 64245 |
| 0048 | L | C | R | U | 1971 | 65821 |
| 0048 | L | D | R | N | 1972 | 66602 |
| 0048 | L | C | E | U | 1971 | 67354 |
| 0048 | L | M | E | N | 1972 | 69342 |
| 0048 | L | D | R | N | 1972 | 69364 |
| 0048 | L | D | E | N | 1972 | 69365 |
| 0048 | L | D | E | N | 1972 | 69793 |
| 0048 | L | M | E | N | 1971 | 70814 |
| 0048 | L | D | E | N | 1973 | 70917 |
| 0048 | L | D | R | N | 1974 | 74281 |
| 0048 | L | D | E | N | 1973 | 75574 |
| 0048 | L | D | E | N | 1973 | 75782 |
| 0048 | L | D | E | N | 1974 | 76260 |
| 0057 | L | D | R | N | 1973 | 70588 |
| 0057 | L | D | E | N | 1973 | 72553 |
| 0057 | L | M | R | N | 1973 | 73064 |
| 0057 | L | M | E | N | 1973 | 75430 |
| 0061 | G | C | E | N | 1958 | 70660 |
| 0064 | L | E | E | N | 1971 | 65173 |
| 0064 | L | E | E | N | 1972 | 65842 |
| 0065 | L | D | R | N | 1972 | 69654 |
| 0065 | L | D | E | N | 1972 | 69655 |
| 0067 | L | M | R | N | 1971 | 64245 |
| 0067 | L | C | E | U | 1972 | 68536 |
| 0067 | L | D | R | N | 1972 | 69362 |
| 0067 | L | D | E | N | 1972 | 69363 |
| 0067 | L | D | R | N | 1972 | 70000 |
| 0067 | L | D | E | N | 1972 | 70682 |
| 0067 | L | M | E | N | 1971 | 70814 |
| 0067 | L | D | R | N | 1971 | 75493 |
| 0067 | L | D | E | N | 1971 | 75494 |
| 0067 | L | C | R | N | 1973 | 75656 |
| 0067 | L | C | E | N | 1973 | 75657 |
| 0070 | L | C | R | N | 1970 | 60855 |
| 0070 | L | D | E | N | 1974 | 76260 |
| 0084 | L | C | E | U | 1972 | 68536 |
| 0089 | L | C | R | U | 1971 | 65821 |
| 0089 | L | C | E | U | 1971 | 67354 |
| 0089 | L | C | E | U | 1972 | 68536 |
| 0089 | L | C | R | N | 1973 | 71801 |
| 0089 | L | D | R | N | 1973 | 73009 |
| 0089 | L | D | E | N | 1973 | 75855 |
| 0099 | L | C | R | N | 1970 | 60855 |
| 0099 | L | D | R | N | 1971 | 63183 |
| 0099 | L | D | R | N | 1971 | 63550 |
| 0099 | L | D | R | N | 1971 | 63767 |
| 0099 | L | D | R | N | 1971 | 64393 |
| 0099 | L | D | E | N | 1971 | 64394 |

**Phys. State:** **D.** Doped; **E.** Expanded; **F.** Fibrous or Whisker; **G.** Gas; **L.** Liquid; **M.** Multiphase; **P.** Powder; **S.** Solid
**Subject:** **C.** Calculated Data + Theory; **D.** Data; **E.** Experiment; **G.** General (Data + Theory + Experiment); **M.** Measured Data + Experiment; **S.** Survey (Review, Compendium, etc.); **T.** Theory
**Language:** **C.** Czech; **D.** Dutch; **E.** English; **F.** French; **G.** German; **I.** Italian; **J.** Japanese; **O.** Other Languages; **P.** Polish; **R.** Russian; **S.** Spanish
**Temperature:** **F.** Full Range (Low + Normal + High); **L.** Low (0 to 75K) + Overlap into Normal; **N.** Normal (75 to 1273K); **H.** High (above 1273K) + Overlap into Normal; **U.** Undefined

| Substance Number | Phys. State | Subject | Language | Temperature | Year | TPRC Number |
|---|---|---|---|---|---|---|
| 515-0099 | L | D | E | N | 1971 | 67243 |
| 0099 | L | D | E | N | 1971 | 67244 |
| 0099 | L | D | R | N | 1973 | 75611 |
| 0099 | L | D | E | N | 1973 | 75612 |
| 0099 | L | D | E | N | 1974 | 76260 |
| 0100 | L | C | R | N | 1970 | 60855 |
| 0100 | L | D | E | N | 1973 | 75574 |
| 0100 | L | D | E | N | 1974 | 76260 |
| 0101 | L | C | R | N | 1970 | 60855 |
| 0101 | L | D | E | N | 1974 | 76260 |
| 0118 | L | D | E | N | 1971 | 61648 |
| 0118 | L | G | E | N | 1971 | 66156 |
| 0121 | L | D | E | N | 1972 | 67822 |
| 0127 | L | D | E | N | 1973 | 73415 |
| 0141 | L | D | R | N | 1972 | 69364 |
| 0141 | L | D | E | N | 1972 | 69365 |
| 0144 | L | C | R | N | 1970 | 60855 |
| 0144 | L | C | R | U | 1971 | 65821 |
| 0144 | L | C | E | U | 1971 | 67354 |
| 0144 | L | D | E | N | 1973 | 72373 |
| 0144 | L | D | E | N | 1974 | 76260 |
| 0152 | G | T | R | N | 1971 | 65016 |
| 0152 | G | C | E | N | 1958 | 70660 |
| 0182 | L | C | E | U | 1972 | 68536 |
| 0182 | L | D | R | N | 1973 | 70588 |
| 0182 | L | D | E | N | 1973 | 72553 |
| 0182 | L | M | R | N | 1973 | 73064 |
| 0182 | L | M | E | N | 1973 | 75430 |
| 0183 | L | M | R | N | 1973 | 73064 |
| 0183 | L | M | E | N | 1973 | 75430 |
| 0184 | L | M | R | N | 1972 | 69716 |
| 0189 | L | D | E | N | 1971 | 63114 |
| 0189 | L | C | E | U | 1972 | 68536 |
| 0191 | L | C | R | N | 1965 | 45340 |
| 0194 | L | C | E | U | 1972 | 68536 |
| 0194 | L | D | E | N | 1974 | 76260 |
| 0194 | L | S | E | N | 1960 | 94260 |
| 0208 | L | D | E | N | 1973 | 75781 |
| 0211 | L | D | E | N | 1973 | 73415 |
| 0213 | L | D | E | N | 1974 | 74951 |
| 0220 | L | D | E | N | 1972 | 60206 |
| 0220 | L | D | E | N | 1971 | 61648 |
| 0221 | L | G | E | N | 1971 | 66156 |
| 0221 | L | C | R | N | 1973 | 71887 |
| 0221 | L | C | E | N | 1973 | 73461 |
| 0237 | L | D | R | N | 1971 | 63183 |
| 0258 | L | C | R | N | 1970 | 60855 |
| 0258 | L | D | E | N | 1974 | 76260 |
| 0259 | L | C | R | N | 1970 | 60855 |
| 0259 | L | D | E | N | 1973 | 75574 |
| 0259 | L | D | E | N | 1974 | 76260 |
| 0263 | L | D | E | N | 1972 | 60206 |
| 0280 | L | C | R | N | 1970 | 60855 |
| 0280 | L | D | E | N | 1974 | 76260 |
| 0281 | L | C | R | N | 1970 | 60855 |
| 0288 | L | D | R | N | 1969 | 57532 |
| 0288 | L | D | E | N | 1969 | 70689 |
| 0290 | L | C | R | N | 1973 | 71801 |
| 0290 | L | S | R | N | 1973 | 72954 |
| 0311 | L | D | R | N | 1963 | 37245 |
| 0311 | L | M | R | N | 1971 | 63260 |
| 0311 | L | M | E | N | 1971 | 66485 |
| 0311 | L | C | E | U | 1972 | 68536 |
| 0357 | L | G | E | N | 1971 | 66156 |
| 0357 | L | D | E | N | 1972 | 66926 |
| 0387 | L | D | R | N | 1970 | 63898 |
| 0387 | L | C | R | U | 1971 | 65821 |
| 0387 | L | D | R | N | 1972 | 66602 |
| 0387 | L | C | E | U | 1971 | 67354 |
| 0387 | L | C | R | N | 1971 | 67818 |
| 0387 | L | M | E | N | 1972 | 69342 |
| 0387 | L | D | E | N | 1972 | 69793 |
| 0387 | L | D | E | N | 1973 | 70917 |
| 0387 | L | C | R | N | 1973 | 71801 |
| 0387 | L | D | R | N | 1973 | 73009 |
| 0387 | L | C | R | N | 1973 | 75656 |
| 0387 | L | C | E | N | 1973 | 75657 |
| 0387 | L | D | E | N | 1973 | 75855 |
| 0408 | L | D | R | N | 1974 | 75958 |
| 0408 | L | D | E | N | 1974 | 76461 |
| 0412 | L | D | R | N | 1969 | 56222 |
| 0412 | L | D | R | N | 1970 | 65456 |
| 0433 | L | D | R | N | 1971 | 64576 |
| 0433 | L | D | E | N | 1971 | 69784 |
| 0460 | L | C | R | N | 1973 | 71801 |
| 0460 | L | D | R | N | 1973 | 73009 |
| 0460 | L | C | R | N | 1973 | 75656 |
| 0460 | L | C | E | N | 1973 | 75657 |
| 0460 | L | D | E | N | 1973 | 75855 |
| 0465 | M | C | R | N | 1965 | 42555 |
| 0465 | M | D | R | N | 1964 | 43252 |
| 0465 | S | D | R | N | 1968 | 60916 |
| 0472 | L | D | E | N | 1972 | 66024 |
| 0472 | M | D | R | N | 1971 | 73714 |
| 0472 | M | D | E | N | 1971 | 73715 |
| 0493 | L | M | E | N | 1972 | 69342 |
| 0493 | L | D | E | N | 1973 | 71555 |
| 0493 | L | D | R | N | 1973 | 72963 |
| 0494 | L | G | R | N | 1974 | 75961 |
| 0494 | L | G | E | N | 1974 | 76464 |
| 0496 | L | D | R | N | 1971 | 64788 |
| 515-0496 | L | D | E | N | 1971 | 64789 |
| 0496 | L | C | R | N | 1973 | 71801 |
| 0498 | L | C | R | N | 1970 | 60855 |
| 0498 | L | D | E | N | 1974 | 76260 |
| 0499 | L | C | R | N | 1970 | 60855 |
| 0502 | L | C | R | N | 1970 | 60855 |
| 0502 | L | D | E | N | 1974 | 76260 |
| 0504 | L | D | R | N | 1972 | 70000 |
| 0504 | L | D | E | N | 1972 | 70682 |
| 0506 | L | M | R | N | 1971 | 64245 |
| 0506 | L | D | R | N | 1972 | 70000 |
| 0506 | L | D | E | N | 1972 | 70682 |
| 0506 | L | M | E | N | 1971 | 70814 |
| 0507 | L | M | E | N | 1971 | 65173 |
| 0507 | L | D | E | N | 1973 | 73415 |
| 0509 | L | M | R | N | 1971 | 64245 |
| 0509 | L | D | R | N | 1972 | 70000 |
| 0509 | L | D | E | N | 1972 | 70682 |
| 0509 | L | M | E | N | 1971 | 70814 |
| 0526 | L | D | E | N | 1974 | 74951 |
| 0544 | L | D | E | N | 1974 | 74951 |
| 0560 | L | D | C | N | 1970 | 62412 |
| 0577 | L | D | E | N | 1972 | 70476 |
| 0578 | L | C | F | N | 1971 | 65091 |
| 0585 | L | D | E | N | 1973 | 72373 |
| 0595 | L | G | R | N | 1974 | 75961 |
| 0595 | L | G | E | N | 1974 | 76464 |
| 0597 | L | G | R | N | 1974 | 75961 |
| 0597 | L | G | E | N | 1974 | 76464 |
| 0599 | L | C | R | N | 1970 | 60855 |
| 0599 | L | D | R | N | 1971 | 65349 |
| 0599 | L | D | E | N | 1971 | 65350 |
| 0600 | L | M | R | N | 1971 | 64245 |
| 0600 | L | D | R | N | 1972 | 70000 |
| 0600 | L | D | E | N | 1972 | 70682 |
| 0600 | L | M | E | N | 1971 | 70814 |
| 0600 | L | D | E | N | 1973 | 72373 |
| 0600 | L | C | R | N | 1973 | 75656 |
| 0600 | L | C | E | N | 1973 | 75657 |
| 0601 | L | C | R | N | 1973 | 71887 |
| 0601 | L | C | E | N | 1973 | 73461 |
| 0603 | L | C | R | N | 1970 | 60855 |
| 0603 | L | D | E | N | 1974 | 76260 |
| 0605 | L | D | R | N | 1970 | 62433 |
| 0605 | L | C | R | N | 1971 | 64152 |
| 0607 | L | D | R | N | 1970 | 62433 |
| 0607 | L | C | R | N | 1971 | 64152 |
| 0609 | L | G | R | N | 1974 | 75961 |
| 0609 | L | G | E | N | 1974 | 76464 |
| 0610 | L | G | R | N | 1974 | 75961 |
| 0610 | L | G | E | N | 1974 | 76464 |
| 0612 | L | D | R | N | 1970 | 62433 |
| 0612 | L | C | R | N | 1971 | 64152 |
| 0613 | L | D | R | N | 1970 | 62433 |
| 0613 | L | C | R | N | 1971 | 64152 |
| 0616 | L | D | R | N | 1970 | 62433 |
| 0616 | L | C | R | N | 1971 | 64152 |
| 0618 | L | D | R | N | 1971 | 65349 |
| 0618 | L | D | E | N | 1971 | 65350 |
| 0619 | L | D | R | N | 1970 | 63898 |
| 0619 | L | D | R | N | 1971 | 64788 |
| 0619 | L | D | E | N | 1971 | 64789 |
| 0619 | L | C | R | N | 1973 | 71801 |
| 0620 | L | G | E | N | 1971 | 70973 |
| 0621 | L | G | E | N | 1971 | 70973 |
| 0622 | L | G | E | N | 1971 | 70973 |
| 0623 | L | G | E | N | 1971 | 70973 |
| 0624 | L | G | E | N | 1971 | 70973 |
| 0625 | L | G | E | N | 1971 | 70973 |
| 0626 | L | G | E | N | 1971 | 70973 |
| 0627 | L | G | E | N | 1971 | 70973 |
| 0628 | L | D | R | N | 1973 | 75539 |
| 0628 | L | D | E | N | 1973 | 75540 |
| 0632 | L | C | R | N | 1970 | 60855 |
| 0632 | L | D | E | N | 1974 | 76260 |
| 0638 | L | G | E | N | 1971 | 70973 |
| 0639 | L | C | R | N | 1970 | 60855 |
| 0640 | L | C | R | N | 1970 | 60855 |
| 0640 | L | D | E | N | 1974 | 76260 |
| 0641 | L | C | R | N | 1970 | 60855 |
| 0642 | L | D | R | N | 1971 | 64576 |
| 0642 | L | D | E | N | 1971 | 69784 |
| 0642 | L | D | R | N | 1973 | 72532 |
| 0642 | L | D | E | N | 1973 | 74820 |
| 0642 | L | D | R | N | 1977 | 88635 |
| 0642 | L | E | R | N | 1977 | 88635 |
| 0642 | L | D | E | N | 1977 | 93697 |
| 0642 | L | E | E | N | 1977 | 93697 |
| 0643 | L | D | R | N | 1971 | 64576 |
| 0643 | L | C | E | U | 1972 | 68536 |
| 0643 | L | D | R | N | 1972 | 69076 |
| 0643 | L | D | E | N | 1971 | 69784 |
| 0644 | L | C | E | U | 1972 | 68536 |
| 0647 | L | D | E | N | 1973 | 75574 |
| 0648 | L | G | E | N | 1971 | 70973 |
| 0650 | L | C | R | N | 1970 | 60855 |
| 0651 | L | C | R | N | 1970 | 60855 |
| 0651 | L | D | E | N | 1974 | 76260 |
| 0663 | L | C | E | U | 1972 | 68536 |
| 0663 | L | M | E | N | 1972 | 69342 |
| 0663 | L | D | E | N | 1973 | 70917 |
| 515-0674 | L | G | E | N | 1971 | 70973 |
| 0675 | L | G | E | N | 1971 | 70973 |
| 0676 | L | G | E | N | 1971 | 70973 |
| 0677 | L | G | E | N | 1971 | 70973 |
| 0678 | L | G | E | N | 1971 | 70973 |
| 0681 | L | M | E | N | 1972 | 69342 |
| 0681 | L | D | R | N | 1972 | 69362 |
| 0681 | L | D | E | N | 1972 | 69363 |
| 0681 | L | D | E | N | 1973 | 70917 |
| 0681 | L | M | R | N | 1973 | 73064 |
| 0681 | L | M | E | N | 1973 | 75430 |
| 0701 | L | D | E | N | 1973 | 73415 |
| 0717 | L | D | E | N | 1973 | 73415 |
| 0721 | L | D | E | N | 1973 | 72373 |
| 0737 | L | D | R | N | 1972 | 69077 |
| 0742 | L | D | E | N | 1973 | 73415 |
| 0757 | L | D | E | N | 1973 | 72373 |
| 0757 | L | D | E | N | 1973 | 72430 |
| 0795 | L | D | E | N | 1972 | 72751 |
| 0826 | L | D | R | N | 1973 | 75539 |
| 0826 | L | D | E | N | 1973 | 75540 |
| 0829 | L | D | E | N | 1973 | 72373 |
| 0830 | L | D | E | N | 1973 | 72373 |
| 0831 | L | D | E | N | 1973 | 72373 |
| 0834 | L | D | E | N | 1973 | 73415 |
| 0835 | L | D | E | N | 1973 | 73415 |
| 0836 | L | D | E | N | 1973 | 73415 |
| 0837 | L | D | E | N | 1973 | 73415 |
| 0838 | L | D | E | N | 1973 | 73415 |
| 0839 | L | D | E | N | 1973 | 73415 |
| 0840 | L | D | E | N | 1973 | 73415 |
| 0841 | L | D | E | N | 1973 | 73415 |
| 0842 | L | D | E | N | 1973 | 73415 |
| 0843 | L | D | E | N | 1973 | 73415 |
| 0844 | L | D | E | N | 1973 | 73415 |
| 0845 | L | D | E | N | 1973 | 73415 |
| 0846 | L | D | E | N | 1973 | 73415 |
| 0847 | L | D | E | N | 1973 | 73415 |
| 0848 | L | D | E | N | 1973 | 73415 |
| 0849 | L | D | E | N | 1973 | 73415 |
| 0850 | L | D | E | N | 1973 | 73415 |
| 0851 | L | D | E | N | 1973 | 73415 |
| 0852 | L | D | E | N | 1973 | 73415 |
| 0860 | L | D | R | N | 1974 | 75958 |
| 0860 | L | D | E | N | 1974 | 76461 |
| 0861 | L | D | R | N | 1974 | 75958 |
| 0861 | L | D | E | N | 1974 | 76461 |
| 0862 | L | D | R | N | 1974 | 75958 |
| 0862 | L | D | E | N | 1974 | 76461 |
| 0863 | L | D | R | N | 1974 | 75958 |
| 0863 | L | D | E | N | 1974 | 76461 |
| 0864 | L | D | R | N | 1974 | 75958 |
| 0864 | L | D | E | N | 1974 | 76461 |
| 0865 | L | D | R | N | 1974 | 75958 |
| 0865 | L | D | E | N | 1974 | 76461 |
| 0866 | L | D | R | N | 1974 | 75958 |
| 0866 | L | D | E | N | 1974 | 76461 |
| 0868 | L | G | R | N | 1974 | 75961 |
| 0868 | L | G | E | N | 1974 | 76464 |
| 0869 | M | D | R | N | 1971 | 73714 |
| 0869 | M | D | E | N | 1971 | 73715 |
| 0870 | L | D | R | N | 1973 | 75539 |
| 0870 | L | D | E | N | 1973 | 75540 |
| 0871 | L | D | R | N | 1973 | 75539 |
| 0871 | L | D | E | N | 1973 | 75540 |
| 0872 | L | D | R | N | 1973 | 75539 |
| 0872 | L | D | E | N | 1973 | 75540 |
| 0873 | L | D | R | N | 1973 | 75539 |
| 0873 | L | D | E | N | 1973 | 75540 |
| 0874 | L | D | R | N | 1973 | 75539 |
| 0874 | L | D | E | N | 1973 | 75540 |
| 0885 | L | D | E | L | 1974 | 74450 |
| 0898 | L | D | E | N | 1973 | 75574 |
| 0901 | L | D | R | N | 1972 | 75349 |
| 0901 | L | D | E | N | 1974 | 75350 |
| 0912 | L | D | E | N | 1974 | 74951 |
| 0913 | L | D | E | N | 1974 | 74951 |
| 0914 | L | D | E | N | 1974 | 74951 |
| 0915 | L | D | E | N | 1974 | 74951 |
| 0916 | L | D | E | N | 1974 | 74951 |
| 0934 | L | C | R | N | 1973 | 73002 |
| 0936 | L | S | E | N | 1960 | 94260 |
| 0955 | L | S | E | N | 1960 | 94260 |
| 0971 | L | D | E | N | 1973 | 75781 |
| 0972 | L | D | E | N | 1973 | 75781 |
| 0973 | L | D | E | N | 1973 | 75781 |
| 0974 | L | D | E | N | 1973 | 75781 |
| 0978 | L | D | E | N | 1973 | 75800 |
| 0979 | L | D | E | N | 1973 | 75800 |
| 0980 | L | D | E | N | 1973 | 75800 |
| 0981 | L | D | E | N | 1973 | 75800 |
| 0982 | L | D | E | N | 1973 | 75800 |
| 0983 | L | D | E | N | 1973 | 75800 |
| 0984 | L | D | E | N | 1973 | 75800 |
| 0985 | L | D | E | N | 1973 | 75800 |
| 0986 | L | D | E | N | 1973 | 75800 |
| 0987 | L | D | E | N | 1973 | 75800 |
| 0988 | L | D | E | N | 1973 | 75800 |
| 1083 | L | S | E | N | 1960 | 94260 |
| 517-0001 | G | D | G | N | 1972 | 68888 |

**Phys. State:** **D.** Doped; **E.** Expanded; **F.** Fibrous or Whisker; **G.** Gas; **L.** Liquid; **M.** Multiphase; **P.** Powder; **S.** Solid
**Subject:** **C.** Calculated Data + Theory; **D.** Data; **E.** Experiment; **G.** General (Data + Theory + Experiment); **M.** Measured Data + Experiment; **S.** Survey (Review, Compendium, etc.); **T.** Theory
**Language:** **C.** Czech; **D.** Dutch; **E.** English; **F.** French; **G.** German; **I.** Italian; **J.** Japanese; **O.** Other Languages; **P.** Polish; **R.** Russian; **S.** Spanish
**Temperature:** **F.** Full Range (Low + Normal + High); **L.** Low (0 to 75K) + Overlap into Normal; **N.** Normal (75 to 1273K); **H.** High (above 1273K) + Overlap into Normal; **U.** Undefined

| Substance Number | Phys. State | Sub-ject | Lan-guage | Temper-ature | Year | TPRC Number |
|---|---|---|---|---|---|---|
| 517-0003 | G | D | G | N | 1972 | 68888 |
| 0011 | L | M | R | N | 1971 | 64245 |
| 0011 | L | D | R | N | 1972 | 70000 |
| 0011 | L | D | E | N | 1972 | 70682 |
| 0011 | L | M | E | N | 1971 | 70814 |
| 0011 | L | D | R | N | 1971 | 75493 |
| 0011 | L | D | E | N | 1971 | 75494 |
| 0030 | L | C | R | N | 1968 | 51848 |
| 0030 | L | M | R | N | 1971 | 64245 |
| 0030 | L | C | E | N | 1972 | 66212 |
| 0030 | L | D | R | N | 1972 | 70000 |
| 0030 | L | D | E | N | 1972 | 70682 |
| 0030 | L | M | E | N | 1971 | 70814 |
| 0030 | L | D | R | N | 1971 | 75493 |
| 0030 | L | D | E | N | 1971 | 75494 |
| 0048 | L | C | R | N | 1968 | 51848 |
| 0048 | L | C | E | N | 1972 | 66212 |
| 0048 | L | D | R | N | 1972 | 69364 |
| 0048 | L | D | E | N | 1972 | 69365 |
| 0048 | L | D | R | N | 1974 | 74281 |
| 0057 | L | C | R | N | 1968 | 51848 |
| 0057 | L | C | E | N | 1972 | 66212 |
| 0063 | L | D | E | N | 1971 | 61648 |
| 0065 | L | D | R | N | 1971 | 64893 |
| 0065 | L | D | E | N | 1971 | 69787 |
| 0067 | L | M | R | N | 1971 | 64245 |
| 0067 | L | D | R | N | 1971 | 64893 |
| 0067 | L | D | E | N | 1971 | 69787 |
| 0067 | L | M | E | N | 1971 | 70814 |
| 0067 | L | D | R | N | 1971 | 75493 |
| 0089 | L | D | R | N | 1973 | 73009 |
| 0089 | L | D | E | N | 1973 | 75855 |
| 0099 | L | D | R | N | 1971 | 63183 |
| 0113 | L | D | E | N | 1971 | 61648 |
| 0118 | L | D | E | N | 1971 | 61648 |
| 0136 | L | D | E | N | 1971 | 61648 |
| 0141 | L | D | R | N | 1972 | 69364 |
| 0141 | L | D | E | N | 1972 | 69365 |
| 0141 | L | D | R | N | 1974 | 74281 |
| 0158 | G | D | G | N | 1972 | 68888 |
| 0182 | L | C | R | N | 1968 | 51848 |
| 0182 | L | C | E | N | 1972 | 66212 |
| 0193 | G | D | G | N | 1972 | 68888 |
| 0220 | L | D | E | N | 1971 | 61648 |
| 0237 | L | D | E | N | 1971 | 61648 |
| 0237 | L | D | R | N | 1971 | 63183 |
| 0387 | L | C | R | N | 1968 | 51848 |
| 0387 | L | D | R | N | 1970 | 63898 |
| 0387 | L | D | R | N | 1971 | 65349 |
| 0387 | L | D | E | N | 1971 | 65350 |
| 0387 | L | C | E | N | 1972 | 66212 |
| 0387 | L | D | R | N | 1973 | 73009 |
| 0387 | L | D | E | N | 1973 | 75855 |
| 0408 | L | D | R | N | 1974 | 75958 |
| 0408 | L | D | E | N | 1974 | 76461 |
| 0433 | L | D | R | N | 1968 | 51196 |
| 0433 | L | D | R | N | 1970 | 58349 |
| 0433 | L | D | R | N | 1971 | 64576 |
| 0433 | L | D | R | N | 1971 | 64893 |
| 0433 | L | D | E | N | 1970 | 66456 |
| 0433 | L | D | E | N | 1968 | 67255 |
| 0433 | L | D | E | N | 1971 | 69784 |
| 0433 | L | D | E | N | 1971 | 69787 |
| 0460 | L | D | R | N | 1973 | 73009 |
| 0460 | L | D | R | N | 1971 | 75493 |
| 0460 | L | D | E | N | 1971 | 75494 |
| 0460 | L | D | E | N | 1973 | 75855 |
| 0467 | G | D | G | H | 1965 | 42486 |
| 0469 | G | D | R | H | 1970 | 63896 |
| 0470 | G | D | G | H | 1965 | 42486 |
| 0471 | G | D | G | H | 1965 | 42486 |
| 0492 | L | M | J | N | 1972 | 67978 |
| 0492 | L | D | E | N | 1968 | 70167 |
| 0492 | L | D | R | N | 1973 | 72963 |
| 0492 | L | D | R | N | 1973 | 72964 |
| 0492 | L | D | E | N | 1973 | 74677 |
| 0496 | L | D | R | N | 1971 | 64788 |
| 0496 | L | D | E | N | 1971 | 64789 |
| 0496 | L | D | R | N | 1971 | 65349 |
| 0496 | L | D | E | N | 1971 | 65350 |
| 0506 | L | M | R | N | 1971 | 64245 |
| 0506 | L | D | R | N | 1972 | 70000 |
| 0506 | L | D | E | N | 1972 | 70682 |
| 0506 | L | M | E | N | 1971 | 70814 |
| 0509 | L | M | R | N | 1971 | 64245 |
| 0509 | L | M | E | N | 1971 | 70814 |
| 0509 | L | D | R | N | 1971 | 75493 |
| 0509 | L | D | E | N | 1971 | 75494 |
| 0600 | L | M | R | N | 1971 | 64245 |
| 0600 | L | D | R | N | 1972 | 70000 |
| 0600 | L | D | E | N | 1972 | 70682 |
| 0600 | L | M | E | N | 1971 | 70814 |
| 0600 | L | D | R | N | 1971 | 75493 |
| 0600 | L | D | E | N | 1971 | 75494 |
| 0619 | L | D | R | N | 1970 | 63898 |
| 0619 | L | D | R | N | 1971 | 64788 |
| 0619 | L | D | E | N | 1971 | 64789 |
| 0655 | L | D | R | N | 1968 | 51196 |
| 0655 | L | D | R | N | 1970 | 58349 |
| 0655 | L | D | R | N | 1971 | 64576 |
| 0655 | L | D | R | N | 1971 | 64893 |
| 517-0655 | L | D | E | N | 1970 | 66456 |
| 0655 | L | D | E | N | 1968 | 67255 |
| 0655 | L | D | E | N | 1971 | 69784 |
| 0655 | L | D | E | N | 1971 | 69787 |
| 0663 | L | C | R | N | 1968 | 51848 |
| 0663 | L | C | E | N | 1972 | 66212 |
| 0663 | L | D | R | N | 1972 | 66602 |
| 0663 | L | D | E | N | 1972 | 69793 |
| 0679 | L | D | R | N | 1968 | 51196 |
| 0679 | L | D | R | N | 1971 | 64576 |
| 0679 | L | D | R | N | 1971 | 64893 |
| 0679 | L | D | E | N | 1968 | 67255 |
| 0679 | L | D | E | N | 1971 | 69784 |
| 0679 | L | D | E | N | 1971 | 69787 |
| 0681 | L | C | R | N | 1968 | 51848 |
| 0681 | L | C | E | N | 1972 | 66212 |
| 0681 | L | D | R | N | 1972 | 66602 |
| 0681 | L | D | E | N | 1972 | 69793 |
| 0681 | L | D | R | N | 1971 | 75493 |
| 0681 | L | D | E | N | 1971 | 75494 |
| 0857 | L | D | R | N | 1973 | 74394 |
| 0866 | L | D | R | N | 1974 | 75958 |
| 0866 | L | D | E | N | 1974 | 76461 |
| 0886 | L | C | R | N | 1971 | 75503 |
| 0886 | L | C | E | N | 1971 | 75504 |
| 0887 | L | C | R | N | 1971 | 75503 |
| 0887 | L | C | E | N | 1971 | 75504 |
| 0888 | L | C | R | N | 1973 | 73002 |
| 0888 | L | C | R | N | 1971 | 75503 |
| 0888 | L | C | E | N | 1971 | 75504 |
| 0890 | G | D | R | H | 1974 | 75075 |
| 0890 | G | D | E | H | 1974 | 75076 |
| 0893 | G | C | R | N | 1974 | 75361 |
| 0934 | L | C | R | N | 1973 | 73002 |
| 0935 | L | C | R | N | 1973 | 73002 |
| 0936 | L | C | R | N | 1973 | 73002 |
| 0937 | L | C | R | N | 1973 | 73002 |
| 0938 | L | C | R | N | 1973 | 73002 |
| 0939 | L | C | R | N | 1973 | 73002 |
| 0940 | L | C | R | N | 1973 | 73002 |
| 0941 | L | C | R | N | 1973 | 73002 |
| 0942 | L | C | R | N | 1973 | 73002 |
| 0943 | L | C | R | N | 1973 | 73002 |
| 0944 | L | C | R | N | 1973 | 73002 |
| 0945 | L | C | R | N | 1973 | 73002 |
| 0946 | L | C | R | N | 1973 | 73002 |
| 0947 | L | C | R | N | 1973 | 73002 |
| 0948 | L | C | R | N | 1973 | 73002 |
| 0949 | L | C | R | N | 1973 | 73002 |
| 0951 | L | C | R | N | 1973 | 73002 |
| 0952 | L | C | R | N | 1973 | 73002 |
| 0953 | L | C | R | N | 1973 | 73002 |
| 0954 | L | C | R | N | 1973 | 73002 |
| 0955 | L | C | R | N | 1973 | 73002 |
| 0956 | L | C | R | N | 1973 | 73002 |
| 1014 | G | D | E | H | 1965 | 78046 |
| 1015 | G | D | E | H | 1965 | 78046 |
| 1016 | G | D | E | H | 1965 | 78046 |
| 1017 | G | D | E | H | 1965 | 78046 |
| 1018 | G | D | E | H | 1965 | 78046 |
| 1032 | G | S | R | H | 1974 | 78662 |
| 1033 | G | S | R | H | 1974 | 78662 |
| 1034 | G | S | R | H | 1974 | 78662 |
| 1035 | G | S | R | H | 1974 | 78662 |
| 1036 | G | S | R | H | 1974 | 78662 |
| 1037 | G | S | R | H | 1974 | 78662 |
| 1038 | G | S | R | H | 1974 | 78662 |
| 1039 | G | S | R | H | 1974 | 78662 |
| 1052 | G | D | E | H | 1964 | 70137 |

**Phys. State:** **D.** Doped; **E.** Expanded; **F.** Fibrous or Whisker; **G.** Gas; **L.** Liquid; **M.** Multiphase; **P.** Powder; **S.** Solid
**Subject:** **C.** Calculated Data + Theory; **D.** Data; **E.** Experiment; **G.** General (Data + Theory + Experiment); **M.** Measured Data + Experiment; **S.** Survey (Review, Compendium, etc.); **T.** Theory
**Language:** **C.** Czech; **D.** Dutch; **E.** English; **F.** French; **G.** German; **I.** Italian; **J.** Japanese; **O.** Other Languages; **P.** Polish; **R.** Russian; **S.** Spanish
**Temperature:** **F.** Full Range (Low + Normal + High); **L.** Low (0 to 75K) + Overlap into Normal; **N.** Normal (75 to 1273K); **H.** High (above 1273K) + Overlap into Normal; **U.** Undefined

# Chapter 4 Viscosity

| Substance Number | Phys. State | Subject | Language | Temperature | Year | TPRC Number |
|---|---|---|---|---|---|---|
| 512-0001 | G | C | E | N | 1964 | 38316 |
| 0001 | G | D | R | N | 1968 | 50752 |
| 0001 | G | C | R | N | 1970 | 59056 |
| 0001 | G | D | E | N | 1971 | 63989 |
| 0001 | G | C | E | N | 1971 | 64301 |
| 0001 | G | G | E | N | 1970 | 65568 |
| 0001 | G | M | G | N | 1972 | 68391 |
| 0001 | G | G | E | N | 1972 | 71409 |
| 0001 | G | C | E | N | 1973 | 71949 |
| 0001 | G | D | E | N | 1973 | 72144 |
| 0001 | G | S | E | N | 1977 | 93622 |
| 0001 | L | D | J | N | 1968 | 50613 |
| 0001 | L | G | E | N | 1970 | 65568 |
| 0002 | G | C | R | N | 1968 | 54413 |
| 0002 | G | C | E | H | 1971 | 63977 |
| 0002 | G | C | E | N | 1971 | 64301 |
| 0002 | G | C | E | N | 1972 | 66200 |
| 0003 | G | C | E | H | 1972 | 38084 |
| 0003 | G | C | E | N | 1966 | 41190 |
| 0003 | G | C | R | L | 1968 | 54413 |
| 0003 | G | M | R | N | 1968 | 54591 |
| 0003 | G | C | R | N | 1970 | 59056 |
| 0003 | G | D | E | N | 1971 | 63430 |
| 0003 | G | C | E | H | 1971 | 63977 |
| 0003 | G | G | E | N | 1972 | 64103 |
| 0003 | G | C | E | N | 1971 | 64301 |
| 0003 | G | C | E | L | 1972 | 66200 |
| 0003 | G | M | R | N | 1971 | 68806 |
| 0003 | G | G | E | N | 1972 | 71409 |
| 0003 | G | C | E | N | 1974 | 73619 |
| 0003 | G | C | E | H | 1973 | 74430 |
| 0003 | G | D | E | N | 1977 | 93611 |
| 0003 | G | T | E | N | 1977 | 93611 |
| 0003 | L | C | R | L | 1968 | 54413 |
| 0003 | L | C | E | L | 1972 | 66200 |
| 0003 | L | G | G | N | 1972 | 66352 |
| 0004 | L | D | G | N | 1970 | 61267 |
| 0005 | L | D | G | N | 1970 | 61267 |
| 0006 | L | C | E | N | 1966 | 42568 |
| 0006 | L | C | R | N | 1965 | 44572 |
| 0006 | L | C | J | N | 1968 | 54484 |
| 0006 | L | D | E | N | 1968 | 55833 |
| 0006 | L | C | R | N | 1969 | 56923 |
| 0006 | L | D | S | N | 1970 | 58973 |
| 0006 | L | D | R | N | 1970 | 59007 |
| 0006 | L | D | E | N | 1971 | 61008 |
| 0006 | L | G | F | N | 1971 | 61394 |
| 0006 | L | D | E | N | 1971 | 61461 |
| 0006 | L | C | E | N | 1971 | 63573 |
| 0006 | L | G | E | N | 1971 | 63708 |
| 0006 | L | D | E | N | 1971 | 64211 |
| 0006 | L | D | S | N | 1971 | 65213 |
| 0006 | L | D | E | N | 1971 | 65840 |
| 0006 | L | G | E | N | 1971 | 65945 |
| 0006 | L | D | E | N | 1971 | 66023 |
| 0006 | L | C | E | N | 1972 | 66056 |
| 0006 | L | C | E | N | 1972 | 66353 |
| 0006 | L | D | E | N | 1970 | 66459 |
| 0006 | L | D | E | N | 1972 | 66920 |
| 0006 | L | D | E | N | 1972 | 67063 |
| 0006 | L | D | E | N | 1972 | 67064 |
| 0006 | L | D | E | N | 1971 | 67567 |
| 0006 | L | D | E | N | 1972 | 67965 |
| 0006 | L | M | E | N | 1972 | 68545 |
| 0006 | L | D | E | N | 1972 | 68667 |
| 0006 | L | D | E | N | 1972 | 68796 |
| 0006 | L | D | R | N | 1972 | 70041 |
| 0006 | L | D | E | N | 1973 | 70901 |
| 0006 | L | C | E | N | 1974 | 75021 |
| 0006 | L | D | R | N | 1971 | 75499 |
| 0006 | L | D | E | N | 1971 | 75500 |
| 0006 | L | D | R | N | 1973 | 75638 |
| 0006 | L | D | E | N | 1973 | 75639 |
| 0006 | L | D | E | N | 1974 | 77413 |
| 0008 | G | D | E | N | 1977 | 93630 |
| 0011 | G | C | E | H | 1972 | 38084 |
| 0011 | G | D | E | N | 1971 | 63910 |
| 0011 | G | C | E | H | 1971 | 63977 |
| 0011 | G | C | E | N | 1971 | 64301 |
| 0011 | G | C | E | N | 1972 | 70184 |
| 0011 | G | C | E | N | 1973 | 71949 |
| 0011 | G | C | E | H | 1973 | 74430 |
| 0011 | L | D | E | N | 1964 | 34654 |
| 0011 | L | D | F | N | 1969 | 56421 |
| 0011 | L | C | F | N | 1971 | 63200 |
| 0011 | L | D | F | N | 1971 | 63202 |
| 0011 | L | D | R | N | 1973 | 75644 |
| 0011 | L | D | E | N | 1973 | 75645 |
| 0011 | L | D | J | N | 1974 | 78428 |
| 0012 | G | D | E | N | 1964 | 37222 |
| 0012 | G | G | E | N | 1969 | 50258 |
| 0012 | G | C | R | N | 1968 | 54413 |
| 0012 | G | C | R | N | 1970 | 59056 |
| 0012 | G | C | E | N | 1971 | 61334 |
| 0012 | G | D | E | N | 1970 | 63745 |

| Substance Number | Phys. State | Subject | Language | Temperature | Year | TPRC Number |
|---|---|---|---|---|---|---|
| 512-0012 | G | C | E | H | 1971 | 63977 |
| 0012 | G | C | E | N | 1971 | 64301 |
| 0012 | G | C | E | N | 1972 | 66200 |
| 0012 | G | C | E | N | 1972 | 66545 |
| 0012 | G | D | E | H | 1972 | 68099 |
| 0012 | G | C | E | N | 1972 | 70184 |
| 0012 | G | D | E | N | 1973 | 71090 |
| 0012 | G | C | E | N | 1973 | 71949 |
| 0012 | G | D | E | N | 1973 | 72144 |
| 0012 | G | C | E | H | 1973 | 74430 |
| 0012 | G | D | E | N | 1977 | 93611 |
| 0012 | G | T | E | N | 1977 | 93611 |
| 0012 | G | D | E | N | 1977 | 93622 |
| 0012 | G | S | E | N | 1977 | 93622 |
| 0012 | L | G | G | N | 1972 | 66352 |
| 0012 | L | D | E | N | 1973 | 76830 |
| 0014 | G | D | E | N | 1971 | 61145 |
| 0014 | G | C | E | N | 1971 | 62561 |
| 0014 | G | C | E | N | 1971 | 64301 |
| 0015 | G | D | E | N | 1971 | 61145 |
| 0015 | G | C | E | N | 1971 | 62561 |
| 0015 | G | D | E | N | 1973 | 72144 |
| 0018 | G | D | E | N | 1964 | 37222 |
| 0018 | G | C | R | N | 1968 | 54413 |
| 0018 | G | C | R | N | 1970 | 59056 |
| 0018 | G | C | E | N | 1971 | 61334 |
| 0018 | G | D | E | N | 1972 | 66090 |
| 0018 | G | C | E | N | 1972 | 66200 |
| 0018 | G | C | E | N | 1972 | 66533 |
| 0018 | G | C | E | N | 1972 | 66545 |
| 0018 | G | G | E | N | 1972 | 71386 |
| 0018 | G | D | E | N | 1973 | 72144 |
| 0018 | G | D | E | N | 1977 | 93611 |
| 0018 | G | S | E | N | 1977 | 93622 |
| 0018 | L | G | G | N | 1972 | 66352 |
| 0019 | G | G | E | N | 1971 | 63879 |
| 0019 | G | G | E | N | 1970 | 65568 |
| 0019 | G | D | E | N | 1974 | 75220 |
| 0019 | L | G | E | N | 1970 | 65568 |
| 0021 | G | C | E | N | 1964 | 38316 |
| 0021 | L | D | J | N | 1968 | 50613 |
| 0022 | L | M | E | N | 1969 | 55678 |
| 0023 | G | C | E | N | 1971 | 64301 |
| 0026 | G | C | E | H | 1972 | 68719 |
| 0026 | G | D | E | H | 1972 | 68848 |
| 0026 | G | C | E | H | 1974 | 76631 |
| 0026 | G | D | E | H | 1977 | 89755 |
| 0026 | G | T | E | H | 1977 | 89755 |
| 0026 | G | D | E | H | 1976 | 91157 |
| 0026 | G | T | E | H | 1976 | 91157 |
| 0028 | G | M | G | N | 1972 | 68391 |
| 0028 | G | M | R | N | 1971 | 68806 |
| 0028 | L | D | R | N | 1968 | 50148 |
| 0028 | L | M | G | N | 1968 | 54509 |
| 0028 | L | D | R | N | 1969 | 55656 |
| 0028 | L | D | E | N | 1969 | 58796 |
| 0028 | L | D | R | N | 1970 | 59007 |
| 0028 | L | D | R | N | 1969 | 59060 |
| 0028 | L | D | E | N | 1972 | 60206 |
| 0028 | L | D | R | N | 1970 | 60923 |
| 0028 | L | D | R | N | 1970 | 62486 |
| 0028 | L | M | E | N | 1971 | 63173 |
| 0028 | L | M | G | N | 1971 | 63267 |
| 0028 | L | D | E | N | 1971 | 63984 |
| 0028 | L | D | E | N | 1972 | 64802 |
| 0028 | L | D | R | N | 1971 | 65789 |
| 0028 | L | G | E | N | 1971 | 65945 |
| 0028 | L | D | E | N | 1972 | 66132 |
| 0028 | L | D | E | N | 1970 | 66459 |
| 0028 | L | D | E | N | 1970 | 66465 |
| 0028 | L | D | E | N | 1972 | 66505 |
| 0028 | L | M | E | N | 1972 | 66775 |
| 0028 | L | D | S | N | 1972 | 66854 |
| 0028 | L | D | E | N | 1972 | 66920 |
| 0028 | L | D | E | N | 1972 | 67064 |
| 0028 | L | D | E | N | 1971 | 67344 |
| 0028 | L | D | E | N | 1972 | 67965 |
| 0028 | L | D | E | N | 1970 | 69014 |
| 0028 | L | D | R | N | 1972 | 69271 |
| 0028 | L | D | O | N | 1971 | 69609 |
| 0028 | L | D | R | N | 1972 | 70041 |
| 0028 | L | D | E | N | 1974 | 74929 |
| 0028 | L | G | R | N | 1974 | 75203 |
| 0028 | L | D | R | N | 1972 | 75513 |
| 0028 | L | D | E | N | 1972 | 75514 |
| 0028 | L | D | R | N | 1971 | 75605 |
| 0028 | L | D | E | N | 1971 | 75606 |
| 0028 | L | D | E | N | 1973 | 75791 |
| 0028 | L | D | E | N | 1966 | 76530 |
| 0029 | L | M | G | N | 1968 | 54509 |
| 0029 | L | D | R | N | 1969 | 55656 |
| 0029 | L | D | E | N | 1970 | 62536 |
| 0029 | L | M | G | N | 1971 | 63267 |
| 0029 | L | D | E | N | 1972 | 65339 |
| 0029 | L | D | R | N | 1971 | 65756 |

| Substance Number | Phys. State | Subject | Language | Temperature | Year | TPRC Number |
|---|---|---|---|---|---|---|
| 512-0029 | L | D | E | N | 1972 | 66505 |
| 0029 | L | D | R | N | 1972 | 68527 |
| 0029 | L | D | R | N | 1972 | 69271 |
| 0029 | L | D | E | N | 1973 | 72514 |
| 0029 | L | M | F | N | 1974 | 79079 |
| 0030 | G | D | E | N | 1972 | 65575 |
| 0030 | L | D | R | N | 1965 | 42060 |
| 0030 | L | M | G | N | 1968 | 54509 |
| 0030 | L | D | E | N | 1971 | 62773 |
| 0030 | L | M | E | N | 1971 | 63173 |
| 0030 | L | M | G | N | 1971 | 63267 |
| 0030 | L | D | R | N | 1971 | 63713 |
| 0030 | L | D | E | N | 1972 | 65575 |
| 0030 | L | D | O | N | 1970 | 66348 |
| 0030 | L | D | R | N | 1972 | 66919 |
| 0030 | L | D | E | N | 1971 | 67376 |
| 0030 | L | D | R | N | 1972 | 68527 |
| 0030 | L | D | E | H | 1973 | 73932 |
| 0031 | L | M | G | N | 1968 | 54509 |
| 0031 | L | D | R | H | 1969 | 61933 |
| 0031 | L | M | G | N | 1971 | 63267 |
| 0031 | L | D | R | N | 1971 | 63713 |
| 0031 | L | D | E | N | 1971 | 67376 |
| 0032 | L | D | E | N | 1970 | 62127 |
| 0032 | L | D | E | N | 1972 | 64802 |
| 0032 | L | D | E | N | 1971 | 66385 |
| 0032 | L | D | E | N | 1972 | 66508 |
| 0036 | G | G | E | N | 1969 | 50258 |
| 0036 | G | C | E | N | 1971 | 61334 |
| 0036 | G | D | E | N | 1972 | 66090 |
| 0036 | G | D | E | H | 1972 | 68099 |
| 0036 | G | D | E | N | 1973 | 71090 |
| 0036 | G | G | E | N | 1972 | 71386 |
| 0036 | G | D | E | N | 1973 | 72144 |
| 0036 | G | D | E | N | 1977 | 93611 |
| 0036 | G | T | E | N | 1977 | 93611 |
| 0036 | G | S | E | N | 1977 | 93622 |
| 0036 | L | D | E | N | 1973 | 76830 |
| 0037 | G | C | E | N | 1971 | 61334 |
| 0037 | G | C | E | U | 1972 | 65366 |
| 0037 | G | D | E | N | 1973 | 72144 |
| 0038 | G | T | E | H | 1971 | 61380 |
| 0038 | G | T | E | H | 1971 | 63982 |
| 0038 | G | T | E | H | 1971 | 64129 |
| 0038 | G | T | E | H | 1970 | 65541 |
| 0038 | G | T | R | H | 1972 | 67514 |
| 0038 | G | T | R | H | 1972 | 68453 |
| 0038 | G | C | E | H | 1972 | 68719 |
| 0038 | G | T | E | H | 1973 | 69252 |
| 0038 | G | T | E | H | 1972 | 69670 |
| 0038 | G | T | E | H | 1972 | 69683 |
| 0038 | G | T | E | H | 1972 | 69808 |
| 0039 | L | C | J | N | 1968 | 54484 |
| 0039 | L | D | R | N | 1970 | 59007 |
| 0039 | L | D | R | N | 1970 | 60922 |
| 0039 | L | D | E | N | 1971 | 61461 |
| 0039 | L | D | R | N | 1970 | 62486 |
| 0039 | L | C | E | N | 1971 | 63573 |
| 0039 | L | D | E | N | 1971 | 63819 |
| 0039 | L | G | E | N | 1970 | 65550 |
| 0039 | L | D | E | N | 1971 | 65840 |
| 0039 | L | D | E | N | 1972 | 66132 |
| 0039 | L | D | E | N | 1970 | 66459 |
| 0039 | L | D | E | N | 1970 | 66463 |
| 0039 | L | D | E | N | 1972 | 67063 |
| 0039 | L | D | E | N | 1972 | 67064 |
| 0039 | L | D | E | N | 1971 | 67567 |
| 0039 | L | D | E | N | 1972 | 67576 |
| 0039 | L | D | E | N | 1972 | 68667 |
| 0039 | L | D | E | N | 1972 | 68796 |
| 0039 | L | D | E | N | 1970 | 69014 |
| 0039 | L | D | R | N | 1972 | 69860 |
| 0039 | L | D | E | N | 1973 | 70901 |
| 0039 | L | C | E | U | 1973 | 74736 |
| 0040 | L | D | E | N | 1971 | 63819 |
| 0040 | L | C | E | N | 1973 | 70212 |
| 0041 | L | D | R | N | 1971 | 63380 |
| 0041 | L | D | E | N | 1971 | 63381 |
| 0041 | L | D | E | N | 1971 | 63819 |
| 0041 | L | D | R | N | 1971 | 66417 |
| 0041 | L | D | E | N | 1971 | 66418 |
| 0041 | L | D | R | N | 1971 | 66644 |
| 0041 | L | D | E | N | 1971 | 66645 |
| 0041 | L | D | E | N | 1971 | 67344 |
| 0041 | L | D | R | N | 1972 | 69560 |
| 0041 | L | D | E | N | 1972 | 72680 |
| 0041 | L | D | R | N | 1972 | 75513 |
| 0041 | L | D | E | N | 1972 | 75514 |
| 0041 | L | D | R | N | 1974 | 75684 |
| 0041 | L | D | E | N | 1974 | 75685 |
| 0042 | L | D | R | N | 1969 | 55656 |
| 0042 | L | D | S | N | 1970 | 58973 |
| 0042 | L | D | R | N | 1970 | 59007 |
| 0042 | L | D | R | N | 1969 | 59060 |
| 0042 | L | D | E | N | 1972 | 61093 |

**Phys. State:** **D.** Doped; **E.** Expanded; **F.** Fibrous or Whisker; **G.** Gas; **L.** Liquid; **M.** Multiphase; **P.** Powder; **S.** Solid
**Subject:** **C.** Calculated Data + Theory; **D.** Data; **E.** Experiment; **G.** General (Data + Theory + Experiment); **M.** Measured Data + Experiment; **S.** Survey (Review, Compendium, etc.); **T.** Theory
**Language:** **C.** Czech; **D.** Dutch; **E.** English; **F.** French; **G.** German; **I.** Italian; **J.** Japanese; **O.** Other Languages; **P.** Polish; **R.** Russian; **S.** Spanish
**Temperature:** **F.** Full Range (Low + Normal + High); **L.** Low (0 to 75K) + Overlap into Normal; **N.** Normal (75 to 1273K); **H.** High (above 1273K) + Overlap into Normal; **U.** Undefined

| Substance Number | Phys. State | Subject | Language | Temperature | Year | TPRC Number |
|---|---|---|---|---|---|---|
| 512-0042 | L | D | E | N | 1971 | 61461 |
| 0042 | L | C | E | N | 1971 | 63598 |
| 0042 | L | D | R | N | 1971 | 65599 |
| 0042 | L | D | E | N | 1971 | 65600 |
| 0042 | L | G | E | N | 1971 | 65945 |
| 0042 | L | D | E | N | 1970 | 66459 |
| 0042 | L | D | E | N | 1972 | 66505 |
| 0042 | L | M | E | N | 1972 | 66775 |
| 0042 | L | D | S | N | 1972 | 66854 |
| 0042 | L | D | R | N | 1972 | 67845 |
| 0042 | L | C | O | N | 1971 | 69611 |
| 0042 | L | D | E | N | 1974 | 74929 |
| 0042 | L | D | R | N | 1971 | 75605 |
| 0042 | L | D | E | N | 1971 | 75606 |
| 0043 | L | D | R | N | 1969 | 55656 |
| 0043 | L | D | S | N | 1970 | 58973 |
| 0043 | L | D | E | N | 1972 | 61093 |
| 0043 | L | D | E | N | 1971 | 61866 |
| 0043 | L | C | E | N | 1971 | 63598 |
| 0043 | L | G | E | N | 1971 | 63708 |
| 0043 | L | D | E | N | 1972 | 64802 |
| 0043 | L | D | R | N | 1971 | 65599 |
| 0043 | L | D | E | N | 1971 | 65600 |
| 0043 | L | G | E | N | 1971 | 65945 |
| 0043 | L | D | E | N | 1972 | 66505 |
| 0043 | L | M | E | N | 1972 | 66775 |
| 0043 | L | D | S | N | 1972 | 66854 |
| 0043 | L | D | R | N | 1972 | 67845 |
| 0043 | L | D | E | N | 1974 | 74929 |
| 0043 | L | D | R | N | 1971 | 75605 |
| 0043 | L | D | E | N | 1971 | 75606 |
| 0043 | L | D | E | N | 1974 | 77413 |
| 0044 | L | D | E | U | 1968 | 54571 |
| 0044 | L | D | R | N | 1969 | 55656 |
| 0044 | L | D | S | N | 1970 | 58973 |
| 0044 | L | D | R | N | 1969 | 59485 |
| 0044 | L | D | E | N | 1972 | 61093 |
| 0044 | L | C | E | N | 1968 | 62352 |
| 0044 | L | D | E | N | 1971 | 63123 |
| 0044 | L | C | E | N | 1971 | 63598 |
| 0044 | L | D | E | N | 1972 | 64802 |
| 0044 | L | D | R | N | 1971 | 65756 |
| 0044 | L | G | E | N | 1971 | 65945 |
| 0044 | L | D | E | N | 1972 | 66505 |
| 0044 | L | D | S | N | 1972 | 66854 |
| 0044 | L | D | E | N | 1973 | 69031 |
| 0044 | L | D | E | N | 1974 | 74929 |
| 0044 | L | C | E | N | 1974 | 75339 |
| 0044 | L | D | R | N | 1971 | 75605 |
| 0044 | L | D | E | N | 1971 | 75606 |
| 0045 | L | C | R | N | 1969 | 56923 |
| 0045 | L | D | E | N | 1972 | 61093 |
| 0045 | L | D | E | N | 1972 | 64802 |
| 0045 | L | G | E | N | 1971 | 65945 |
| 0045 | L | C | E | N | 1972 | 66056 |
| 0045 | L | D | E | N | 1974 | 74929 |
| 0045 | L | D | R | N | 1973 | 75591 |
| 0045 | L | D | E | N | 1973 | 75592 |
| 0046 | L | D | E | N | 1972 | 61093 |
| 0046 | L | D | E | N | 1972 | 64802 |
| 0046 | L | G | E | N | 1971 | 65945 |
| 0047 | L | C | R | N | 1964 | 36663 |
| 0047 | L | D | R | N | 1970 | 59007 |
| 0047 | L | D | E | N | 1971 | 61461 |
| 0047 | L | D | E | N | 1971 | 63123 |
| 0047 | L | G | E | N | 1971 | 63708 |
| 0047 | L | D | E | N | 1971 | 64211 |
| 0047 | L | D | E | N | 1971 | 65840 |
| 0047 | L | G | E | N | 1971 | 65945 |
| 0047 | L | D | E | N | 1970 | 66459 |
| 0047 | L | D | E | N | 1972 | 67063 |
| 0047 | L | D | E | N | 1971 | 67567 |
| 0047 | L | D | E | N | 1972 | 67576 |
| 0047 | L | D | E | N | 1973 | 70901 |
| 0047 | L | C | E | N | 1974 | 75021 |
| 0047 | L | D | E | N | 1974 | 77413 |
| 0048 | L | D | R | N | 1965 | 42060 |
| 0048 | L | D | R | N | 1970 | 61935 |
| 0048 | L | D | E | N | 1970 | 62536 |
| 0048 | L | D | E | N | 1972 | 65357 |
| 0048 | L | D | O | N | 1970 | 66348 |
| 0048 | L | D | R | N | 1971 | 66364 |
| 0048 | L | D | R | N | 1972 | 66919 |
| 0048 | L | D | R | N | 1972 | 68527 |
| 0048 | L | M | F | N | 1974 | 79079 |
| 0052 | L | M | G | N | 1964 | 34078 |
| 0052 | L | D | R | N | 1963 | 37245 |
| 0052 | L | D | R | N | 1968 | 50147 |
| 0052 | L | D | E | N | 1971 | 61265 |
| 0054 | L | D | E | N | 1971 | 62773 |
| 0054 | L | D | E | N | 1972 | 65357 |
| 0054 | L | D | E | N | 1972 | 66094 |
| 0054 | L | M | J | N | 1973 | 72914 |
| 0055 | L | S | E | L | 1964 | 38454 |
| 0055 | L | G | E | L | 1971 | 62985 |
| 0055 | L | M | E | L | 1971 | 63554 |
| 0055 | L | D | E | L | 1972 | 64803 |
| 0055 | L | M | E | L | 1972 | 66080 |
| 0055 | L | D | E | L | 1972 | 70797 |
| 0055 | L | M | R | L | 1973 | 73015 |
| 0055 | L | C | E | L | 1974 | 74745 |

| Substance Number | Phys. State | Subject | Language | Temperature | Year | TPRC Number |
|---|---|---|---|---|---|---|
| 512-0055 | L | G | E | L | 1973 | 74780 |
| 0055 | L | G | E | L | 1973 | 76343 |
| 0056 | L | S | E | L | 1964 | 38454 |
| 0056 | L | G | E | L | 1971 | 62985 |
| 0056 | L | M | E | L | 1971 | 63554 |
| 0056 | L | D | E | L | 1972 | 64803 |
| 0056 | L | M | E | L | 1972 | 66080 |
| 0056 | L | D | E | L | 1972 | 70797 |
| 0056 | L | M | R | L | 1973 | 73015 |
| 0056 | L | C | E | L | 1974 | 74745 |
| 0056 | L | G | E | L | 1973 | 74780 |
| 0056 | L | G | E | L | 1973 | 76343 |
| 0057 | L | C | F | N | 1969 | 57596 |
| 0057 | L | D | E | N | 1971 | 61330 |
| 0057 | L | C | E | N | 1971 | 61478 |
| 0057 | L | G | G | N | 1971 | 62442 |
| 0057 | L | D | E | N | 1971 | 63914 |
| 0057 | L | G | E | N | 1969 | 65879 |
| 0057 | L | D | R | N | 1971 | 70101 |
| 0057 | L | D | R | N | 1972 | 70280 |
| 0057 | L | D | E | N | 1972 | 71431 |
| 0057 | L | D | E | N | 1972 | 71432 |
| 0057 | L | D | R | N | 1973 | 74280 |
| 0058 | L | D | E | N | 1971 | 61330 |
| 0058 | L | C | E | N | 1971 | 61478 |
| 0058 | L | D | E | N | 1971 | 63914 |
| 0058 | L | D | R | N | 1971 | 70101 |
| 0058 | L | D | R | N | 1972 | 70280 |
| 0058 | L | D | E | N | 1972 | 71431 |
| 0058 | L | D | E | N | 1972 | 71432 |
| 0058 | L | D | R | N | 1973 | 75644 |
| 0058 | L | D | E | N | 1973 | 75645 |
| 0061 | G | D | E | N | 1964 | 37222 |
| 0061 | G | C | E | H | 1972 | 38084 |
| 0061 | G | C | O | N | 1964 | 43155 |
| 0061 | G | G | E | N | 1969 | 50258 |
| 0061 | G | C | R | N | 1968 | 54413 |
| 0061 | G | C | R | N | 1970 | 59056 |
| 0061 | G | C | E | N | 1971 | 61334 |
| 0061 | G | G | E | N | 1971 | 63879 |
| 0061 | G | D | E | N | 1971 | 63910 |
| 0061 | G | C | E | H | 1971 | 63977 |
| 0061 | G | C | E | U | 1972 | 65366 |
| 0061 | G | D | E | N | 1972 | 66089 |
| 0061 | G | C | E | N | 1972 | 66200 |
| 0061 | G | C | E | N | 1972 | 66533 |
| 0061 | G | G | E | N | 1972 | 71386 |
| 0061 | G | D | E | N | 1973 | 73805 |
| 0061 | G | C | E | H | 1973 | 74430 |
| 0061 | G | D | E | N | 1974 | 76378 |
| 0061 | G | D | E | N | 1977 | 93611 |
| 0061 | G | T | E | N | 1977 | 93611 |
| 0061 | G | S | E | N | 1977 | 93622 |
| 0061 | L | G | G | N | 1972 | 66352 |
| 0062 | L | D | R | N | 1964 | 36255 |
| 0063 | L | D | R | N | 1964 | 36255 |
| 0063 | L | C | R | N | 1972 | 69535 |
| 0064 | G | T | O | N | 1964 | 43155 |
| 0064 | G | T | G | U | 1969 | 56735 |
| 0064 | G | T | E | U | 1962 | 64901 |
| 0064 | G | C | F | U | 1971 | 65623 |
| 0064 | G | C | E | N | 1972 | 66545 |
| 0064 | G | T | E | N | 1972 | 67501 |
| 0064 | G | T | R | N | 1972 | 70272 |
| 0064 | G | C | R | F | 1952 | 72020 |
| 0064 | G | S | E | N | 1973 | 72366 |
| 0064 | G | T | R | N | 1973 | 72971 |
| 0064 | G | S | E | N | 1973 | 73278 |
| 0064 | G | T | E | U | 1974 | 74827 |
| 0064 | G | T | E | U | 1974 | 74952 |
| 0064 | G | T | R | U | 1974 | 75069 |
| 0064 | G | T | E | U | 1974 | 75070 |
| 0064 | G | T | E | U | 1974 | 76389 |
| 0064 | L | T | E | U | 1964 | 38250 |
| 0064 | L | C | R | N | 1968 | 53467 |
| 0064 | L | C | J | N | 1968 | 54484 |
| 0064 | L | T | R | U | 1970 | 58511 |
| 0064 | L | S | J | N | 1970 | 58989 |
| 0064 | L | D | R | N | 1970 | 62486 |
| 0064 | L | D | E | N | 1972 | 66557 |
| 0064 | L | T | E | N | 1972 | 67501 |
| 0064 | L | C | E | N | 1972 | 68203 |
| 0064 | L | D | E | N | 1970 | 69014 |
| 0064 | L | T | O | U | 1972 | 69124 |
| 0064 | L | S | E | N | 1972 | 69620 |
| 0064 | L | S | E | N | 1971 | 69767 |
| 0064 | L | T | E | U | 1970 | 69886 |
| 0064 | L | T | E | N | 1973 | 72040 |
| 0064 | L | S | E | N | 1973 | 72366 |
| 0064 | L | T | E | U | 1973 | 73174 |
| 0064 | L | S | E | N | 1914 | 73602 |
| 0064 | L | T | E | U | 1974 | 74827 |
| 0064 | L | T | E | U | 1974 | 74952 |
| 0064 | L | T | E | U | 1974 | 75396 |
| 0065 | L | M | R | N | 1965 | 38887 |
| 0066 | L | M | R | N | 1965 | 38887 |
| 0067 | L | M | R | N | 1965 | 38887 |
| 0068 | G | D | E | N | 1971 | 61145 |
| 0069 | G | C | E | H | 1972 | 38084 |
| 0069 | G | C | R | L | 1968 | 54413 |
| 0069 | G | C | E | L | 1972 | 66200 |

| Substance Number | Phys. State | Subject | Language | Temperature | Year | TPRC Number |
|---|---|---|---|---|---|---|
| 512-0069 | G | C | E | H | 1973 | 74430 |
| 0069 | L | C | R | L | 1968 | 54413 |
| 0069 | L | C | E | L | 1972 | 66200 |
| 0070 | L | D | R | N | 1965 | 42060 |
| 0071 | L | D | R | N | 1965 | 42060 |
| 0071 | L | D | R | N | 1971 | 66364 |
| 0072 | L | G | F | N | 1971 | 61394 |
| 0072 | L | D | R | N | 1972 | 69652 |
| 0072 | L | D | E | N | 1972 | 69653 |
| 0073 | L | G | F | N | 1971 | 61394 |
| 0074 | L | D | E | N | 1971 | 63169 |
| 0075 | L | D | E | N | 1971 | 63169 |
| 0076 | L | D | E | N | 1971 | 63169 |
| 0077 | L | D | E | N | 1971 | 63169 |
| 0078 | L | D | E | N | 1971 | 63169 |
| 0079 | L | D | E | N | 1971 | 63169 |
| 0080 | L | D | E | N | 1971 | 63169 |
| 0081 | L | D | E | N | 1971 | 63169 |
| 0084 | L | D | E | N | 1971 | 61008 |
| 0084 | L | D | E | N | 1971 | 63123 |
| 0084 | L | D | E | U | 1971 | 65066 |
| 0084 | L | G | E | N | 1970 | 65550 |
| 0084 | L | C | E | U | 1973 | 74736 |
| 0085 | L | D | E | N | 1971 | 62626 |
| 0085 | L | D | E | N | 1971 | 63123 |
| 0085 | L | C | E | N | 1974 | 75339 |
| 0086 | L | D | R | N | 1969 | 59485 |
| 0087 | L | M | R | H | 1968 | 59478 |
| 0088 | L | M | R | H | 1968 | 59478 |
| 0089 | L | D | R | H | 1969 | 61933 |
| 0089 | L | D | E | N | 1973 | 72514 |
| 0089 | L | D | E | N | 1974 | 74778 |
| 0095 | G | C | G | H | 1965 | 47645 |
| 0095 | G | G | G | H | 1970 | 61549 |
| 0095 | G | D | E | U | 1971 | 62797 |
| 0095 | G | G | E | H | 1971 | 63637 |
| 0095 | G | D | R | H | 1971 | 64641 |
| 0095 | G | D | E | H | 1971 | 64642 |
| 0095 | G | D | E | H | 1971 | 64753 |
| 0095 | G | C | E | H | 1970 | 64975 |
| 0095 | G | C | E | H | 1973 | 71182 |
| 0095 | G | D | E | H | 1976 | 91157 |
| 0095 | G | T | E | H | 1976 | 91157 |
| 0096 | G | G | G | H | 1970 | 61549 |
| 0096 | G | D | G | H | 1971 | 64571 |
| 0096 | G | D | E | H | 1971 | 64753 |
| 0096 | G | C | E | H | 1970 | 64975 |
| 0096 | G | T | R | H | 1972 | 68678 |
| 0096 | G | T | E | H | 1972 | 68679 |
| 0096 | G | D | E | H | 1977 | 89755 |
| 0096 | G | T | E | H | 1977 | 89755 |
| 0097 | L | D | R | N | 1970 | 61935 |
| 0097 | L | D | R | N | 1971 | 63380 |
| 0097 | L | D | E | N | 1971 | 63381 |
| 0097 | L | D | R | N | 1971 | 66417 |
| 0097 | L | D | E | N | 1971 | 66418 |
| 0098 | L | D | R | N | 1970 | 58348 |
| 0098 | L | D | E | N | 1971 | 61867 |
| 0098 | L | G | G | N | 1971 | 62442 |
| 0098 | L | D | E | N | 1971 | 62471 |
| 0098 | L | D | E | N | 1971 | 64137 |
| 0098 | L | D | R | N | 1971 | 66417 |
| 0098 | L | D | E | N | 1971 | 66418 |
| 0098 | L | D | E | N | 1970 | 67239 |
| 0098 | L | D | E | N | 1972 | 68398 |
| 0098 | L | D | E | N | 1972 | 68399 |
| 0098 | L | D | E | N | 1972 | 68796 |
| 0098 | L | D | F | N | 1973 | 71842 |
| 0098 | L | M | E | N | 1974 | 76685 |
| 0099 | L | D | E | N | 1972 | 64802 |
| 0099 | L | D | R | N | 1971 | 65756 |
| 0099 | L | D | R | N | 1971 | 66417 |
| 0099 | L | D | E | N | 1971 | 66418 |
| 0099 | L | C | R | N | 1974 | 75680 |
| 0099 | L | C | E | N | 1974 | 75681 |
| 0100 | L | D | F | N | 1969 | 56421 |
| 0100 | L | D | E | N | 1972 | 64802 |
| 0100 | L | D | R | N | 1971 | 66417 |
| 0100 | L | D | E | N | 1971 | 66418 |
| 0100 | L | C | R | N | 1972 | 69535 |
| 0100 | L | D | E | N | 1973 | 72514 |
| 0101 | L | D | R | N | 1971 | 66417 |
| 0101 | L | D | E | N | 1971 | 66418 |
| 0101 | L | D | E | N | 1973 | 72514 |
| 0101 | L | M | E | N | 1974 | 76685 |
| 0102 | L | D | R | N | 1971 | 63380 |
| 0102 | L | D | E | N | 1971 | 63381 |
| 0102 | L | D | R | N | 1971 | 66417 |
| 0102 | L | D | E | N | 1971 | 66418 |
| 0103 | L | D | R | N | 1963 | 36636 |
| 0103 | L | G | G | N | 1971 | 62442 |
| 0103 | L | D | R | N | 1971 | 66417 |
| 0103 | L | D | E | N | 1971 | 66418 |
| 0103 | L | D | E | N | 1971 | 67344 |
| 0103 | L | D | R | N | 1972 | 68527 |
| 0104 | G | C | R | L | 1968 | 54413 |
| 0104 | G | C | E | L | 1972 | 66200 |
| 0104 | L | C | R | L | 1968 | 54413 |
| 0104 | L | C | E | L | 1972 | 66200 |
| 0106 | G | C | E | N | 1966 | 41190 |
| 0106 | G | C | O | N | 1964 | 43155 |

**Phys. State:** **D.** Doped; **E.** Expanded; **F.** Fibrous or Whisker; **G.** Gas; **L.** Liquid; **M.** Multiphase; **P.** Powder; **S.** Solid
**Subject:** **C.** Calculated Data + Theory; **D.** Data; **E.** Experiment; **G.** General (Data + Theory + Experiment); **M.** Measured Data + Experiment; **S.** Survey (Review, Compendium, etc.); **T.** Theory
**Language:** **C.** Czech; **D.** Dutch; **E.** English; **F.** French; **G.** German; **I.** Italian; **J.** Japanese; **O.** Other Languages; **P.** Polish; **R.** Russian; **S.** Spanish
**Temperature:** **F.** Full Range (Low + Normal + High); **L.** Low (0 to 75K) + Overlap into Normal; **N.** Normal (75 to 1273K); **H.** High (above 1273K) + Overlap into Normal; **U.** Undefined

| Substance Number | Phys. State | Subject | Language | Temperature | Year | TPRC Number |
|---|---|---|---|---|---|---|
| 512-0106 | G | C | I | N | 1965 | 46295 |
| 0106 | G | D | R | N | 1968 | 50752 |
| 0106 | G | C | R | N | 1968 | 54413 |
| 0106 | G | M | R | N | 1968 | 54591 |
| 0106 | G | C | R | N | 1970 | 59056 |
| 0106 | G | D | E | N | 1970 | 63745 |
| 0106 | G | G | E | N | 1971 | 63879 |
| 0106 | G | C | E | H | 1971 | 63977 |
| 0106 | G | D | E | N | 1971 | 63989 |
| 0106 | G | C | E | N | 1971 | 64301 |
| 0106 | G | D | E | N | 1972 | 66089 |
| 0106 | G | C | E | N | 1972 | 66200 |
| 0106 | G | C | E | N | 1972 | 66533 |
| 0106 | G | D | E | H | 1972 | 68099 |
| 0106 | G | M | R | N | 1971 | 68806 |
| 0106 | G | D | R | H | 1972 | 70621 |
| 0106 | G | D | E | H | 1972 | 70622 |
| 0106 | G | D | E | N | 1973 | 71090 |
| 0106 | G | G | E | N | 1972 | 71386 |
| 0106 | G | C | E | H | 1973 | 74430 |
| 0106 | G | S | E | N | 1977 | 93622 |
| 0106 | L | G | G | N | 1972 | 66352 |
| 0108 | L | D | S | N | 1972 | 66854 |
| 0108 | L | C | O | N | 1971 | 69611 |
| 0109 | G | T | E | L | 1972 | 65367 |
| 0110 | G | G | E | N | 1972 | 71404 |
| 0110 | L | D | E | N | 1972 | 65339 |
| 0111 | L | D | F | N | 1971 | 63202 |
| 0111 | L | D | E | N | 1972 | 65339 |
| 0113 | L | D | S | N | 1971 | 65213 |
| 0113 | L | D | R | N | 1972 | 69758 |
| 0113 | L | D | E | N | 1973 | 70901 |
| 0113 | L | D | E | N | 1974 | 74929 |
| 0114 | L | D | R | N | 1970 | 59007 |
| 0114 | L | D | E | N | 1972 | 60206 |
| 0114 | L | D | E | N | 1971 | 61461 |
| 0114 | L | C | E | N | 1971 | 63598 |
| 0114 | L | G | E | N | 1971 | 63708 |
| 0114 | L | D | E | N | 1971 | 65840 |
| 0114 | L | M | R | N | 1971 | 65899 |
| 0114 | L | D | E | N | 1970 | 66459 |
| 0114 | L | M | E | N | 1971 | 66488 |
| 0114 | L | D | E | N | 1972 | 67063 |
| 0114 | L | D | E | N | 1972 | 67064 |
| 0114 | L | D | E | N | 1973 | 73180 |
| 0114 | L | D | E | N | 1974 | 77413 |
| 0117 | L | C | J | N | 1968 | 54484 |
| 0117 | L | D | R | N | 1970 | 59007 |
| 0117 | L | D | E | N | 1971 | 61008 |
| 0117 | L | D | F | N | 1971 | 63201 |
| 0117 | L | C | E | N | 1971 | 63573 |
| 0117 | L | G | E | N | 1971 | 63708 |
| 0117 | L | D | E | N | 1971 | 63835 |
| 0117 | L | D | E | U | 1971 | 65066 |
| 0117 | L | D | S | N | 1971 | 65213 |
| 0117 | L | D | E | N | 1971 | 66023 |
| 0117 | L | D | E | N | 1970 | 66459 |
| 0117 | L | D | E | N | 1972 | 66526 |
| 0117 | L | D | E | N | 1972 | 66773 |
| 0117 | L | D | E | N | 1972 | 67063 |
| 0117 | L | D | S | N | 1972 | 69065 |
| 0117 | L | D | R | N | 1972 | 70041 |
| 0117 | L | C | E | N | 1973 | 70212 |
| 0117 | L | D | E | N | 1973 | 70901 |
| 0117 | L | D | E | N | 1973 | 72220 |
| 0117 | L | D | E | N | 1973 | 72221 |
| 0117 | L | G | E | N | 1974 | 75020 |
| 0117 | L | C | E | N | 1974 | 75021 |
| 0117 | L | D | R | N | 1971 | 75605 |
| 0117 | L | D | E | N | 1971 | 75606 |
| 0117 | L | D | E | N | 1966 | 76530 |
| 0117 | L | D | E | N | 1974 | 77413 |
| 0118 | L | D | J | N | 1968 | 55653 |
| 0118 | L | D | E | N | 1972 | 60206 |
| 0118 | L | D | E | N | 1972 | 66526 |
| 0119 | L | D | R | N | 1969 | 59060 |
| 0121 | L | D | R | N | 1971 | 62378 |
| 0121 | L | D | E | N | 1971 | 62379 |
| 0121 | L | D | R | N | 1972 | 69650 |
| 0121 | L | D | E | N | 1972 | 69651 |
| 0121 | L | D | R | N | 1972 | 69652 |
| 0121 | L | D | E | N | 1972 | 69653 |
| 0121 | L | D | R | N | 1971 | 74763 |
| 0121 | L | D | R | N | 1973 | 75591 |
| 0121 | L | D | E | N | 1973 | 75592 |
| 0121 | L | D | R | N | 1973 | 75654 |
| 0121 | L | D | E | N | 1973 | 75655 |
| 0122 | L | D | R | N | 1971 | 62378 |
| 0122 | L | D | E | N | 1971 | 62379 |
| 0122 | L | D | E | N | 1972 | 67576 |
| 0122 | L | D | R | N | 1972 | 75513 |
| 0122 | L | D | E | N | 1972 | 75514 |
| 0123 | L | D | R | N | 1971 | 62378 |
| 0123 | L | D | E | N | 1971 | 62379 |
| 0126 | L | G | E | N | 1971 | 63708 |
| 0126 | L | D | E | N | 1971 | 65840 |
| 0126 | L | D | E | N | 1972 | 68398 |
| 0126 | L | D | E | N | 1974 | 77413 |
| 0127 | L | D | E | N | 1972 | 68667 |
| 0127 | L | D | S | N | 1972 | 69065 |
| 0128 | G | C | E | N | 1972 | 66545 |

| Substance Number | Phys. State | Subject | Language | Temperature | Year | TPRC Number |
|---|---|---|---|---|---|---|
| 512-0130 | L | S | E | N | 1972 | 68175 |
| 0131 | L | D | R | N | 1971 | 64575 |
| 0131 | L | D | R | N | 1971 | 65599 |
| 0131 | L | D | E | N | 1971 | 65600 |
| 0131 | L | D | E | N | 1971 | 66479 |
| 0131 | L | D | E | N | 1971 | 67567 |
| 0134 | L | D | E | N | 1971 | 65840 |
| 0135 | L | D | E | N | 1971 | 63835 |
| 0135 | L | D | E | N | 1971 | 65840 |
| 0135 | L | D | E | N | 1972 | 68667 |
| 0135 | L | D | S | N | 1972 | 69065 |
| 0136 | L | D | R | N | 1963 | 36636 |
| 0136 | L | D | R | N | 1970 | 59007 |
| 0136 | L | D | E | N | 1971 | 61866 |
| 0136 | L | D | R | N | 1971 | 63380 |
| 0136 | L | D | E | N | 1971 | 63381 |
| 0136 | L | D | R | N | 1969 | 65051 |
| 0136 | L | D | E | N | 1971 | 65840 |
| 0136 | L | D | E | N | 1970 | 66459 |
| 0136 | L | D | E | N | 1972 | 66508 |
| 0136 | L | D | S | N | 1972 | 69065 |
| 0136 | L | D | R | N | 1972 | 69860 |
| 0136 | L | D | E | N | 1973 | 75791 |
| 0141 | L | D | S | N | 1970 | 58973 |
| 0141 | L | D | E | N | 1972 | 61093 |
| 0141 | L | D | E | N | 1970 | 62127 |
| 0141 | L | D | E | N | 1972 | 64802 |
| 0141 | L | M | E | N | 1972 | 66775 |
| 0141 | L | D | S | N | 1972 | 66854 |
| 0141 | L | D | E | N | 1974 | 74929 |
| 0141 | L | D | E | N | 1966 | 76530 |
| 0143 | L | G | G | N | 1971 | 62442 |
| 0143 | L | D | R | N | 1971 | 63713 |
| 0143 | L | D | E | N | 1971 | 67376 |
| 0144 | L | G | G | N | 1971 | 62442 |
| 0145 | L | D | R | N | 1971 | 63380 |
| 0145 | L | D | E | N | 1971 | 63381 |
| 0146 | L | D | R | N | 1971 | 63380 |
| 0146 | L | D | E | N | 1971 | 63381 |
| 0146 | L | D | E | N | 1971 | 66385 |
| 0146 | L | D | E | N | 1973 | 72514 |
| 0147 | L | D | R | N | 1971 | 63380 |
| 0147 | L | D | E | N | 1971 | 63381 |
| 0147 | L | D | E | N | 1971 | 66385 |
| 0148 | L | M | E | N | 1971 | 64086 |
| 0148 | L | D | E | N | 1970 | 65883 |
| 0148 | L | C | E | U | 1972 | 66405 |
| 0148 | L | D | E | N | 1972 | 66530 |
| 0148 | L | D | E | N | 1973 | 71151 |
| 0148 | L | C | E | U | 1973 | 74736 |
| 0148 | L | C | E | U | 1972 | 77761 |
| 0148 | L | C | E | U | 1972 | 77762 |
| 0149 | L | M | E | N | 1971 | 64086 |
| 0149 | L | D | E | N | 1970 | 65883 |
| 0149 | L | C | E | U | 1972 | 66405 |
| 0149 | L | C | E | N | 1973 | 70212 |
| 0149 | L | D | E | N | 1973 | 71151 |
| 0149 | L | D | E | N | 1974 | 73605 |
| 0149 | L | C | E | U | 1973 | 74736 |
| 0149 | L | C | E | U | 1972 | 77761 |
| 0149 | L | C | E | U | 1972 | 77762 |
| 0151 | L | C | E | N | 1971 | 63598 |
| 0151 | L | D | E | N | 1971 | 65840 |
| 0151 | L | D | E | U | 1973 | 72280 |
| 0152 | G | D | E | N | 1970 | 63745 |
| 0152 | G | G | E | N | 1971 | 63879 |
| 0152 | G | C | E | H | 1971 | 63977 |
| 0152 | G | C | E | N | 1971 | 64301 |
| 0152 | G | D | E | N | 1973 | 71090 |
| 0152 | G | D | E | N | 1974 | 76378 |
| 0152 | G | S | E | N | 1977 | 93622 |
| 0153 | L | D | E | N | 1971 | 62471 |
| 0154 | L | D | E | N | 1971 | 62471 |
| 0154 | L | D | E | N | 1973 | 72514 |
| 0155 | L | M | O | N | 1970 | 60754 |
| 0155 | L | D | R | N | 1972 | 75521 |
| 0155 | L | D | E | N | 1972 | 75522 |
| 0156 | L | M | O | N | 1970 | 60754 |
| 0157 | L | M | O | N | 1970 | 60754 |
| 0158 | G | C | I | N | 1965 | 46295 |
| 0158 | G | C | R | N | 1968 | 54413 |
| 0158 | G | C | E | N | 1972 | 66200 |
| 0159 | L | D | R | N | 1971 | 66644 |
| 0159 | L | D | E | N | 1971 | 66645 |
| 0160 | L | D | R | N | 1971 | 66644 |
| 0160 | L | D | E | N | 1971 | 66645 |
| 0161 | L | D | R | N | 1971 | 66644 |
| 0161 | L | D | E | N | 1971 | 66645 |
| 0162 | L | G | E | N | 1971 | 63708 |
| 0162 | L | D | E | U | 1971 | 65066 |
| 0162 | L | D | E | N | 1972 | 66773 |
| 0162 | L | D | E | N | 1973 | 75791 |
| 0162 | L | D | E | N | 1974 | 77413 |
| 0163 | L | D | E | N | 1972 | 66773 |
| 0164 | L | D | E | N | 1972 | 66773 |
| 0165 | L | D | E | N | 1971 | 66358 |
| 0165 | L | D | E | N | 1972 | 66773 |
| 0166 | L | M | E | N | 1972 | 66775 |
| 0167 | L | M | E | N | 1972 | 66775 |
| 0168 | L | M | E | N | 1972 | 66775 |
| 0173 | L | D | E | N | 1973 | 73180 |

| Substance Number | Phys. State | Subject | Language | Temperature | Year | TPRC Number |
|---|---|---|---|---|---|---|
| 512-0174 | G | T | E | H | 1973 | 73019 |
| 0176 | L | D | R | H | 1969 | 59058 |
| 0178 | L | D | R | H | 1969 | 59058 |
| 0179 | L | D | R | H | 1969 | 59058 |
| 0180 | L | C | F | N | 1969 | 57596 |
| 0181 | L | C | F | N | 1969 | 57596 |
| 0181 | L | D | R | N | 1973 | 74280 |
| 0182 | L | C | F | N | 1969 | 57596 |
| 0182 | L | G | E | N | 1969 | 65879 |
| 0182 | L | D | E | N | 1972 | 68056 |
| 0182 | L | D | R | N | 1971 | 70101 |
| 0182 | L | D | E | N | 1972 | 71431 |
| 0182 | L | D | R | N | 1973 | 75644 |
| 0182 | L | D | E | N | 1973 | 75645 |
| 0183 | L | C | F | N | 1969 | 57596 |
| 0183 | L | D | R | N | 1971 | 70101 |
| 0183 | L | D | E | N | 1972 | 71431 |
| 0183 | L | D | R | N | 1974 | 74365 |
| 0183 | L | D | R | N | 1973 | 75644 |
| 0183 | L | D | E | N | 1973 | 75645 |
| 0184 | L | D | F | N | 1971 | 63201 |
| 0184 | L | D | E | N | 1972 | 67134 |
| 0185 | L | D | F | N | 1971 | 63201 |
| 0186 | L | D | F | N | 1969 | 56421 |
| 0186 | L | C | F | N | 1971 | 63200 |
| 0187 | L | D | F | N | 1969 | 56421 |
| 0187 | L | D | F | N | 1971 | 63202 |
| 0188 | G | G | E | N | 1972 | 71404 |
| 0189 | L | D | R | N | 1968 | 51467 |
| 0189 | L | D | R | N | 1970 | 58348 |
| 0189 | L | D | R | N | 1969 | 65051 |
| 0189 | L | D | E | N | 1970 | 67239 |
| 0189 | L | D | E | N | 1972 | 67576 |
| 0189 | L | D | E | N | 1973 | 72221 |
| 0189 | L | G | R | N | 1974 | 75203 |
| 0189 | L | D | R | N | 1971 | 75605 |
| 0189 | L | D | E | N | 1971 | 75606 |
| 0189 | L | G | E | N | 1974 | 76465 |
| 0192 | G | C | E | H | 1973 | 74430 |
| 0193 | G | G | E | N | 1971 | 63879 |
| 0193 | G | C | E | N | 1971 | 64301 |
| 0193 | G | D | R | H | 1972 | 70621 |
| 0193 | G | D | E | H | 1972 | 70622 |
| 0193 | G | S | E | N | 1977 | 93622 |
| 0193 | L | G | G | N | 1972 | 66352 |
| 0194 | G | C | E | N | 1971 | 64301 |
| 0194 | G | C | E | H | 1973 | 74430 |
| 0195 | L | D | E | N | 1971 | 64211 |
| 0196 | L | D | E | N | 1971 | 64211 |
| 0196 | L | D | E | N | 1972 | 66095 |
| 0197 | L | M | E | N | 1971 | 62627 |
| 0198 | L | M | E | N | 1971 | 62627 |
| 0199 | L | D | E | U | 1968 | 54571 |
| 0199 | L | C | E | N | 1968 | 62352 |
| 0199 | L | D | E | N | 1971 | 62626 |
| 0199 | L | D | R | N | 1972 | 69650 |
| 0199 | L | D | E | N | 1972 | 69651 |
| 0200 | L | D | E | N | 1971 | 62626 |
| 0200 | L | C | E | N | 1974 | 75339 |
| 0200 | L | D | R | N | 1973 | 75591 |
| 0200 | L | D | E | N | 1973 | 75592 |
| 0201 | L | D | E | N | 1971 | 65840 |
| 0201 | L | D | E | N | 1972 | 67063 |
| 0202 | G | C | E | H | 1972 | 38084 |
| 0202 | G | D | E | N | 1971 | 63430 |
| 0202 | G | G | E | N | 1972 | 64103 |
| 0202 | G | C | E | N | 1974 | 73619 |
| 0202 | G | C | E | H | 1973 | 74430 |
| 0205 | G | C | E | U | 1972 | 65366 |
| 0205 | G | S | E | N | 1977 | 93622 |
| 0206 | L | D | R | N | 1971 | 64575 |
| 0206 | L | D | E | U | 1971 | 65066 |
| 0206 | L | D | E | N | 1971 | 66023 |
| 0206 | L | D | E | N | 1971 | 66479 |
| 0207 | L | D | R | N | 1970 | 58348 |
| 0207 | L | D | E | N | 1970 | 67239 |
| 0208 | L | D | R | N | 1970 | 58552 |
| 0208 | L | D | R | N | 1971 | 65018 |
| 0208 | L | D | E | N | 1970 | 67240 |
| 0209 | L | D | R | N | 1970 | 58552 |
| 0209 | L | D | E | N | 1970 | 67240 |
| 0211 | L | D | R | N | 1968 | 53203 |
| 0212 | L | D | R | N | 1968 | 53203 |
| 0213 | L | D | E | N | 1972 | 68056 |
| 0213 | L | D | E | N | 1973 | 72217 |
| 0214 | L | D | E | N | 1972 | 61093 |
| 0214 | L | D | R | N | 1972 | 69758 |
| 0214 | L | D | E | N | 1974 | 74929 |
| 0215 | L | D | E | N | 1972 | 61093 |
| 0216 | L | D | R | N | 1970 | 59007 |
| 0216 | L | D | E | N | 1972 | 61093 |
| 0216 | L | D | E | N | 1970 | 66459 |
| 0216 | L | D | R | N | 1973 | 75638 |
| 0216 | L | D | E | N | 1973 | 75639 |
| 0217 | L | D | R | N | 1969 | 58481 |
| 0217 | L | D | E | N | 1969 | 67238 |
| 0218 | L | D | R | N | 1969 | 58481 |
| 0218 | L | D | E | N | 1969 | 67238 |
| 0219 | L | D | E | N | 1971 | 61867 |
| 0220 | L | D | E | N | 1972 | 60206 |
| 0220 | L | D | E | N | 1971 | 61867 |

**Phys. State:** **D.** Doped; **E.** Expanded; **F.** Fibrous or Whisker; **G.** Gas; **L.** Liquid; **M.** Multiphase; **P.** Powder; **S.** Solid
**Subject:** **C.** Calculated Data + Theory; **D.** Data; **E.** Experiment; **G.** General (Data + Theory + Experiment); **M.** Measured Data + Experiment; **S.** Survey (Review, Compendium, etc.); **T.** Theory
**Language:** **C.** Czech; **D.** Dutch; **E.** English; **F.** French; **G.** German; **I.** Italian; **J.** Japanese; **O.** Other Languages; **P.** Polish; **R.** Russian; **S.** Spanish
**Temperature:** **F.** Full Range (Low + Normal + High); **L.** Low (0 to 75K) + Overlap into Normal; **N.** Normal (75 to 1273K); **H.** High (above 1273K) + Overlap into Normal; **U.** Undefined

| Substance Number | Phys. State | Subject | Language | Temperature | Year | TPRC Number |
|---|---|---|---|---|---|---|
| 512-0220 | L | D | E | N | 1973 | 71442 |
| 0221 | L | D | E | N | 1971 | 61867 |
| 0221 | L | M | E | N | 1971 | 63173 |
| 0221 | L | D | R | U | 1973 | 72529 |
| 0221 | L | D | E | U | 1973 | 73455 |
| 0222 | G | C | E | H | 1970 | 64975 |
| 0223 | L | D | E | N | 1971 | 61867 |
| 0224 | L | D | E | N | 1971 | 61867 |
| 0227 | L | D | E | N | 1973 | 70901 |
| 0227 | L | D | E | N | 1973 | 72220 |
| 0227 | L | D | E | U | 1973 | 72280 |
| 0227 | L | C | E | N | 1974 | 75021 |
| 0228 | L | C | E | N | 1971 | 63573 |
| 0230 | G | D | E | N | 1974 | 75220 |
| 0235 | L | D | E | N | 1971 | 65840 |
| 0235 | L | D | S | N | 1972 | 69065 |
| 0236 | L | D | E | N | 1971 | 65840 |
| 0237 | G | D | E | N | 1972 | 65575 |
| 0237 | L | D | R | N | 1968 | 50148 |
| 0237 | L | D | J | N | 1968 | 55653 |
| 0237 | L | D | E | N | 1971 | 63984 |
| 0237 | L | D | E | N | 1972 | 65575 |
| 0237 | L | D | E | N | 1971 | 65840 |
| 0237 | L | C | O | N | 1971 | 69611 |
| 0237 | L | D | E | N | 1973 | 72217 |
| 0238 | L | M | E | N | 1969 | 55678 |
| 0238 | L | M | R | N | 1971 | 65899 |
| 0238 | L | M | E | N | 1971 | 66488 |
| 0238 | L | D | E | N | 1971 | 67567 |
| 0239 | L | D | E | U | 1968 | 54571 |
| 0239 | L | D | R | N | 1970 | 60922 |
| 0239 | L | D | E | N | 1970 | 66463 |
| 0239 | L | D | E | N | 1973 | 70901 |
| 0239 | L | C | E | N | 1974 | 75021 |
| 0239 | L | D | R | N | 1973 | 75591 |
| 0239 | L | D | E | N | 1973 | 75592 |
| 0241 | L | D | O | N | 1970 | 66348 |
| 0241 | L | D | E | H | 1973 | 73932 |
| 0243 | G | T | R | H | 1972 | 73479 |
| 0243 | G | T | E | H | 1973 | 73480 |
| 0245 | L | D | E | N | 1972 | 66095 |
| 0248 | L | D | E | N | 1972 | 66094 |
| 0256 | G | T | E | H | 1972 | 70286 |
| 0258 | L | D | R | N | 1971 | 65756 |
| 0258 | L | D | R | N | 1971 | 65789 |
| 0259 | L | D | R | N | 1970 | 61935 |
| 0260 | L | D | E | N | 1970 | 62127 |
| 0260 | L | D | R | N | 1971 | 65756 |
| 0260 | L | D | E | N | 1974 | 74929 |
| 0260 | L | C | R | N | 1974 | 75680 |
| 0260 | L | C | E | N | 1974 | 75681 |
| 0263 | L | D | E | N | 1972 | 60206 |
| 0264 | L | D | E | N | 1972 | 60206 |
| 0265 | L | D | E | N | 1972 | 60206 |
| 0269 | L | D | E | N | 1968 | 55833 |
| 0269 | L | D | E | N | 1972 | 68399 |
| 0271 | L | C | E | N | 1972 | 66353 |
| 0272 | L | C | E | N | 1972 | 66353 |
| 0273 | L | C | E | N | 1972 | 66353 |
| 0274 | L | C | E | N | 1972 | 66353 |
| 0275 | L | C | E | N | 1972 | 66353 |
| 0276 | L | C | E | N | 1972 | 66353 |
| 0277 | L | C | E | N | 1972 | 66353 |
| 0280 | L | D | E | N | 1973 | 74721 |
| 0281 | L | D | E | H | 1973 | 73932 |
| 0282 | L | G | E | N | 1971 | 63708 |
| 0282 | L | D | E | N | 1972 | 68667 |
| 0282 | L | C | E | N | 1974 | 75021 |
| 0282 | L | D | E | N | 1974 | 77413 |
| 0283 | L | D | E | N | 1972 | 68667 |
| 0284 | L | D | E | N | 1972 | 68667 |
| 0285 | L | C | R | N | 1965 | 44572 |
| 0286 | L | C | R | N | 1965 | 44572 |
| 0288 | L | D | R | N | 1971 | 74763 |
| 0289 | L | D | E | N | 1973 | 69031 |
| 0290 | L | S | O | U | 1970 | 60879 |
| 0290 | L | T | E | N | 1971 | 62926 |
| 0290 | L | T | R | N | 1972 | 69270 |
| 0292 | G | C | E | H | 1972 | 38084 |
| 0292 | G | C | E | H | 1973 | 74430 |
| 0293 | G | C | E | H | 1972 | 38084 |
| 0293 | G | C | E | H | 1973 | 74430 |
| 0294 | G | C | E | H | 1972 | 38084 |
| 0294 | G | C | E | H | 1973 | 74430 |
| 0295 | G | C | E | H | 1972 | 38084 |
| 0295 | G | C | E | H | 1973 | 74430 |
| 0296 | G | C | E | H | 1972 | 38084 |
| 0296 | G | C | E | H | 1973 | 74430 |
| 0297 | G | C | E | H | 1972 | 38084 |
| 0297 | G | C | E | H | 1973 | 74430 |
| 0300 | L | E | E | N | 1971 | 65197 |
| 0302 | L | M | G | N | 1964 | 34078 |
| 0303 | L | D | E | N | 1972 | 64802 |
| 0304 | L | C | E | N | 1968 | 62352 |
| 0304 | L | D | E | N | 1972 | 64802 |
| 0305 | L | D | E | N | 1972 | 64802 |
| 0306 | L | D | E | N | 1972 | 64802 |
| 0310 | L | D | E | N | 1969 | 58796 |
| 0310 | L | D | E | N | 1971 | 64141 |
| 0310 | L | D | R | N | 1974 | 74365 |
| 0311 | L | D | R | N | 1963 | 37245 |
| 512-0311 | L | D | R | N | 1968 | 50147 |
| 0311 | L | D | E | N | 1971 | 61265 |
| 0317 | L | D | E | N | 1971 | 67567 |
| 0321 | L | G | G | N | 1972 | 66352 |
| 0321 | L | D | E | N | 1972 | 66530 |
| 0322 | L | G | E | N | 1974 | 75020 |
| 0330 | L | D | E | N | 1971 | 64137 |
| 0330 | L | D | E | N | 1972 | 65835 |
| 0331 | L | D | E | N | 1971 | 64141 |
| 0331 | L | D | R | N | 1974 | 74365 |
| 0332 | L | D | E | N | 1972 | 64466 |
| 0333 | L | D | E | N | 1972 | 64466 |
| 0334 | L | D | E | N | 1972 | 64466 |
| 0335 | L | D | E | N | 1972 | 64466 |
| 0336 | L | D | E | N | 1972 | 64466 |
| 0337 | L | D | E | N | 1972 | 64466 |
| 0338 | L | D | E | N | 1972 | 64466 |
| 0339 | L | D | E | N | 1972 | 64466 |
| 0344 | L | D | E | N | 1971 | 66023 |
| 0345 | L | D | E | N | 1971 | 66023 |
| 0346 | L | D | E | N | 1971 | 66023 |
| 0347 | L | M | E | N | 1971 | 63173 |
| 0347 | L | D | R | U | 1973 | 72529 |
| 0347 | L | D | E | U | 1973 | 73455 |
| 0348 | L | D | E | N | 1971 | 66358 |
| 0349 | L | D | E | N | 1971 | 66358 |
| 0350 | L | D | E | N | 1971 | 66358 |
| 0351 | L | D | E | N | 1971 | 66358 |
| 0352 | L | D | E | N | 1971 | 66358 |
| 0355 | L | D | R | N | 1970 | 60923 |
| 0355 | L | D | E | N | 1970 | 66465 |
| 0355 | L | D | R | N | 1974 | 75684 |
| 0355 | L | D | E | N | 1974 | 75685 |
| 0356 | G | T | R | N | 1973 | 72971 |
| 0359 | L | C | E | N | 1966 | 42568 |
| 0360 | L | C | E | N | 1966 | 42568 |
| 0361 | L | C | E | N | 1966 | 42568 |
| 0362 | L | C | E | N | 1966 | 42568 |
| 0363 | L | C | E | N | 1966 | 42568 |
| 0364 | L | C | E | N | 1966 | 42568 |
| 0365 | G | D | E | N | 1977 | 93630 |
| 0366 | L | D | G | N | 1972 | 67600 |
| 0367 | L | D | G | N | 1972 | 67600 |
| 0370 | G | C | E | H | 1973 | 74430 |
| 0370 | L | M | G | N | 1967 | 47301 |
| 0370 | L | D | R | N | 1971 | 65018 |
| 0370 | L | D | R | N | 1972 | 66917 |
| 0371 | L | D | R | N | 1972 | 66917 |
| 0372 | L | D | R | N | 1972 | 66917 |
| 0373 | L | D | E | N | 1972 | 65835 |
| 0374 | L | D | E | N | 1972 | 65835 |
| 0375 | L | D | E | N | 1972 | 65835 |
| 0376 | L | D | E | N | 1971 | 66385 |
| 0377 | L | D | E | N | 1971 | 66385 |
| 0378 | L | D | R | N | 1970 | 60921 |
| 0378 | L | D | E | N | 1970 | 66462 |
| 0379 | L | D | R | N | 1970 | 60921 |
| 0379 | L | D | E | N | 1970 | 66462 |
| 0380 | L | D | E | H | 1973 | 73932 |
| 0380 | L | D | E | N | 1974 | 74778 |
| 0383 | L | M | R | N | 1966 | 44540 |
| 0384 | L | M | R | N | 1966 | 44540 |
| 0385 | L | D | R | N | 1968 | 51467 |
| 0387 | L | D | R | N | 1972 | 66919 |
| 0388 | L | D | R | N | 1972 | 66919 |
| 0392 | L | D | E | N | 1972 | 67576 |
| 0392 | L | D | E | N | 1973 | 72220 |
| 0392 | L | C | E | N | 1974 | 75021 |
| 0393 | L | D | E | N | 1972 | 67134 |
| 0393 | L | D | E | N | 1972 | 67576 |
| 0393 | L | D | E | N | 1973 | 72221 |
| 0394 | L | D | E | N | 1972 | 67576 |
| 0395 | G | T | E | H | 1972 | 67737 |
| 0396 | L | D | R | N | 1970 | 61935 |
| 0397 | L | D | R | N | 1970 | 61935 |
| 0397 | L | M | J | N | 1973 | 72914 |
| 0398 | L | D | R | N | 1970 | 61935 |
| 0399 | L | C | R | N | 1964 | 36663 |
| 0400 | L | C | R | N | 1964 | 36663 |
| 0401 | L | C | R | N | 1964 | 36663 |
| 0402 | L | C | R | N | 1964 | 36663 |
| 0402 | L | D | E | N | 1973 | 70901 |
| 0403 | L | C | R | N | 1969 | 56923 |
| 0403 | L | C | E | N | 1972 | 66056 |
| 0403 | L | D | R | N | 1973 | 75815 |
| 0403 | L | D | E | N | 1973 | 75816 |
| 0404 | L | C | R | N | 1969 | 56923 |
| 0404 | L | C | E | N | 1972 | 66056 |
| 0405 | L | C | R | N | 1969 | 56923 |
| 0405 | L | C | E | N | 1972 | 66056 |
| 0405 | L | D | R | N | 1972 | 67845 |
| 0406 | L | C | R | N | 1969 | 56923 |
| 0406 | L | C | E | N | 1972 | 66056 |
| 0407 | L | C | R | N | 1969 | 56923 |
| 0407 | L | C | E | N | 1972 | 66056 |
| 0408 | L | C | R | N | 1969 | 56923 |
| 0408 | L | C | E | N | 1972 | 66056 |
| 0416 | L | D | R | N | 1972 | 69650 |
| 0416 | L | D | E | N | 1972 | 69651 |
| 0417 | L | D | R | N | 1972 | 69652 |
| 0417 | L | D | E | N | 1972 | 69653 |
| 512-0424 | L | G | E | N | 1971 | 63708 |
| 0424 | L | D | E | N | 1974 | 77413 |
| 0429 | G | C | E | H | 1970 | 64975 |
| 0433 | L | D | E | H | 1973 | 73932 |
| 0451 | G | S | E | N | 1977 | 93622 |
| 0452 | L | D | R | N | 1972 | 70280 |
| 0452 | L | D | E | N | 1972 | 71432 |
| 0453 | L | D | R | N | 1972 | 70280 |
| 0453 | L | D | E | N | 1972 | 71432 |
| 0453 | L | D | R | N | 1973 | 75644 |
| 0453 | L | D | E | N | 1973 | 75645 |
| 0458 | L | D | R | N | 1974 | 74365 |
| 0462 | L | M | E | N | 1972 | 68545 |
| 0462 | L | D | E | N | 1974 | 74929 |
| 0476 | G | T | R | U | 1969 | 56446 |
| 0476 | G | T | E | U | 1970 | 64427 |
| 0500 | L | M | E | N | 1974 | 76685 |
| 0507 | L | D | R | N | 1973 | 75654 |
| 0507 | L | D | E | N | 1973 | 75655 |
| 0510 | L | M | E | N | 1974 | 76685 |
| 0523 | L | D | E | N | 1973 | 72221 |
| 0529 | L | D | F | N | 1973 | 71842 |
| 0537 | L | D | R | N | 1972 | 69560 |
| 0537 | L | D | E | N | 1972 | 72680 |
| 0540 | L | D | E | N | 1973 | 72221 |
| 0600 | L | M | E | N | 1974 | 76685 |
| 0633 | L | D | R | N | 1973 | 74280 |
| 0644 | L | D | R | N | 1973 | 75819 |
| 0644 | L | D | E | N | 1973 | 75820 |
| 0647 | L | D | R | N | 1972 | 75521 |
| 0647 | L | D | E | N | 1972 | 75522 |
| 0653 | L | D | R | N | 1973 | 74280 |
| 0688 | L | D | R | N | 1971 | 65766 |
| 0688 | L | D | E | N | 1971 | 69786 |
| 0689 | L | D | R | N | 1971 | 65766 |
| 0689 | L | D | E | N | 1971 | 69786 |
| 0693 | L | D | R | H | 1969 | 59058 |
| 0693 | L | D | R | H | 1969 | 62116 |
| 0694 | L | D | R | H | 1969 | 59058 |
| 0694 | L | D | R | H | 1969 | 62116 |
| 0695 | L | D | R | H | 1969 | 59058 |
| 0695 | L | D | R | H | 1969 | 62116 |
| 0696 | L | D | R | H | 1969 | 62116 |
| 0697 | L | D | O | N | 1971 | 69609 |
| 0699 | G | C | E | H | 1973 | 74430 |
| 0701 | G | G | E | N | 1972 | 71404 |
| 0708 | L | M | G | N | 1967 | 47301 |
| 0717 | L | D | E | N | 1973 | 71442 |
| 0718 | L | D | E | N | 1973 | 72220 |
| 0718 | L | C | E | N | 1974 | 75021 |
| 0724 | L | D | E | N | 1973 | 72221 |
| 0733 | L | D | R | N | 1973 | 75563 |
| 0733 | L | D | E | N | 1973 | 75564 |
| 0738 | G | C | E | H | 1971 | 63977 |
| 0739 | G | C | E | H | 1971 | 63977 |
| 0739 | G | T | E | H | 1973 | 73019 |
| 0742 | L | D | E | N | 1973 | 71442 |
| 0766 | L | D | R | N | 1972 | 75521 |
| 0766 | L | D | E | N | 1972 | 75522 |
| 0796 | L | D | E | N | 1974 | 73605 |
| 0804 | L | D | E | N | 1973 | 74721 |
| 0806 | G | D | E | N | 1973 | 73805 |
| 0806 | G | M | F | N | 1973 | 73961 |
| 0806 | G | G | F | N | 1973 | 73962 |
| 0812 | L | D | R | U | 1973 | 72529 |
| 0812 | L | D | E | U | 1973 | 73455 |
| 0853 | L | T | E | N | 1973 | 74817 |
| 0880 | L | D | R | N | 1971 | 75499 |
| 0880 | L | D | E | N | 1971 | 75500 |
| 0882 | G | C | E | H | 1973 | 74430 |
| 0883 | G | C | E | H | 1973 | 74430 |
| 0884 | G | C | E | H | 1973 | 74430 |
| 0889 | L | D | R | N | 1972 | 75521 |
| 0889 | L | D | E | N | 1972 | 75522 |
| 0896 | L | D | R | N | 1973 | 75563 |
| 0896 | L | D | E | N | 1973 | 75564 |
| 0897 | L | D | R | N | 1973 | 75563 |
| 0897 | L | D | E | N | 1973 | 75564 |
| 0899 | L | D | E | H | 1973 | 73932 |
| 0900 | L | D | E | H | 1973 | 73932 |
| 0902 | L | D | R | N | 1973 | 75815 |
| 0902 | L | D | E | N | 1973 | 75816 |
| 0903 | L | D | R | N | 1973 | 75815 |
| 0903 | L | D | E | N | 1973 | 75816 |
| 0904 | L | D | R | N | 1973 | 75819 |
| 0904 | L | D | E | N | 1973 | 75820 |
| 0905 | L | D | R | N | 1973 | 75819 |
| 0905 | L | D | E | N | 1973 | 75820 |
| 0911 | L | T | E | N | 1972 | 67762 |
| 0931 | G | M | F | N | 1973 | 73961 |
| 0932 | G | M | F | N | 1973 | 73961 |
| 0933 | G | G | F | N | 1973 | 73962 |
| 0995 | L | D | R | N | 1974 | 75684 |
| 0995 | L | D | E | N | 1974 | 75685 |
| 0999 | L | D | E | N | 1974 | 77413 |
| 1006 | L | D | E | N | 1964 | 34654 |
| 1044 | L | D | J | N | 1974 | 78428 |
| 1046 | L | T | R | N | 1973 | 74471 |
| 1046 | L | T | E | N | 1973 | 77385 |
| 1047 | G | T | E | H | 1974 | 77458 |

**Phys. State:** **D.** Doped; **E.** Expanded; **F.** Fibrous or Whisker; **G.** Gas; **L.** Liquid; **M.** Multiphase; **P.** Powder; **S.** Solid
**Subject:** **C.** Calculated Data + Theory; **D.** Data; **E.** Experiment; **G.** General (Data + Theory + Experiment); **M.** Measured Data + Experiment; **S.** Survey (Review, Compendium, etc.); **T.** Theory
**Language:** **C.** Czech; **D.** Dutch; **E.** English; **F.** French; **G.** German; **I.** Italian; **J.** Japanese; **O.** Other Languages; **P.** Polish; **R.** Russian; **S.** Spanish
**Temperature:** **F.** Full Range (Low + Normal + High); **L.** Low (0 to 75K) + Overlap into Normal; **N.** Normal (75 to 1273K); **H.** High (above 1273K) + Overlap into Normal; **U.** Undefined

| Substance Number | Phys. State | Sub-ject | Lan-guage | Temper-ature | Year | TPRC Number |
|---|---|---|---|---|---|---|
| 514-0001 | G | D | R | N | 1968 | 50752 |
| 0001 | G | C | R | N | 1970 | 59056 |
| 0001 | G | D | E | N | 1971 | 63989 |
| 0001 | G | C | E | N | 1974 | 76851 |
| 0003 | G | D | R | N | 1968 | 50752 |
| 0003 | G | C | R | N | 1970 | 59056 |
| 0003 | G | C | E | H | 1971 | 63977 |
| 0003 | G | D | E | N | 1971 | 63989 |
| 0006 | L | D | E | N | 1972 | 67064 |
| 0006 | L | D | E | N | 1972 | 68796 |
| 0006 | L | D | R | N | 1972 | 70041 |
| 0006 | L | D | E | N | 1973 | 75783 |
| 0008 | G | C | E | H | 1971 | 63977 |
| 0011 | L | C | F | N | 1971 | 63200 |
| 0012 | G | C | E | H | 1971 | 63977 |
| 0012 | G | D | E | N | 1973 | 70794 |
| 0012 | G | C | E | N | 1974 | 74390 |
| 0012 | G | C | E | N | 1974 | 76851 |
| 0012 | G | S | E | N | 1977 | 93622 |
| 0014 | G | D | E | N | 1971 | 61145 |
| 0015 | G | D | E | N | 1971 | 61145 |
| 0018 | G | C | E | H | 1971 | 63977 |
| 0018 | G | D | E | N | 1973 | 70794 |
| 0018 | G | C | E | N | 1974 | 74390 |
| 0018 | G | C | E | N | 1974 | 76851 |
| 0018 | G | S | E | N | 1977 | 93622 |
| 0020 | G | G | E | N | 1965 | 34063 |
| 0020 | G | S | E | F | 1971 | 62671 |
| 0020 | G | G | E | N | 1971 | 63597 |
| 0020 | G | G | E | H | 1970 | 65551 |
| 0020 | G | C | R | H | 1971 | 67807 |
| 0020 | G | S | E | H | 1972 | 68186 |
| 0020 | G | M | R | N | 1971 | 68806 |
| 0020 | G | S | E | H | 1972 | 69811 |
| 0020 | G | D | E | U | 1972 | 70027 |
| 0020 | G | D | E | N | 1973 | 71090 |
| 0020 | G | S | E | N | 1971 | 71643 |
| 0020 | G | C | R | H | 1973 | 72523 |
| 0020 | G | C | R | N | 1970 | 73694 |
| 0020 | G | C | E | N | 1970 | 73695 |
| 0020 | G | C | E | H | 1974 | 74346 |
| 0020 | G | C | R | H | 1973 | 74573 |
| 0020 | G | C | E | H | 1973 | 74574 |
| 0020 | G | D | E | N | 1973 | 76337 |
| 0020 | G | C | E | H | 1974 | 76376 |
| 0020 | G | C | E | N | 1974 | 76851 |
| 0028 | L | C | E | N | 1971 | 63598 |
| 0028 | L | D | R | N | 1972 | 70041 |
| 0028 | L | D | E | N | 1973 | 75791 |
| 0030 | L | D | O | N | 1970 | 66348 |
| 0030 | L | D | R | N | 1971 | 66364 |
| 0030 | L | D | E | N | 1974 | 74778 |
| 0036 | G | D | E | N | 1973 | 70794 |
| 0036 | G | C | E | N | 1974 | 74390 |
| 0036 | G | S | E | N | 1977 | 93622 |
| 0039 | L | D | E | N | 1972 | 67064 |
| 0039 | L | D | E | N | 1972 | 68796 |
| 0041 | L | D | R | N | 1971 | 63380 |
| 0041 | L | D | E | N | 1971 | 63381 |
| 0041 | L | D | R | N | 1971 | 66417 |
| 0041 | L | D | E | N | 1971 | 66418 |
| 0041 | L | D | R | N | 1971 | 66644 |
| 0041 | L | D | E | N | 1971 | 66645 |
| 0041 | L | D | E | N | 1973 | 75783 |
| 0042 | L | C | E | N | 1971 | 63598 |
| 0043 | L | C | E | N | 1971 | 63598 |
| 0044 | L | D | E | U | 1968 | 54571 |
| 0044 | L | C | E | N | 1968 | 62352 |
| 0044 | L | C | E | N | 1971 | 63598 |
| 0044 | L | C | E | N | 1974 | 75339 |
| 0046 | L | C | E | N | 1968 | 62352 |
| 0047 | L | D | E | N | 1973 | 70901 |
| 0048 | L | D | O | N | 1970 | 66348 |
| 0048 | L | D | R | N | 1971 | 66364 |
| 0048 | L | D | R | N | 1973 | 75654 |
| 0048 | L | D | E | N | 1973 | 75655 |
| 0051 | G | M | R | N | 1966 | 44588 |
| 0051 | G | C | R | N | 1969 | 56170 |
| 0051 | G | C | E | H | 1972 | 65518 |
| 0051 | G | S | R | N | 1971 | 66284 |
| 0051 | G | S | E | N | 1971 | 67539 |
| 0052 | G | M | R | N | 1966 | 44588 |
| 0052 | G | C | R | N | 1969 | 56170 |
| 0052 | G | C | E | H | 1972 | 65518 |
| 0052 | G | S | R | N | 1971 | 66284 |
| 0052 | G | S | E | N | 1971 | 67539 |
| 0061 | G | C | E | N | 1974 | 76851 |
| 0064 | G | T | R | U | 1969 | 55647 |
| 0064 | G | C | E | H | 1971 | 63977 |
| 0064 | G | T | R | N | 1973 | 72971 |
| 0064 | G | S | E | N | 1973 | 73278 |
| 0064 | G | T | G | H | 1972 | 73814 |
| 0064 | G | T | R | U | 1974 | 75069 |
| 0064 | G | T | E | U | 1974 | 75070 |
| 0068 | G | D | E | N | 1971 | 61145 |
| 0071 | L | D | R | N | 1971 | 66364 |
| 0085 | L | C | E | N | 1974 | 75339 |
| 0089 | L | D | E | N | 1974 | 74778 |
| 0097 | L | D | R | N | 1971 | 63380 |
| 0097 | L | D | E | N | 1971 | 63381 |
| 0097 | L | D | R | N | 1971 | 66417 |
| 514-0097 | L | D | E | N | 1971 | 66418 |
| 0098 | L | D | R | N | 1971 | 66417 |
| 0098 | L | D | E | N | 1971 | 66418 |
| 0098 | L | D | E | N | 1972 | 68796 |
| 0099 | L | D | R | N | 1971 | 66417 |
| 0099 | L | D | E | N | 1971 | 66418 |
| 0100 | L | D | R | N | 1971 | 66417 |
| 0100 | L | D | E | N | 1971 | 66418 |
| 0101 | L | D | R | N | 1971 | 66417 |
| 0101 | L | D | E | N | 1971 | 66418 |
| 0102 | L | D | R | N | 1971 | 63380 |
| 0102 | L | D | E | N | 1971 | 63381 |
| 0102 | L | D | R | N | 1971 | 66417 |
| 0102 | L | D | E | N | 1971 | 66418 |
| 0103 | L | D | R | N | 1971 | 66417 |
| 0103 | L | D | E | N | 1971 | 66418 |
| 0106 | G | D | R | N | 1968 | 50752 |
| 0106 | G | C | R | N | 1970 | 59056 |
| 0106 | G | D | E | N | 1971 | 63989 |
| 0113 | L | D | E | N | 1973 | 70901 |
| 0113 | L | D | E | N | 1973 | 72220 |
| 0114 | L | C | E | N | 1971 | 63598 |
| 0114 | L | D | E | N | 1972 | 67064 |
| 0114 | L | D | E | N | 1974 | 76146 |
| 0116 | L | D | E | N | 1974 | 76146 |
| 0117 | L | D | R | N | 1972 | 70041 |
| 0117 | L | D | E | N | 1973 | 72220 |
| 0117 | L | C | E | N | 1974 | 75021 |
| 0118 | L | D | R | N | 1972 | 70041 |
| 0121 | L | D | R | N | 1971 | 74763 |
| 0121 | L | D | R | N | 1973 | 75654 |
| 0121 | L | D | E | N | 1973 | 75655 |
| 0129 | G | T | R | H | 1971 | 66074 |
| 0129 | G | T | E | H | 1971 | 66075 |
| 0129 | G | D | R | H | 1972 | 70621 |
| 0129 | G | D | E | H | 1972 | 70622 |
| 0129 | G | S | E | H | 1973 | 72551 |
| 0135 | L | D | E | N | 1973 | 72220 |
| 0136 | L | D | R | N | 1971 | 63380 |
| 0136 | L | D | E | N | 1971 | 63381 |
| 0136 | L | D | E | N | 1973 | 75791 |
| 0145 | L | D | R | N | 1971 | 63380 |
| 0145 | L | D | E | N | 1971 | 63381 |
| 0146 | L | D | R | N | 1971 | 63380 |
| 0146 | L | D | E | N | 1971 | 63381 |
| 0147 | L | D | R | N | 1971 | 63380 |
| 0147 | L | D | E | N | 1971 | 63381 |
| 0152 | G | M | R | N | 1966 | 44588 |
| 0152 | G | C | R | N | 1969 | 56170 |
| 0152 | G | C | R | N | 1970 | 59056 |
| 0152 | G | C | E | H | 1972 | 65518 |
| 0152 | G | S | R | N | 1971 | 66284 |
| 0152 | G | S | E | N | 1971 | 67539 |
| 0155 | L | M | O | N | 1970 | 60754 |
| 0155 | L | D | R | N | 1972 | 75521 |
| 0156 | L | M | O | N | 1970 | 60754 |
| 0157 | L | M | O | N | 1970 | 60754 |
| 0159 | L | D | R | N | 1971 | 66644 |
| 0159 | L | D | E | N | 1971 | 66645 |
| 0160 | L | D | R | N | 1971 | 66644 |
| 0160 | L | D | E | N | 1971 | 66645 |
| 0161 | L | D | R | N | 1971 | 66644 |
| 0161 | L | D | E | N | 1971 | 66645 |
| 0162 | L | D | E | N | 1973 | 75791 |
| 0171 | G | C | E | H | 1967 | 47652 |
| 0171 | G | T | R | H | 1972 | 68678 |
| 0171 | G | T | E | H | 1972 | 68679 |
| 0186 | L | C | F | N | 1971 | 63200 |
| 0188 | L | C | F | N | 1971 | 63200 |
| 0193 | G | C | R | N | 1970 | 59056 |
| 0193 | G | C | E | H | 1971 | 63977 |
| 0193 | G | C | E | N | 1974 | 76851 |
| 0199 | L | D | E | U | 1968 | 54571 |
| 0199 | L | C | E | N | 1968 | 62352 |
| 0200 | L | C | E | N | 1974 | 75339 |
| 0227 | L | D | E | N | 1973 | 70901 |
| 0227 | L | D | E | N | 1973 | 72220 |
| 0227 | L | C | E | N | 1974 | 75021 |
| 0239 | L | D | E | U | 1968 | 54571 |
| 0280 | L | D | O | N | 1970 | 66348 |
| 0281 | L | D | E | N | 1973 | 74721 |
| 0281 | L | D | E | N | 1974 | 74778 |
| 0282 | L | C | E | N | 1974 | 75021 |
| 0288 | L | D | R | N | 1971 | 74763 |
| 0312 | L | M | F | N | 1964 | 37547 |
| 0312 | L | M | I | N | 1965 | 49040 |
| 0313 | L | M | F | N | 1964 | 37547 |
| 0313 | L | M | I | N | 1965 | 49040 |
| 0314 | L | M | F | N | 1964 | 37547 |
| 0314 | L | M | I | N | 1965 | 49040 |
| 0315 | L | M | I | N | 1965 | 49040 |
| 0316 | L | M | F | N | 1964 | 37547 |
| 0316 | L | M | I | N | 1965 | 49040 |
| 0356 | G | T | R | N | 1973 | 72971 |
| 0380 | L | D | O | N | 1970 | 66348 |
| 0380 | L | D | E | N | 1974 | 74778 |
| 0381 | L | D | O | N | 1970 | 66348 |
| 0386 | G | C | R | N | 1969 | 55396 |
| 0386 | G | C | G | H | 1972 | 73814 |
| 0387 | L | D | R | N | 1971 | 66364 |
| 0391 | G | M | R | N | 1966 | 44588 |
| 514-0391 | G | C | R | N | 1969 | 56170 |
| 0391 | G | C | E | H | 1972 | 65518 |
| 0391 | G | S | R | N | 1971 | 66284 |
| 0391 | G | S | E | N | 1971 | 67539 |
| 0392 | L | D | E | N | 1973 | 72220 |
| 0425 | L | D | E | N | 1972 | 65094 |
| 0426 | L | D | E | N | 1972 | 65094 |
| 0447 | L | D | R | N | 1971 | 74763 |
| 0454 | L | D | E | N | 1973 | 74721 |
| 0475 | G | C | R | H | 1971 | 67807 |
| 0476 | G | T | R | U | 1969 | 55647 |
| 0507 | L | D | R | N | 1973 | 75654 |
| 0507 | L | D | E | N | 1973 | 75655 |
| 0647 | L | D | R | N | 1972 | 75521 |
| 0718 | L | D | E | N | 1973 | 72220 |
| 0718 | L | C | E | N | 1974 | 75021 |
| 0766 | L | D | R | N | 1972 | 75521 |
| 0779 | L | D | R | N | 1973 | 72445 |
| 0780 | L | D | R | N | 1973 | 72445 |
| 0781 | L | D | R | N | 1973 | 72445 |
| 0782 | L | D | R | N | 1973 | 72445 |
| 0783 | L | D | R | N | 1973 | 72445 |
| 0784 | L | D | R | N | 1973 | 72445 |
| 0785 | L | D | R | N | 1973 | 72445 |
| 0786 | L | D | R | N | 1973 | 72445 |
| 0787 | L | D | R | N | 1973 | 72445 |
| 0788 | L | D | R | N | 1973 | 72445 |
| 0789 | L | D | R | N | 1973 | 72445 |
| 0790 | L | D | R | N | 1973 | 72445 |
| 0791 | L | D | R | N | 1973 | 72445 |
| 0792 | L | D | R | N | 1973 | 72445 |
| 0793 | L | D | R | N | 1973 | 72445 |
| 0805 | L | D | E | N | 1973 | 74721 |
| 0867 | L | D | E | N | 1974 | 76146 |
| 0889 | L | D | R | N | 1972 | 75521 |
| 0919 | L | D | R | N | 1972 | 75619 |
| 0919 | L | D | E | N | 1972 | 75620 |
| 0920 | L | D | R | N | 1972 | 75619 |
| 0920 | L | D | E | N | 1972 | 75620 |
| 0921 | L | D | R | N | 1972 | 75619 |
| 0921 | L | D | E | N | 1972 | 75620 |
| 0922 | L | D | R | N | 1972 | 75619 |
| 0922 | L | D | E | N | 1972 | 75620 |
| 0923 | L | D | R | N | 1972 | 75619 |
| 0923 | L | D | E | N | 1972 | 75620 |
| 0975 | L | D | E | N | 1973 | 75783 |
| 515-0003 | G | C | R | N | 1971 | 65016 |
| 0003 | G | C | E | N | 1973 | 70014 |
| 0005 | L | D | R | N | 1973 | 75575 |
| 0005 | L | D | E | N | 1973 | 75576 |
| 0006 | L | D | E | N | 1972 | 60206 |
| 0006 | L | D | R | N | 1970 | 62455 |
| 0010 | L | D | R | N | 1972 | 69566 |
| 0011 | L | D | R | N | 1971 | 62768 |
| 0011 | L | D | E | N | 1971 | 66477 |
| 0011 | L | D | E | N | 1972 | 67824 |
| 0011 | L | D | R | N | 1973 | 70859 |
| 0011 | L | D | E | N | 1973 | 71138 |
| 0013 | L | M | E | N | 1974 | 75328 |
| 0028 | D | G | S | N | 1971 | 66853 |
| 0028 | G | D | E | N | 1971 | 61145 |
| 0028 | G | C | E | N | 1971 | 62561 |
| 0028 | L | D | R | N | 1970 | 59007 |
| 0028 | L | D | G | N | 1970 | 61243 |
| 0028 | L | D | E | N | 1971 | 62472 |
| 0028 | L | D | E | N | 1970 | 66459 |
| 0028 | L | G | S | N | 1971 | 66853 |
| 0028 | L | D | E | N | 1971 | 66855 |
| 0028 | L | D | E | N | 1973 | 69219 |
| 0028 | L | G | R | N | 1974 | 75203 |
| 0028 | L | G | E | N | 1974 | 76465 |
| 0029 | L | M | E | N | 1964 | 38472 |
| 0029 | L | C | R | N | 1971 | 62600 |
| 0029 | L | D | E | N | 1971 | 63359 |
| 0029 | L | D | R | N | 1971 | 65462 |
| 0029 | L | C | E | N | 1971 | 66475 |
| 0029 | L | D | E | N | 1972 | 69723 |
| 0029 | L | D | E | N | 1971 | 69788 |
| 0029 | L | D | R | N | 1973 | 70583 |
| 0029 | L | D | R | N | 1972 | 71590 |
| 0029 | L | D | E | N | 1973 | 72340 |
| 0029 | L | D | E | N | 1973 | 72975 |
| 0029 | L | D | F | N | 1974 | 74976 |
| 0030 | G | D | E | N | 1972 | 65575 |
| 0030 | L | M | E | N | 1964 | 38472 |
| 0030 | L | C | R | N | 1971 | 62600 |
| 0030 | L | M | E | N | 1971 | 63173 |
| 0030 | L | M | R | N | 1970 | 63600 |
| 0030 | L | D | E | N | 1972 | 65575 |
| 0030 | L | C | E | N | 1971 | 66475 |
| 0030 | L | D | E | N | 1972 | 66662 |
| 0030 | L | D | R | N | 1973 | 70583 |
| 0030 | L | D | R | N | 1972 | 71590 |
| 0030 | L | D | R | N | 1973 | 71995 |
| 0030 | L | D | E | N | 1973 | 72340 |
| 0030 | L | D | E | N | 1973 | 72975 |
| 0030 | L | D | F | N | 1974 | 74976 |
| 0030 | L | D | R | N | 1973 | 75573 |
| 0030 | L | S | E | N | 1960 | 94260 |
| 0031 | L | M | E | N | 1964 | 38472 |

**Phys. State:** **D.** Doped; **E.** Expanded; **F.** Fibrous or Whisker; **G.** Gas; **L.** Liquid; **M.** Multiphase; **P.** Powder; **S.** Solid
**Subject:** **C.** Calculated Data + Theory; **D.** Data; **E.** Experiment; **G.** General (Data + Theory + Experiment); **M.** Measured Data + Experiment; **S.** Survey (Review, Compendium, etc.); **T.** Theory
**Language:** **C.** Czech; **D.** Dutch; **E.** English; **F.** French; **G.** German; **I.** Italian; **J.** Japanese; **O.** Other Languages; **P.** Polish; **R.** Russian; **S.** Spanish
**Temperature:** **F.** Full Range (Low + Normal + High); **L.** Low (0 to 75K) + Overlap into Normal; **N.** Normal (75 to 1273K); **H.** High (above 1273K) + Overlap into Normal; **U.** Undefined

| Substance Number | Phys. State | Subject | Language | Temperature | Year | TPRC Number |
|---|---|---|---|---|---|---|
| 515-0031 | L | D | E | N | 1972 | 69723 |
| 0031 | L | D | R | N | 1972 | 71590 |
| 0031 | L | D | E | N | 1973 | 72975 |
| 0032 | D | D | E | U | 1972 | 68367 |
| 0032 | L | D | E | N | 1972 | 42002 |
| 0032 | L | D | R | N | 1970 | 60759 |
| 0032 | L | C | E | N | 1971 | 61312 |
| 0032 | L | M | E | N | 1971 | 64054 |
| 0032 | L | D | E | N | 1972 | 65613 |
| 0032 | L | D | R | N | 1971 | 66363 |
| 0032 | L | D | E | N | 1971 | 66855 |
| 0032 | L | D | E | N | 1972 | 68039 |
| 0032 | L | C | E | N | 1972 | 68203 |
| 0032 | L | D | E | U | 1972 | 68367 |
| 0032 | L | D | E | N | 1970 | 68707 |
| 0032 | L | D | E | N | 1973 | 69219 |
| 0032 | L | D | R | N | 1970 | 75459 |
| 0032 | L | D | E | N | 1970 | 75460 |
| 0033 | L | D | R | N | 1973 | 71082 |
| 0041 | L | D | G | N | 1970 | 61243 |
| 0041 | L | D | E | N | 1971 | 63878 |
| 0041 | L | D | R | N | 1972 | 67139 |
| 0041 | L | D | E | N | 1971 | 67344 |
| 0041 | L | D | R | U | 1972 | 68163 |
| 0041 | L | D | E | U | 1972 | 68164 |
| 0041 | L | D | E | N | 1973 | 69219 |
| 0041 | L | D | E | N | 1972 | 69794 |
| 0042 | D | G | S | N | 1971 | 66853 |
| 0042 | G | D | E | N | 1971 | 61145 |
| 0042 | G | C | E | N | 1971 | 62561 |
| 0042 | L | M | E | N | 1965 | 41259 |
| 0042 | L | C | R | N | 1968 | 53467 |
| 0042 | L | D | R | N | 1970 | 59007 |
| 0042 | L | D | G | N | 1970 | 61243 |
| 0042 | L | C | E | N | 1971 | 63598 |
| 0042 | L | D | R | N | 1971 | 65593 |
| 0042 | L | D | E | N | 1971 | 65594 |
| 0042 | L | D | E | N | 1970 | 66459 |
| 0042 | L | G | S | N | 1971 | 66853 |
| 0042 | L | D | E | N | 1971 | 66855 |
| 0042 | L | D | G | N | 1972 | 67710 |
| 0042 | L | D | E | N | 1973 | 69219 |
| 0042 | L | D | R | N | 1974 | 74964 |
| 0042 | L | D | R | N | 1970 | 75459 |
| 0042 | L | D | E | N | 1970 | 75460 |
| 0043 | D | G | S | N | 1971 | 66853 |
| 0043 | L | D | R | N | 1967 | 55527 |
| 0043 | L | C | E | N | 1971 | 63598 |
| 0043 | L | D | E | N | 1972 | 66063 |
| 0043 | L | G | S | N | 1971 | 66853 |
| 0043 | L | D | E | N | 1971 | 66855 |
| 0043 | L | D | E | N | 1973 | 69219 |
| 0044 | L | D | E | N | 1973 | 69219 |
| 0048 | L | M | E | N | 1964 | 38472 |
| 0048 | L | D | R | N | 1968 | 50828 |
| 0048 | L | D | R | N | 1969 | 56940 |
| 0048 | L | D | G | N | 1970 | 61243 |
| 0048 | L | D | E | N | 1972 | 68218 |
| 0048 | L | D | R | N | 1973 | 70583 |
| 0048 | L | D | R | N | 1973 | 71995 |
| 0048 | L | D | E | N | 1973 | 72340 |
| 0048 | L | D | E | N | 1973 | 72975 |
| 0048 | L | T | R | N | 1972 | 75525 |
| 0048 | L | T | E | N | 1972 | 75526 |
| 0048 | L | D | R | N | 1973 | 75573 |
| 0054 | L | D | R | N | 1969 | 55597 |
| 0054 | L | D | E | N | 1972 | 65502 |
| 0057 | L | M | R | N | 1971 | 62128 |
| 0057 | L | M | E | N | 1971 | 66472 |
| 0057 | L | D | R | N | 1972 | 68167 |
| 0057 | L | D | E | N | 1972 | 68168 |
| 0057 | L | D | G | N | 1973 | 72511 |
| 0057 | L | D | R | N | 1973 | 75555 |
| 0057 | L | D | E | N | 1973 | 75556 |
| 0057 | L | D | R | N | 1973 | 75664 |
| 0057 | L | D | E | N | 1973 | 75665 |
| 0057 | L | C | R | N | 1974 | 75678 |
| 0057 | L | C | E | N | 1974 | 75679 |
| 0057 | L | D | R | N | 1973 | 75830 |
| 0057 | L | D | E | N | 1973 | 75831 |
| 0058 | L | D | R | N | 1973 | 75826 |
| 0058 | L | D | E | N | 1973 | 75827 |
| 0063 | D | D | E | N | 1974 | 75983 |
| 0063 | L | C | E | N | 1970 | 61062 |
| 0063 | L | D | E | N | 1971 | 62467 |
| 0063 | L | D | E | N | 1971 | 66855 |
| 0063 | L | C | E | N | 1972 | 68203 |
| 0063 | L | D | E | N | 1973 | 69219 |
| 0063 | L | D | E | N | 1974 | 75983 |
| 0064 | L | C | E | N | 1972 | 68551 |
| 0064 | L | S | E | N | 1971 | 69767 |
| 0064 | L | S | E | N | 1973 | 72366 |
| 0064 | L | T | E | U | 1973 | 73174 |
| 0064 | L | S | E | N | 1914 | 73602 |
| 0065 | L | D | R | N | 1971 | 64917 |
| 0065 | L | D | E | N | 1971 | 64918 |
| 0065 | L | D | F | N | 1972 | 66356 |
| 0067 | L | D | E | N | 1970 | 59461 |
| 0070 | L | M | E | N | 1964 | 38472 |
| 0070 | L | D | E | N | 1972 | 69723 |
| 0085 | L | D | R | N | 1968 | 56795 |

| Substance Number | Phys. State | Subject | Language | Temperature | Year | TPRC Number |
|---|---|---|---|---|---|---|
| 515-0085 | L | D | E | N | 1972 | 65496 |
| 0086 | L | D | E | N | 1971 | 73917 |
| 0087 | L | D | R | N | 1971 | 66303 |
| 0087 | L | D | E | N | 1971 | 69790 |
| 0089 | L | D | F | N | 1974 | 74976 |
| 0089 | L | D | R | N | 1973 | 75660 |
| 0089 | L | D | E | N | 1973 | 75661 |
| 0098 | L | D | G | N | 1970 | 61243 |
| 0099 | L | M | E | N | 1964 | 38472 |
| 0099 | L | G | G | N | 1971 | 62442 |
| 0099 | L | D | E | N | 1972 | 69723 |
| 0099 | L | D | E | N | 1973 | 72975 |
| 0100 | L | M | E | N | 1964 | 38472 |
| 0100 | L | D | G | N | 1970 | 61243 |
| 0100 | L | G | G | N | 1971 | 62442 |
| 0100 | L | D | E | N | 1972 | 64799 |
| 0100 | L | D | E | N | 1972 | 64802 |
| 0100 | L | D | E | N | 1972 | 68218 |
| 0100 | L | D | E | N | 1972 | 69723 |
| 0100 | L | D | R | N | 1973 | 70583 |
| 0100 | L | D | E | N | 1973 | 72340 |
| 0100 | L | T | R | N | 1972 | 75525 |
| 0100 | L | T | E | N | 1972 | 75526 |
| 0100 | L | D | R | N | 1973 | 75573 |
| 0100 | L | D | R | N | 1973 | 75575 |
| 0100 | L | D | E | N | 1973 | 75576 |
| 0101 | L | D | E | N | 1972 | 64802 |
| 0101 | L | D | E | N | 1972 | 69723 |
| 0101 | L | D | E | N | 1973 | 72975 |
| 0103 | L | D | G | N | 1970 | 61243 |
| 0103 | L | D | E | N | 1971 | 67344 |
| 0103 | L | D | E | N | 1973 | 69219 |
| 0113 | L | D | E | N | 1973 | 69219 |
| 0117 | L | D | R | N | 1970 | 62455 |
| 0118 | L | D | E | N | 1972 | 43271 |
| 0118 | L | D | J | N | 1968 | 55653 |
| 0118 | L | G | G | N | 1971 | 62442 |
| 0118 | L | G | E | N | 1970 | 65885 |
| 0118 | L | D | E | N | 1971 | 66238 |
| 0118 | L | D | E | N | 1972 | 66511 |
| 0118 | L | D | E | N | 1972 | 69723 |
| 0118 | L | D | E | N | 1973 | 72975 |
| 0119 | L | D | R | N | 1969 | 59060 |
| 0121 | L | D | O | N | 1968 | 55109 |
| 0121 | L | D | R | N | 1971 | 62768 |
| 0121 | L | D | R | N | 1971 | 65686 |
| 0121 | L | D | R | N | 1969 | 65730 |
| 0121 | L | D | E | N | 1971 | 66477 |
| 0121 | L | D | R | N | 1973 | 71672 |
| 0121 | L | D | E | N | 1973 | 72559 |
| 0122 | L | D | R | N | 1968 | 56795 |
| 0122 | L | D | E | N | 1972 | 65496 |
| 0136 | L | D | R | N | 1968 | 56795 |
| 0136 | L | D | R | N | 1970 | 59007 |
| 0136 | L | D | G | N | 1970 | 61243 |
| 0136 | L | D | R | N | 1970 | 62455 |
| 0136 | L | D | E | N | 1971 | 62472 |
| 0136 | L | D | E | N | 1972 | 65496 |
| 0136 | L | D | E | N | 1970 | 66459 |
| 0136 | L | D | E | N | 1972 | 66508 |
| 0136 | L | D | G | N | 1973 | 72511 |
| 0136 | L | D | R | N | 1971 | 75601 |
| 0136 | L | D | E | N | 1971 | 75602 |
| 0140 | L | C | E | N | 1972 | 68203 |
| 0141 | D | G | S | N | 1971 | 66853 |
| 0141 | L | D | R | N | 1968 | 56795 |
| 0141 | L | D | E | N | 1972 | 65496 |
| 0141 | L | G | S | N | 1971 | 66853 |
| 0141 | L | D | E | N | 1973 | 69219 |
| 0141 | L | D | J | N | 1973 | 71050 |
| 0143 | L | D | G | N | 1970 | 61243 |
| 0144 | L | D | G | N | 1970 | 61243 |
| 0144 | L | G | G | N | 1971 | 62442 |
| 0144 | L | D | E | N | 1973 | 72975 |
| 0146 | L | D | E | N | 1971 | 64305 |
| 0146 | L | D | E | N | 1972 | 69723 |
| 0147 | L | D | R | N | 1970 | 62455 |
| 0147 | L | D | E | N | 1972 | 69723 |
| 0152 | G | C | R | N | 1971 | 65016 |
| 0152 | G | C | E | N | 1973 | 70014 |
| 0154 | L | D | E | N | 1974 | 74255 |
| 0154 | L | G | E | N | 1973 | 74787 |
| 0155 | L | D | R | N | 1971 | 67734 |
| 0155 | L | D | R | N | 1972 | 68213 |
| 0155 | L | D | E | N | 1972 | 68214 |
| 0155 | L | D | R | N | 1971 | 75489 |
| 0155 | L | D | E | N | 1971 | 75490 |
| 0155 | L | D | R | N | 1973 | 75664 |
| 0155 | L | D | E | N | 1973 | 75665 |
| 0157 | L | D | O | N | 1967 | 45354 |
| 0157 | L | G | G | N | 1971 | 62442 |
| 0157 | L | D | R | N | 1971 | 75489 |
| 0157 | L | D | E | N | 1971 | 75490 |
| 0157 | L | D | R | N | 1973 | 75541 |
| 0157 | L | D | E | N | 1973 | 75542 |
| 0181 | L | D | R | N | 1972 | 68167 |
| 0181 | L | D | E | N | 1972 | 68168 |
| 0182 | L | D | E | N | 1972 | 67640 |
| 0182 | L | D | R | N | 1972 | 68169 |
| 0182 | L | D | E | N | 1972 | 68170 |
| 0182 | L | D | R | N | 1972 | 68213 |

| Substance Number | Phys. State | Subject | Language | Temperature | Year | TPRC Number |
|---|---|---|---|---|---|---|
| 515-0182 | L | D | E | N | 1972 | 68214 |
| 0182 | L | D | G | N | 1973 | 72511 |
| 0182 | L | D | R | N | 1973 | 75555 |
| 0182 | L | D | E | N | 1973 | 75556 |
| 0182 | L | D | R | N | 1973 | 75664 |
| 0182 | L | D | E | N | 1973 | 75665 |
| 0182 | L | C | R | N | 1974 | 75678 |
| 0182 | L | C | E | N | 1974 | 75679 |
| 0183 | L | G | G | N | 1971 | 62442 |
| 0183 | L | D | R | N | 1971 | 64892 |
| 0183 | L | D | R | N | 1972 | 68169 |
| 0183 | L | D | E | N | 1972 | 68170 |
| 0183 | L | D | E | N | 1971 | 69785 |
| 0183 | L | D | R | N | 1973 | 75664 |
| 0183 | L | D | E | N | 1973 | 75665 |
| 0183 | L | C | R | N | 1974 | 75678 |
| 0183 | L | C | E | N | 1974 | 75679 |
| 0183 | L | D | E | N | 1974 | 75867 |
| 0189 | L | D | E | N | 1971 | 63114 |
| 0189 | L | M | G | N | 1972 | 68390 |
| 0189 | L | D | E | N | 1973 | 72221 |
| 0189 | L | D | G | N | 1973 | 74497 |
| 0189 | L | G | R | N | 1974 | 75203 |
| 0189 | L | G | E | N | 1974 | 76465 |
| 0193 | L | D | E | N | 1971 | 62467 |
| 0194 | L | G | P | N | 1964 | 38366 |
| 0194 | L | D | R | N | 1969 | 65730 |
| 0194 | L | D | O | N | 1973 | 70048 |
| 0194 | L | D | R | N | 1972 | 71590 |
| 0200 | L | D | R | N | 1968 | 55473 |
| 0207 | L | D | E | N | 1973 | 69219 |
| 0213 | L | M | E | N | 1971 | 63921 |
| 0213 | L | D | E | N | 1972 | 64802 |
| 0213 | L | D | R | N | 1970 | 67941 |
| 0213 | L | D | E | N | 1972 | 70107 |
| 0213 | L | G | R | N | 1973 | 72537 |
| 0213 | L | G | E | N | 1973 | 74825 |
| 0214 | L | D | E | N | 1973 | 69219 |
| 0220 | L | D | E | N | 1972 | 60206 |
| 0221 | L | D | R | N | 1969 | 56919 |
| 0221 | L | D | R | N | 1968 | 57029 |
| 0221 | L | M | E | N | 1971 | 63173 |
| 0221 | L | D | R | N | 1971 | 65317 |
| 0221 | L | D | E | N | 1971 | 65318 |
| 0221 | L | D | E | N | 1972 | 66503 |
| 0221 | L | D | R | N | 1970 | 67731 |
| 0221 | L | D | E | N | 1974 | 74255 |
| 0237 | L | D | E | N | 1967 | 49931 |
| 0237 | L | D | G | N | 1970 | 61243 |
| 0237 | L | D | E | N | 1971 | 61585 |
| 0237 | L | D | E | N | 1972 | 67640 |
| 0237 | L | D | E | N | 1972 | 68266 |
| 0237 | L | D | R | N | 1971 | 75601 |
| 0237 | L | D | E | N | 1971 | 75602 |
| 0258 | L | G | G | N | 1971 | 62442 |
| 0258 | L | D | R | N | 1971 | 65462 |
| 0258 | L | G | E | N | 1970 | 65885 |
| 0258 | L | D | E | N | 1972 | 69723 |
| 0258 | L | D | E | N | 1971 | 69788 |
| 0259 | L | M | E | N | 1964 | 38472 |
| 0259 | L | D | E | N | 1972 | 64799 |
| 0259 | L | D | R | N | 1973 | 70583 |
| 0259 | L | D | E | N | 1973 | 72340 |
| 0259 | L | T | R | N | 1972 | 75525 |
| 0259 | L | T | E | N | 1972 | 75526 |
| 0259 | L | D | R | N | 1973 | 75573 |
| 0264 | L | D | E | N | 1972 | 60206 |
| 0280 | L | D | E | N | 1972 | 69723 |
| 0280 | L | D | E | N | 1973 | 72975 |
| 0281 | L | D | E | N | 1972 | 69723 |
| 0288 | L | D | R | N | 1973 | 71672 |
| 0288 | L | D | E | N | 1973 | 72559 |
| 0290 | L | M | R | N | 1971 | 62128 |
| 0290 | L | C | R | N | 1971 | 62600 |
| 0290 | L | M | E | N | 1971 | 66472 |
| 0290 | L | C | E | N | 1971 | 66475 |
| 0290 | L | T | R | N | 1972 | 68030 |
| 0290 | L | T | R | N | 1972 | 68031 |
| 0290 | L | C | E | N | 1972 | 68551 |
| 0290 | L | T | R | N | 1972 | 69270 |
| 0290 | L | T | E | N | 1972 | 70603 |
| 0290 | L | T | E | N | 1972 | 70604 |
| 0290 | L | T | R | N | 1973 | 72530 |
| 0290 | L | D | E | N | 1974 | 74351 |
| 0290 | L | G | E | N | 1973 | 74787 |
| 0290 | L | T | E | N | 1973 | 74817 |
| 0290 | L | T | R | N | 1972 | 75523 |
| 0290 | L | T | E | N | 1972 | 75524 |
| 0290 | L | T | R | N | 1972 | 75525 |
| 0290 | L | T | E | N | 1972 | 75526 |
| 0290 | L | T | R | N | 1972 | 75527 |
| 0290 | L | T | E | N | 1972 | 75528 |
| 0305 | L | D | E | N | 1969 | 62583 |
| 0310 | L | D | E | N | 1969 | 58796 |
| 0311 | L | D | R | N | 1963 | 37245 |
| 0311 | L | D | R | N | 1969 | 65730 |
| 0319 | L | D | R | N | 1972 | 68663 |
| 0319 | L | D | E | N | 1972 | 68664 |
| 0344 | L | D | R | N | 1971 | 75601 |
| 0344 | L | D | E | N | 1971 | 75602 |
| 0347 | L | G | G | N | 1971 | 62442 |

**Phys. State:** **D.** Doped; **E.** Expanded; **F.** Fibrous or Whisker; **G.** Gas; **L.** Liquid; **M.** Multiphase; **P.** Powder; **S.** Solid
**Subject:** **C.** Calculated Data + Theory; **D.** Data; **E.** Experiment; **G.** General (Data + Theory + Experiment); **M.** Measured Data + Experiment; **S.** Survey (Review, Compendium, etc.); **T.** Theory
**Language:** **C.** Czech; **D.** Dutch; **E.** English; **F.** French; **G.** German; **I.** Italian; **J.** Japanese; **O.** Other Languages; **P.** Polish; **R.** Russian; **S.** Spanish
**Temperature:** **F.** Full Range (Low + Normal + High); **L.** Low (0 to 75K) + Overlap into Normal; **N.** Normal (75 to 1273K); **H.** High (above 1273K) + Overlap into Normal; **U.** Undefined

| Substance Number | Phys. State | Subject | Language | Temperature | Year | TPRC Number |
|---|---|---|---|---|---|---|
| 515-0347 | L | M | E | N | 1971 | 63173 |
| 0357 | L | G | E | N | 1970 | 65885 |
| 0357 | L | D | E | N | 1972 | 66511 |
| 0357 | L | D | E | N | 1972 | 69723 |
| 0357 | L | D | E | N | 1973 | 72975 |
| 0376 | L | D | E | N | 1972 | 69723 |
| 0377 | L | D | E | N | 1971 | 64305 |
| 0377 | L | D | E | N | 1972 | 69723 |
| 0387 | L | C | R | N | 1967 | 53326 |
| 0387 | L | C | R | N | 1968 | 53467 |
| 0387 | L | D | R | N | 1968 | 57029 |
| 0387 | L | C | E | N | 1972 | 66038 |
| 0387 | L | D | O | N | 1973 | 70048 |
| 0387 | L | D | R | N | 1972 | 71590 |
| 0387 | L | D | R | N | 1973 | 73022 |
| 0387 | L | G | R | N | 1973 | 73061 |
| 0387 | L | D | F | N | 1974 | 74976 |
| 0387 | L | G | E | N | 1973 | 75844 |
| 0389 | L | D | R | N | 1970 | 67941 |
| 0389 | M | D | J | N | 1969 | 64774 |
| 0390 | L | G | E | N | 1971 | 66136 |
| 0390 | L | D | R | N | 1970 | 67941 |
| 0390 | L | D | R | N | 1970 | 71214 |
| 0390 | M | D | J | N | 1969 | 64774 |
| 0390 | M | M | E | N | 1972 | 71210 |
| 0393 | L | D | R | N | 1968 | 53364 |
| 0393 | L | D | E | N | 1972 | 66039 |
| 0393 | L | D | E | N | 1973 | 72221 |
| 0396 | L | D | R | N | 1972 | 69832 |
| 0396 | L | D | E | N | 1972 | 69833 |
| 0412 | L | M | R | N | 1970 | 64359 |
| 0433 | L | D | E | N | 1971 | 69851 |
| 0447 | L | D | R | N | 1967 | 50605 |
| 0447 | L | D | R | N | 1971 | 62475 |
| 0447 | L | D | E | N | 1971 | 62476 |
| 0447 | L | C | R | N | 1971 | 62600 |
| 0447 | L | D | E | N | 1971 | 63245 |
| 0447 | L | G | R | N | 1970 | 63262 |
| 0447 | L | G | E | N | 1970 | 66470 |
| 0447 | L | C | E | N | 1971 | 66475 |
| 0447 | L | D | R | N | 1973 | 71577 |
| 0447 | L | D | R | N | 1974 | 75672 |
| 0447 | L | D | E | N | 1974 | 75673 |
| 0447 | L | C | R | N | 1974 | 75678 |
| 0447 | L | C | E | N | 1974 | 75679 |
| 0448 | L | D | R | N | 1973 | 75541 |
| 0448 | L | D | E | N | 1973 | 75542 |
| 0452 | L | D | R | N | 1973 | 75833 |
| 0452 | L | D | E | N | 1973 | 75834 |
| 0453 | L | D | E | N | 1971 | 64864 |
| 0453 | L | D | O | N | 1971 | 66967 |
| 0453 | L | D | R | N | 1971 | 67734 |
| 0453 | L | D | R | N | 1973 | 75833 |
| 0453 | L | D | E | N | 1973 | 75834 |
| 0460 | L | D | R | N | 1967 | 53325 |
| 0460 | L | C | R | N | 1968 | 53467 |
| 0460 | L | D | R | N | 1969 | 56919 |
| 0460 | L | D | G | N | 1970 | 61937 |
| 0460 | L | D | E | N | 1972 | 65514 |
| 0460 | L | D | E | N | 1972 | 66503 |
| 0460 | L | D | S | N | 1973 | 71315 |
| 0460 | L | D | R | N | 1972 | 71590 |
| 0460 | L | D | R | N | 1973 | 73022 |
| 0460 | L | D | F | N | 1974 | 74976 |
| 0460 | L | D | R | N | 1974 | 75674 |
| 0460 | L | D | E | N | 1974 | 75675 |
| 0475 | L | C | R | N | 1968 | 53467 |
| 0477 | L | D | R | N | 1970 | 68958 |
| 0477 | L | D | R | N | 1970 | 68959 |
| 0478 | L | C | E | N | 1972 | 68551 |
| 0479 | L | D | C | N | 1972 | 68960 |
| 0480 | L | D | R | N | 1971 | 68954 |
| 0481 | L | D | G | N | 1971 | 68964 |
| 0483 | L | D | C | N | 1972 | 68960 |
| 0485 | L | D | R | N | 1969 | 68957 |
| 0486 | L | D | R | N | 1969 | 68957 |
| 0487 | M | D | J | N | 1969 | 64774 |
| 0488 | M | D | J | N | 1969 | 64774 |
| 0489 | M | D | J | N | 1969 | 64774 |
| 0489 | M | M | E | N | 1972 | 71210 |
| 0490 | L | D | R | N | 1969 | 68957 |
| 0491 | L | M | E | N | 1971 | 66414 |
| 0493 | L | M | E | N | 1972 | 67500 |
| 0493 | L | M | E | N | 1972 | 69342 |
| 0493 | L | M | E | N | 1973 | 72969 |
| 0493 | L | D | E | N | 1973 | 74678 |
| 0494 | L | G | R | N | 1969 | 56930 |
| 0495 | L | D | R | N | 1970 | 62769 |
| 0495 | L | D | E | N | 1970 | 66469 |
| 0497 | L | G | R | N | 1969 | 56930 |
| 0498 | L | D | E | N | 1972 | 69723 |
| 0500 | L | D | O | N | 1967 | 45354 |
| 0500 | L | G | E | N | 1970 | 65885 |
| 0502 | L | D | E | N | 1972 | 69723 |
| 0503 | L | G | G | N | 1971 | 62442 |
| 0503 | L | D | F | N | 1974 | 74976 |
| 0504 | L | D | R | N | 1973 | 75830 |
| 0504 | L | D | E | N | 1973 | 75831 |
| 0505 | L | D | G | N | 1970 | 61243 |
| 0505 | L | G | G | N | 1971 | 62442 |
| 0506 | L | D | R | N | 1973 | 70859 |
| 515-0506 | L | D | E | N | 1973 | 71138 |
| 0507 | L | D | O | N | 1967 | 45354 |
| 0507 | L | D | R | N | 1969 | 56940 |
| 0507 | L | D | R | N | 1971 | 62744 |
| 0507 | L | D | E | N | 1971 | 62745 |
| 0507 | L | D | R | N | 1971 | 64729 |
| 0507 | L | D | E | N | 1971 | 64730 |
| 0507 | L | D | R | N | 1972 | 69022 |
| 0507 | L | D | E | N | 1972 | 69023 |
| 0507 | L | D | E | N | 1973 | 69219 |
| 0507 | L | D | R | N | 1970 | 75461 |
| 0507 | L | D | E | N | 1970 | 75462 |
| 0507 | L | D | E | N | 1974 | 75671 |
| 0508 | L | D | R | N | 1971 | 66117 |
| 0508 | L | D | E | N | 1971 | 66118 |
| 0509 | L | D | R | N | 1973 | 70859 |
| 0509 | L | D | E | N | 1973 | 71138 |
| 0510 | L | G | G | N | 1971 | 62442 |
| 0510 | L | D | R | N | 1971 | 64913 |
| 0510 | L | D | E | N | 1971 | 64914 |
| 0511 | L | G | G | N | 1971 | 62442 |
| 0511 | L | D | E | N | 1972 | 67824 |
| 0512 | L | D | R | N | 1972 | 68165 |
| 0512 | L | D | E | N | 1972 | 68166 |
| 0513 | L | D | R | N | 1972 | 67140 |
| 0513 | L | D | E | N | 1972 | 69795 |
| 0515 | L | G | G | N | 1971 | 62442 |
| 0516 | L | D | R | N | 1972 | 69760 |
| 0516 | L | D | E | N | 1972 | 70681 |
| 0518 | L | D | E | N | 1970 | 60208 |
| 0518 | L | D | E | N | 1973 | 69219 |
| 0519 | L | D | E | N | 1971 | 63834 |
| 0519 | L | D | E | N | 1973 | 69219 |
| 0520 | L | D | E | N | 1973 | 69219 |
| 0521 | L | D | E | N | 1971 | 65423 |
| 0523 | L | D | E | N | 1973 | 72221 |
| 0524 | L | D | E | N | 1973 | 69219 |
| 0525 | L | D | E | N | 1971 | 63834 |
| 0526 | L | D | E | N | 1973 | 69219 |
| 0527 | L | D | E | N | 1973 | 69219 |
| 0529 | L | D | F | N | 1972 | 62151 |
| 0530 | L | D | E | N | 1973 | 69219 |
| 0532 | D | D | E | L | 1972 | 66431 |
| 0532 | D | M | E | N | 1974 | 73285 |
| 0532 | L | D | E | N | 1971 | 63113 |
| 0532 | L | D | E | L | 1972 | 66431 |
| 0532 | L | M | E | N | 1974 | 73285 |
| 0532 | L | C | E | U | 1973 | 74736 |
| 0533 | L | D | E | N | 1973 | 69219 |
| 0534 | L | D | E | N | 1971 | 63834 |
| 0535 | L | D | E | N | 1973 | 69219 |
| 0536 | L | D | E | N | 1973 | 69219 |
| 0537 | L | D | R | N | 1967 | 49914 |
| 0537 | L | D | E | N | 1972 | 66199 |
| 0538 | L | D | O | N | 1971 | 65003 |
| 0538 | L | D | E | N | 1972 | 69723 |
| 0540 | L | D | R | N | 1968 | 53364 |
| 0540 | L | D | E | N | 1972 | 66039 |
| 0541 | L | D | E | N | 1973 | 69219 |
| 0542 | L | D | E | N | 1973 | 69219 |
| 0543 | L | D | E | N | 1973 | 69219 |
| 0544 | L | D | E | N | 1971 | 63836 |
| 0544 | L | D | E | N | 1973 | 69219 |
| 0545 | L | D | R | N | 1970 | 60759 |
| 0545 | L | D | E | N | 1970 | 68707 |
| 0545 | L | D | G | N | 1972 | 68955 |
| 0546 | L | D | E | N | 1971 | 64305 |
| 0547 | L | D | O | N | 1971 | 64146 |
| 0548 | L | D | E | N | 1971 | 64305 |
| 0549 | L | D | E | N | 1973 | 69219 |
| 0550 | L | D | R | N | 1971 | 62129 |
| 0550 | L | D | E | N | 1971 | 66474 |
| 0551 | L | D | R | N | 1971 | 65317 |
| 0551 | L | D | E | N | 1971 | 65318 |
| 0551 | L | D | R | N | 1972 | 67139 |
| 0551 | L | D | E | N | 1972 | 69794 |
| 0552 | L | D | F | N | 1968 | 50472 |
| 0552 | L | C | E | N | 1972 | 68203 |
| 0552 | L | D | G | N | 1972 | 71225 |
| 0553 | L | D | F | N | 1968 | 50472 |
| 0553 | L | C | E | H | 1968 | 55140 |
| 0553 | L | D | G | N | 1972 | 71225 |
| 0554 | L | D | G | N | 1972 | 71225 |
| 0555 | L | D | E | N | 1971 | 63834 |
| 0556 | L | D | E | N | 1971 | 64305 |
| 0557 | L | D | G | N | 1972 | 68955 |
| 0557 | L | D | G | N | 1972 | 71225 |
| 0558 | L | D | E | N | 1971 | 64305 |
| 0559 | L | D | F | N | 1973 | 70408 |
| 0560 | L | D | C | N | 1970 | 62412 |
| 0560 | L | D | E | N | 1973 | 71785 |
| 0561 | L | D | E | N | 1970 | 64935 |
| 0562 | L | D | E | N | 1971 | 63970 |
| 0562 | L | D | E | N | 1971 | 65230 |
| 0562 | L | D | E | N | 1972 | 66538 |
| 0562 | L | D | E | N | 1972 | 66539 |
| 0562 | L | C | E | U | 1973 | 74736 |
| 0564 | L | D | E | N | 1971 | 64305 |
| 0566 | L | D | E | N | 1971 | 64305 |
| 0567 | L | D | E | N | 1971 | 64305 |
| 0568 | L | D | E | N | 1971 | 64305 |
| 515-0569 | L | D | E | N | 1971 | 64305 |
| 0570 | L | D | E | N | 1971 | 64305 |
| 0571 | L | G | E | N | 1970 | 65885 |
| 0571 | L | D | E | N | 1972 | 66511 |
| 0571 | L | D | E | N | 1972 | 69723 |
| 0571 | L | D | E | N | 1973 | 72975 |
| 0572 | L | D | O | N | 1971 | 65003 |
| 0572 | L | D | E | N | 1972 | 69723 |
| 0573 | L | D | E | N | 1971 | 64305 |
| 0574 | L | D | R | N | 1972 | 69020 |
| 0574 | L | D | E | N | 1972 | 69021 |
| 0575 | L | D | R | N | 1972 | 69020 |
| 0575 | L | D | E | N | 1972 | 69021 |
| 0576 | L | D | R | N | 1972 | 69020 |
| 0576 | L | D | E | N | 1972 | 69021 |
| 0577 | L | D | E | N | 1972 | 70476 |
| 0578 | L | M | E | N | 1963 | 36908 |
| 0578 | L | M | E | N | 1965 | 41259 |
| 0578 | L | C | E | H | 1968 | 55140 |
| 0578 | L | M | R | N | 1970 | 63600 |
| 0578 | L | C | F | N | 1971 | 65091 |
| 0578 | L | D | R | N | 1971 | 65734 |
| 0578 | L | C | E | N | 1972 | 68203 |
| 0578 | L | D | E | N | 1972 | 68223 |
| 0578 | L | D | E | N | 1972 | 68939 |
| 0578 | L | D | E | N | 1973 | 69219 |
| 0578 | L | D | G | N | 1972 | 71225 |
| 0578 | L | D | G | N | 1973 | 73987 |
| 0578 | L | S | E | N | 1974 | 74962 |
| 0579 | L | D | E | N | 1971 | 61577 |
| 0579 | L | D | E | N | 1971 | 61579 |
| 0580 | L | D | E | N | 1972 | 66776 |
| 0581 | L | G | E | N | 1970 | 65885 |
| 0581 | L | D | E | N | 1972 | 66511 |
| 0581 | L | D | E | N | 1972 | 69723 |
| 0581 | L | D | E | N | 1973 | 72975 |
| 0582 | L | D | O | N | 1971 | 65003 |
| 0582 | L | D | E | N | 1972 | 69723 |
| 0584 | L | D | E | N | 1972 | 66776 |
| 0585 | L | D | O | N | 1971 | 65003 |
| 0585 | L | D | E | N | 1972 | 69723 |
| 0586 | L | D | R | N | 1971 | 63034 |
| 0586 | L | D | E | N | 1971 | 66473 |
| 0587 | L | D | E | N | 1972 | 66776 |
| 0588 | L | D | E | N | 1970 | 64935 |
| 0590 | L | D | E | N | 1972 | 66776 |
| 0591 | L | D | E | N | 1972 | 66776 |
| 0593 | L | C | E | N | 1971 | 61312 |
| 0595 | L | D | R | N | 1969 | 56919 |
| 0595 | L | D | E | N | 1972 | 66503 |
| 0595 | L | D | R | N | 1971 | 67733 |
| 0595 | L | D | R | N | 1970 | 76333 |
| 0595 | L | D | E | N | 1973 | 76334 |
| 0596 | L | D | R | N | 1970 | 62769 |
| 0596 | L | D | E | N | 1970 | 66469 |
| 0597 | L | G | R | N | 1969 | 56930 |
| 0598 | L | D | R | N | 1970 | 62769 |
| 0598 | L | D | E | N | 1970 | 66469 |
| 0600 | L | G | G | N | 1971 | 62442 |
| 0600 | L | D | R | N | 1973 | 70859 |
| 0600 | L | D | E | N | 1973 | 71138 |
| 0603 | L | M | E | N | 1964 | 38472 |
| 0603 | L | D | E | N | 1972 | 69723 |
| 0603 | L | D | E | N | 1973 | 72975 |
| 0604 | L | D | R | N | 1971 | 63382 |
| 0604 | L | D | E | N | 1971 | 63383 |
| 0604 | L | G | R | N | 1974 | 74999 |
| 0604 | L | G | E | N | 1974 | 76460 |
| 0606 | L | D | R | N | 1971 | 66119 |
| 0606 | L | D | E | N | 1971 | 66120 |
| 0606 | L | D | R | N | 1972 | 71590 |
| 0608 | L | D | R | N | 1968 | 50466 |
| 0608 | L | D | R | N | 1969 | 55597 |
| 0608 | L | D | E | N | 1972 | 65502 |
| 0608 | L | D | R | N | 1972 | 69996 |
| 0608 | L | D | E | N | 1972 | 70683 |
| 0608 | L | D | R | N | 1972 | 71119 |
| 0609 | L | G | R | N | 1969 | 56930 |
| 0610 | L | G | R | N | 1969 | 56930 |
| 0610 | L | D | R | N | 1968 | 57029 |
| 0610 | L | G | G | N | 1971 | 62442 |
| 0610 | L | D | R | N | 1971 | 67733 |
| 0611 | L | D | R | N | 1971 | 63382 |
| 0611 | L | D | E | N | 1971 | 63383 |
| 0611 | L | D | R | N | 1972 | 71119 |
| 0611 | L | D | R | N | 1972 | 71590 |
| 0613 | L | D | R | N | 1970 | 67731 |
| 0613 | L | D | R | N | 1972 | 69022 |
| 0613 | L | D | E | N | 1972 | 69023 |
| 0614 | L | D | R | N | 1971 | 66119 |
| 0614 | L | D | E | N | 1971 | 66120 |
| 0614 | L | D | R | N | 1972 | 69832 |
| 0614 | L | D | E | N | 1972 | 69833 |
| 0615 | L | D | G | N | 1971 | 62771 |
| 0616 | L | G | R | N | 1969 | 56930 |
| 0617 | L | D | G | N | 1971 | 62771 |
| 0619 | L | D | F | N | 1974 | 74976 |
| 0630 | L | D | R | N | 1971 | 62744 |
| 0630 | L | D | E | N | 1971 | 62745 |
| 0631 | L | D | R | N | 1971 | 64729 |
| 0631 | L | D | E | N | 1971 | 64730 |

**Phys. State:** **D.** Doped; **E.** Expanded; **F.** Fibrous or Whisker; **G.** Gas; **L.** Liquid; **M.** Multiphase; **P.** Powder; **S.** Solid
**Subject:** **C.** Calculated Data + Theory; **D.** Data; **E.** Experiment; **G.** General (Data + Theory + Experiment); **M.** Measured Data + Experiment; **S.** Survey (Review, Compendium, etc.); **T.** Theory
**Language:** **C.** Czech; **D.** Dutch; **E.** English; **F.** French; **G.** German; **I.** Italian; **J.** Japanese; **O.** Other Languages; **P.** Polish; **R.** Russian; **S.** Spanish
**Temperature:** **F.** Full Range (Low + Normal + High); **L.** Low (0 to 75K) + Overlap into Normal; **N.** Normal (75 to 1273K); **H.** High (above 1273K) + Overlap into Normal; **U.** Undefined

| Substance Number | Phys. State | Subject | Language | Temperature | Year | TPRC Number |
|---|---|---|---|---|---|---|
| 515-0631 | L | D | R | N | 1971 | 67732 |
| 0632 | L | M | E | N | 1964 | 38472 |
| 0632 | L | D | E | N | 1972 | 69723 |
| 0632 | L | D | R | N | 1972 | 75609 |
| 0632 | L | D | E | N | 1972 | 75610 |
| 0633 | L | D | R | N | 1968 | 57029 |
| 0633 | L | D | R | N | 1970 | 62769 |
| 0633 | L | D | E | N | 1970 | 66469 |
| 0635 | L | D | E | N | 1971 | 64864 |
| 0635 | L | D | R | N | 1970 | 75461 |
| 0635 | L | D | E | N | 1970 | 75462 |
| 0636 | L | D | E | N | 1970 | 60208 |
| 0636 | L | D | E | N | 1973 | 74024 |
| 0636 | L | D | E | N | 1974 | 75671 |
| 0640 | L | M | E | N | 1964 | 38472 |
| 0640 | L | D | E | N | 1972 | 69723 |
| 0642 | L | D | R | N | 1970 | 58553 |
| 0642 | L | D | R | N | 1970 | 59317 |
| 0642 | L | M | R | N | 1971 | 62128 |
| 0642 | L | D | R | N | 1969 | 65730 |
| 0642 | L | D | F | N | 1972 | 66356 |
| 0642 | L | D | E | N | 1970 | 66457 |
| 0642 | L | D | E | N | 1970 | 66458 |
| 0642 | L | M | E | N | 1971 | 66472 |
| 0642 | L | D | R | N | 1972 | 68151 |
| 0642 | L | D | E | N | 1972 | 68152 |
| 0642 | L | D | R | N | 1973 | 71669 |
| 0642 | L | D | E | N | 1973 | 72555 |
| 0642 | L | C | R | N | 1974 | 75678 |
| 0642 | L | C | E | N | 1974 | 75679 |
| 0643 | L | D | R | N | 1967 | 51061 |
| 0643 | L | D | R | N | 1969 | 56288 |
| 0643 | L | D | R | N | 1969 | 65730 |
| 0643 | L | D | R | N | 1972 | 68151 |
| 0643 | L | D | E | N | 1972 | 68152 |
| 0643 | L | D | E | N | 1971 | 69851 |
| 0643 | L | D | E | N | 1972 | 70492 |
| 0643 | L | D | R | N | 1973 | 71669 |
| 0643 | L | D | E | N | 1973 | 72555 |
| 0643 | L | C | R | N | 1974 | 75678 |
| 0643 | L | C | E | N | 1974 | 75679 |
| 0644 | L | D | O | N | 1968 | 55109 |
| 0644 | L | D | R | N | 1971 | 62475 |
| 0644 | L | D | E | N | 1971 | 62476 |
| 0644 | L | D | R | N | 1971 | 65294 |
| 0644 | L | D | R | N | 1969 | 65730 |
| 0644 | L | D | E | N | 1971 | 66480 |
| 0644 | L | D | R | N | 1971 | 75489 |
| 0644 | L | D | E | N | 1971 | 75490 |
| 0644 | L | D | R | N | 1972 | 75621 |
| 0644 | L | D | E | N | 1972 | 75622 |
| 0644 | L | D | R | N | 1973 | 75646 |
| 0644 | L | D | E | N | 1973 | 75647 |
| 0644 | L | D | R | N | 1974 | 75924 |
| 0644 | L | D | E | N | 1974 | 76441 |
| 0646 | L | D | R | N | 1971 | 62543 |
| 0646 | L | D | E | N | 1971 | 66471 |
| 0647 | L | D | O | N | 1967 | 45354 |
| 0647 | L | D | R | N | 1972 | 70102 |
| 0647 | L | D | E | N | 1972 | 71433 |
| 0647 | L | D | R | N | 1973 | 75573 |
| 0650 | L | D | R | N | 1971 | 65462 |
| 0650 | L | D | E | N | 1971 | 69788 |
| 0651 | L | D | E | N | 1972 | 69723 |
| 0652 | L | D | E | N | 1972 | 64799 |
| 0653 | L | D | R | N | 1970 | 62769 |
| 0653 | L | D | E | N | 1970 | 66469 |
| 0654 | L | D | R | N | 1971 | 64913 |
| 0654 | L | D | E | N | 1971 | 64914 |
| 0654 | L | D | R | N | 1971 | 64917 |
| 0654 | L | D | E | N | 1971 | 64918 |
| 0655 | L | D | R | N | 1967 | 48539 |
| 0656 | L | D | R | N | 1971 | 65697 |
| 0656 | L | D | E | N | 1971 | 69339 |
| 0657 | L | D | R | N | 1971 | 66303 |
| 0657 | L | D | E | N | 1971 | 69790 |
| 0658 | L | D | R | N | 1972 | 69830 |
| 0658 | L | D | E | N | 1972 | 69831 |
| 0659 | L | D | R | N | 1969 | 53690 |
| 0659 | L | D | R | N | 1970 | 62104 |
| 0659 | L | D | R | N | 1970 | 62769 |
| 0659 | L | D | E | N | 1970 | 66469 |
| 0659 | L | D | E | N | 1969 | 67237 |
| 0660 | L | G | G | N | 1971 | 62442 |
| 0660 | L | D | R | N | 1974 | 75672 |
| 0660 | L | D | E | N | 1974 | 75673 |
| 0660 | L | C | R | N | 1974 | 75678 |
| 0660 | L | C | E | N | 1974 | 75679 |
| 0661 | L | D | R | N | 1970 | 62769 |
| 0661 | L | D | E | N | 1970 | 66469 |
| 0662 | L | D | E | N | 1971 | 64864 |
| 0662 | L | D | R | N | 1971 | 67732 |
| 0662 | L | D | R | N | 1972 | 69830 |
| 0662 | L | D | E | N | 1972 | 69831 |
| 0662 | L | G | R | N | 1973 | 73061 |
| 0662 | L | G | E | N | 1973 | 75844 |
| 0663 | L | C | R | N | 1967 | 53326 |
| 0663 | L | D | G | N | 1969 | 56950 |
| 0663 | L | D | G | N | 1970 | 61243 |
| 0663 | L | C | E | N | 1972 | 66038 |
| 0663 | L | G | R | N | 1973 | 73061 |

| Substance Number | Phys. State | Subject | Language | Temperature | Year | TPRC Number |
|---|---|---|---|---|---|---|
| 515-0663 | L | G | E | N | 1973 | 75844 |
| 0664 | L | D | E | N | 1971 | 63967 |
| 0664 | L | D | R | N | 1972 | 71119 |
| 0664 | L | D | R | N | 1971 | 75599 |
| 0664 | L | D | E | N | 1971 | 75600 |
| 0664 | L | D | R | N | 1971 | 75603 |
| 0664 | L | D | E | N | 1971 | 75604 |
| 0665 | L | D | R | N | 1967 | 50605 |
| 0665 | L | D | E | N | 1971 | 63245 |
| 0665 | L | D | R | N | 1972 | 71119 |
| 0665 | L | D | R | N | 1973 | 75557 |
| 0665 | L | D | E | N | 1973 | 75558 |
| 0666 | L | D | R | N | 1970 | 66274 |
| 0666 | L | D | R | N | 1972 | 68165 |
| 0666 | L | D | E | N | 1972 | 68166 |
| 0667 | L | D | R | N | 1969 | 53690 |
| 0667 | L | D | R | N | 1970 | 60759 |
| 0667 | L | D | R | N | 1970 | 62769 |
| 0667 | L | D | E | N | 1970 | 66469 |
| 0667 | L | D | E | N | 1969 | 67237 |
| 0667 | L | D | E | N | 1970 | 68707 |
| 0668 | L | D | R | N | 1970 | 62769 |
| 0668 | L | D | E | N | 1970 | 66469 |
| 0669 | L | D | R | N | 1970 | 62769 |
| 0669 | L | D | E | N | 1970 | 66469 |
| 0670 | L | D | R | N | 1972 | 68167 |
| 0670 | L | D | E | N | 1972 | 68168 |
| 0670 | L | D | R | N | 1972 | 68169 |
| 0670 | L | D | E | N | 1972 | 68170 |
| 0670 | L | D | R | N | 1973 | 75826 |
| 0670 | L | D | E | N | 1973 | 75827 |
| 0670 | L | D | R | N | 1973 | 75833 |
| 0670 | L | D | E | N | 1973 | 75834 |
| 0671 | L | D | R | N | 1970 | 62769 |
| 0671 | L | D | E | N | 1970 | 66469 |
| 0672 | L | D | R | N | 1969 | 64328 |
| 0672 | L | D | R | N | 1971 | 66117 |
| 0672 | L | D | E | N | 1971 | 66118 |
| 0673 | L | D | G | N | 1971 | 62771 |
| 0680 | L | D | R | N | 1968 | 50725 |
| 0681 | L | G | R | N | 1970 | 63262 |
| 0681 | L | G | E | N | 1970 | 66470 |
| 0681 | L | D | R | N | 1972 | 69020 |
| 0681 | L | D | E | N | 1972 | 69021 |
| 0681 | L | D | R | N | 1973 | 71995 |
| 0681 | L | D | R | N | 1974 | 75672 |
| 0681 | L | D | E | N | 1974 | 75673 |
| 0681 | L | C | R | N | 1974 | 75678 |
| 0681 | L | C | E | N | 1974 | 75679 |
| 0681 | L | D | R | N | 1974 | 75867 |
| 0682 | L | D | R | N | 1971 | 65293 |
| 0682 | L | D | E | N | 1972 | 66777 |
| 0682 | L | D | E | N | 1971 | 69789 |
| 0684 | L | D | G | N | 1971 | 62771 |
| 0686 | L | D | R | N | 1971 | 64727 |
| 0686 | L | D | E | N | 1971 | 64728 |
| 0686 | L | D | R | N | 1973 | 75559 |
| 0686 | L | D | E | N | 1973 | 75560 |
| 0691 | L | D | E | N | 1972 | 69724 |
| 0692 | L | D | E | N | 1972 | 69724 |
| 0698 | L | D | R | N | 1973 | 75652 |
| 0698 | L | D | E | N | 1973 | 75653 |
| 0709 | L | D | R | N | 1970 | 76333 |
| 0709 | L | D | E | N | 1973 | 76334 |
| 0715 | L | D | F | N | 1974 | 74976 |
| 0716 | L | D | F | N | 1974 | 74976 |
| 0717 | L | D | E | N | 1973 | 72052 |
| 0719 | L | D | E | N | 1973 | 72221 |
| 0720 | L | D | E | N | 1973 | 72221 |
| 0721 | L | D | E | N | 1973 | 72221 |
| 0721 | L | D | R | N | 1972 | 72505 |
| 0721 | L | D | E | N | 1974 | 74351 |
| 0722 | L | D | E | N | 1973 | 72221 |
| 0723 | L | D | E | N | 1973 | 72221 |
| 0732 | L | D | R | N | 1969 | 56940 |
| 0732 | L | D | R | N | 1973 | 71995 |
| 0733 | L | D | R | N | 1972 | 69078 |
| 0733 | L | G | E | N | 1973 | 72958 |
| 0734 | M | M | R | U | 1971 | 69961 |
| 0737 | L | D | R | N | 1973 | 71669 |
| 0737 | L | D | E | N | 1973 | 72555 |
| 0737 | L | C | R | N | 1974 | 75678 |
| 0737 | L | C | E | N | 1974 | 75679 |
| 0743 | L | D | R | N | 1972 | 71119 |
| 0744 | L | D | R | N | 1972 | 71119 |
| 0745 | L | D | R | N | 1972 | 71119 |
| 0746 | L | D | R | N | 1972 | 71119 |
| 0747 | L | D | R | N | 1972 | 71119 |
| 0747 | L | G | R | N | 1974 | 74999 |
| 0747 | L | G | E | N | 1974 | 76460 |
| 0748 | L | D | R | N | 1972 | 71119 |
| 0749 | L | D | F | N | 1968 | 50472 |
| 0752 | L | D | E | N | 1972 | 73919 |
| 0758 | L | D | E | N | 1973 | 72406 |
| 0759 | L | D | E | N | 1973 | 72406 |
| 0765 | L | D | O | N | 1971 | 66967 |
| 0769 | L | D | R | N | 1970 | 62104 |
| 0770 | L | D | R | N | 1973 | 71577 |
| 0771 | L | D | E | N | 1973 | 71785 |
| 0799 | L | D | S | N | 1973 | 71315 |
| 0800 | L | D | S | N | 1973 | 71315 |

| Substance Number | Phys. State | Subject | Language | Temperature | Year | TPRC Number |
|---|---|---|---|---|---|---|
| 515-0801 | L | D | S | N | 1973 | 71315 |
| 0802 | L | D | S | N | 1973 | 71315 |
| 0803 | L | D | S | N | 1973 | 71315 |
| 0813 | L | D | E | N | 1974 | 74351 |
| 0814 | L | D | E | N | 1974 | 74351 |
| 0815 | L | D | E | N | 1974 | 74351 |
| 0816 | L | D | E | N | 1974 | 74351 |
| 0817 | L | D | E | N | 1974 | 74351 |
| 0818 | L | D | E | N | 1974 | 74351 |
| 0819 | L | D | E | N | 1974 | 74351 |
| 0820 | L | D | E | N | 1974 | 74351 |
| 0821 | L | D | E | N | 1974 | 74351 |
| 0822 | L | D | E | N | 1974 | 74351 |
| 0823 | L | D | E | N | 1974 | 74351 |
| 0824 | L | D | E | N | 1974 | 74351 |
| 0825 | L | D | E | N | 1974 | 74351 |
| 0827 | L | D | R | N | 1970 | 76333 |
| 0827 | L | D | E | N | 1973 | 76334 |
| 0832 | L | D | E | N | 1973 | 72399 |
| 0854 | L | D | R | N | 1972 | 72505 |
| 0855 | L | D | R | N | 1972 | 72505 |
| 0856 | L | D | R | N | 1972 | 72505 |
| 0858 | L | D | F | N | 1974 | 74976 |
| 0859 | L | D | F | N | 1974 | 74976 |
| 0871 | L | D | E | N | 1974 | 75329 |
| 0876 | L | D | R | N | 1971 | 75489 |
| 0876 | L | D | E | N | 1971 | 75490 |
| 0880 | L | D | R | N | 1967 | 76987 |
| 0880 | L | D | E | N | 1967 | 76988 |
| 0894 | L | D | R | N | 1973 | 75557 |
| 0894 | L | D | E | N | 1973 | 75558 |
| 0895 | L | D | R | N | 1973 | 75559 |
| 0895 | L | D | E | N | 1973 | 75560 |
| 0897 | L | G | E | N | 1973 | 72958 |
| 0898 | L | D | R | N | 1973 | 75573 |
| 0906 | L | D | R | N | 1973 | 75826 |
| 0906 | L | D | E | N | 1973 | 75827 |
| 0917 | L | D | E | N | 1974 | 75204 |
| 0918 | L | D | E | N | 1974 | 75204 |
| 0926 | L | D | E | N | 1974 | 75329 |
| 0927 | L | D | E | N | 1974 | 75329 |
| 0928 | L | D | E | N | 1974 | 75329 |
| 0929 | L | D | E | N | 1974 | 75329 |
| 0930 | L | D | E | N | 1974 | 75329 |
| 0957 | L | D | R | N | 1973 | 73022 |
| 0958 | L | D | R | N | 1973 | 73022 |
| 0969 | L | D | R | N | 1973 | 75660 |
| 0969 | L | D | E | N | 1973 | 75661 |
| 0970 | L | D | R | N | 1974 | 75674 |
| 0970 | L | D | E | N | 1974 | 75675 |
| 0971 | L | D | E | N | 1973 | 75781 |
| 0972 | L | D | E | N | 1973 | 75781 |
| 0973 | L | D | E | N | 1973 | 75781 |
| 0974 | L | D | E | N | 1973 | 75781 |
| 1019 | L | G | E | N | 1973 | 72958 |
| 1020 | L | G | E | N | 1973 | 72958 |
| 1021 | L | M | R | N | 1973 | 74765 |
| 1022 | L | M | R | N | 1973 | 74765 |
| 1023 | L | M | R | N | 1973 | 74765 |
| 1024 | L | M | R | N | 1973 | 74765 |
| 517-0001 | G | D | G | N | 1972 | 68888 |
| 0003 | G | D | G | N | 1972 | 68888 |
| 0005 | L | D | R | N | 1973 | 75575 |
| 0005 | L | D | E | N | 1973 | 75576 |
| 0006 | L | D | R | N | 1970 | 62455 |
| 0011 | L | D | R | N | 1973 | 75652 |
| 0011 | L | D | E | N | 1973 | 75653 |
| 0011 | L | D | R | N | 1974 | 75674 |
| 0011 | L | D | E | N | 1974 | 75675 |
| 0020 | G | C | R | N | 1970 | 73694 |
| 0020 | G | C | E | N | 1970 | 73695 |
| 0023 | L | D | E | N | 1972 | 67825 |
| 0028 | L | C | R | N | 1969 | 56638 |
| 0028 | L | D | G | N | 1969 | 56950 |
| 0028 | L | D | E | N | 1969 | 58796 |
| 0028 | L | D | G | N | 1970 | 61243 |
| 0028 | L | D | G | N | 1970 | 61937 |
| 0028 | L | M | E | N | 1971 | 63173 |
| 0028 | L | C | E | N | 1972 | 65495 |
| 0029 | L | D | R | N | 1971 | 66115 |
| 0029 | L | D | E | N | 1971 | 66116 |
| 0029 | L | D | E | N | 1973 | 72975 |
| 0029 | L | D | E | N | 1974 | 74832 |
| 0029 | L | D | E | N | 1974 | 75983 |
| 0030 | G | D | E | N | 1972 | 65575 |
| 0030 | L | C | R | N | 1967 | 53326 |
| 0030 | L | D | E | N | 1971 | 61577 |
| 0030 | L | D | R | N | 1969 | 62595 |
| 0030 | L | M | E | N | 1971 | 63173 |
| 0030 | L | D | R | N | 1969 | 63252 |
| 0030 | L | D | E | N | 1972 | 65575 |
| 0030 | L | D | R | N | 1971 | 65734 |
| 0030 | L | C | E | N | 1972 | 66038 |
| 0030 | L | D | E | N | 1972 | 66662 |
| 0030 | L | D | E | N | 1972 | 68266 |
| 0030 | L | D | R | N | 1971 | 68954 |
| 0030 | L | D | E | N | 1973 | 72975 |
| 0030 | L | D | G | N | 1973 | 73821 |
| 0030 | L | D | R | N | 1974 | 75674 |
| 0030 | L | D | E | N | 1974 | 75675 |

**Phys. State:** **D.** Doped; **E.** Expanded; **F.** Fibrous or Whisker; **G.** Gas; **L.** Liquid; **M.** Multiphase; **P.** Powder; **S.** Solid
**Subject:** **C.** Calculated Data + Theory; **D.** Data; **E.** Experiment; **G.** General (Data + Theory + Experiment); **M.** Measured Data + Experiment; **S.** Survey (Review, Compendium, etc.); **T.** Theory
**Language:** **C.** Czech; **D.** Dutch; **E.** English; **F.** French; **G.** German; **I.** Italian; **J.** Japanese; **O.** Other Languages; **P.** Polish; **R.** Russian; **S.** Spanish
**Temperature:** **F.** Full Range (Low + Normal + High); **L.** Low (0 to 75K) + Overlap into Normal; **N.** Normal (75 to 1273K); **H.** High (above 1273K) + Overlap intc Normal; **U.** Undefined

| Substance Number | Phys. State | Sub-ject | Lan-guage | Temper-ature | Year | TPRC Number |
|---|---|---|---|---|---|---|
| 517-0030 | L | D | E | N | 1974 | 75983 |
| 0031 | L | D | E | N | 1973 | 72975 |
| 0032 | L | D | R | N | 1971 | 66363 |
| 0032 | L | D | E | N | 1974 | 72185 |
| 0032 | L | D | R | N | 1971 | 75603 |
| 0032 | L | D | E | N | 1971 | 75604 |
| 0039 | L | C | R | N | 1974 | 75682 |
| 0039 | L | C | E | N | 1974 | 75683 |
| 0041 | L | D | G | N | 1970 | 61243 |
| 0042 | D | D | G | N | 1972 | 67710 |
| 0042 | L | D | R | N | 1969 | 56229 |
| 0042 | L | C | R | N | 1969 | 56638 |
| 0042 | L | D | R | N | 1970 | 60854 |
| 0042 | L | D | G | N | 1970 | 61243 |
| 0042 | L | D | R | N | 1970 | 62518 |
| 0042 | L | C | E | N | 1971 | 63598 |
| 0042 | L | T | R | N | 1971 | 63783 |
| 0042 | L | C | E | N | 1972 | 65495 |
| 0042 | L | D | E | N | 1972 | 65532 |
| 0042 | L | D | R | N | 1972 | 66881 |
| 0042 | L | D | G | N | 1972 | 67710 |
| 0042 | L | D | E | N | 1974 | 76330 |
| 0043 | L | D | O | N | 1971 | 65003 |
| 0044 | L | C | E | N | 1971 | 63598 |
| 0048 | L | D | R | N | 1969 | 56940 |
| 0048 | L | D | G | N | 1970 | 61243 |
| 0048 | L | D | R | N | 1971 | 65686 |
| 0048 | L | D | E | N | 1972 | 68218 |
| 0048 | L | D | E | N | 1972 | 68266 |
| 0048 | L | D | E | N | 1973 | 72975 |
| 0048 | L | D | R | N | 1973 | 75652 |
| 0048 | L | D | E | N | 1973 | 75653 |
| 0048 | L | D | E | N | 1974 | 75983 |
| 0054 | L | D | R | N | 1968 | 50466 |
| 0057 | L | D | E | N | 1971 | 64154 |
| 0057 | L | D | R | N | 1971 | 65686 |
| 0057 | L | D | R | N | 1972 | 68167 |
| 0057 | L | D | E | N | 1972 | 68168 |
| 0057 | L | D | E | N | 1973 | 70729 |
| 0057 | L | D | G | N | 1973 | 72511 |
| 0057 | L | D | R | N | 1973 | 75555 |
| 0057 | L | D | E | N | 1973 | 75556 |
| 0057 | L | D | R | N | 1973 | 75830 |
| 0057 | L | D | E | N | 1973 | 75831 |
| 0058 | L | D | R | N | 1973 | 75826 |
| 0058 | L | D | E | N | 1973 | 75827 |
| 0063 | L | D | E | N | 1971 | 62467 |
| 0063 | L | C | E | N | 1972 | 68203 |
| 0063 | L | D | E | N | 1974 | 75983 |
| 0065 | L | D | R | N | 1971 | 64917 |
| 0065 | L | D | E | N | 1971 | 64918 |
| 0065 | L | D | F | N | 1972 | 66356 |
| 0065 | L | D | E | N | 1972 | 67825 |
| 0067 | L | D | R | N | 1974 | 75674 |
| 0067 | L | D | E | N | 1974 | 75675 |
| 0087 | L | D | E | N | 1973 | 72406 |
| 0099 | L | D | E | N | 1973 | 72975 |
| 0099 | L | D | E | N | 1974 | 75983 |
| 0100 | L | D | G | N | 1970 | 61243 |
| 0100 | L | D | E | N | 1972 | 64799 |
| 0100 | L | D | E | N | 1972 | 68218 |
| 0100 | L | D | R | N | 1973 | 75575 |
| 0100 | L | D | E | N | 1973 | 75576 |
| 0100 | L | D | E | N | 1974 | 75983 |
| 0101 | L | D | E | N | 1973 | 72975 |
| 0103 | L | D | G | N | 1970 | 61243 |
| 0115 | L | D | E | N | 1970 | 64935 |
| 0117 | L | D | R | N | 1970 | 62455 |
| 0118 | L | D | J | N | 1968 | 55653 |
| 0118 | L | D | R | N | 1971 | 65686 |
| 0118 | L | D | E | N | 1973 | 72975 |
| 0121 | L | D | O | N | 1968 | 55109 |
| 0121 | L | D | R | N | 1971 | 65686 |
| 0121 | L | D | R | N | 1969 | 65730 |
| 0121 | L | D | R | N | 1973 | 71672 |
| 0121 | L | D | E | N | 1973 | 72559 |
| 0121 | L | D | R | N | 1971 | 75491 |
| 0121 | L | D | E | N | 1971 | 75492 |
| 0121 | L | D | R | N | 1973 | 75652 |
| 0121 | L | D | E | N | 1973 | 75653 |
| 0135 | L | D | E | N | 1974 | 76330 |
| 0136 | L | D | R | N | 1970 | 62455 |
| 0136 | L | D | G | N | 1973 | 72511 |
| 0140 | L | C | E | N | 1972 | 68203 |
| 0141 | L | D | J | N | 1973 | 71050 |
| 0144 | L | D | G | N | 1970 | 61243 |
| 0144 | L | D | E | N | 1972 | 68266 |
| 0144 | L | D | E | N | 1973 | 72975 |
| 0144 | L | D | E | N | 1974 | 75983 |
| 0147 | L | D | R | N | 1970 | 62455 |
| 0155 | L | D | R | N | 1971 | 62543 |
| 0155 | L | D | R | N | 1971 | 65686 |
| 0155 | L | D | E | N | 1971 | 66471 |
| 0155 | L | D | R | N | 1971 | 67734 |
| 0155 | L | D | R | N | 1972 | 68213 |
| 0155 | L | D | E | N | 1972 | 68214 |
| 0155 | L | D | E | N | 1972 | 75522 |
| 0155 | L | D | R | N | 1973 | 75646 |
| 0155 | L | D | E | N | 1973 | 75647 |
| 0155 | L | C | R | N | 1974 | 75682 |
| 0155 | L | C | E | N | 1974 | 75683 |
| 517-0157 | L | D | O | N | 1967 | 45354 |
| 0157 | L | D | R | N | 1973 | 75541 |
| 0157 | L | D | E | N | 1973 | 75542 |
| 0157 | L | D | R | N | 1973 | 75646 |
| 0157 | L | D | E | N | 1973 | 75647 |
| 0157 | L | C | R | N | 1974 | 75682 |
| 0157 | L | C | E | N | 1974 | 75683 |
| 0181 | L | D | R | N | 1972 | 68167 |
| 0181 | L | D | E | N | 1972 | 68168 |
| 0182 | L | D | E | N | 1971 | 64154 |
| 0182 | L | D | E | N | 1972 | 67640 |
| 0182 | L | D | R | N | 1972 | 68169 |
| 0182 | L | D | E | N | 1972 | 68170 |
| 0182 | L | D | R | N | 1972 | 68213 |
| 0182 | L | D | E | N | 1972 | 68214 |
| 0182 | L | D | R | N | 1972 | 69372 |
| 0182 | L | D | E | N | 1972 | 69373 |
| 0182 | L | D | G | N | 1973 | 72511 |
| 0182 | L | D | R | N | 1973 | 75555 |
| 0182 | L | D | E | N | 1973 | 75556 |
| 0182 | L | C | R | N | 1974 | 75682 |
| 0182 | L | C | E | N | 1974 | 75683 |
| 0183 | L | D | R | N | 1971 | 64892 |
| 0183 | L | D | R | N | 1972 | 68169 |
| 0183 | L | D | E | N | 1972 | 68170 |
| 0183 | L | D | E | N | 1971 | 69785 |
| 0189 | L | D | E | N | 1971 | 73917 |
| 0193 | L | D | E | N | 1971 | 62467 |
| 0194 | L | G | P | N | 1964 | 38366 |
| 0194 | L | D | R | N | 1969 | 55892 |
| 0194 | L | D | R | N | 1969 | 65730 |
| 0194 | L | D | R | N | 1971 | 66119 |
| 0194 | L | D | E | N | 1971 | 66120 |
| 0194 | L | D | E | N | 1969 | 66455 |
| 0194 | L | D | E | N | 1972 | 67823 |
| 0194 | L | D | R | N | 1972 | 68663 |
| 0194 | L | D | E | N | 1972 | 68664 |
| 0194 | L | D | O | N | 1973 | 70048 |
| 0196 | L | D | E | N | 1971 | 73917 |
| 0200 | L | D | R | N | 1968 | 55473 |
| 0200 | L | T | R | N | 1971 | 63783 |
| 0213 | L | D | E | N | 1974 | 74832 |
| 0216 | L | C | E | N | 1971 | 63598 |
| 0221 | L | D | R | N | 1969 | 56919 |
| 0221 | L | D | R | N | 1968 | 57029 |
| 0221 | L | D | R | N | 1969 | 62595 |
| 0221 | L | M | E | N | 1971 | 63173 |
| 0221 | L | D | R | N | 1969 | 63252 |
| 0221 | L | D | R | N | 1971 | 65317 |
| 0221 | L | D | E | N | 1971 | 65318 |
| 0221 | L | D | E | N | 1972 | 66503 |
| 0221 | L | D | R | N | 1970 | 67731 |
| 0237 | G | D | E | N | 1972 | 65575 |
| 0237 | L | D | E | N | 1967 | 49931 |
| 0237 | L | D | R | N | 1968 | 55473 |
| 0237 | L | D | J | N | 1968 | 55653 |
| 0237 | L | D | G | N | 1970 | 61243 |
| 0237 | L | D | E | N | 1971 | 64154 |
| 0237 | L | D | E | N | 1972 | 65575 |
| 0237 | L | D | E | N | 1972 | 66662 |
| 0237 | L | D | E | N | 1972 | 67640 |
| 0237 | L | D | E | N | 1972 | 68266 |
| 0237 | L | D | E | N | 1973 | 72975 |
| 0238 | L | D | E | N | 1970 | 64935 |
| 0259 | L | D | E | N | 1972 | 64799 |
| 0259 | L | D | E | N | 1972 | 68266 |
| 0259 | L | D | E | N | 1974 | 74832 |
| 0259 | L | D | E | N | 1974 | 75983 |
| 0260 | L | D | R | N | 1969 | 56229 |
| 0260 | L | D | E | N | 1972 | 65532 |
| 0280 | L | D | E | N | 1973 | 72975 |
| 0281 | L | D | E | N | 1974 | 75983 |
| 0288 | L | D | R | N | 1973 | 71672 |
| 0288 | L | D | E | N | 1973 | 72559 |
| 0310 | L | D | E | N | 1969 | 58796 |
| 0311 | L | D | R | N | 1969 | 63174 |
| 0311 | L | D | R | N | 1969 | 65730 |
| 0311 | L | C | R | N | 1974 | 75682 |
| 0311 | L | C | E | N | 1974 | 75683 |
| 0319 | L | D | R | N | 1972 | 68663 |
| 0319 | L | D | E | N | 1972 | 68664 |
| 0319 | L | C | R | N | 1974 | 75682 |
| 0319 | L | C | E | N | 1974 | 75683 |
| 0347 | L | M | E | N | 1971 | 63173 |
| 0357 | L | D | E | N | 1973 | 72975 |
| 0387 | L | D | R | N | 1968 | 50466 |
| 0387 | L | C | R | N | 1967 | 53326 |
| 0387 | L | D | R | N | 1968 | 57029 |
| 0387 | L | C | E | N | 1972 | 66038 |
| 0387 | L | D | O | N | 1973 | 70048 |
| 0387 | L | D | R | N | 1973 | 73022 |
| 0396 | L | D | R | N | 1972 | 69832 |
| 0396 | L | D | E | N | 1972 | 69833 |
| 0412 | L | D | R | N | 1970 | 62440 |
| 0412 | L | D | R | N | 1971 | 66115 |
| 0412 | L | D | E | N | 1971 | 66116 |
| 0433 | L | D | R | N | 1967 | 48539 |
| 0433 | L | D | R | N | 1967 | 51061 |
| 0433 | L | D | R | N | 1968 | 51196 |
| 0433 | L | D | E | N | 1968 | 67255 |
| 0433 | L | D | E | N | 1971 | 69851 |
| 517-0433 | L | D | E | N | 1972 | 70492 |
| 0433 | L | D | R | N | 1973 | 75828 |
| 0433 | L | D | E | N | 1973 | 75829 |
| 0447 | L | D | R | N | 1967 | 50605 |
| 0447 | L | D | R | N | 1971 | 62475 |
| 0447 | L | D | E | N | 1971 | 62476 |
| 0447 | L | D | E | N | 1971 | 63245 |
| 0447 | L | D | R | N | 1973 | 71577 |
| 0447 | L | D | R | N | 1973 | 71672 |
| 0447 | L | D | E | N | 1973 | 72559 |
| 0448 | L | D | R | N | 1973 | 75541 |
| 0448 | L | D | E | N | 1973 | 75542 |
| 0452 | L | D | R | N | 1973 | 75833 |
| 0452 | L | D | E | N | 1973 | 75834 |
| 0453 | L | D | E | N | 1971 | 64864 |
| 0453 | L | D | O | N | 1971 | 66967 |
| 0453 | L | D | R | N | 1971 | 67734 |
| 0453 | L | D | R | N | 1973 | 75833 |
| 0453 | L | D | E | N | 1973 | 75834 |
| 0460 | L | D | R | N | 1969 | 56919 |
| 0460 | L | D | G | N | 1970 | 61937 |
| 0460 | L | D | R | N | 1971 | 65686 |
| 0460 | L | D | R | N | 1971 | 66363 |
| 0460 | L | D | E | N | 1972 | 66503 |
| 0460 | L | D | R | N | 1973 | 73022 |
| 0460 | L | D | E | N | 1974 | 74832 |
| 0467 | G | C | R | H | 1971 | 67807 |
| 0468 | G | T | R | H | 1968 | 55010 |
| 0468 | G | T | E | H | 1968 | 64428 |
| 0469 | G | C | R | H | 1968 | 55010 |
| 0469 | G | D | R | H | 1970 | 63896 |
| 0469 | G | D | R | H | 1971 | 63955 |
| 0469 | G | D | E | H | 1971 | 63956 |
| 0469 | G | C | E | H | 1968 | 64428 |
| 0480 | L | D | R | N | 1971 | 68954 |
| 0492 | L | D | E | N | 1968 | 70168 |
| 0492 | L | M | E | N | 1973 | 72969 |
| 0492 | L | D | E | N | 1973 | 74678 |
| 0498 | L | D | E | N | 1974 | 74832 |
| 0500 | L | D | O | N | 1967 | 45354 |
| 0501 | L | D | E | N | 1971 | 64154 |
| 0502 | L | D | E | N | 1974 | 74832 |
| 0504 | L | D | E | N | 1972 | 67825 |
| 0504 | L | D | R | N | 1973 | 75830 |
| 0504 | L | D | E | N | 1973 | 75831 |
| 0505 | L | D | G | N | 1970 | 61243 |
| 0506 | L | D | E | N | 1973 | 72406 |
| 0507 | L | D | O | N | 1967 | 45354 |
| 0507 | L | D | R | N | 1969 | 56940 |
| 0507 | L | D | R | N | 1971 | 62744 |
| 0507 | L | D | E | N | 1971 | 62745 |
| 0507 | L | D | R | N | 1971 | 64729 |
| 0507 | L | D | E | N | 1971 | 64730 |
| 0507 | L | D | O | N | 1971 | 66368 |
| 0507 | L | D | E | N | 1972 | 68218 |
| 0507 | L | D | R | N | 1972 | 69022 |
| 0507 | L | D | E | N | 1972 | 69023 |
| 0507 | L | D | R | N | 1970 | 75461 |
| 0507 | L | D | E | N | 1970 | 75462 |
| 0507 | L | D | R | N | 1973 | 75658 |
| 0507 | L | D | E | N | 1973 | 75659 |
| 0507 | L | D | E | N | 1974 | 75671 |
| 0508 | L | D | R | N | 1971 | 66117 |
| 0508 | L | D | E | N | 1971 | 66118 |
| 0510 | L | D | R | N | 1971 | 64913 |
| 0510 | L | D | E | N | 1971 | 64914 |
| 0512 | L | D | R | N | 1972 | 68165 |
| 0512 | L | D | E | N | 1972 | 68166 |
| 0517 | L | D | R | N | 1972 | 66881 |
| 0519 | L | D | E | N | 1972 | 68218 |
| 0522 | L | D | R | N | 1972 | 66881 |
| 0523 | L | D | R | N | 1970 | 62518 |
| 0528 | L | D | R | N | 1972 | 66881 |
| 0531 | L | D | R | N | 1972 | 66881 |
| 0538 | L | D | O | N | 1971 | 65003 |
| 0539 | L | D | R | N | 1970 | 62518 |
| 0547 | L | D | J | N | 1973 | 71050 |
| 0550 | L | C | R | N | 1974 | 75682 |
| 0550 | L | C | E | N | 1974 | 75683 |
| 0551 | L | D | R | N | 1971 | 65317 |
| 0551 | L | D | E | N | 1971 | 65318 |
| 0552 | L | D | F | N | 1968 | 50472 |
| 0552 | L | D | G | N | 1972 | 71225 |
| 0553 | L | D | F | N | 1968 | 50472 |
| 0553 | L | D | G | N | 1972 | 71225 |
| 0554 | L | D | G | N | 1972 | 71225 |
| 0561 | L | D | E | N | 1970 | 64935 |
| 0563 | L | G | P | N | 1964 | 38366 |
| 0565 | L | D | R | N | 1970 | 60854 |
| 0571 | L | D | E | N | 1973 | 72975 |
| 0572 | L | D | O | N | 1971 | 65003 |
| 0578 | L | D | R | N | 1971 | 65734 |
| 0578 | L | D | G | N | 1972 | 71225 |
| 0578 | L | D | R | N | 1971 | 75599 |
| 0578 | L | D | E | N | 1971 | 75600 |
| 0579 | L | D | E | N | 1971 | 61577 |
| 0580 | L | D | R | U | 1969 | 56961 |
| 0580 | L | D | E | U | 1971 | 63986 |
| 0580 | L | C | R | N | 1974 | 75682 |
| 0581 | L | D | E | N | 1973 | 72975 |
| 0582 | L | D | O | N | 1971 | 65003 |

**Phys. State:** **D.** Doped; **E.** Expanded; **F.** Fibrous or Whisker; **G.** Gas; **L.** Liquid; **M.** Multiphase; **P.** Powder; **S.** Solid
**Subject:** **C.** Calculated Data + Theory; **D.** Data; **E.** Experiment; **G.** General (Data + Theory + Experiment); **M.** Measured Data + Experiment; **S.** Survey (Review, Compendium, etc.); **T.** Theory
**Language:** **C.** Czech; **D.** Dutch; **E.** English; **F.** French; **G.** German; **I.** Italian; **J.** Japanese; **O.** Other Languages; **P.** Polish; **R.** Russian; **S.** Spanish
**Temperature:** **F.** Full Range (Low + Normal + High); **L.** Low (0 to 75K) + Overlap into Normal; **N.** Normal (75 to 1273K); **H.** High (above 1273K) + Overlap into Normal; **U.** Undefined

| Substance Number | Phys. State | Subject | Language | Temperature | Year | TPRC Number |
|---|---|---|---|---|---|---|
| 517-0583 | L | G | P | N | 1964 | 38366 |
| 0585 | L | D | O | N | 1971 | 65003 |
| 0586 | L | C | E | N | 1974 | 75683 |
| 0588 | L | D | E | N | 1970 | 64935 |
| 0589 | L | G | P | N | 1964 | 38366 |
| 0593 | L | C | E | N | 1971 | 61312 |
| 0595 | L | D | R | N | 1969 | 56919 |
| 0595 | L | D | E | N | 1972 | 66503 |
| 0595 | L | D | R | N | 1971 | 67733 |
| 0595 | L | D | R | N | 1970 | 76333 |
| 0595 | L | D | E | N | 1973 | 76334 |
| 0600 | L | D | R | N | 1971 | 65686 |
| 0601 | L | D | R | N | 1969 | 62595 |
| 0601 | L | D | R | N | 1969 | 63252 |
| 0603 | L | D | E | N | 1973 | 72975 |
| 0604 | L | D | R | N | 1971 | 63382 |
| 0604 | L | D | E | N | 1971 | 63383 |
| 0606 | L | D | R | N | 1971 | 66119 |
| 0606 | L | D | E | N | 1971 | 66120 |
| 0606 | L | D | E | N | 1972 | 67823 |
| 0608 | L | D | R | N | 1968 | 50466 |
| 0610 | L | D | R | N | 1968 | 57029 |
| 0610 | L | D | R | N | 1971 | 67733 |
| 0611 | L | D | R | N | 1971 | 63382 |
| 0611 | L | D | E | N | 1971 | 63383 |
| 0613 | L | D | R | N | 1970 | 67731 |
| 0613 | L | D | R | N | 1972 | 69022 |
| 0613 | L | D | E | N | 1972 | 69023 |
| 0614 | L | D | R | N | 1971 | 66119 |
| 0614 | L | D | E | N | 1971 | 66120 |
| 0614 | L | D | R | N | 1972 | 69832 |
| 0614 | L | D | E | N | 1972 | 69833 |
| 0628 | L | D | R | U | 1969 | 56961 |
| 0628 | L | D | E | U | 1971 | 63986 |
| 0629 | L | D | R | N | 1969 | 55892 |
| 0629 | L | D | E | N | 1969 | 66455 |
| 0630 | L | D | R | N | 1971 | 62744 |
| 0630 | L | D | E | N | 1971 | 62745 |
| 0631 | L | D | R | N | 1971 | 64729 |
| 0631 | L | D | E | N | 1971 | 64730 |
| 0631 | L | D | R | N | 1971 | 67732 |
| 0633 | L | D | R | N | 1968 | 57029 |
| 0635 | L | D | E | N | 1971 | 64864 |
| 0635 | L | D | R | N | 1970 | 75461 |
| 0635 | L | D | E | N | 1970 | 75462 |
| 0636 | L | D | E | N | 1974 | 75671 |
| 0642 | L | D | R | N | 1969 | 65730 |
| 0642 | L | D | F | N | 1972 | 66356 |
| 0643 | L | D | R | N | 1967 | 51061 |
| 0643 | L | D | R | N | 1969 | 65730 |
| 0643 | L | D | E | N | 1971 | 69851 |
| 0643 | L | D | E | N | 1972 | 70492 |
| 0644 | L | D | O | N | 1968 | 55109 |
| 0644 | L | D | R | N | 1969 | 55892 |
| 0644 | L | D | R | N | 1971 | 62475 |
| 0644 | L | D | E | N | 1971 | 62476 |
| 0644 | L | D | R | N | 1969 | 65730 |
| 0644 | L | D | E | N | 1969 | 66455 |
| 0644 | L | D | R | N | 1973 | 75646 |
| 0644 | L | D | E | N | 1973 | 75647 |
| 0644 | L | C | R | N | 1974 | 75682 |
| 0644 | L | C | E | N | 1974 | 75683 |
| 0645 | L | D | R | N | 1970 | 62440 |
| 0646 | L | D | R | N | 1971 | 62543 |
| 0646 | L | D | E | N | 1971 | 66471 |
| 0647 | L | D | O | N | 1967 | 45354 |
| 0647 | L | D | O | N | 1971 | 66368 |
| 0647 | L | D | E | N | 1972 | 75522 |
| 0647 | L | D | R | N | 1973 | 75658 |
| 0647 | L | D | E | N | 1973 | 75659 |
| 0649 | L | D | E | N | 1971 | 64154 |
| 0652 | L | D | E | N | 1972 | 64799 |
| 0654 | L | D | R | N | 1971 | 64913 |
| 0654 | L | D | E | N | 1971 | 64914 |
| 0654 | L | D | R | N | 1971 | 64917 |
| 0654 | L | D | E | N | 1971 | 64918 |
| 0655 | L | D | R | N | 1967 | 48539 |
| 0655 | L | D | R | N | 1968 | 51196 |
| 0655 | L | D | E | N | 1968 | 67255 |
| 0658 | L | D | R | N | 1972 | 69830 |
| 0658 | L | D | E | N | 1972 | 69831 |
| 0659 | L | D | R | N | 1969 | 53690 |
| 0659 | L | D | R | N | 1970 | 62104 |
| 0659 | L | D | E | N | 1969 | 67237 |
| 0662 | L | D | R | N | 1969 | 63174 |
| 0662 | L | D | R | N | 1969 | 64328 |
| 0662 | L | D | E | N | 1971 | 64864 |
| 0662 | L | D | R | N | 1971 | 64892 |
| 0662 | L | D | R | N | 1971 | 67732 |
| 0662 | L | D | E | N | 1971 | 69785 |
| 0662 | L | D | R | N | 1972 | 69830 |
| 0662 | L | D | E | N | 1972 | 69831 |
| 0663 | L | C | R | N | 1967 | 53326 |
| 0663 | L | D | G | N | 1969 | 56950 |
| 0663 | L | D | G | N | 1970 | 61243 |
| 0663 | L | C | E | N | 1972 | 66038 |
| 0664 | L | D | R | N | 1971 | 75599 |
| 0664 | L | D | E | N | 1971 | 75600 |
| 0664 | L | D | R | N | 1971 | 75603 |
| 0664 | L | D | E | N | 1971 | 75604 |
| 0665 | L | D | R | N | 1967 | 50605 |
| 517-0665 | L | D | E | N | 1971 | 63245 |
| 0665 | L | D | R | N | 1973 | 75557 |
| 0665 | L | D | E | N | 1973 | 75558 |
| 0666 | L | D | R | N | 1972 | 68165 |
| 0666 | L | D | E | N | 1972 | 68166 |
| 0667 | L | D | R | N | 1969 | 53690 |
| 0667 | L | D | E | N | 1969 | 67237 |
| 0667 | L | D | R | N | 1972 | 69372 |
| 0667 | L | D | E | N | 1972 | 69373 |
| 0667 | L | D | E | N | 1973 | 70729 |
| 0670 | L | D | R | N | 1972 | 68167 |
| 0670 | L | D | E | N | 1972 | 68168 |
| 0670 | L | D | R | N | 1972 | 68169 |
| 0670 | L | D | E | N | 1972 | 68170 |
| 0670 | L | C | R | N | 1974 | 75682 |
| 0670 | L | C | E | N | 1974 | 75683 |
| 0670 | L | D | R | N | 1973 | 75826 |
| 0670 | L | D | E | N | 1973 | 75827 |
| 0670 | L | D | R | N | 1973 | 75833 |
| 0670 | L | D | E | N | 1973 | 75834 |
| 0672 | L | D | R | N | 1969 | 63174 |
| 0672 | L | D | R | N | 1969 | 64328 |
| 0672 | L | D | R | N | 1971 | 66117 |
| 0672 | L | D | E | N | 1971 | 66118 |
| 0679 | L | D | R | N | 1968 | 51196 |
| 0679 | L | D | E | N | 1968 | 67255 |
| 0681 | L | C | R | N | 1967 | 53326 |
| 0681 | L | C | E | N | 1972 | 66038 |
| 0681 | L | D | R | N | 1973 | 75646 |
| 0681 | L | D | E | N | 1973 | 75647 |
| 0682 | L | D | E | N | 1967 | 49931 |
| 0683 | L | D | R | N | 1967 | 51061 |
| 0685 | L | D | R | N | 1969 | 55892 |
| 0685 | L | D | E | N | 1969 | 66455 |
| 0709 | L | D | R | N | 1970 | 76333 |
| 0709 | L | D | E | N | 1973 | 76334 |
| 0713 | L | D | E | N | 1974 | 72185 |
| 0721 | L | D | E | N | 1973 | 72406 |
| 0732 | L | D | R | N | 1969 | 56940 |
| 0749 | L | D | F | N | 1968 | 50472 |
| 0750 | L | D | F | N | 1968 | 50472 |
| 0751 | L | D | F | N | 1968 | 50472 |
| 0758 | L | D | E | N | 1973 | 72406 |
| 0759 | L | D | E | N | 1973 | 72406 |
| 0760 | L | D | E | N | 1973 | 72406 |
| 0761 | L | D | E | N | 1973 | 72406 |
| 0762 | L | D | E | N | 1973 | 72406 |
| 0763 | L | D | E | N | 1973 | 72406 |
| 0764 | L | D | E | N | 1973 | 72406 |
| 0765 | L | D | O | N | 1971 | 66967 |
| 0765 | L | C | R | N | 1974 | 75682 |
| 0765 | L | C | E | N | 1974 | 75683 |
| 0766 | L | D | O | N | 1971 | 66368 |
| 0766 | L | D | E | N | 1972 | 75522 |
| 0766 | L | D | R | N | 1973 | 75658 |
| 0766 | L | D | E | N | 1973 | 75659 |
| 0769 | L | D | R | N | 1970 | 62104 |
| 0770 | L | D | R | N | 1973 | 71577 |
| 0807 | L | D | G | N | 1973 | 73821 |
| 0826 | L | D | E | N | 1974 | 74832 |
| 0827 | L | D | R | N | 1970 | 76333 |
| 0827 | L | D | E | N | 1973 | 76334 |
| 0830 | L | D | R | N | 1973 | 75646 |
| 0830 | L | D | E | N | 1973 | 75647 |
| 0857 | L | C | R | N | 1972 | 70524 |
| 0877 | L | D | R | N | 1971 | 75491 |
| 0877 | L | D | E | N | 1971 | 75492 |
| 0878 | L | D | R | N | 1971 | 75491 |
| 0878 | L | D | E | N | 1971 | 75492 |
| 0879 | L | D | R | N | 1971 | 75491 |
| 0879 | L | D | E | N | 1971 | 75492 |
| 0889 | L | D | E | N | 1972 | 75522 |
| 0890 | G | D | R | H | 1974 | 75751 |
| 0890 | G | D | E | H | 1974 | 75752 |
| 0894 | L | D | R | N | 1973 | 75557 |
| 0894 | L | D | E | N | 1973 | 75558 |
| 0906 | L | D | R | N | 1973 | 75826 |
| 0906 | L | D | E | N | 1973 | 75827 |
| 0907 | L | D | R | N | 1973 | 75828 |
| 0907 | L | D | E | N | 1973 | 75829 |
| 0908 | L | D | R | N | 1973 | 75828 |
| 0908 | L | D | E | N | 1973 | 75829 |
| 0950 | L | D | R | N | 1973 | 75652 |
| 0950 | L | D | E | N | 1973 | 75653 |
| 0957 | L | D | R | N | 1973 | 73022 |
| 0958 | L | D | R | N | 1973 | 73022 |
| 0968 | L | D | R | N | 1973 | 75652 |
| 0968 | L | D | E | N | 1973 | 75653 |
| 0989 | L | C | R | N | 1974 | 75682 |
| 0989 | L | C | E | N | 1974 | 75683 |
| 0990 | L | C | R | N | 1974 | 75682 |
| 0990 | L | C | E | N | 1974 | 75683 |
| 0991 | L | C | R | N | 1974 | 75682 |
| 0991 | L | C | E | N | 1974 | 75683 |
| 0992 | L | C | R | N | 1974 | 75682 |
| 0992 | L | C | E | N | 1974 | 75683 |
| 0993 | L | C | R | N | 1974 | 75682 |
| 0993 | L | C | E | N | 1974 | 75683 |
| 0994 | L | C | R | N | 1974 | 75682 |
| 0994 | L | C | E | N | 1974 | 75683 |

**Phys. State:** **D.** Doped; **E.** Expanded; **F.** Fibrous or Whisker; **G.** Gas; **L.** Liquid; **M.** Multiphase; **P.** Powder; **S.** Solid
**Subject:** **C.** Calculated Data + Theory; **D.** Data; **E.** Experiment; **G.** General (Data + Theory + Experiment); **M.** Measured Data + Experiment; **S.** Survey (Review, Compendium, etc.); **T.** Theory
**Language:** **C.** Czech; **D.** Dutch; **E.** English; **F.** French; **G.** German; **I.** Italian; **J.** Japanese; **O.** Other Languages; **P.** Polish; **R.** Russian; **S.** Spanish
**Temperature:** **F.** Full Range (Low + Normal + High); **L.** Low (0 to 75K) + Overlap into Normal; **N.** Normal (75 to 1273K); **H.** High (above 1273K) + Overlap into Normal; **U.** Undefined

# Chapter 5 Emittance

| *Substance Number* | *Phys. State* | *Subject* | *Language* | *Temperature* | *Year* | *TPRC Number* |
|---|---|---|---|---|---|---|
| 512-0027 | G | M | G | H | 1966 | 42186 |
| 0027 | G | C | E | H | 1973 | 70475 |
| 0038 | G | C | E | H | 1964 | 37296 |
| 0038 | G | T | E | H | 1967 | 47656 |
| 0038 | G | C | R | H | 1967 | 49045 |
| 0038 | G | C | E | H | 1967 | 49046 |
| 0090 | S | M | R | H | 1972 | 69824 |
| 0090 | S | M | E | H | 1972 | 69825 |
| 0091 | S | M | R | H | 1972 | 69824 |
| 0091 | S | M | E | H | 1972 | 69825 |
| 0095 | G | G | E | H | 1967 | 47657 |
| 0172 | G | D | R | H | 1972 | 67143 |
| 0172 | G | D | E | H | 1972 | 76925 |
| 0174 | S | M | E | H | 1969 | 77573 |
| 0193 | M | C | R | N | 1969 | 73771 |
| 0193 | M | C | E | N | 1969 | 73772 |
| 0262 | S | M | R | H | 1973 | 73345 |
| 0262 | S | M | E | H | 1973 | 73346 |
| 0711 | G | T | E | H | 1967 | 47235 |
| 0778 | S | M | R | H | 1973 | 73345 |
| 0778 | S | M | E | H | 1973 | 73346 |
| 1048 | S | M | E | H | 1969 | 77573 |
| 514-0020 | G | D | E | N | 1971 | 61868 |
| 0020 | G | M | E | N | 1972 | 65359 |
| 0020 | G | C | E | U | 1972 | 66342 |
| 0020 | G | G | E | H | 1956 | 70593 |
| 0091 | S | D | E | H | 1977 | 93601 |
| 0174 | S | M | E | H | 1969 | 77573 |
| 0175 | S | M | E | H | 1969 | 77573 |
| 0233 | S | D | E | H | 1977 | 93601 |
| 0262 | S | D | E | H | 1977 | 93601 |
| 0308 | M | C | R | N | 1969 | 73771 |
| 0308 | M | C | E | N | 1969 | 73772 |
| 1048 | S | M | E | H | 1969 | 77573 |
| 515-0389 | M | D | J | N | 1969 | 64774 |
| 0390 | M | D | J | N | 1969 | 64774 |
| 0476 | M | D | E | H | 1973 | 70750 |
| 0484 | M | D | E | H | 1973 | 70750 |
| 0488 | M | D | J | N | 1969 | 64774 |
| 517-0106 | G | D | E | H | 1973 | 72550 |
| 0193 | G | D | E | H | 1973 | 72550 |
| 0308 | M | C | R | H | 1969 | 73771 |
| 0308 | M | C | E | H | 1969 | 73772 |
| 0466 | G | C | E | U | 1971 | 63414 |
| 0466 | G | D | E | U | 1971 | 65020 |
| 0466 | G | D | E | U | 1971 | 65038 |
| 0466 | G | D | E | H | 1973 | 72152 |
| 0466 | G | D | E | U | 1973 | 72409 |
| 0466 | M | S | R | N | 1969 | 57577 |
| 0466 | M | M | E | N | 1972 | 65359 |
| 0466 | M | D | E | N | 1974 | 77706 |
| 0466 | M | D | E | N | 1974 | 78814 |
| 0466 | M | T | E | N | 1974 | 78814 |
| 0466 | M | D | E | N | 1974 | 78855 |
| 0466 | M | T | E | N | 1974 | 78855 |
| 0470 | G | E | E | N | 1961 | 83560 |
| 0470 | G | G | E | N | 1961 | 83560 |
| 0473 | G | C | R | H | 1971 | 44544 |
| 0473 | G | C | E | H | 1972 | 62496 |
| 0473 | G | C | E | H | 1974 | 75740 |
| 0473 | G | D | E | H | 1961 | 83558 |
| 0473 | G | T | E | H | 1961 | 83558 |
| 0474 | G | D | E | H | 1972 | 76997 |
| 0474 | M | M | F | H | 1970 | 64521 |
| 0492 | L | T | F | N | 1972 | 64937 |
| 0492 | L | C | E | N | 1971 | 66552 |
| 0492 | L | T | E | N | 1972 | 68979 |
| 0700 | L | D | E | N | 1972 | 72552 |
| 0833 | G | D | R | H | 1971 | 73370 |
| 0833 | G | D | E | H | 1971 | 73371 |
| 0833 | G | E | E | H | 1961 | 83558 |
| 0890 | G | C | E | N | 1974 | 74356 |
| 0996 | G | C | E | H | 1974 | 75740 |
| 0997 | G | C | E | H | 1974 | 75740 |
| 0998 | G | C | E | H | 1974 | 75740 |
| 1080 | G | G | E | H | 1961 | 83558 |

**Phys. State:** **D.** Doped; **E.** Expanded; **F.** Fibrous or Whisker; **G.** Gas; **L.** Liquid; **M.** Multiphase; **P.** Powder; **S.** Solid
**Subject:** **C.** Calculated Data + Theory; **D.** Data; **E.** Experiment; **G.** General (Data + Theory + Experiment); **M.** Measured Data + Experiment; **S.** Survey (Review, Compendium, etc.); **T.** Theory
**Language:** **C.** Czech; **D.** Dutch; **E.** English; **F.** French; **G.** German; **I.** Italian; **J.** Japanese; **O.** Other Languages; **P.** Polish; **R.** Russian; **S.** Spanish
**Temperature:** **F.** Full Range (Low + Normal + High); **L.** Low (0 to 75K) + Overlap into Normal; **N.** Normal (75 to 1273K); **H.** High (above 1273K) + Overlap into Normal; **U.** Undefined

# Chapter 6 Reflectance

| Substance Number | Phys. State | Subject | Language | Temperature | Year | TPRC Number |
|---|---|---|---|---|---|---|
| 512-0027 | G | M | E | H | 1974 | 76169 |
| 0030 | M | D | E | N | 1972 | 65399 |
| 0032 | M | D | E | N | 1974 | 78825 |
| 0032 | M | E | E | N | 1974 | 78825 |
| 0038 | G | T | E | H | 1971 | 61020 |
| 0038 | G | T | E | H | 1971 | 61155 |
| 0038 | G | C | E | U | 1971 | 61455 |
| 0038 | G | T | E | U | 1971 | 62602 |
| 0038 | G | T | E | H | 1971 | 64038 |
| 0038 | G | T | E | U | 1972 | 66829 |
| 0038 | G | C | E | H | 1973 | 68987 |
| 0038 | G | T | E | H | 1973 | 72548 |
| 0038 | G | T | R | H | 1969 | 73373 |
| 0038 | G | T | E | H | 1970 | 73374 |
| 0038 | G | T | R | H | 1973 | 73469 |
| 0038 | G | T | E | H | 1973 | 73470 |
| 0038 | G | T | R | H | 1973 | 73473 |
| 0038 | G | T | E | H | 1973 | 73474 |
| 0114 | L | C | R | N | 1972 | 70515 |
| 0138 | S | D | R | N | 1972 | 68627 |
| 0138 | S | D | E | N | 1972 | 77027 |
| 0174 | S | C | E | N | 1972 | 66692 |
| 0175 | S | C | E | N | 1972 | 66692 |
| 0194 | L | D | E | N | 1972 | 70303 |
| 0203 | L | G | E | N | 1970 | 62479 |
| 0203 | L | M | E | N | 1972 | 65365 |
| 0204 | L | G | E | N | 1970 | 62479 |
| 0204 | L | M | E | N | 1972 | 65365 |
| 0222 | G | D | E | H | 1974 | 75862 |
| 0237 | L | D | E | N | 1972 | 70303 |
| 0240 | S | D | E | N | 1965 | 36324 |
| 0241 | S | D | E | N | 1965 | 36324 |
| 0241 | S | D | E | U | 1971 | 61603 |
| 0256 | G | T | E | U | 1971 | 62400 |
| 0256 | G | T | E | H | 1972 | 68489 |
| 0256 | G | T | R | H | 1972 | 73467 |
| 0256 | G | T | E | H | 1972 | 73468 |
| 0257 | S | D | E | U | 1971 | 61603 |
| 0353 | S | D | E | N | 1971 | 63577 |
| 0354 | S | D | E | N | 1971 | 63577 |
| 0392 | L | C | R | N | 1972 | 70515 |
| 0422 | S | D | E | N | 1971 | 63737 |
| 0422 | S | D | R | N | 1972 | 68627 |
| 0422 | S | D | E | N | 1973 | 69943 |
| 0422 | S | D | E | N | 1972 | 77027 |
| 0423 | S | D | E | N | 1971 | 63737 |
| 0436 | S | D | E | N | 1970 | 62505 |
| 0438 | S | D | E | N | 1970 | 62505 |
| 0443 | S | D | E | N | 1970 | 62505 |
| 0449 | S | D | E | N | 1973 | 69943 |
| 0702 | S | G | E | L | 1972 | 71413 |
| 0703 | S | G | E | L | 1972 | 71413 |
| 0704 | S | G | E | H | 1972 | 71413 |
| 0706 | S | D | G | N | 1972 | 77500 |
| 0767 | S | D | R | N | 1971 | 64869 |
| 0768 | S | D | R | N | 1971 | 64869 |
| 0910 | G | D | E | H | 1974 | 75862 |
| 1011 | S | D | G | N | 1972 | 77500 |
| 1078 | M | D | E | N | 1974 | 78825 |
| 1078 | M | E | E | N | 1974 | 78825 |
| 514-0020 | G | D | E | N | 1972 | 53540 |
| 0020 | G | C | E | U | 1971 | 62404 |
| 0020 | G | D | E | N | 1970 | 65083 |
| 0020 | G | C | E | H | 1971 | 73089 |
| 0020 | G | T | E | U | 1974 | 74528 |
| 0020 | G | D | E | N | 1972 | 77039 |
| 0020 | M | D | E | N | 1972 | 65399 |
| 0057 | L | D | E | N | 1977 | 93596 |
| 0129 | G | T | E | U | 1972 | 67419 |
| 0129 | M | M | E | N | 1972 | 66031 |
| 0182 | L | D | E | N | 1977 | 93596 |
| 0249 | G | C | E | N | 1966 | 47643 |
| 0308 | M | D | E | N | 1972 | 65399 |
| 0309 | G | D | E | U | 1968 | 50438 |
| 0309 | G | C | E | H | 1972 | 69085 |
| 0667 | L | D | E | N | 1977 | 93596 |
| 0698 | S | M | G | U | 1963 | 69326 |
| 0698 | S | M | E | U | 1972 | 69327 |
| 0753 | G | D | E | N | 1973 | 73943 |
| 1001 | G | D | E | U | 1974 | 76406 |
| 1002 | G | D | E | U | 1974 | 76407 |
| 515-0030 | L | M | E | N | 1972 | 67433 |
| 0030 | L | C | E | N | 1973 | 68974 |
| 0048 | L | D | E | N | 1974 | 72596 |
| 0048 | S | G | E | N | 1971 | 66171 |
| 0070 | L | D | E | N | 1974 | 72596 |
| 0070 | S | G | E | N | 1971 | 66171 |
| 0100 | L | D | E | N | 1974 | 72596 |
| 0100 | S | G | E | N | 1971 | 66171 |
| 0144 | L | D | E | N | 1974 | 72596 |
| 0144 | S | G | E | N | 1971 | 66171 |
| 0182 | L | M | E | N | 1973 | 76384 |
| 0194 | L | D | E | N | 1974 | 75373 |
| 515-0194 | S | G | E | N | 1971 | 66171 |
| 0258 | L | D | E | N | 1974 | 72596 |
| 0258 | S | G | E | N | 1971 | 66171 |
| 0259 | L | D | E | N | 1974 | 72596 |
| 0259 | S | G | E | N | 1971 | 66171 |
| 0366 | L | C | E | N | 1971 | 63740 |
| 0387 | L | D | E | N | 1971 | 63689 |
| 0387 | L | G | E | N | 1971 | 65942 |
| 0402 | L | C | E | N | 1974 | 75059 |
| 0447 | L | D | E | N | 1974 | 73964 |
| 0482 | M | D | E | U | 1972 | 73832 |
| 0504 | L | D | E | N | 1971 | 63689 |
| 0504 | L | G | E | N | 1971 | 65942 |
| 0506 | L | M | E | N | 1973 | 76384 |
| 0592 | L | C | E | N | 1971 | 63740 |
| 0594 | L | C | E | N | 1971 | 63740 |
| 0642 | L | D | E | N | 1974 | 75373 |
| 0642 | S | G | E | N | 1971 | 66171 |
| 0643 | L | D | E | N | 1974 | 75373 |
| 0643 | S | G | E | N | 1971 | 66171 |
| 0644 | L | D | E | N | 1974 | 73964 |
| 0644 | L | D | R | U | 1973 | 75837 |
| 0644 | L | D | E | U | 1973 | 75838 |
| 0644 | L | D | E | N | 1975 | 76211 |
| 0644 | L | M | E | N | 1973 | 76384 |
| 0647 | L | D | E | N | 1974 | 73964 |
| 0663 | L | D | E | N | 1971 | 63689 |
| 0663 | L | G | E | N | 1971 | 65942 |
| 1013 | L | M | E | N | 1973 | 76384 |
| 1049 | L | D | E | N | 1974 | 77784 |
| 517-0408 | L | D | R | N | 1972 | 70878 |
| 0408 | S | D | E | N | 1972 | 70879 |
| 0466 | G | D | E | U | 1971 | 61407 |
| 0466 | G | D | E | U | 1973 | 70706 |
| 0466 | G | D | E | H | 1973 | 72152 |
| 0466 | G | D | E | U | 1973 | 72409 |
| 0466 | G | M | E | U | 1968 | 77523 |
| 0466 | M | D | E | U | 1968 | 50438 |
| 0466 | M | S | R | N | 1969 | 57577 |
| 0466 | M | C | R | N | 1972 | 69064 |
| 0466 | M | D | E | N | 1974 | 77706 |
| 0466 | M | D | E | N | 1974 | 78855 |
| 0466 | M | T | E | N | 1974 | 78855 |
| 0492 | G | D | E | N | 1972 | 77039 |
| 0492 | L | C | E | N | 1971 | 63398 |
| 0492 | L | D | E | N | 1971 | 63689 |
| 0492 | L | G | E | N | 1971 | 65942 |
| 0492 | L | C | E | N | 1971 | 66552 |
| 0492 | L | D | E | U | 1972 | 66757 |
| 0492 | L | D | E | N | 1972 | 67551 |
| 0492 | L | D | E | N | 1973 | 70783 |
| 0492 | L | T | E | N | 1973 | 72026 |
| 0492 | L | C | E | N | 1974 | 75063 |
| 0492 | L | C | E | N | 1975 | 76568 |
| 0492 | L | D | E | N | 1972 | 77039 |
| 0492 | L | D | E | N | 1974 | 77784 |
| 0602 | L | D | R | N | 1972 | 70878 |
| 0602 | S | D | E | N | 1972 | 70879 |
| 0700 | L | S | E | N | 1971 | 69328 |
| 0700 | L | D | E | U | 1972 | 69340 |
| 1010 | L | E | E | N | 1973 | 75325 |
| 1010 | L | M | E | N | 1973 | 76384 |
| 1010 | L | M | E | N | 1974 | 78687 |

**Phys. State:** **D.** Doped; **E.** Expanded; **F.** Fibrous or Whisker; **G.** Gas; **L.** Liquid; **M.** Multiphase; **P.** Powder; **S.** Solid
**Subject:** **C.** Calculated Data + Theory; **D.** Data; **E.** Experiment; **G.** General (Data + Theory + Experiment); **M.** Measured Data + Experiment; **S.** Survey (Review, Compendium, etc.); **T.** Theory
**Language:** **C.** Czech; **D.** Dutch; **E.** English; **F.** French; **G.** German; **I.** Italian; **J.** Japanese; **O.** Other Languages; **P.** Polish; **R.** Russian; **S.** Spanish
**Temperature:** **F.** Full Range (Low + Normal + High); **L.** Low (0 to 75K) + Overlap into Normal; **N.** Normal (75 to 1273K); **H.** High (above 1273K) + Overlap into Normal; **U.** Undefined

# Chapter 7 Absorptance

| Substance Number | Phys. State | Subject | Language | Temperature | Year | TPRC Number |
|---|---|---|---|---|---|---|
| 512-0023 | D | D | J | N | 1972 | 67595 |
| 0036 | S | D | E | L | 1974 | 75023 |
| 0037 | S | D | E | L | 1974 | 75023 |
| 0117 | S | D | E | N | 1970 | 65044 |
| 0158 | S | D | E | N | 1970 | 65044 |
| 0194 | S | D | E | N | 1970 | 65044 |
| 0261 | P | D | R | N | 1970 | 63227 |
| 0262 | P | D | R | N | 1970 | 63227 |
| 0288 | D | D | J | N | 1972 | 67595 |
| 0891 | S | D | E | L | 1974 | 75023 |
| 514-0012 | S | D | E | L | 1971 | 68236 |
| 0106 | S | D | E | L | 1971 | 68236 |
| 0293 | S | D | E | L | 1971 | 68236 |
| 0388 | S | D | E | L | 1971 | 68236 |
| 0461 | S | D | E | L | 1971 | 68236 |

**Phys. State:** **D.** Doped; **E.** Expanded; **F.** Fibrous or Whisker; **G.** Gas; **L.** Liquid; **M.** Multiphase; **P.** Powder; **S.** Solid
**Subject:** **C.** Calculated Data + Theory; **D.** Data; **E.** Experiment; **G.** General (Data + Theory + Experiment); **M.** Measured Data + Experiment; **S.** Survey (Review, Compendium, etc.); **T.** Theory
**Language:** **C.** Czech; **D.** Dutch; **E.** English; **F.** French; **G.** German; **I.** Italian; **J.** Japanese; **O.** Other Languages; **P.** Polish; **R.** Russian; **S.** Spanish
**Temperature:** **F.** Full Range (Low + Normal + High); **L.** Low (0 to 75K) + Overlap into Normal; **N.** Normal (75 to 1273K); **H.** High (above 1273K) + Overlap into Normal; **U.** Undefined

# Chapter 8 Transmittance

| Substance Number | Phys. State | Sub-ject | Lan-guage | Temper-ature | Year | TPRC Number |
|---|---|---|---|---|---|---|
| 512-0003 | S | D | E | L | 1972 | 66088 |
| 0011 | S | D | E | L | 1971 | 63052 |
| 0012 | D | D | E | L | 1972 | 69701 |
| 0012 | S | D | E | L | 1972 | 69701 |
| 0023 | S | D | E | L | 1973 | 69028 |
| 0029 | D | D | R | L | 1973 | 73627 |
| 0029 | D | D | E | L | 1973 | 73628 |
| 0030 | D | D | R | L | 1973 | 73627 |
| 0030 | D | D | E | L | 1973 | 73628 |
| 0030 | S | D | E | N | 1972 | 78292 |
| 0031 | S | D | E | N | 1972 | 78292 |
| 0048 | D | D | R | L | 1973 | 73627 |
| 0048 | D | D | E | L | 1973 | 73628 |
| 0048 | S | D | E | N | 1972 | 78292 |
| 0069 | S | D | E | L | 1972 | 66088 |
| 0082 | S | D | E | N | 1972 | 64583 |
| 0083 | S | D | E | N | 1972 | 64583 |
| 0111 | S | D | E | L | 1971 | 63661 |
| 0111 | S | D | E | L | 1971 | 63662 |
| 0111 | S | D | E | L | 1973 | 69028 |
| 0139 | S | D | R | N | 1970 | 62991 |
| 0144 | S | D | E | N | 1972 | 78292 |
| 0158 | D | D | E | L | 1972 | 69701 |
| 0158 | S | D | E | L | 1972 | 69701 |
| 0175 | S | D | E | N | 1971 | 65832 |
| 0187 | S | D | E | L | 1971 | 63661 |
| 0187 | S | D | E | L | 1971 | 63662 |
| 0192 | S | D | E | L | 1971 | 63052 |
| 0240 | S | D | E | N | 1965 | 36324 |
| 0240 | S | D | E | L | 1973 | 71361 |
| 0240 | S | G | E | N | 1970 | 77936 |
| 0241 | S | D | E | N | 1965 | 36324 |
| 0241 | S | D | E | U | 1971 | 61603 |
| 0241 | S | D | E | L | 1973 | 71361 |
| 0241 | S | G | E | N | 1970 | 77936 |
| 0242 | S | D | E | L | 1973 | 72903 |
| 0246 | S | D | R | N | 1969 | 59066 |
| 0247 | S | D | R | N | 1969 | 59066 |
| 0250 | S | G | R | N | 1971 | 63862 |
| 0251 | S | G | R | N | 1971 | 63862 |
| 0252 | S | G | R | N | 1971 | 63862 |
| 0253 | S | G | R | N | 1971 | 63862 |
| 0254 | S | G | R | N | 1971 | 63862 |
| 0255 | S | G | R | N | 1971 | 63862 |
| 0257 | S | D | E | U | 1971 | 61603 |
| 0257 | S | M | E | N | 1972 | 68340 |
| 0259 | S | D | E | N | 1972 | 78292 |
| 0280 | S | D | E | U | 1971 | 67368 |
| 0281 | S | D | E | U | 1971 | 67368 |
| 0281 | S | M | E | N | 1972 | 68340 |
| 0281 | S | D | E | L | 1973 | 71132 |
| 0298 | S | D | E | N | 1973 | 68977 |
| 0299 | S | D | E | N | 1973 | 68977 |
| 0382 | S | D | R | N | 1970 | 62991 |
| 0434 | S | D | E | N | 1971 | 65832 |
| 0435 | S | D | E | U | 1971 | 67368 |
| 0442 | S | D | E | L | 1971 | 63661 |
| 0448 | P | D | G | N | 1974 | 75417 |
| 0454 | S | D | E | L | 1973 | 71132 |
| 0603 | S | D | E | N | 1972 | 78292 |
| 0683 | S | D | R | N | 1971 | 75483 |
| 0683 | S | D | E | N | 1971 | 75484 |
| 0729 | S | D | E | L | 1971 | 63661 |
| 0730 | S | D | E | L | 1971 | 63661 |
| 0730 | S | D | E | L | 1971 | 63662 |
| 0731 | S | D | E | L | 1971 | 63662 |
| 0731 | S | D | E | L | 1973 | 72903 |
| 0875 | S | D | R | N | 1971 | 75483 |
| 0875 | S | D | E | N | 1971 | 75484 |
| 0892 | P | D | G | N | 1974 | 75417 |
| 0959 | S | D | F | N | 1974 | 73509 |
| 0960 | S | D | F | N | 1974 | 73509 |
| 0961 | S | D | F | N | 1974 | 73509 |
| 0962 | S | D | F | N | 1974 | 73509 |
| 0963 | S | D | F | N | 1974 | 73509 |
| 0964 | S | D | F | N | 1974 | 73509 |
| 0965 | S | D | F | N | 1974 | 73509 |
| 0966 | S | D | F | N | 1974 | 73509 |
| 0967 | S | D | F | N | 1974 | 73509 |
| 0976 | S | D | E | L | 1974 | 75784 |
| 0977 | S | D | E | L | 1974 | 75784 |
| 1004 | S | D | E | N | 1956 | 76857 |
| 1005 | S | D | E | N | 1956 | 76857 |
| 514-0048 | S | M | E | N | 1972 | 68340 |
| 0177 | S | D | F | N | 1974 | 73509 |
| 0257 | S | M | E | N | 1972 | 68340 |
| 0278 | S | M | E | N | 1972 | 68340 |
| 0279 | S | M | E | N | 1972 | 68340 |
| 0280 | S | M | E | N | 1972 | 68340 |
| 0281 | S | M | E | N | 1972 | 68340 |
| 0308 | P | D | E | U | 1973 | 71767 |
| 0412 | S | D | E | L | 1968 | 69016 |
| 0413 | S | D | E | L | 1968 | 69016 |
| 0414 | S | D | E | L | 1968 | 69016 |
| 514-0415 | S | D | E | L | 1968 | 69016 |
| 0418 | S | D | F | N | 1974 | 73509 |
| 0959 | S | D | F | N | 1974 | 73509 |
| 515-0158 | S | D | E | L | 1972 | 69701 |
| 0643 | S | D | E | N | 1971 | 61308 |
| 0643 | S | D | E | L | 1971 | 61671 |

**Phys. State:** **D.** Doped; **E.** Expanded; **F.** Fibrous or Whisker; **G.** Gas; **L.** Liquid; **M.** Multiphase; **P.** Powder; **S.** Solid
**Subject:** **C.** Calculated Data + Theory; **D.** Data; **E.** Experiment; **G.** General (Data + Theory + Experiment); **M.** Measured Data + Experiment; **S.** Survey (Review, Compendium, etc.); **T.** Theory
**Language:** **C.** Czech; **D.** Dutch; **E.** English; **F.** French; **G.** German; **I.** Italian; **J.** Japanese; **O.** Other Languages; **P.** Polish; **R.** Russian; **S.** Spanish
**Temperature:** **F.** Full Range (Low + Normal + High); **L.** Low (0 to 75K) + Overlap into Normal; **N.** Normal (75 to 1273K); **H.** High (above 1273K) + Overlap into Normal; **U.** Undefined

# Chapter 9 Prandtl Number

| Substance Number | Phys. State | Sub-ject | Lan-guage | Temper-ature | Year | TPRC Number |
|---|---|---|---|---|---|---|
| 512-0001 | G | C | G | N | 1972 | 66889 |
| 0003 | G | C | G | N | 1972 | 66889 |
| 0008 | G | D | E | N | 1977 | 93634 |
| 0008 | G | T | E | N | 1977 | 93634 |
| 0008 | L | D | E | N | 1977 | 93634 |
| 0008 | L | T | E | N | 1977 | 93634 |
| 0012 | G | C | G | N | 1972 | 66889 |
| 0018 | G | C | G | N | 1972 | 66889 |
| 0036 | G | C | G | N | 1972 | 66889 |
| 0037 | G | C | G | N | 1972 | 66889 |
| 0051 | G | G | E | N | 1971 | 65727 |
| 0052 | G | G | E | N | 1971 | 65727 |
| 0061 | G | C | G | N | 1972 | 66889 |
| 0096 | G | D | E | H | 1971 | 64330 |
| 0105 | L | D | E | N | 1977 | 93634 |
| 0105 | L | T | E | N | 1977 | 93634 |
| 0106 | G | D | R | H | 1972 | 70621 |
| 0106 | G | D | E | H | 1972 | 70622 |
| 0152 | G | C | G | N | 1972 | 66889 |
| 0158 | G | C | G | N | 1972 | 66889 |
| 0193 | G | C | G | N | 1972 | 66889 |
| 0193 | G | D | R | H | 1972 | 70621 |
| 0193 | G | D | E | H | 1972 | 70622 |
| 0291 | L | D | E | N | 1977 | 93634 |
| 0291 | L | T | E | N | 1977 | 93634 |
| 0329 | L | D | E | N | 1977 | 93634 |
| 0329 | L | T | E | N | 1977 | 93634 |
| 0365 | G | D | E | N | 1977 | 93634 |
| 0365 | G | T | E | N | 1977 | 93634 |
| 0365 | L | D | E | N | 1977 | 93634 |
| 0365 | L | T | E | N | 1977 | 93634 |
| 0476 | G | G | E | N | 1971 | 65727 |
| 1082 | L | D | E | N | 1977 | 93634 |
| 1082 | L | T | E | N | 1977 | 93634 |
| 514-0020 | G | S | E | F | 1971 | 62671 |
| 0020 | G | G | E | N | 1971 | 65727 |
| 0020 | G | S | E | N | 1971 | 71643 |
| 0020 | G | C | E | N | 1974 | 76376 |
| 0020 | G | D | E | N | 1977 | 93634 |
| 0020 | G | T | E | N | 1977 | 93634 |
| 0020 | L | D | E | N | 1977 | 93634 |
| 0020 | L | T | E | N | 1977 | 93634 |
| 0129 | G | D | R | H | 1972 | 70621 |
| 0129 | G | D | E | H | 1972 | 70622 |
| 0171 | G | C | E | H | 1967 | 47652 |
| 0386 | G | T | R | U | 1970 | 73706 |
| 0386 | G | T | E | U | 1970 | 73707 |
| 515-0032 | L | D | R | N | 1970 | 75459 |
| 0032 | L | D | E | N | 1970 | 75460 |
| 0042 | L | D | R | N | 1970 | 75459 |
| 0042 | L | D | E | N | 1970 | 75460 |
| 0311 | L | D | R | N | 1963 | 37245 |
| 517-0001 | G | D | G | N | 1972 | 68888 |
| 0003 | G | D | G | N | 1972 | 68888 |
| 0158 | G | D | G | N | 1972 | 68888 |
| 0193 | G | D | G | N | 1972 | 68888 |
| 1053 | G | D | E | H | 1968 | 36538 |

**Phys. State:** **D.** Doped; **E.** Expanded; **F.** Fibrous or Whisker; **G.** Gas; **L.** Liquid; **M.** Multiphase; **P.** Powder; **S.** Solid
**Subject:** **C.** Calculated Data + Theory; **D.** Data; **E.** Experiment; **G.** General (Data + Theory + Experiment); **M.** Measured Data + Experiment; **S.** Survey (Review, Compendium, etc.); **T.** Theory
**Language:** **C.** Czech; **D.** Dutch; **E.** English; **F.** French; **G.** German; **I.** Italian; **J.** Japanese; **O.** Other Languages; **P.** Polish; **R.** Russian; **S.** Spanish
**Temperature:** **F.** Full Range (Low + Normal + High); **L.** Low (0 to 75K) + Overlap into Normal; **N.** Normal (75 to 1273K); **H.** High (above 1273K) + Overlap into Normal; **U.** Undefined

# Chapter 10 Thermal Linear Expansion

| Substance Number | Phys. State | Subject | Language | Temperature | Year | TPRC Number |
|---|---|---|---|---|---|---|
| 512-0030 | S | D | E | N | 1972 | 67004 |
| 0031 | S | D | E | N | 1972 | 67004 |
| 0048 | S | D | R | N | 1970 | 64874 |
| 0048 | S | D | E | N | 1972 | 67004 |
| 0050 | S | D | E | U | 1973 | 70322 |
| 0070 | S | D | R | N | 1971 | 68604 |
| 0090 | S | M | G | N | 1964 | 36882 |
| 0091 | S | M | G | N | 1964 | 36882 |
| 0091 | S | D | R | N | 1970 | 62763 |
| 0091 | S | M | E | H | 1972 | 68014 |
| 0091 | S | S | E | H | 1974 | 76377 |
| 0091 | S | M | E | H | 1973 | 77262 |
| 0091 | S | D | E | H | 1962 | 77305 |
| 0092 | S | M | G | N | 1964 | 36882 |
| 0092 | S | D | R | N | 1970 | 62763 |
| 0093 | S | M | G | N | 1964 | 36882 |
| 0093 | S | D | R | N | 1970 | 62763 |
| 0093 | S | M | E | H | 1973 | 77262 |
| 0093 | S | D | E | H | 1962 | 77305 |
| 0094 | S | M | G | N | 1964 | 36882 |
| 0094 | S | D | E | H | 1969 | 56631 |
| 0094 | S | D | R | N | 1970 | 62763 |
| 0094 | S | D | E | H | 1965 | 77249 |
| 0094 | S | M | E | H | 1973 | 77262 |
| 0094 | S | D | E | H | 1968 | 78345 |
| 0099 | S | D | R | N | 1971 | 65414 |
| 0099 | S | D | E | N | 1971 | 65415 |
| 0120 | S | D | R | N | 1971 | 65414 |
| 0120 | S | D | E | N | 1971 | 65415 |
| 0174 | S | D | E | H | 1969 | 56631 |
| 0174 | S | D | E | H | 1970 | 77563 |
| 0174 | S | M | E | H | 1969 | 77573 |
| 0175 | S | S | E | H | 1974 | 76377 |
| 0175 | S | D | E | H | 1962 | 77305 |
| 0175 | S | D | E | N | 1964 | 77354 |
| 0175 | S | M | E | H | 1974 | 77928 |
| 0233 | S | M | E | H | 1972 | 68014 |
| 0233 | S | D | E | H | 1971 | 70372 |
| 0240 | S | D | E | L | 1972 | 65400 |
| 0241 | S | D | E | L | 1972 | 65400 |
| 0259 | S | D | R | N | 1970 | 64874 |
| 0262 | S | D | E | N | 1963 | 70328 |
| 0262 | S | D | R | H | 1970 | 74076 |
| 0262 | S | D | E | H | 1970 | 74077 |
| 0262 | S | D | E | N | 1974 | 74519 |
| 0262 | S | S | E | H | 1974 | 76377 |
| 0262 | S | D | E | H | 1965 | 77249 |
| 0262 | S | M | E | H | 1973 | 77262 |
| 0262 | S | D | E | H | 1962 | 77305 |
| 0262 | S | M | E | H | 1974 | 77928 |
| 0262 | S | D | E | H | 1968 | 78345 |
| 0281 | S | D | F | N | 1971 | 65620 |
| 0287 | S | D | E | N | 1963 | 70328 |
| 0287 | S | D | E | N | 1974 | 74519 |
| 0301 | S | D | F | N | 1971 | 65620 |
| 0418 | S | D | E | U | 1973 | 70322 |
| 0428 | S | D | E | H | 1971 | 70372 |
| 0430 | S | D | E | N | 1974 | 77377 |
| 0430 | S | D | J | U | 1973 | 77946 |
| 0430 | S | D | E | U | 1973 | 77947 |
| 0431 | S | D | E | N | 1974 | 77377 |
| 0431 | S | D | J | U | 1973 | 77946 |
| 0431 | S | D | E | U | 1973 | 77947 |
| 0433 | S | D | E | N | 1974 | 77377 |
| 0441 | S | D | G | H | 1972 | 70392 |
| 0442 | S | D | G | H | 1972 | 70392 |
| 0445 | S | D | R | H | 1970 | 74076 |
| 0445 | S | D | E | H | 1970 | 74077 |
| 0447 | S | D | E | N | 1973 | 71230 |
| 0448 | S | D | E | N | 1973 | 71230 |
| 0454 | S | D | R | N | 1971 | 68604 |
| 0694 | S | D | R | N | 1972 | 74097 |
| 0694 | S | D | E | N | 1972 | 74098 |
| 0696 | S | D | R | N | 1972 | 74097 |
| 0696 | S | D | E | N | 1972 | 74098 |
| 0705 | S | M | E | L | 1969 | 63999 |
| 0706 | S | M | E | L | 1969 | 63999 |
| 0740 | S | S | E | H | 1974 | 76377 |
| 0756 | S | D | R | H | 1970 | 74076 |
| 0756 | S | D | E | H | 1970 | 74077 |
| 1009 | S | D | E | N | 1964 | 77354 |
| 1028 | S | S | E | H | 1974 | 76377 |
| 1042 | S | D | E | H | 1970 | 77563 |
| 1043 | S | D | E | H | 1970 | 77563 |
| 1048 | S | M | E | H | 1969 | 77573 |
| 514-0024 | S | D | E | N | 1972 | 67004 |
| 0031 | S | D | E | N | 1972 | 67004 |
| 0048 | S | D | E | N | 1972 | 67004 |
| 0050 | S | D | E | U | 1973 | 70322 |
| 0174 | S | D | E | H | 1970 | 77561 |
| 0174 | S | D | E | H | 1970 | 77563 |
| 0174 | S | M | E | H | 1969 | 77573 |
| 0175 | S | D | E | H | 1970 | 77561 |
| 0175 | S | D | E | H | 1970 | 77563 |
| 514-0175 | S | M | E | H | 1969 | 77573 |
| 0262 | S | D | R | H | 1970 | 74072 |
| 0262 | S | D | E | H | 1970 | 74073 |
| 0262 | S | D | R | H | 1970 | 74076 |
| 0262 | S | D | E | H | 1970 | 74077 |
| 0287 | S | D | R | H | 1970 | 74072 |
| 0287 | S | D | E | H | 1970 | 74073 |
| 0419 | S | D | E | U | 1973 | 70322 |
| 0420 | S | D | E | U | 1973 | 70322 |
| 0432 | S | D | G | H | 1972 | 70392 |
| 0437 | S | D | G | H | 1972 | 70392 |
| 0439 | S | D | G | H | 1972 | 70392 |
| 0440 | S | D | G | H | 1972 | 70392 |
| 0445 | S | D | R | H | 1970 | 74072 |
| 0445 | S | D | E | H | 1970 | 74073 |
| 0445 | S | D | R | H | 1970 | 74076 |
| 0445 | S | D | E | H | 1970 | 74077 |
| 0445 | S | D | E | H | 1970 | 77561 |
| 0740 | S | D | R | H | 1970 | 74072 |
| 0740 | S | D | E | H | 1970 | 74073 |
| 0740 | S | D | R | H | 1970 | 74076 |
| 0740 | S | D | E | H | 1970 | 74077 |
| 0741 | S | D | R | H | 1970 | 74072 |
| 0741 | S | D | E | H | 1970 | 74073 |
| 1042 | S | D | E | H | 1970 | 77561 |
| 1042 | S | D | E | H | 1970 | 77563 |
| 1043 | S | D | E | H | 1970 | 77561 |
| 1043 | S | D | E | H | 1970 | 77563 |
| 1048 | S | M | E | H | 1969 | 77573 |
| 1051 | S | D | E | N | 1973 | 78311 |

**Phys. State:** **D.** Doped; **E.** Expanded; **F.** Fibrous or Whisker; **G.** Gas; **L.** Liquid; **M.** Multiphase; **P.** Powder; **S.** Solid
**Subject:** **C.** Calculated Data + Theory; **D.** Data; **E.** Experiment; **G.** General (Data + Theory + Experiment); **M.** Measured Data + Experiment; **S.** Survey (Review, Compendium, etc.); **T.** Theory
**Language:** **C.** Czech; **D.** Dutch; **E.** English; **F.** French; **G.** German; **I.** Italian; **J.** Japanese; **O.** Other Languages; **P.** Polish; **R.** Russian; **S.** Spanish
**Temperature:** **F.** Full Range (Low + Normal + High); **L.** Low (0 to 75K) + Overlap into Normal; **N.** Normal (75 to 1273K); **H.** High (above 1273K) + Overlap into Normal; **U.** Undefined

# Chapter 11 Thermal Volumetric Expansion

| Substance Number | Phys. State | Sub-ject | Lan-guage | Temper-ature | Year | TPRC Number |
|---|---|---|---|---|---|---|
| 512-0030 | S | D | E | N | 1972 | 67004 |
| 0031 | S | D | E | N | 1972 | 67004 |
| 0048 | S | D | E | N | 1972 | 67004 |

**Phys. State:** **D.** Doped; **E.** Expanded; **F.** Fibrous or Whisker; **G.** Gas; **L.** Liquid; **M.** Multiphase; **P.** Powder; **S.** Solid
**Subject:** **C.** Calculated Data + Theory; **D.** Data; **E.** Experiment; **G.** General (Data + Theory + Experiment); **M.** Measured Data + Experiment; **S.** Survey (Review, Compendium, etc.); **T.** Theory
**Language:** **C.** Czech; **D.** Dutch; **E.** English; **F.** French; **G.** German; **I.** Italian; **J.** Japanese; **O.** Other Languages; **P.** Polish; **R.** Russian; **S.** Spanish
**Temperature:** **F.** Full Range (Low + Normal + High); **L.** Low (0 to 75K) + Overlap into Normal; **N.** Normal (75 to 1273K); **H.** High (above 1273K) + Overlap into Normal; **U.** Undefined

# Part C
# BIBLIOGRAPHY

PART C

# USE OF THE BIBLIOGRAPHY

## A. GENERAL CONSIDERATIONS

Because of the wide variety of literature sources (i.e., journal publications, dissertations, government and industrial reports, books, etc.), three specific formats for the citations are used in the *Bibliography*. CINDAS' procedures are described below:

1. Titles reported in the *Bibliography* are taken from an abstract or from the original document. In the case of translated titles, discrepancies often exist between various sources. In general, CINDAS makes no special effort to check the accuracy of titles.
2. The names of authors reported in abstracts, and even in original publications, are at times misspelled and/or incomplete. CINDAS attempts to report the correct names of the authors.
3. The names of scientific and technical journals are abbreviated according to the notations used by the abstracting journals. In those cases where a journal name is not applicable, the name of the publisher, symposium, or disseminating agency is entered in place of the journal name, depending upon the reference work.
4. Foreign language documents and their English translations, if any, are reported in the *Bibliography*. The translation refers to the original document by the words "For original article, see TPRC No. . . ."; the original article refers to the translation by "For English translation, see TPRC No. . . . ." In some cases, translations may refer to original articles reported in the six-volume Supplement I of the Retrieval Guide.
5. Subscripts are either spelled out or written on the line. Examples: $C_P \equiv$ specific heat at constant pressure; $Si_3Fe_5 \equiv$ Si(3) Fe(5); etc.
6. Superscripts are spelled out. Examples: $V^2 \equiv$ V squared; $\sqrt{2} \equiv$ square root of two; etc.

## B. ABBREVIATIONS AND ACRONYMS USED IN BIBLIOGRAPHIC CITATIONS

AAAS. . . . . . . . . American Association for the Advancement of Science
AAIE. . . . . . . . . .American Association of Industrial Engineers
AD. . . . . . . . . . . .Prefix Catalog Codes for Defense Documentation Center (DDC) (Formerly ASTIA)
AEC. . . . . . . . . . .Atomic Energy Commission
AEDC. . . . . . . . .Arnold Engineering and Development Center
AF. . . . . . . . . . . . Air Force
AFB. . . . . . . . . . .Air Force Base
AFML. . . . . . . . .Air Force Materials Laboratory
AFWL. . . . . . . . .Air Force Weapons Laboratory
AGARD. . . . . . .Advisory Group for Aeronautical Research and Development
AIAA. . . . . . . . . American Institute of Aeronautics and Astronautics
AIAE. . . . . . . . . .Associate Institute of Automobile Engineering
AICE. . . . . . . . . .American Institute of Chemical Engineering
AIEE. . . . . . . . . .American Institute of Electrical Engineers (Now IEEE)
AIIE. . . . . . . . . . American Institute of Industrial Engineers
AIME. . . . . . . . . American Institute of Mining and Metallurgical Engineers
AIP. . . . . . . . . . . American Institute of Physics
AISI. . . . . . . . . . .American Iron and Steel Institute
AM. . . . . . . . . . . .Applied Mechanics Reviews
AMRA. . . . . . . . Army Materials Research Agency
ANL. . . . . . . . . . .Argonne National Laboratory
AOA. . . . . . . . . . American Ordnance Association
APDA. . . . . . . . .Atomic Power Development Association, Inc.
API. . . . . . . . . . . American Petroleum Institute
ARC. . . . . . . . . . Aeronautical Research Council
ARDE. . . . . . . . .Armament Research and Development Establishment
ARF. . . . . . . . . . .Armour Research Foundation
ARL. . . . . . . . . . .Aeronautical Research Laboratories
ARS. . . . . . . . . . .American Rocket Society (Now AIAA)
ASHRAE. . . . . . American Society of Heating, Refrigeration, Air Conditioning Engineers
ASM. . . . . . . . . . American Society for Metals
ASME. . . . . . . . .American Society for Mechanical Engineers
ASR. . . . . . . . . . .American Society of Rocketry
ASSE. . . . . . . . . .American Society of Safety Engineers
ASTE. . . . . . . . . .American Society of Tool Engineers
ASTIA. . . . . . . . .Armed Services Technical Information Agency (Now DDC)
ASTM. . . . . . . . .American Society for Testing and Materials
ASTME. . . . . . . American Society of Tool and Manufacturing Engineers
BC. . . . . . . . . . . . British Ceramic Society
BISI. . . . . . . . . . . British Iron and Steel Industry
BR. . . . . . . . . . . . Battelle Technical Review
BU ORD. . . . . . .Bureau of Ordnance
BU WEPS. . . . . .Bureau of Weapons
CFSTI. . . . . . . . .Clearinghouse for Federal Scientific and Technical Information (Now NTIS)
CINDAS–EPIC. . . . . . . .Center for Information and Numerical Data Analysis and Synthesis—Electronic Properties Information Center

CINDAS–
TPRC....... Center for Information and Numerical Data Analysis and Synthesis—Thermophysical Properties Research Center
CNEN......... Comitato Nazionale per L'Energia Nucleare (National Nuclear Energy Committee)
CNDR......... Consiglio Nazionale Delle Ricerche (National Research Council of Italy)
CNES......... Centre National D'Etudes Spatiales (National Center for Space Studies)
COSATI....... Committee on Scientific and Technical Information (Formerly COSI)
COSI.......... Committee on Scientific Information (Now COSATI)
CPIA.......... Chemical Propulsion Information Agency
CRREL........ Cold Regions Research and Engineering Laboratory
DASA......... Defense Atomic Support Agency
DDC.......... Defense Documentation Center
DDR&E....... Directorate of Defense Research and Engineering
DMIC......... Defense Metals Information Center
DOD.......... U. S. Department of Defense
DOFL......... Diamond Ordnance Fuze Laboratory (HDL)
DPGR......... Dugway Proving Ground
ECOM......... Electronics Command, U. S. Army
EPIC.......... Electronic Properties Information Center
EURATOM.... European Atomic Energy Community
FAI........... Federation Aeronautique Internationale
FID........... International Federation of Documentation
FPL........... Forest Production Laboratory
FTD........... Foreign Technology Division
GRA.......... Government Research Announcements (Formerly UGAR)
HDL.......... Harry Diamond Laboratory
IAA........... International Academy of Astronautics
IAS........... Institute of Aero/Space Science (Now AIAA)
IEE........... Institute of Electrical and Electronic Engineers (Formerly AIEE & IRE)
IEEE.......... Institute of Electrical and Electronic Engineers
IIT............ Illinois Institute of Technology Research Institute
IPST.......... Israel Program for Scientific Translation, Ltd.
IRE........... Institute of Radio Engineers (Now IEEE)
ISA........... Instrument Society of America
JAERI......... Japan Atomic Energy Research Institute
JANAF........ Joint Army–Navy–Air Force
JPL........... Jet Propulsion Laboratory
KAPL......... Knolls Atomic Power Plant
MR............ ASM Review of Current Metal Literature (in Metals Review)
NADC......... Naval Air Development Center
NAS........... National Academy of Science
NASA......... National Aeronautics and Space Administration
NASC......... National Aeronautics and Space Council
NAS/NRC..... National Academy of Science—National Research Council
NATO......... North Atlantic Treaty Organization
NBS........... National Bureau of Standards
NDAC......... National Defense Advisory Committee
NDRC......... National Defense Research Council
NOTS......... Naval Ordnance Test Station
NPL........... National Physical Laboratory (England)
NRC.......... National Research Council (Canada)
NRD.......... Naval Research and Development (Now ONR)
NRL........... Naval Research Laboratory
NS............ Nuclear Science Abstracts
NSF........... National Science Foundation
NTIS.......... National Technical Information Service (Formerly CFSTI)
OAR.......... Office Aerospace Research
ONERA....... Office National D'Etudes et de Recherches Aerospatiales (France)
ONR.......... Office of Naval Research
ORNL......... Oak Ridge National Laboratory
OSI........... Office of Scientific Information
OSR.......... Office of Scientific Research
OTS.......... Office of Technological Service
REIC.......... Radiation Effects Information Center
RIA........... Rock Island Arsenal
RM............ ASM Review of Metal Literature
RPL........... Rocket Propulsion Laboratory
RR............ U. S. Government Research Reports (OTS)
SIPRE......... Snow, Ice and Permafrost Research Establishment
TML.......... Titanium Metallurgical Laboratory
TPRC......... Thermophysical Properties Research Center
TT............ Technical Translations (OTS)
UCRL......... University of California Radiation Laboratory
UGAR........ United States Government Announcements Reports (Now GRA)
USAEC........ U. S. Atomic Energy Commission
USBM......... U. S. Bureau of Mines
USDA......... U. S. Department of Agriculture
USJPRS....... U. S. Joint Publication Research Service
USL........... Underwater Sound Laboratory
WADC........ Wright Air Development Center
WADD........ Wright Air Development Division
WRAC........ Willow Run Aeronautical Center

# BIBLIOGRAPHY


*TPRC Number* — *Bibliographic Citation*

34063 TEMPERATURE AND PRESSURE DEPENDENCE OF THE VISCOSITY OF GASES.
CRONIN, D. J.
AM. J. PHYS.
33 ( 10 ), 835–40, 1965.

34078 THE SYSTEM NITROGEN TETROXIDE-TETRANITROMETHANE.
BACHMAIER, F.
DEUT. LUFT RAUMFAHRT FORSCHUNGSBER.
64–29, 28PP., 1964.

34150 WORK IN THE FIELD OF PLASMA PHYSICS.
MICHEL, A.
SIEMENS- Z.
39 ( 5 ), 428–33, 1965.

34644 THE JOULE - THOMSON ( J - T ) EFFECT IN TWO HYDROCARBON - HYDROGEN MIXTURES.
AYBER, R.
KAELTETECHNIK
17 ( 9 ), 276–81, 1965.

34654 PHYSICAL PROPERTIES OF CALCIUM-AMMONIA SOLUTIONS.
WONG, W.-S.
NAVAL ORDANCE LAB. SYMP. AMMONIA BATTERIES, 6TH
1–7PP., 1964.
( N64–22881, AD–433973, NTIS )

35381 THE PROPERTIES OF COMPRESSED GASES.
VODAR, B. SAUREL, J.
HIGH PRESSURE PHYS. CHEM.
1, 51–143, 1963.

35673 CALORIMETRIC MEASUREMENTS ON MIXTURES OF NONELECTROLYTES. I. MOLECULAR HEAT CAPACITY OF THE 1,2- DIBROMOETHANE- BENZENE SYSTEM.
FINDENEGG, G. H. GRUBER, K. PEREIRA, J. F.
KOHLER, F.
MONATSH. CHEM.
96 ( 2 ), 669–78, 1965.

35709 EVAPORATION ENTHALPY AND MEAN SPECIFIC HEAT OF TERNARY POSITIVE-NEGATIVE AZEOTROPES.
PAWLAK, J. ZIELENKIEWICZ, A.
ROCZNIKI CHEM.
39 ( 3 ), 419–23, 1965.

36022 BINARY SALT SYSTEMS STUDY BY THE METHOD OF TEMPERATURE CONDUCTIVITY.
BERG, L. G. SIDOROVA, E. E.
FIZ.-KHIM. ANALIZ, AKAD. NAUK SSSR, SIBIRSK. OTD., INST. NEORGAN. KHIM., TR. YUBILEINOI KONF., NOVOSIBIRSK
201–8, 1963.

36064 EFFECT OF A SULFUR IMPURITY ON HEAT CONDUCTIVITY OF CRYSTALLINE SELENIUM.
SAMEDOVA-BEIBUTOVA, Z. A.
TR. AZERB. POLITEKHN. INST.
( 9 ), 178–84, 1963.

36129 THERMODYNAMIC INVESTIGATION OF THE SILVER AND CUPROUS HALIDE SYSTEMS. I. THE SILVER-CUPROUS ION SYSTEM.
NOELTING, J.
BER. BUNSENGES. PHYSIK. CHEM.
68 ( 10 ), 932–9, 1964.

36130 THERMODYNAMIC INVESTIGATIONS OF SILVER AND CUPROUS HALIDE SYSTEMS. II. THE CONTRIBUTION OF DEFECTS TO SPECIFIC HEAT.
NOELTING, J.
BER. BUNSENGESELL. PHYS. CHEM.
68 ( 10 ), 939–44, 1964.

36131 THERMODYNAMIC INVESTIGATIONS OF THE SILVER AND COPPER( I ) HALIDE SYSTEMS. III. THE PHASE DIAGRAM OF SILVER BROMIDE - COPPER BROMIDE.
NOELTING, J.
BER. BUNSENGES. PHY. CHEM. ( GERMANY )
68 ( 10 ), 944–6, 1964.

36223 HEAT CONDUCTIVITY IN GAS MIXTURES.
SCHRAMM, B. SCHAEFER, K.
BER. BUNSENGES. PHYS. CHEM.
69, 110–4, 1965.

36255 DEPENDENCE OF THE VISCOSITY OF LIQUIDS ON THE PRESSURE UP TO 10,000 KILOGRAM PER SQUARE CENTIMETER.
ZOLOTYKH, E. V.
TR. INST. GOS. KOM. STANDARTOV, MER I IZMERITELN. PRIBOROV SSSR
75, 123–33, 1964.

36324 THE REFLECTION AND TRANSMISSION OF INFRARED MATERIALS. III, SPECTRA FROM 2 TO 50 MICRONS.
MC CARTHY D E
APPL OPT
4 3 317–20 1965 CA 62 14052

36336 THERMAL CONDUCTIVITY OF STARCH PARTICLES.
ROTH, D. D.
IOWA STATE UNIVERSITY, AMES, IOWA, M.S. THESIS.
43PP., 1969.

36538 STUDIES OF ABLATIVE MATERIAL PERFORMANCE FOR SOLID ROCKET NOZZLE APPLICATIONS. FINAL REPT.
SCHAEFER J W DAHM T J RODRIGUEZ D A
REESE J J JR WOOL M R AEROTHERM CORP PALO ALTO CALIF
NASA AND CFSTI
NASA–CR–72429 N68–31937
1–150 1968 PA N68–6–19 3360

36636 SOME PHYSICOCHEMICAL PROPERTIES OF THE FORMAMIDE - ACETONE SYSTEM.
BLANK, M. G.
DOKL. LVOVSK. POLITEKHN. INST.
5 ( 1–2 ), 114–9, 1963.

36663 EFFECT OF ALKALI METAL CATIONS ON THE VISCOSITY OF TOLUENE AND XYLENE SOLUTIONS.
GORENBEIN, E. YA. RUSIN, G. G.
UKR. KHIM. ZH.
30 ( 6 ), 582–9, 1964.

36882 PRODUCTION OF SOLID SOLUTIONS OF CARBIDES AND THE EXAMINATION OF THEIR PHYSICAL PROPERTIES.
SAMSONOV, G. V. PADERNO, V. N.
PLANSEEBER. PULVERMET.
12 ( 1 ), 19–30, 1964.

36908 VISCOMETRY OF MOLTEN METALS.
ANDERSON, A. J.
TRANS. AMER. FOUND. SOC.
71, 305–14, 1963.

37009 CALCULATION OF THE RELATION OF THE THERMAL CONDUCTIVITY TO COMPOSITION IN HEAT AND MASS TRANSFER DURING VAPORIZATION OF LIQUID IN A HOT GAS.
VARSHAVSKII, G. A.
TR. ODESSK. GOS. UNIV., SER. FIZ. NAUK
152 ( 8 ), 43–50, 1962.

37222 VISCOSITY OF NEON - HELIUM AND NEON - ARGON MIXTURES AT 20 AND 30 DEGREES.
KESTIN, J. NAGASHIMA, A.
J. CHEM. PHYS.
40 ( 12 ), 3648–54, 1964.

37245 DIAGRAMS FOR THE DETERMINATION OF THE PHYSICAL PROPERTIES OF NITRIC ACID.
CHERNYSHEV, A. K.
VESTN. TEKHN. I EKON. INFORM. NAUCHN.-ISSLED. TEKHN.-EKON. ISSLED. GOS. KOM. KHIM. I NEFT. PROM. PRI GOSPLANE SSSR
( 4 ), 13–15, 1963.

37296 THE MICROWAVE EMISSIVITY OF TURBULENT PLASMA.
WORT, D. J. H.
J. NUCL. ENERGY
6 ( PT. C ), 237–50, 1964.

37547 APPARATUS FOR THE MEASUREMENT OF LIQUID VISCOSITIES AT HIGH TEMPERATURE UP TO 500 DEGREES C AND UNDER PRESSURE UP TO 20 KILOGRAM PER SQUARE CENTIMETER.
BESSOUAT, R. ELBERG, S.
J. PHYS. ( FRANCE )
25 ( 3 ), 23–30 A, 1964.

37602 THERMAL DIFFUSION IN A HIGHLY-IONIZED PLASMA COLUMN.
STEFANOV, B.
PRINCETON UNIV., N. J. PLASMA PHYSICS LAB.
21PP., 1970.

37642 THE SPECIFIC HEAT OF SOME URANIUM AND THORIUM CARBIDES BETWEEN 1.8 AND 4.2 DEGREES K.
HARNESS, J. B. MATTHEWS, J. C. MORTON, N.
BRIT. J. APPL. PHYS.
15 ( 8 ), 963–6, 1964.

38084 TRANSPORT PROPERTIES OF SELECTED ELEMENTS AND COMPOUNDS IN THE GASEOUS STATE.
LILEY, P. E.
TPRC, PURDUE UNIVERSITY, WEST LAFAYETTE, IND.
75PP., 1972.

38250 RELATION BETWEEN VISCOSITIES AND THE COMPOSITION OF BINARY SOLUTIONS.
SHASTRY, V. R. ASAWA, H. G. SAHANI, N. S.
J. INDIAN CHEM. SOC.
41 ( 10 ), 733–4, 1964.

38291 EXPERIMENTAL STUDY OF THE THERMAL CONDUCTIVITY OF MONATOMIC GASES AND THEIR MIXTURES.
SHASHKOV, A. G. MARCHENKOV, E. I.
FOREIGN TECHNOLOGY DIVISION
27PP., 1972.
( ENGLISH TRANSLATION OF PROBL. TEPLO - MASSOPERENOSA 297–317, 1970., FOR ORIGINAL SEE TPRC NO. 63876 )
( FTD–HT–23–528–72, AVAIL NTIS )

| TPRC Number | Bibliographic Citation |
|---|---|
| 38316 | GAS-PHASE VISCOSITY OF HYDROCARBON MIXTURES.<br>LEE, A. L. EAKIN, B. E.<br>SOC. PETROL. ENGRS. J.<br>4 ( 3 ), 247-9, 1964. |
| 38366 | PHYSIOCHEMICAL STUDY OF THE SYSTEMS ALKYL SULFATES-P-TOLUIDINE HYDROCHLORIDE. I. SPECIFIC CONDUCTIVITY. II. THE TEMPERATURE COEFFICIENT AND THE ACTIVATION ENERGY OF CONDUCTIVITY. III. VISCOSITY. IV. THE TEMPERATURE COEFFICIENT AND THE ACTIVATION ENERGY OF VISCOSITY.<br>ZWIERZYKOWSKI, W. NIEWIADOMSKI, H.<br>ZESZYTY NAUK. POLITECH. GDANSK., CHEM.<br>( 7 ), 65-80, 81-94, 95-103, 105-18, 1964. |
| 38454 | EQUILIBRIUM PROPERTIES OF LIQUID AND SOLID MIXTURES OF HELIUM 3 AND 4.<br>TACONIS, K. W. DE BRUYN OUBOTER, R.<br>PROGR. LOW TEMP. PHYS.<br>4, 38-96, 1964. |
| 38472 | VISCOSITY OF AQUEOUS ALKALI HALIDE SOLUTIONS.<br>LENGYEL, S. TAMAS, J. GIBER, J.<br>HOLDERITH, V.<br>ACTA CHIM. ACAD. SCI. HUNG.<br>40 ( 2 ), 125-43, 1964. |
| 38887 | VISCOSITY OF FUSED CARBONATES OF ALKALI METALS. I. SYSTEMS SODIUM CARBONATE - POTASSIUM CARBONATE, LITHIUM CARBONATE - SODIUM CARBONATE, AND LITHIUM CARBONATE - POTASSIUM CARBONATE.<br>VOROBEV, G. V. PALGUEV, S. F.<br>KARPACHEV, S. V.<br>TR. INST. ELEKTROKHIM., AKAD. NAUK S S S R, URALSK. FILIAL<br>( 6 ), 39-45, 1965. |
| 38895 | THERMODYNAMIC PROPERTIES OF A METHANE-PROPANE MIXTURE AT LOW TEMPERATURES AND HIGH PRESSURES.<br>MANKER, E. A.<br>UNIV. OF MICHIGAN, PH.D. THESIS<br>135PP., 1965.<br>( UM 65-5928, AVAIL. UNIV. MICROFILMS ) |
| 38921 | THE CONTINUOUS DETERMINATION OF THE THERMAL CONDUCTIVITY OF A GASEOUS BINARY MIXTURE.<br>SHASHKOV, A. G. ABRAMENKO, T. N.<br>VESTI AKAD. NAVUK BELARUSK. SSR, SER. FIZ.-TEKHN. NAVUK<br>( 1 ), 64-8, 1965. |
| 41190 | PROPERTIES OF HYDROCARBON MIXTURES FROM THE CORRESPONDING STATES PRINCIPLE.<br>LELAND, T. W., JR. CHAPPELEAR, P. S.<br>LEACH, J. W.<br>AM. PETROL. INST.<br>PREPRINT NO. 36-66, 38PP., 1966. |
| 41259 | BIOPHYSICAL STUDIES ON SUBCELLULAR PARTICLES. I. ROTATIONAL VISCOMETRY—CONSTRUCTION OF A CONE-GLASS PLATE VISCOMETER.<br>MUKOHATA, Y. MASAYASU, M. ISEMURA, T.<br>ANN. REPT. BIOL. WORKS, FAC. SCI., OSAKA UNIV.<br>13, 93-106, 1965. |
| 41432 | SPECIFIC HEAT OF LIQUID HYDROCARBONS.<br>HADDEN, S. T.<br>HYDROCARBON PROCESS.<br>45 ( 7 ), 137-42, 1966. |
| 42002 | VISCOSITY EFFECTS ON THE PHOTOHYDRATION OF PYRIMIDINES.<br>SUMMERS, W. A., JR. BURR, J. G.<br>J. PHYS. CHEM.<br>76 ( 22 ), 3137-41, 1972. |
| 42060 | PHYSICAL - CHEMICAL PROPERTIES OF FUSED SYSTEMS OF SODIUM, POTASSIUM, AND ZIRCONIUM FLUORIDES AND CHLORIDES.<br>SHEIKO, I. N.<br>TR. VSES. SOVESHCH. PO FIZ. KHIM. RASPLAVLEN. SOLEI, 2ND, KIEV 1963<br>79-87, 1965. |
| 42140 | EXPERIMENTAL DETERMINATION OF THE HEAT AND TEMPERATURE CONDUCTIVITIES OF THE SYSTEM SODIUM CHLORIDE-CALCIUM CHLORIDE.<br>VENERAKI, I. E. CHEBOTAREV, V. A.<br>TOPOL'NITSKII, G. G. ROMAN'KO, K. S.<br>VESTN. KIEV. POLITEKH. INST., SER. TEPLOENERG.<br>( 2 ), 140-6, 1965. |
| 42186 | SPECTROPHOTOMETRIC PROPERTIES OF THE LOW- CURRENT CARBON ARC.<br>MAGDEBURG, H. SCHLEY, U.<br>Z. ANGEW. PHYS.<br>20 ( 5 ), 465-73, 1966. |
| 42261 | THE SPECIFIC HEAT AT CONSTANT OSMOTIC PRESSURE OF HELIUM-3 IN SUPERFLUID HELIUM-4. FROM PROCEEDINGS OF THE 12TH INTERNATIONAL CONFERENCE ON LOW TEMPERATURE PHYSICS. KYOTO JAPAN, SEPTEMBER 4-10, 1970.<br>RADEBAUGH, R. SIEGWARTH, J. D.<br>KEIGAKU PUBLISHING CO., CIYODA-KU, TOKYO, JAPAN<br>163-5, 1971. |

| TPRC Number | Bibliographic Citation |
|---|---|
| 42486 | ALLOWANCE FOR THE TEMPERATURE DEPENDENCE OF SPECIFIC HEAT OF INDIVIDUAL GASES AND GAS MIXTURES IN THERMODYNAMIC CALCULATIONS FOR GAS FLOW MACHINES.<br>FISCHER, S.<br>FREIBERGER FORSCHUNGSH., A<br>381, 117, 1965. |
| 42555 | DETERMINATION OF THE SPECIFIC HEAT CAPACITY OF CEMENT MIXES.<br>KULIEV, S. M. ABDINOV, M. A.<br>IZV. AKAD. NAUK AZERB. SSR, SER. GEOL.-GEORGR. NAUK<br>( 3 ), 66-71, 1965. |
| 42568 | RELAXATION TIMES, MUTUAL AND AVERAGED MUTUAL VISCOSITIES OF SOME SUBSTITUTED XYLENES.<br>MEHROTRA, N. K. SAXENA, M. C.<br>INDIAN J. PURE APPL. PHYS.<br>4 ( 11 ), 438-40, 1966. |
| 43155 | DYNAMIC VISCOSITY COEFFICIENT OF GAS MIXTURES.<br>MAKAVECKAS, R.<br>SILUMINE ENERG., KAUNO POLITECHN. INST., MOKSLINES-TECHN. KONF. DARBAI, 14TH, KAUNAS<br>5-9, 1965. |
| 43208 | THERMAL CONDUCTIVITY OF HELIUM- ARGON MIXTURES UP TO 330 DEGREES C AND 200 BAR.<br>TUFEU, R. GARRABOS, Y. LE NEINDRE, B.<br>COMPT. REND.<br>273 B ( 16 ), 696-9, 1971. |
| 43236 | RELATIONS BETWEEN THE THERMAL CONDUCTIVITY OF SILICON CARBIDE STOPPER HEAD AND CORROSION OF THE STOPPER ROD.<br>NIWA, S. HAGHIHARA, J.<br>TAIKABUTSU<br>15 ( 77 ), 338-40, 1964. |
| 43252 | DETERMINATION OF THE THERMOPHYSICAL CHARACTERISTICS OF RAW MATERIAL WITH CONSIDERATION FOR CHEMICAL CONVERSIONS AND HEATING RATE.<br>VISHNEVSKII, K. P. KITOV, R. N.<br>STEPUKHIN, A. S.<br>TR., GOLOVN. NAUCHN.-ISSLED. INST. TSEMENTN. MASHINOSTR.<br>( 3 ), 8-12, 1964. |
| 43271 | FLUIDITY AND CONDUCTANCE IN AQUEOUS ELECTROLYTE SOLUTIONS. AN APPROACH FROM THE GLASSY STATE AND HIGH-CONCENTRATION LIMIT. I. CALCIUM NITRATE SOLUTIONS.<br>ANGELL, C. A. BRESSEL, R. D.<br>J. PHYS. CHEM.<br>76 ( 22 ), 3244-53, 1972. |
| 43277 | JOULE- THOMSON EFFECT IN METHANE- HYDROGEN AND ETHYLENE- HYDROGEN MIXTURES.<br>AYBER, R.<br>V D I FORSCHUNGSH.<br>( 511 ), 35PP., 1965. |
| 43291 | CALCULATION OF THERMAL CONDUCTIVITY OF BINARY GAS MIXTURES.<br>SHASHKOV, A. G. ABRAMENKO, T. N.<br>VESTSI AKAD. NAVUK BELARUSK. SSR, SER. FIZ-TEKHN. NAVUK<br>( 3 ), 49-54, 1965. |
| 44126 | THERMODYNAMIC PROPERTIES OF AZEOTROPIC MIXTURES OF FREON 22 AND FREON 115.<br>BADYLKES, I. S.<br>BULL. INST. INTERN. FROID, ANNEXE<br>( 4 ), 195-202, 1965. |
| 44432 | EXPERIMENTAL HEAT CONDUCTIVITIES OF PURE LIQUIDS AND MIXTURES.<br>PURANASAMRIDDHI, D.<br>KAELTETECHNIK<br>18 ( 12 ), 445-50, 1966. |
| 44504 | HEATS OF MIXING, EXCESS VOLUMES OF MIXING AND EXCESS HEAT CAPACITY OF THE SYSTEMS FORMED BY TERT- BUTYL ALCOHOL AND NORMAL HYDROCARBONS.<br>RECKO, W. M. SADOWSKA, K. W.<br>WOYCICKA, M. K.<br>BULL. ACAD. POLON. SCI., SER. SCI. CHIM.<br>19 ( 8 ), 475-9, 1971. |
| 44540 | PHYSICAL PROPERTIES OF THE ACETONE-METHYL ETHER-BORON FLUORIDE COMPLEX.<br>KARAMYAN, A. T. KAMINSKII, V. A.<br>BOCHIKASHVILI, T. P.<br>SOOBSHCH. AKAD. NAUK GRUZ. SSR<br>44 ( 3 ), 585-8, 1966. |
| 44544 | CALCULATION OF SPECTRAL PROPERTIES OF EMITTERS AT CALCULATION OF OUTWARD HEAT TRANSFER IN FLAMING KILNS.<br>DONETS, V. I. KLEKL, A. E.<br>IZV. VYSSH. UCHEB. ZAVED., CHERN. MET.<br>14 ( 11 ), 170-2, 1971. |

**TPRC Number** | ***Bibliographic Citation***

44572 **LINEAR CORRELATIONS OF VARIOUS PROPERTIES OF CHEMICAL COMPOUNDS.**
MASLOV, P. G.
UCH. ZAP., LENINGRAD. GOS. PEDAGOG. INST.
265, 399–408, 1965.

44588 **THE VISCOSITY OF NITROGEN TETROXIDE UNDER ATMOSPHERIC PRESSURE.**
TIMOFEEV, B. D.
VESTSI AKAD NAVUK BELARUS. SSR, SER. FIZ.–TEKH. NAVUK
( 4 ), 27–31, 1966.

44589 **EXPERIMENTAL STUDY OF SPECIFIC HEAT OF DISSOCIATING NITROGEN TETROXIDE EQUILIBRIUM.**
NESTERENKO V. B.   TIMOFEEV, B. D.
IL'YUKHIN, YU. D.
VESTSI AKAD. NAVUK BELARUS. SSR, SER. FIZ.–TEKH. NAVUK
( 4 ), 123–5, 1966.

44787 **SOME PROPERTIES OF COMPOUNDS OF GROUPS II, IV AND V TYPE.**
CHIANG, P.–H.   TYCHINA, I. I.   OSMANOV, E.
GORYUNOVA, N. A.
DOKL. NAUCH KONF., LENINGRAD. INZH.–STROIT. INST., 21ST, LENINGRAD
8–11, 1963.

45152 **EFFECT OF FORMULATION VARIABLES ON THE THERMAL CONDUCTIVITY OF GRAPHITE. FROM PROCEEDINGS OF THE 5TH CONFERENCE ON THERMAL CONDUCTIVITY. DENVER HILTON HOTEL, DENVER, COLORADO. OCTOBER 20–22, 1965. VOLUME 2.**
FAIR, F. V.
SPEER CARBON COMPANY
PP. V–C–1–V–C–19, 1965.

45340 **THERMODYNAMIC CHARACTERISTICS OF COMPOUNDS CONTAINING ZINC.**
MASLOV, P. G.
UCH. ZAP., LENINGRAD. GOS. PEDAGOG. INST.
265, 362–72, 1965.

45354 **PHYSICAL AND CHEMICAL PROPERTIES OF SOME WATER – UREA – AMMONIUM SALT SYSTEMS.**
BARKAN, A. S.   TUPCHY, I. A.
VESTSI AKAD. NAVUK BELARUS. SSR, SER. KHIM. NAVUK
( 1 ), 44–50, 1967.

45651 **THERMODYNAMIC PROPERTIES OF GASES AND LIQUIDS. VOL. I. THERMODYNAMIC PROPERTIES OF AIR IN TEMPERATURE RANGE BETWEEN MINUS 210 C AND 1250 C WITH PRESSURES TO 4500 BAR.**
BAEHR, H. D.   SCHWIER, K.
SPRINGER–VERLAG, BERLIN, HEIDELBERG, NEW YORK
1, 136PP., 1961.

46165 **A NEW HOT–WIRE MODIFICATION OF THE THERMAL CONDUCTIVITY CELL FOR LIQUIDS AND GASES.**
RASTORGUEV, YU. L.   GELLER, V. Z.
INZH.–FIZ. ZH.
13 ( 1 ), 16–23, 1967.
( FOR ENGLISH TRANSLATION SEE TPRC NO. 64176 )

46295 **VISCOSITY AND DIFFUSION OF MATERIAL IN A MIXED GASEOUS AND DENSE STATE.**
PARATELLA, A.
ATTI IST. VENETO SCI., LETT. ARTI, CL. SCI. MAT. NATUR.
128, 403–14, 1965–1966.

46359 **EFFECT OF PARTICLE SPATIAL DISTRIBUTION ON THERMAL CONDUCTIVITY OF TWO–PHASE SYSTEMS.**
DESHPANDE, P. B.
UNIVERSITY OF ARKANSAS, PH.D. THESIS
135PP., 1970.
( UM 70–17185, N71–31252, AVAIL. UNIV. MICROFILMS )

47005 **SHOCK REFLECTION AND SURFACE EFFECTS IN THE SHOCK TUBE.**
BUSING, J. R.   CLARKE, J. F.
CONF. PROC.
( 12 ), 165–81, 1967.

47235 **THE EMISSIVITY AND THE ABSORPTION COEFFICIENT OF A HOMOGENEOUS NON–EQUILIBRIUM PLASMA.**
BEL, N.   MANGENEY, A.
ANN. ASTROPHYS.
30, 649–657, 1967.

47301 **HYDRAZINE NITRATE–HYDRAZINE AS A MONERGOLIC PROPELLANT. II.**
BACHMAIER, F.
DEUT. LUFT– RAUMFAHRT, FORSCHUNGSBER.
22PP., 1967.
DLR–FB–67–22

47643 **ULTRAVIOLET REFLECTIVITY OF VENUS AND JUPITER.**
EVANS, D. C.
7TH INTERN. SPACE SCI. SYMP.
135–49, 1967.

47645 **TRANSPORT PROPERTY DETERMINATION FROM ARC PLASMAS.**
UHLENBUSCH, J.   FISCHER, E.
PROCEEDINGS OF INTERN. CONF. ON PHENOMENA IN IONIZED GASES, 7TH BELGRADE, YUGOSLAVIA, AUG. 22–7, 1965.
703–9, 1966.
( A67–35067 19–25 )

47652 **TRANSPORT PROPERTIES OF THERMALLY IONIZED AIR.**
KULIK, P. P.   PANEVIN, Y. G.
8TH INT. CONF. PHENOMENA IN IONIZ. GASES
306, 1967.

47653 **PLASMA PROPERTIES AND COLLISION CROSS SECTIONS FROM ARC MEASUREMENTS.**
HERMANN, W.
INTERN. CONF. PHENOMENA IN IONIZED GASES, 8TH
308, 1967.

47654 **PROPERTIES OF THE HYDROGEN – PLASMA UP TO 24,000 DEGREES K.**
PLANTIKOW, U.   STEINBERGER, S.
8TH INT. CONF. PHENOMENA IN IONIZ. GASES
315, 1967.

47655 **PROPERTIES OF THE NITROGEN PLASMA UP TO 25,000 DEGREES K.**
SCHADE, E.
8TH INT. CONF. PHENOMENA IN IONIZ. GASES
316, 1967.

47656 **THE EMISSIVITY AND ABSORPTION COEFFICIENT OF AN HOMOGENEOUS NON–EQUILIBRIUM PLASMA.**
BEL, N.   MANGENEY, A.
8TH INT. CONF. PHENOMENA IN IONIZ. GASES
445, 1967.

47657 **EMISSION OF AN ARGON PLASMA JET IN THE NEIGHBOURHOOD OF THE PLASMA FREQUENCY 50 CM( –1 ) LESS THAN SIGMA LESS THAN 200 C M( –1 ).**
CHAPELLE, J.
8TH INT. CONF. PHENOMENA IN IONIZ. GASES
446, 1967.

47658 **SHOCK WAVE STRUCTURE IN RECOMBINING LOW–DENSITY PLASMAS.**
CHRISTIANSEN, W. H.
8TH INT. CONF. PHENOMENA IN IONIZ. GASES
467, 1967.

48010 **MAGNETIC FIELD MEASUREMENT OF ELECTRON THERMAL CONDUCTIVITY AND ELECTRON – ATOM CROSS SECTION IN CESIUM PLASMA.**
GERASIMOV, N.   STEFANOV, B.   ZARKOVA, L.
PHENOMENA IONIZ. GDSES, INT. CONF., CONTRIB. PAP., 8TH, VIENNA
313, 1967.

48539 **VISCOSITY AND DENSITY OF POTASSIUM ALUMINATE SOLUTIONS.**
LAINER, YU. A.   KITLER, I. N.
MET. TSVET. REDK. METAL., AKAD. NAUK S S S R, INST. MET.
187–91, 1967.

49040 **MEASUREMENT OF VISCOSITY OF MIXTURES OF DIPHENYL, OMP, AND HIGH BOILING RESIDUE.**
TERMINE, G.   ZANELLI, S.
RIV. ING. NUCL.
3, 860–7, 1965.

49045 **DETERMINATION OF THE RADIAL DISTRIBUTION OF THE ABSORPTION AND EMISSION COEFFICIENTS IN OPTICALLY DENSE PLASMA.**
KOLOBOVA, G. A.
IZV. VYSSH. UCHEB. ZAVED., FIZ.
10 ( 12 ), 75–80, 1967.
( FOR ENGLISH TRANSLATION SEE TPRC NO. 49046 )

49046 **DETERMINATION OF THE RADIAL DISTRIBUTION OF THE ABSORPTION AND EMISSION COEFFICIENTS IN OPTICALLY DENSE PLASMA.**
KOLOBOVA, G. A.
SOV. PHYS. J.
10 ( 12 ), 46–50, 1967.
( ENGLISH TRANSLATION OF IZV. VYSSH. UCHEB. ZAVED., FIZ. 10 ( 12 ), 75–80, 1967., FOR ORIGINAL SEE TPRC NO. 49045 )

49226 **THERMAL CONDUCTIVITY OF BINARY GAS MIXTURES AT LOW TEMPERATURES.**
SHASHKOV, A. G.   ABRAMENKO, T. N.
INZH.–FIZ. ZH.
13 ( 4 ), 549–53, 1967.
( FOR ENGLISH TRANSLATION SEE TPRC NO. 64189 )

49504 **DETERMINATION OF THERMAL PROPERTIES OF LOOSE MATERIALS AND THICK BEDS OF VARIOUS MATERIALS.**
MEDVEDEV, N. N.
INZH. – FIZ. ZH.
14 ( 2 ), 329–33, 1968.
( FOR ENGLISH TRANSLATION SEE TPRC NO. 65123 )

| TPRC Number | Bibliographic Citation |
|---|---|
| 49693 | **CALCULATION OF THE ELECTRICAL CONDUCTIVITY, HEAT CONDUCTIVITY, AND EMITTANCE OF AN ARGON PLASMA.**<br>ISMAILOVA, D. MERENKOVA, L. K.<br>NESMACHNAYA, I. P. ENGEL'SHT, V. S.<br>ISSLED. ELEK. DUGI ARGONE<br>3-23, 1966. |
| 49792 | **THERMAL CONDUCTIVITY OF TRANSISTOR CAP FILLER MATERIAL.**<br>STAAR, K.<br>HIRADASTECH. IPARI KUT. INTEZ. KOZLEM.<br>6 ( 4 ), 53-9, 1966. |
| 49902 | **THERMAL CONDUCTIVITY OF GASES OVER A WIDE TEMPERATURE RANGE.**<br>CORREIA, P. SCHRAMM, B. SCHAEFER, K.<br>BER. BUNSENGES. PHYS. CHEM.<br>72 ( 3 ), 393-9, 1968. |
| 49914 | **THE VISCOSITY OF AQUEOUS DIETHYLENE GLYCOL SOLUTIONS.**<br>KLIMENKO, A. P. KOSTYUK, V. I.<br>PETRUSHENKO, A. A. KRAVCHENKO, M. I.<br>GAZOV. PROM.<br>12 ( 11 ), 47-8, 1967.<br>( FOR ENGLISH TRANSLATION SEE TPRC NO. 66199 ) |
| 49920 | **THERMAL CONDUCTIVITY OF POLAR GAS BINARY MIXTURES.**<br>SHASHKOV, A. G. ABRAMENKO, T. N.<br>INZH. FIZ. ZH.<br>14 ( 2 ), 322-8, 1968.<br>( FOR ENGLISH TRANSLATION SEE TPRC NO. 65122 ) |
| 49931 | **VISCOSITY AND CONDUCTIVITY OF NICKEL SULFATE IN DIOXANE-WATER MIXTURES AT 35 DEGREES.**<br>DAS, P. B.<br>J. INST. CHEM. ( INDIA )<br>39 ( 6 ), 245-9, 1967. |
| 50037 | **THERMAL CONDUCTIVITY OF GAS MIXTURES OVER A WIDE TEMPERATURE RANGE. II.**<br>CORREIA, P. SCHAEFER, K.<br>BER. BUNSENGES. PHYS. CHEM.<br>72 ( 4 ), 507-14, 1968. |
| 50147 | **DENSITY AND VISCOSITY OF NITROGEN OXIDE SOLUTIONS IN 65-75 PERCENT NITRIC ACID.**<br>KARAVAEV, M. M. ZHANTALAI, V. A.<br>UKR. KHIM. ZH.<br>34 ( 2 ), 214-6, 1968.<br>( FOR ENGLISH TRANSLATION SEE TPRC NO. 61265 ) |
| 50148 | **ELECTRICAL CONDUCTIVITY AND VISCOSITY OF SODIUM IODIDE SOLUTIONS IN MIXTURES OF METHANOL AND DIOXANE.**<br>SHKODIN, A. M. LEVITSKAYA, N. K.<br>UKR. KHIM. ZH.<br>34 ( 4 ), 330-3, 1968.<br>( FOR ENGLISH TRANSLATION SEE TPRC NO. 63984 ) |
| 50258 | **VISCOSITY OF TWO BINARY, GASEOUS MIXTURES.**<br>KALELKAR, A. S.<br>BROWN UNIV., PROVIDENCE, R. I., PH.D. THESIS<br>114PP., 1969.<br>( UM 70-8741, AVAIL. UNIV. MICROFILMS ) |
| 50438 | **NEAR INFRARED REFLECTIVITY OF VENUS AND ICE CLOUDS.**<br>HANSEN, J. E. CHEYNEY, H.<br>J. ATMOS. SCI.<br>25 ( 4 ), 629-33, 1968. |
| 50466 | **IRON CHLORIDE - MAGNESIUM CHLORIDE - WATER AND IRON CHLORIDE - ZINC CHLORIDE - WATER SYSTEMS.**<br>SHCHEDRINA, A. P. OZEROVA, M. I.<br>ZH. NEORG. KHIM.<br>13 ( 6 ), 1713-16, 1968. |
| 50472 | **THE EFFECT OF VARIOUS SUGAR SOLUTIONS ON BLOOD VISCOSITY, THE ERYTHROCYTE SEDIMENTATION RATE, AND RED CELL MORPHOLOGY.**<br>LARCAN, A. STREIFF, F. PETERS, A.<br>GENETET, B. HELMER, J.<br>PATHOL. BIOL.<br>16 ( 7-8 ), 397-402, 1968. |
| 50605 | **POTASSIUM SULFATE-MANGANESE SULFATE-WATER TERNARY SYSTEM AT 20 DEGREES.**<br>KARNAUKHOV, A. S. RUNOV, N. N.<br>IZV. VYSSH. UCHEB. ZAVED., KHIM. KHIM. TEKHNOL.<br>10 ( 12 ), 1299-302, 1967.<br>( FOR ENGLISH TRANSLATION SEE TPRC NO. 63245 ) |
| 50613 | **PHYSICOCHEMICAL CONSTANTS. VISCOSITY OF FLUID ( METHANE + PROPANE ) UNDER LOW TEMPERATURES AND HIGH PRESSURES.**<br>IGUCHI, A.<br>KAGAKU SOCHI<br>10 ( 6 ), 52-3, 1968. |
| 50725 | **SELF-DIFFUSION OF CALCIUM IN SODIUM METASILICATE SOLUTIONS. I.**<br>BABAYAN, S. G. PETROSYAN, K. A.<br>MANVELYAN, M. G.<br>ARM. KHIM. ZH.<br>21 ( 2 ), 89-98, 1968. |
| 50730 | **CELLULAR STRUCTURE AND PHYSICAL PROPERTIES OF PLASTICS.**<br>BRISCALL, H. THOMAS, C. R.<br>BRIT. PLAST.<br>41 ( 7 ), 79-84, 1968. |
| 50752 | **VISCOSITY OF METHANE-NITROGEN AND METHANE-NITROGEN-HYDROGEN MIXTURES AT 273-473 K AND PRESSURES TO 490.3 X 10( TO THE 5TH POWER ) NEWTONS PER SQUARE METER.**<br>GNEZDILOV, N. E. GOLUBEV, I. F.<br>GAZOV. PROM.<br>13 ( 4 ), 46-8, 1968.<br>( FOR ENGLISH TRANSLATION SEE TPRC NO. 63989 ) |
| 50756 | **THERMAL CONDUCTIVITY OF LIQUID SOLUTIONS.**<br>RASTORGUEV, YU. L. GANIEV, YU. A.<br>INZH. - FIZ. ZH.<br>14 ( 4 ), 689-97, 1968.<br>( FOR ENGLISH TRANSLATION SEE TPRC NO. 65336 ) |
| 50758 | **DEPENDENCE OF THE THERMAL CONDUCTIVITY OF NON - NEWTONIAN FLUIDS ON THE RATE OF SHEAR.**<br>KHABAKHPASHEVA, E. M. CHIRKOV, YU. S.<br>KOSTYLEV, YU. V.<br>INZH. - FIZ. ZH.<br>14 ( 5 ), 918-20, 1968.<br>( FOR ENGLISH TRANSLATION SEE TPRC NO. 65912 ) |
| 50828 | **VISCOSITY OF A LIQUID IN A SYSTEM OF THIN CAPILLARIES.**<br>KHOLODNITSKII, B. A.<br>VESTN. LENINGRAD. UNIV., FIZ., KHIM.<br>23 ( 10 ), 153-4, 1968. |
| 50991 | **EXCESS ISOTHERMAL COMPRESSIBILITY AND EXCESS MOLAR HEAT CAPACITY OF THE SYSTEM BENZENE - N-HEPTANE.**<br>SUBRAHMANYAM, S. V. RAGHUNATH, K.<br>CHEM. PHYS. LETT.<br>64 ( 87 ), 21-55, 1968. |
| 51001 | **THERMAL CONDUCTIVITY OF GAS MIXTURES, REACTING AT A FINITE RATE.**<br>BAKALIN, YU. I. VERBA, M. I. PORTNOV, V. D.<br>INZH. - FIZ. ZH.<br>14 ( 5 ), 925-6, 1968.<br>( FOR ENGLISH TRANSLATION SEE TPRC NO. 65913 ) |
| 51061 | **VISCOSITY OF HIGH - MODULAR ALUMINATE SOLUTIONS.**<br>ZAKHAROVA, M. V. NI, L. P. PONOMAREV, V. D.<br>TR. INST. MET. OBOGASHCH., AKAD. NAUK KAZ. S S R<br>23, 50-7, 1967. |
| 51196 | **SOME PHYSICAL - CHEMICAL PROPERTIES OF SOLUTIONS OF THE SODIUM OXIDE-POTASSIUM OXIDE-ALUMINUM OXIDE-WATER SYSTEM.**<br>MASHOVETS, V. P. KUROCHKINA, V. V.<br>PENKINA, N. V. PUCHKOV, L. V. FEDOROV, M. K.<br>TSVET. METAL.<br>41 ( 7 ), 49-51, 1968.<br>( FOR ENGLISH TRANSLATION SEE TPRC NO. 67255 ) |
| 51467 | **PHYSICAL PROPERTIES OF SOLUTIONS OF THE ACETIC ACID-N,N-DIMETHYLANILINE SYSTEM.**<br>PANCHENKO, M. F.<br>UKR. KHIM. ZH.<br>34 ( 8 ), 783-7, 1968. |
| 51848 | **SOME METHODS FOR CALCULATING THE DENSITY AND HEAT CAPACITY OF MIXED ELECTROLYTE SOLUTIONS.**<br>MIKULIN, G. I.<br>VOP. FIZ. KHIM. RASTVOROV ELEKTROLIT.<br>401-16, 1968.<br>( FOR ENGLISH TRANSLATION SEE TPRC NO. 66212 ) |
| 53203 | **ISOTOPE EQUILIBRIUM IN THE SYSTEM BORON TRIFLUORIDE - TETRAHYDROPYRAN COMPLEX OF BORON TRIFLUORIDE.**<br>KATALNIKOV, S. G. NGO, Q. B.<br>ISOTOPENPRAXIS<br>4 ( 10 ), 386-8, 1968. |
| 53325 | **VISCOSITY OF CALCIUM CHLORIDE SOLUTIONS.**<br>EZROKHI, L. L.<br>TR. VSES. NAUCH.-ISSLED. INST. GALURGII<br>( 52 ), 3-9, 1967.<br>( FOR ENGLISH TRANSLATION SEE TPRC NO. 65514 ) |
| 53326 | **VISCOSITY OF SOLUTIONS OF A SODIUM, MAGNESIUM, CHLORIDE SULFATE IONS-WATER SYSTEM.**<br>EZROKHI, L. L.<br>TR. VSES. NAUCH.-ISSLED. INST. GALURGII<br>( 52 ), 20-36, 1967.<br>( FOR ENGLISH TRANSLATION SEE TPRC NO. 66038 ) |
| 53364 | **PHYSICAL-CHEMICAL PROPERTIES OF PROPIONIC ACID-WATER AND VALERIC ACID-WATER SYSTEMS.**<br>KOVALEVA, A. G.<br>UKR. KHIM. ZH.<br>34 ( 9 ), 871-6, 1968.<br>( FOR ENGLISH TRANSLATION SEE TPRC NO. 66039 ) |

*TPRC Number* | *Bibliographic Citation*

53433 **THERMAL CONDUCTIVITY OF BINARY MIXTURES.**
DULNEV, G. N. ZARICHNYAK, YU. P.
INZH. FIZ. ZH.
15 ( 4 ), 669–77, 1968.
( FOR ENGLISH TRANSLATION SEE TPRC NO. 67624 )

53467 **VISCOSITY OF TWO-COMPONENT SYSTEMS.**
IVANOV, U. I.
LATV. PSR ZINAT. AKAD. VESTIS, FIZ. TEH. ZINAT. SER.
( 4 ), 83–8, 1968.

53540 **SPECTRAL COLOR MATCHING BY MEANS OF MINIMAX APPROXIMATION.**
OHTA, N. URABE, H.
APPL. OPT.
11 ( 11 ), 2551–3, 1972.

53690 **PHYSICAL - CHEMICAL PROPERTIES OF SATURATED SOLUTIONS OF A LITHIUM NITRITE-SODIUM NITRITE WATER SYSTEM.**
RAZUMOVSKAYA, O. N. PROTSENKO, P. I.
IVLEVA, T. I.
ZH. NEORG. KHIM.
14 ( 2 ), 563–6, 1969.
( FOR ENGLISH TRANSLATION SEE TPRC NO. 67237 )

53699 **FUNDAMENTALS OF ENGINEERING HPI ( HYDROCARBON PROCESS INDUSTRY ) PLANTS WITH THE DIGITAL COMPUTER. IV. MATHEMATICS FOR PHYSICAL AND THERMODYNAMIC PROPERTIES. 1.**
HYMAN, M. H.
PETRO/CHEM. ENG.
41 ( 2 ), 30–1, 34–5, 38–40, 42, 1969.

53702 **HEAT CAPACITY OF TERNARY MIXTURES OF ORGANIC LIQUIDS IN THE LIQUID- VAPOR CRITICAL REGION.**
NOZDREV, V. F. GRECHKIN, V. I.
UCH. ZAP. TUL. GOS. PEDAGOG, INST., FIZ.– TEKH.
NAUKI
( 1 ), 16–23, 1967.

53703 **HEAT CAPACITY OF A SUBSTANCE IN THE CRITICAL REGION.**
GRECHKIN, V. I. ZAKURENOV, V. M.
UCH. ZAP. TUL. GOS. PEDAGOG. INST. FIZ.–TEKH.
NAUKI
( 1 ), 24–39, 1967.

54274 **THE INFLUENCE OF AN ELECTRIC FIELD ON THE TRANSPORT PROPERTIES OF POLAR GASES.**
DE GROOT, J. J.
UNIVERSITY OF LEIDEN, HOLLAND, PH.D. THESIS
92PP., 1971.

54290 **TEMPERATURE FIELD MEASUREMENTS IN DARK ZONES OF WALL STABILIZED LOW-CURRENT ARCS BY MEANS OF THE SCHLIERIN METHOD, USED ON HYDROGEN ARC TO DETERMINE HEAT CONDUCTIVITY AND ON MAGNETICALLY DEFLECTED ARGON ARCS.**
KOLLMAR, W.
TECHNISCHE HOCHSCHULE MUNCHEN, WEST GERMANY,
PH.D. THESIS
69PP., 1968.
( AVAIL: CFSTI )

54356 **HELIUM-3-HELIUM-4 MIXTURE.**
SHIGI, T.
BUSSEI
9 ( 7 ), 345–57, 1968.

54378 **CERTAIN METHODS FOR CALCULATING THERMAL CONDUCTIVITY OF MULTICOMPONENT MIXTURES INVOLVING POLYATOMIC GASES.**
GUPTA, G. P. MATHUR, S. SAXENA, S. C.
DEF. SCI. J.
18 ( 4 ), 195–204, 1968.

54413 **TEMPERATURE DPENDENCE OF THE VISCOSITY OF MIXTURES OF RAREFIED GASES AT LOW TEMPERATURES.**
ULYBIN, S. A.
GAZOV. PROM.
13 ( 12 ), 36–9, 1968.
( FOR ENGLISH TRANSLATION SEE TPRC NO. 66200 )

54484 **PHYSIOCHEMICAL CONSTANTS. VISCOSITY OF NORMAL SOLUTIONS.**
IQUCHI, A.
KAGAKU SOCHI
10 ( 12 ), 64–5, 1968.

54509 **VISCOSITY MEASUREMENTS OF ELECTROLYTIC SOLUTIONS.**
FALKENHAGEN, H. GERDES, E. EINFELDT, J.
MONATSBER. DEUT. AKAD. WISS. BERLIN
10 ( 8 ), 613–7, 1968.

54571 **VISCOSITY OF THE TERNARY LIQUID MIXTURES N- BUTYL ACETATE- N- BUTANOL - CYCLOHEXANOL.**
NAGESHWAR, G. D. MENE, P. S.
TRANS. INDIAN INST. CHEM. ENG.
53–5, 1968.

54591 **AUTOMATION OF THE CONTROL OF THE DEGREE OF AMMONIA DISSOCIATION IN FURNACES FOR NITRIDING METALS.**
DEMBOVSKII, V. V.
ZASHCH. POKRYTIYA METAL.
( 2 ), 99–103, 1968.

55010 **THERMODYNAMIC PROPERTIES AND COMBUSTION-PRODUCT TRANSFER PROPERTIES OF HYDROCARBON FUELS AT TEMPERATURES UP TO 4000 K.**
SAMUILOV, E. V. ROZHDESTVENSKII, I. B.
VOSKRSENSKAYA, N. V. TSITELAURI, N. N.
OLEVINSKII, K. K. SHEVELEV, V. P.
ELEC. MHD, PROC. SYMP.
4, 2249–76, 1968.
( FOR ENGLISH TRANSLATION SEE TPRC NO. 64428 )

55109 **DENSITY AND VISCOSITY OF SOME SOLUTIONS FROM THE SYSTEM SULFURIC ACID-PHOSPHORIC ACID-WATER.**
LEONTE, C. GABURICI, M.
BUL. INST. POLITEH. IASI
14 ( 1–2 ), 201–4, 1968.

55140 **FREE VOLUME AND TEMPERATURE DEPENDENCE OF VISCOSITY.**
BERNINI, U. CARRELLI, A. FITTIPALDI, F.
RAGOZZINO, E.
RIC. SCI.
38 ( 11 ), 1017–22, 1968.

55181 **A SEMIEMPIRICAL COLLISION MODEL FOR PLASMAS.**
MEADOR, W. E.
LANGLEY RESEARCH CENTER, LANGLEY STATION,
HAMPTON, VIRGINIA
35PP., 1969.
( N69-23787 NASA-TR-R-310 )

55396 **NONEQUILIBRIUM LAMINAR BOUNDARY LAYER OF DISSOCIATING AIR ON AXISYMMETRIC SOLIDS.**
VORONKIN, V. G. GERASKINA, L. K.
IZV. AKAD. NAUK SSSR, MEKH. ZHIDK. GAZA
( 3 ), 144–50, 1969.

55473 **PHASE EQUILIBRIUMS IN AN ISOAMYL ALCOHOL- WATER- DIOXANE SYSTEM.**
IONIN, M. V. SHERSTNEVA, T. V.
TR. KHIM. KHIM. TEKHNOL.
( 2 ), 18–21, 1968.

55527 **ACOUSTIC AND SOME PHYSICOCHEMICAL PROPERTIES OF PROPYL ALCOHOL-WATER BINARY MIXTURES ALONG THE SATURATION LINE.**
VINITSKII, E. S. ZAKURENOV, V. M.
PRIMEN. UL'TRAAKUST. ISSLED. VESHCHESTVA
( 23 ), 25–30, 1967.
( FOR ENGLISH TRANSLATION SEE TPRC NO. 66063 )

55597 **PHYSICOCHEMICAL STUDY OF THE FERROUS CHLORIDE - ZINC CHLORIDE - WATER SYSTEM.**
SHCHEDRINA, A. P. ZMIEVA, L. V.
KRASNOVA, L. I.
IZV. VYSSH. UCHEB. ZAVED., KHIM. KHIM. TEKHNOL.
12 ( 6 ), 714–7, 1969.
( FOR ENGLISH TRANSLATION SEE TPRC NO. 65502 )

55647 **CALCULATION OF THE VISCOSITY OF CHEMICALLY REACTING MIXTURES IN A DENSE GAS STATE.**
MAKSIMOV, B. G. MISHINA, L. V.
SEREBRYANYI, G. Z.
DOKL. AKAD. NAUK BELORUSS. SSR
13 ( 6 ), 516–8, 1969.

55653 **DISSOCIATION CONSTANT OF TETRA-N-BUTYLAMMONIUM BROMIDE IN P-DIOXANE-WATER MIXTURE.**
HORITA, K. KANO, G. TUBOTA, N.
FUKUI DAIGAKU KOGAKUBU KENKYU HOKOKU
16 ( 2 ), 243–8, 1968.

55656 **VISCOSITY AND ELECTRICAL CONDUCTIVITY OF ALCOHOLIC SOLUTIONS OF LITHIUM CHLORIDE.**
GOLIK, A. Z. ALLANAZAROV, G. CHOLPAI, P. F.
IZV. AKAD. NAUK TURKM. SSR, SER. FIZ.–TEKH., KHIM.
GEOL. NAUK
( 4 ), 112–4, 1969.
( FOR ENGLISH TRANSLATION SEE TPRC NO. 66505 )

55678 **EXPERIMENTAL VISCOSITIES FOR BINARY HYDROCARBON SYSTEMS.**
BENNETT, S. WHITE, B. MADDOX, R. N.
PROC. ANNU. CONV., NATUR. GAS PROCESS. ASS., TECH.
PAP.
( 48 ), 36–40, 1969.

55757 **KINETIC THEORY OF THE THERMAL CONDUCTIVITY OF GAS MIXTURES AT LOW TEMPERATURES. I.**
SHASHKOV, A. G. ABRAMENKO, T. N.
INZH.–FIZ. ZH.
16 ( 6 ), 1009–18, 1969.
( FOR ENGLISH TRANSLATION SEE TPRC NO. 69502 )

55833 **VISCOSITIES OF THE BENZENE-PERFLUOROBENZENE SYSTEM.**
BEDSOLE, A. D. TAYLOR, Z. L., JR.
J. ALA. ACAD. SCI.
39 ( 4 ), 270–8, 1968.

55851 **THERMAL CONDUCTIVITY OF GAS MIXTURES AT LOW TEMPERATURES. II.**
SHASHKOV, A. G. ABRAMENKO, T. N.
INZH.–FIZ. ZH.
17 ( 1 ), 65–71, 1969.
( FOR ENGLISH TRANSLATION SEE TPRC NO. 69503 )

*TPRC Number* | *Bibliographic Citation*

55892 CHANGES OF STATE OF TITANIUM( IV ) DURING ITS HYDROLYSIS IN SULFURIC ACID.
DOMATOV, YU. D.
ZH. PRIKL. KHIM. ( LENINGRAD )
42 ( 8 ), 1725–32, 1969.
( FOR ENGLISH TRANSLATION SEE TPRC NO. 66455 )

56170 NITROGEN TETROXIDE VISCOSITY IN A DENSE GASEOUS STATE.
MAKSIMOV, B. G. MISHINA, L. V.
SEREBRYANYI, G. Z. DUNETS, V. M.
DOKL. AKAD. NAUK BELORUSS. SSR
13 ( 8 ), 691–3, 1969.
( FOR ENGLISH TRANSLATION SEE TPRC NO. 65518 )

56222 ISOBARIC HEAT CAPACITY OF BORIC ACID.
POPOV, V. N. MOROZOVA, N. A.
AT. ENERG.
27 ( 3 ), 233, 1969.

56229 PHYSICOCHEMICAL PROPERTIES OF THREE-COMPONENT SOLUTIONS OF WATER, ETHANOL, AND ISOBUTANOL ( HOMOGENEOUS REGION ).
KHARIN, S. E. KNIGA, A. A. SOROKINA, G. S.
FERMENT. SPIRT. PROM.
35 ( 7 ), 9–12, 1969.
( FOR ENGLISH TRANSLATION SEE TPRC NO. 65532 )

56288 VISCOSITY, DENSITY, AND ELECTRICAL CONDUCTIVITY OF AQUEOUS SODIUM HYDROXIDE SOLUTIONS AT LOW TEMPERATURES.
BARON, N. M. SHCHERBA, M. U.
ZH. PRIKL. KHIM. ( LENINGRAD )
42 ( 9 ), 2128–31, 1969.

56421 COMPARISON OF THE STRUCTURES OF SOLUTIONS OF POTASSIUM, POTASSIUM IODIDE, AND POTASSIUM AMIDE IN LIQUID AMMONIA.
MOREAU, C. DAMAY, P. LEPOUTRE, G.
ANN. CHIM. ( PARIS )
4 ( 4 ), 309–12, 1969.

56446 TRANSFER PHENOMENA IN REACTING GAS MIXTURES.
ALEKSEEV, B. V.
CHISLENNYE METODY TEOR. RAZREZH. GAZOV
140–74, 1969.
( FOR ENGLISH TRANSLATION SEE TPRC NO. 64427 )

56631 FABRICATION AND CHARACTERIZATION OF METAL CARBIDE-GRAPHITE COMPOSITES.
HARADA, Y. BORTZ, S. A. BLUM, S. L.
HIGH TEMP. MATER., PAP. PLANSEE SEMIN., 6TH
608–29, 1969.

56638 VISCOSITY OF THREE - COMPONENT WATER - ETHANOL - METHANOL SOLUTIONS.
KHARIN, S. E. KOKHANOVSKII, N. I.
SOROKINA, G. S.
IZV. VYSSH. UCHEB. ZAVED., KHIM. KHIM. TEKHNOL.
12 ( 9 ), 1219–22, 1969.
( FOR ENGLISH TRANSLATION SEE TPRC NO. 65495 )

56735 DEPENDENCE OF THE SECOND VIRIAL COEFFICIENT AND TRANSPORT PHENOMENA IN GASES ON THE EXPONENTS OF THE LENNARD–JONES POTENTIAL.
SCHRAMM, B.
BER. BUNSENGES. PHYS. CHEM.
73 ( 10 ), 1045–50, 1969.

56795 VISCOSITY AND PROGRESSIVE DIFFUSION IN AQUEOUS KETONE SOLUTIONS.
BAZUNOVA - YAROSHEVSKAYA, L. G.
MORACHEVSKII, A. G. NERONOV, YU. I.
SHAPIL, L. G.
KHIM. TERMODIN. RASTVOROV
( 2 ), 73–7, 1968.
( FOR ENGLISH TRANSLATION SEE TPRC NO. 65496 )

56919 HETEROGENEOUS EQUILIBRIUMS IN TERNARY AQUEOUS SYSTEMS OF CALCIUM AND SODIUM PERCHLORATES AND CHLORIDES AT 40 DEGREES.
IVANOV, S. A.
UCH. ZAP., YAROSLAV. GOS. PEDAGOG. INST.
( 66 ), 82–95, 1969.
( FOR ENGLISH TRANSLATION SEE TPRC NO. 66503 )

56923 ENERGY OF ACTIVATION AND RELATIVE TEMPERATURE COEFFICIENTS OF THE VISCOSITY AND ELECTRICAL CONDUCTIVITY OF BINARY LIQUID SYSTEMS.
FIALKOV, YU. YA. TSENDROVSKAYA, V. A.
UKR. KHIM. ZH.
35 ( 10 ), 1029–35, 1969.
( FOR ENGLISH TRANSLATION SEE TPRC NO. 66056 )

56930 VISCOSITY OF SOLUTIONS OF PERCHLORATES OF GROUP II CATIONS.
LILICH, L. S. KHRIPUN, M. K.
CHURGANOVA, S. S.
VESTN. LENINGRAD. UNIV., FIZ., KHIM.
( 3 ), 91–5, 1969.

56940 PHYSICOCHEMICAL STUDY OF AN AQUEOUS SYSTEM OF UREA AND POTASSIUM AND CHROMIUM CHLORIDES AT 25 C.
KARNAUKHOV, A. S. TATARINOV, V. A.
UCH. ZAP., YAROSLAV. GOS. PEDAGOG. INST.
( 66 ), 162–72, 1969.

*TPRC Number* | *Bibliographic Citation*

56949 APPLICATION OF DIFFERENTIAL SCANNING CALORIMETER ( DSC ).
KIND, R.
WISS. Z. TECH. HOCHSCH. CHEM. LEUNA- MERSEBURG
11 ( 3 ), 225–31, 1969.

56950 SALT- MIXED SOLVENT SYSTEMS. I V. MAGNESIUM SULFATE- METHANOL- WATER SYSTEM.
WINKLER, F. EMONS, H. H.
WISS. Z. TECH. HOCHSCH. CHEM. LEUNA- MERSEBURG
11 ( 3 ), 236–42, 1969.

56961 DISTRIBUTION OF NEODYMIUM AND SARMARIUM CHLORIDES BETWEEN TRIBUTYL PHOSPHATE AND WATER.
MIKHAILICHENKO, A. I. KLIMENKO, M. A.
BOGDANOV, V. P.
ZH. NEORG. KHIM.
14 ( 11 ), 3076–80, 1969.
( FOR ENGLISH TRANSLATION SEE T63986 )

57029 SOLUBILITY IN THE AQUEOUS SYSTEM OF SODIUM AND MAGNESIUM PERCHLORATES AND CHLORIDES AT 90.
KUDRYAKOVA, S. A. LEPESHKOV, I. N.
UCH. ZAP., YAROSLAV. GOS., PEDAGOG. INST.
( 66 ), 40–59, 1969.

57532 STRUCTURE DEPENDENCE OF THE HEAT CAPACITY OF POLYPHOSPHORIC ACIDS.
SMIRNOVA, E. P. MATVEICHUK, N. L.
MIKHAILIN, A. D. ILLARIONOV, V. V.
DOKL. AKAD. NAUK SSSR
189 ( 2 ), 344–6, 1969.
( FOR ENGLISH TRANSLATION SEE TPRC NO. 70689 )

57535 PROPERTIES OF BINARY SOLUTIONS IN THE CRITICAL MIXING REGION.
SALAKHUTDINOV, M. I.
DOKL. AKAD. NAUK TADZH. SSR
12 ( 5 ), 13–16, 1969.

57538 ANALYSIS OF COMPOSITION OF GASES BY THERMAL CONDUCTIVITY. PHYSICAL PRINCIPLES.
KOCHEROV, L. E. SHUTOV, M. D.
ENTSIKL. IZMER., KONTR. AVTOMAT.
( 11 ), 5–6, 1968.
( FOR ENGLISH TRANSLATION SEE TPRC NO. 64047 )

57577 PROBLEMS OF INFRARED ATMOSPHERIC SPECTROSCOPY CONNECTED WITH THE DETERMINATION OF THE UNDERLAYING SURFACE TEMPERATURE.
KONDRATEV, K. YA.
IZV. AKAD. NAUK SSSR, FIZ. ATMOS. OKEANA
5 ( 6 ), 616–30, 1969.

57596 VISCOSITY AND STRUCTURE OF MOLTEN SALTS.
MURGULESCU, I. G.
J. CHIM. PHYS. PHYSICOCHIM. BIOL.
43–56, 1969.

58112 SPECIFIC HEAT OF MIXING OF ALCOHOL–HYDROCARBON SOLUTION SYSTEM.
IGUCHI, A.
KAGAKU SOCHI
11 ( 12 ), 76–7, 1969.

58281 APPLICABILITY OF THE MARON–TORNBULL EQUATION FOR CALCULATION OF THE JOULE–THOMSON COEFFICIENT OF HYDRO–CARBON GASES.
KADYROV, N. K. GASIMOVA, F. A.
GABIBOV, A. B.
IZV. AKAD. NAUK AZERB. SSR, SER. FIZ.–MAT. TEKH. NAUK
( 3 ), 101–8, 1966.

58348 LIQUID - PHASE INTERACTION OF DIMETHYL SULFOXIDE AND DIMETHYLACETAMIDE WITH ACETIC ACID.
KHAZOVA, I. V. CHESNOKOV, V. F.
ZH. OBSHCH. KHIM.
40 ( 1 ), 7–11, 1970.
( FOR ENGLISH TRANSLATION SEE TPRC NO. 67239 )

58349 SPECIFIC HEAT OF POTASSIUM OXIDE–ALUMINUM OXIDE–WATER SYSTEM SOLUTIONS IN THE 25–340 DEGREE RANGE.
MASHOVETS, V. P. PUCHKOV, L. V.
BARANOVA, T. A. LAPIDUS, M. E.
ZH. PRIKL. KHIM. ( LENINGRAD )
43 ( 1 ), 183–6, 1970.
( FOR ENGLISH TRANSLATION SEE TPRC NO. 66456 )

58473 SIMPLIFIED CALCULATION OF ENTHALPY AND ENTROPY OF NITROGEN TO 40,000 DEGREES K.
ANGELIN, L. ZANNIER, E.
QUAD. ING. CHIM. ITAL.
6 ( 1 ), 17–21, 1970.

58481 EXTREMES ON CURVES RELATING TO RETENTION VALUES TO THE FIXED COMPOSITION OF BINARY FIXED PHASES.
SHEVCHUK, I. M. GRANZHAN, V. A.
SAKHAROV, V. M. KOVALENKO, N. V.
ZH. OBSHCH. KHIM.
39 ( 12 ), 2638–41, 1969.
( FOR ENGLISH TRANSLATION SEE TPRC NO. 67238 )

| TPRC Number | Bibliographic Citation |
|---|---|
| 58511 | **SHEAR VISCOSITY OF A BINARY MIXTURE OF SIMPLE LIQUIDS.**<br>BRUK-LEVINSON, E. T. VIKHRENKO, V. S.<br>NEMTSOV, V. B. ROTT, L. A.<br>IZV. VYSSH. UCHEB. ZAVED., FIZ.<br>13 ( 2 ), 70-5, 1970.<br>( FOR ENGLISH TRANSLATION SEE TPRC NO. 69886 ) |
| 58534 | **THERMAL EFFECTS ACCOMPANYING SPONTANEOUS IGNITIONS IN GASES. II. SLOW EXOTHERMIC DECOMPOSITION OF DIETHYL PEROXIDE.**<br>FINE, D. H. GRAY, P. MAC KINVEN, R.<br>PROC. ROY. SOC.<br>316 A ( 1525 ), 223-68, 1970. |
| 58552 | **PHYSICAL PROPERTIES OF SOLUTIONS OF A PHENOL-CHLORO ACETIC ACID SYSTEM.**<br>PANCHENKO, M. F.<br>ZH. OBSHCH. KHIM.<br>40 ( 2 ), 266-70, 1970.<br>( FOR ENGLISH TRANSLATION SEE TPRC NO. 67240 ) |
| 58553 | **SPECIFIC HEAT OF AQUEOUS POTASSIUM HYDROXIDE SOLUTIONS AT TEMPERATURES LESS THAN 400 DEGREES.**<br>PUCHKOV, L. V. BARANOVA, T. A.<br>LAPIDUS, M. E.<br>ZH. PRIKL. KHIM. ( LENINGRAD )<br>43 ( 2 ), 455-7, 1970.<br>( FOR ENGLISH TRANSLATION SEE TPRC NO. 66457 ) |
| 58796 | **ELECTROSTRICTION IN MIXED POLAR SOLVENTS. VISCOSITY OF SILVER NITRATE IN METHANOL - WATER MIXTURES AT 30 DEGREES.**<br>SINGH, D. YADAV, V. S. GOEL, B. K.<br>Z. PHYS. CHEM. ( FRANFURT AM MAIN )<br>68 ( 3-6 ), 242-9, 1969. |
| 58973 | **VISCOSITY OF BINARY MIXTURES. III. BENZENE-ALCOHOL SYSTEMS.**<br>MATO, F. COCA, J.<br>AN. QUIM.<br>66 ( 2 ), 127-32, 1970. |
| 58989 | **RELATIONS BETWEEN ACTIVITY COEFFICIENT AND VISCOSITY OF BINARY LIQUID SOLUTIONS.**<br>IGUCHI, A.<br>KAGAKU SOCHI<br>12 ( 4 ), 78-9, 1970. |
| 59007 | **DENSITY, VISCOSITY, AND SURFACE TENSION OF SOME BINARY SYSTEMS.**<br>KONOBEEV, B. I. LYAPIN, V. V.<br>ZH. PRIKL. KHIM. ( LENINGRAD )<br>43 ( 4 ), 803-11, 1970.<br>( FOR ENGLISH TRANSLATION SEE TPRC NO. 66459 ) |
| 59019 | **CALCULATION OF THE THERMAL CONDUCTIVITY OF BINARY GAS MIXTURES.**<br>DULNEV, G. N. ZARICHNYAK, YU. P.<br>MURATOVA, B. L.<br>INZH.-FIZ. ZH.<br>18 ( 5 ), 849-55, 1970.<br>( FOR ENGLISH TRANSLATION SEE TPRC NO. 71943 ) |
| 59027 | **THERMODYNAMIC PROPERTIES OF FLUOROCARBON PLUS HYDROCARBON MIXTURES. 2. EXCESS ENTHALPIES OF MIXING.**<br>ANDREWS, A. MORCOM, K. W. DUNCAN, W. A.<br>SWINTON, F. L. POLLOCK, J. M.<br>J. CHEM THERMODYN.<br>2 ( 1 ), 95-103, 1970. |
| 59056 | **CALCULATION OF VISCOSITY COEFFICIENTS FOR SOME MULTICOMPONENT GAS MIXTURES.**<br>KURLAPOV, L. I. MARTYNOVA, G. P.<br>TEOR. OSN. KHIM. TEKHNOL.<br>4 ( 3 ), 414-5, 1970. |
| 59058 | **VISCOSITY OF IRON-SULFUR, NICKEL-SULFUR, COBALT-SULFUR MELTS AND ITS RELATION TO THERMODYNAMIC PROPERTIES.**<br>BARMIN, L. N. ESIN, O. A.<br>DOBROVINSKII, I. E. CHUCHMAREV, S. K.<br>TERMODIN. SVOISTVA RASPLAVOV<br>59-65, 1969. |
| 59060 | **DENSITY, VISCOSITY, AND ELECTRICAL CONDUCTIVITY OF AQUEOUS AND ALCOHOLIC SOLUTIONS OF GALLIUM CHLORIDE.**<br>YATSENKO, S. P. DRUZHININA, E. P.<br>RYKOVA, L. N.<br>TR. INST. KHIM., AKAD. NAUK S S S R, URAL. FILIAL<br>( 18 ), 51-9, 1968. |
| 59066 | **INFRARED SPECTROSCOPIC STUDY OF SYNTHETIC AND NATURAL TUNGSTATES AND MOLYBDATES.**<br>PLYUSNINA, I. I. ZAITSEVA, L. A.<br>VESTN. MOSK. UNIV., GEOL.<br>24 ( 4 ), 110-15, 1969. |
| 59295 | **CALCULATION OF THE THERMODYNAMIC PROPERTIES OF CHEMICALLY REACTING SYSTEMS BASED ON THE EQUATION OF STATE.**<br>BAZHIN, M. A. NESTERENKO, V. B.<br>DOKL. AKAD. NAUK BELORUSS. S S R<br>14 ( 4 ), 317-20, 1970. |

| TPRC Number | Bibliographic Citation |
|---|---|
| 59317 | **VISCOSITY OF AQUEOUS POTASSIUM HYDROXIDE SOLUTIONS TO 90 DEGREES.**<br>PUCHKOV, L. V. SARGAEV, P. M. FEDOROV, M. K.<br>ZH. PRIKL. KHIM. ( LENINGRAD )<br>43 ( 3 ), 677-80, 1970.<br>( FOR ENGLISH TRANSLATION SEE TPRC NO. 66458 ) |
| 59461 | **STRUCTURAL DIFFERENCES OF CERTAIN INORGANIC SALT SOLUTIONS IN NATURAL AND HEAVY WATER.**<br>SELECKI, A. TYMINSKI, B. CHMIELEWSKI, A. G.<br>ISOTOPENPRAXIS<br>6 ( 4 ), 138-42, 1970. |
| 59465 | **IMPROVED METHOD FOR MEASURING THE THERMAL CONDUCTIVITY OF ELECTROLYTE SOLUTIONS.**<br>PRUDNIKOV, N. K.<br>IZV. VYSSH. UCHEB. ZAVED., ENERG.<br>13 ( 1 ), 122-6, 1970.<br>( FOR ENGLISH TRANSLATION SEE TPRC NO. 66197 ) |
| 59478 | **VISCOSITY OF MELTS IN THE SODIUM METAPHOSPHATE- SODIUM PYROPHOSPHATE SYSTEM.**<br>POSTNIKOV, N. N. FRENKEL, M. G.<br>MELNIKOVA, N. S.<br>TR., NAUCH.- ISSLED. INST. UDOBR. INSEKTOFUNGITS.<br>( 1 ), 116-20, 1968. |
| 59485 | **BULK VISCOSITY OF BINARY MIXTURES OF ETHYL ACETATE WITH ALCOHOLS.**<br>KOLCHIN, YU. P.<br>UCH. ZAP., KURSK. GOS. PEDAGOG. INST.<br>54, 13-7, 1969. |
| 60119 | **HEAT FLOW IN THE SOUTHWESTERN PACIFIC.**<br>SCLATER, J. G. RITTER, U. G. DIXON, F. S.<br>J. GEOPHYS. RES.<br>77 ( 29 ), 5697-704, 1972. |
| 60128 | **DETERMINATION OF TRANSPORT COEFFICIENTS OF NITROGEN PLASMA UP TO 13,000 DEGREES K.**<br>PLANTIKOW, U.<br>Z. PHYS.<br>237 ( 5 ), 388-402, 1970. |
| 60206 | **A COMPARATIVE STUDY OF TETRABUTYLAMMONIUM BROMIDE, SODIUM TETRAPHENYLBORATE, TETRAPHENYL PHOSPHOROUS CHLORIDE, AND TETRAPHENYL ARSENIC CHLORIDE IN WATER AND METHANOL AT 25 C. PARTIAL MOLAL VOLUMES, HEAT CAPACITIES, AND VISCOSITIES.**<br>JOLICOEUR, C. PHILIP, P. R. PERRON, G.<br>LEDUC, P. A. DESNOYERS, J. E.<br>CAN. J. CHEM.<br>50 ( 19 ), 3167-78, 1972. |
| 60208 | **NEW TECHNIQUE FOR THE STUDY OF COPPER - ACETAMIDE SYSTEM IN AQUEOUS SOLUTION.**<br>AHMED, S. S. KHAN, S. A. KHAN, A. R.<br>PAK. J. SCI. IND. RES.<br>13 ( 1-2 ), 45-7, 1970. |
| 60734 | **THERMAL CONDUCTIVITY OF MIXTURES WITH INTERPENETRATING COMPONENTS.**<br>DULNEV, G. N.<br>INZH.-FIZ. ZH.<br>19 ( 3 ), 562-77, 1970.<br>( FOR ENGLISH TRANSLATION SEE TPRC NO. 72172 ) |
| 60754 | **DETERMINATION OF THE VISCOSITY OF AMMONIUM NITRATE - LIMESTONE MELTS.**<br>ERVACINOVIC, R. NOVOHATNI, G.<br>TEHNIKA ( BELGRADE )<br>25 ( 8 ), 1566-70, 1970. |
| 60759 | **INCREASE IN THE PROTECTIVE PROPERTIES OF SODIUM NITRITE AQUEOUS SOLUTIONS BY THICKENING THEM WITH XYLITOL.**<br>ROZENFEL'D, I. L. PERSIANTSEVA, V. P.<br>KUZNETSOV, YU. I.<br>ZASHCH. METAL.<br>6 ( 5 ), 577-81, 1970.<br>( FOR ENGLISH TRANSLATION SEE TPRC NO. 68707 ) |
| 60796 | **RADIAL DISTRIBUTION OF TEMPERATURE IN A DIRECT CURRENT ARC BURNING IN A NITROGEN ATMOSPHERE UNDER CONDITIONS USED IN SPECTROCHEMICAL EXCITATION.**<br>PAVLOVIC, B. V. IKONOMOV, N. N.<br>VUKANOVIC, V. M.<br>GLAS. HEM. DRUS., BEOGRAD<br>34 ( 5-7 ), 313-26, 1969. |
| 60854 | **EFFECT OF COMPOSITION AND TEMPERATURE ON THE PHYSICOCHEMICAL CONSTANTS OF THE ETHANOL- WATER-ISOAMYL ACETATE SYSTEM.**<br>KHARIN, S. E. SHATALOVA, V. V.<br>IZV. VYSSH. UCHEB. ZAVED., KHIM. KHIM. TECHNOL.<br>13 ( 6 ), 781-5, 1970. |
| 60855 | **SPECIFIC HEATS OF WATER- SALT SYSTEMS CONTAINING ALKALI- METAL HALIDES AT 25 DEGREES.**<br>VASILEV, V. A. SHEVCHENKO, E. YA.<br>IZV. VYSSH. UCHEB. ZAVED., KHIM. KHIM. TEKHNOL.<br>13 ( 6 ), 789-93, 1970. |

| TPRC Number | Bibliographic Citation |
|---|---|
| 60879 | **VISCOSITY OF MOLTEN SALTS.**<br>GALASIU, I.<br>STUD. CERCET. CHIM.<br>18 ( 3 ), 293–313, 1970. |
| 60916 | **MECHANICAL AND THERMOPHYSICAL PROPERTIES OF RAW MATERIALS.**<br>MALYSHEV, A. P.<br>TSEMENT<br>34 ( 2 ), 5–7, 1968. |
| 60917 | **THERMODYNAMIC PROPERTIES OF MATTER IN THE CRITICAL**<br>GRECHKIN, V. I. ZAKURENOV, V. M.<br>UCH. ZAP., KURSK. GOS. PEDAGOG. INST.<br>54, 105–12, 1969. |
| 60921 | **VISCOSITY AND SURFACE TENSION OF CYMENES, THEIR HYDROPEROXIDES, AND DECOMPOSITION PRODUCTS.**<br>SHUB, G. A. ORLOVA, O. S.<br>KHARLAMPOVICH, G. D.<br>ZH. PRIKL. KHIM. ( LENINGRAD )<br>43 ( 5 ), 1175–7, 1970.<br>( FOR ENGLISH TRANSLATION SEE TPRC NO. 66462 ) |
| 60922 | **TEMPERATURE AND CONCENTRATION DEPENDENCE OF DIFFUSION AND VISCOSITY COEFFICIENTS AND DENSITY IN CYCLOHEXANE-CYCLOHEXANOL SYSTEMS.**<br>RAVDEL, A. A. PORAI-KOSHITS, A. B.<br>PENKINA, N. V. OLEVSKII, V. M. IVANOV, V. A.<br>ZH. PRIKL. KHIM. ( LENINGRAD )<br>43 ( 6 ), 1401–2, 1970.<br>( FOR ENGLISH TRANSLATION SEE TPRC NO. 66463 ) |
| 60923 | **PHYSICOCHEMICAL ANALYSIS OF AN N-METHYL-ALPHA-PYRROLIDONE-METHANOL SYSTEM.**<br>GRANZHAN, V. A. KIRILLOVA, O. G.<br>ZH. PRIKL. KHIM. ( LENINGRAD )<br>43 ( 8 ), 1875–7, 1970.<br>( FOR ENGLISH TRANSLATION SEE TPRC NO. 66465 ) |
| 61008 | **STEADY STATE TECHNIQUE FOR MEASURING FLUXES AND DIFFUSIVITIES IN BINARY LIQUID SYSTEMS.**<br>RAO, S. S. BENNETT, C. O.<br>AICHE J.<br>17 ( 1 ), 75–81, 1971. |
| 61020 | **EFFECT OF OBLIQUE INCIDENCE ON OPTICAL ABSORPTION OF LASER LIGHT BY A PLASMA.**<br>SHEARER, J. W.<br>PHYS. FLUIDS<br>14 ( 1 ), 183–5, 1971. |
| 61062 | **EFFECT OF VARIABLE VISCOSITY ON LAMINAR HEAT TRANSFER IN A RECTANGLAR DUCT.**<br>HWANG, S.-T. HONG, S. W.<br>CHEM. ENG. PROGR., SYMP. SER.<br>66 ( 102 ), 100–8, 1970. |
| 61076 | **MEASUREMENT OF THERMAL CONDUCTIVITY IN A DECAYING ARGON PLASMA.**<br>BECKER, E. BEZ, W.<br>INT. CONF. PHENOMENA IONIZ. GASES, CONTRIB. PAP., 9TH<br>418, 1969. |
| 61077 | **TRANSPORT PROPERTIES OF THE NITROGEN PLASMA UP TO 13,000 DEGREES K.**<br>PLANTIKOW, U.<br>INT. CONF. PHENOMENA IONIZ. GASES, CONTRIB. PAP., 9TH<br>421, 1969. |
| 61093 | **DIELECTRIC RELAXATION BEHAVIOUR OF SOME PRIMARY AND SECONDARY ALCOHOLS IN HEPTANE SOLUTIONS.**<br>SAKELLARIDIS, P. HINOPOULOS, B.<br>J. PHYS. D<br>5 ( 10 ), 1815–23, 1972. |
| 61145 | **VISCOSITY AND DIFFUSION COEFFICIENTS OF POLAR GAS MIXTURES.**<br>BAE, J. H. REED, T. M., III<br>IND. ENG. CHEM. FUNDAM.<br>10 ( 1 ), 36–41, 1971. |
| 61155 | **INTERACTION OF AN OBLIQUELY INCIDENT P-POLARIZED PLANE ELECTROMAGNETIC WAVE WITH A HOT PLASMA HALF-SPACE.**<br>BOLDUC, P. E. KLEVANS, E. H.<br>PHYS. FLUIDS<br>14 ( 2 ), 378–86, 1971. |
| 61177 | **THERMAL CONDUCTIVITY OF PHOSPHORIC ACID - WATER MIXTURES AT 25 DEGREES C.**<br>TURNBULL, A. G.<br>J. CHEM. ENG. DATA<br>16 ( 1 ), 79–83, 1971. |
| 61232 | **THERMAL CONDUCTIVITY — METALLIC ELEMENTS AND ALLOYS. VOL. 1 OF THERMOPHYSICAL PROPERTIES OF MATTER-THE TPRC DATA SERIES.**<br>TOULOUKIAN, Y. S. POWELL, R. W. HO, C. Y.<br>KLEMENS, P. G.<br>THERMOPHYSICAL PROPERTIES RES. CTR., PURDUE UNIV., LAFAYETTE, INDIANA<br>1595PP., 1970.<br>( IFI/PLENUM DATA CORP., NEW YORK ) |

| TPRC Number | Bibliographic Citation |
|---|---|
| 61243 | **VISCOSITIES IN MIXTURES OF WATER AND ORGANIC LIQUIDS AS WELL AS SOLUTIONS OF SEVERAL ELECTROLYTES IN THESE SYSTEMS.**<br>EBERT, G. WENDORFF, J.<br>BER. BUNSENGES. PHYS. CHEM.<br>74 ( 10 ), 1071–7, 1970. |
| 61265 | **DENSITY AND VISCOSITY OF NITROGEN OXIDE SOLUTIONS IN 65–75 PERCENT NITRIC ACID.**<br>KARAVAEV, M. M. ZHANTALAI, V. A.<br>FTD-HT-23-1299-71<br>5PP., 1971.<br>( ENGLISH TRANSLATION OF UKR. KHIM. ZH. 34 ( 2 ), 214–6, 1968., FOR ORIGINAL SEE TPRC NO. 50147 ) |
| 61267 | **THERMODYNAMICS OF THE SYSTEM DIETHYLAMINE- TRIETHYLAMINE. V. DENSITY, EXCESS VOLUME, VISCOSITY, AND DIELECTRIC CONSTANTS BETWEEN 20 AND 50 DEGREES.**<br>LANGGUTH, U. BITTRICH, H. J.<br>Z. PHYS. CHEM. ( LEIPZIG )<br>244 ( 5–6 ), 327–39, 1970. |
| 61308 | **OPTICAL ABSORPTION CHARACTERISTICS AND PHOTO-BLEACHING BEHAVIOR OF TRAPPED ELECTRONS IN GAMMA-IRRADIATED ALKALINE ICE.**<br>HASE, H. KEVAN, L.<br>J. CHEM. PHYS.<br>54 ( 3 ), 908–14, 1971. |
| 61312 | **DYNAMIC JAHN-TELLER EFFECTS AND MAGNETIC ANISOTROPIES IN AQUEOUS SOLUTIONS AND WATER-ETHANOL GLASSES OF COPPER ( II ) SOLVATES AND COMPLEXES WITH 2,2 PRIME-DIPYRIDINE.**<br>NOACK, M. KOKOSZKA, G. F. GORDON, G.<br>J. CHEM. PHYS.<br>54 ( 3 ), 1342–50, 1971. |
| 61330 | **INADEQUACIES OF VISCOSITY THEORIES FOR A VITREOUS POTASSIUM NITRATE - CALCIUM NITRATE MELT.**<br>TWEER, H. LABERGE, N. MACEDO, P. B.<br>J. AMER. CERAM. SOC.<br>54 ( 2 ), 121–3, 1971. |
| 61331 | **ON RELATIVISTIC KINETIC GAS THEORY. V. THE COEFFICIENTS OF HEAT CONDUCTIVITY, DIFFUSION AND THERMAL DIFFUSION FOR A BINARY MIXTURE OF RELATIVISTIC MAXWELLIAN MOLECULES.**<br>VAN LEEUWEN, W. A. DE GROOT, S. R.<br>PHYSICA ( UTRECHT )<br>51 ( 1 ), 16–31, 1971. |
| 61334 | **TRANSPORT PROPERTIES OF NOBLE- GAS MIXTURES.**<br>HOGERVORST, W.<br>PHYSICA ( UTRECHT )<br>51 ( 1 ), 90–103, 1971. |
| 61380 | **ANOMALOUS VISCOSITY AS POSSIBLE EXPLANATION FOR AN ANOMALOUS PLASMA SKIN EFFECT.**<br>SCHMIDT, J. YOSHIKAWA, S.<br>PHYS. REV. LETT.<br>26 ( 13 ), 753–6, 1971. |
| 61394 | **THE INFLUENCE OF CERTAIN PHYSIOCHEMICAL FACTORS ON THE DIMERISATION REACTIONS OF FREE CYCLOHEXYL AND SUBSTITUTED CYCLOHEXYL RADICALS.**<br>TESSIER, M. PALLAUD, R.<br>COMPT. REND.<br>272 C ( 6 ), 549–50, 1971. |
| 61407 | **RADIATIVE TRANSFER IN WATER AND ICE CLOUDS IN THE VISIBLE AND INFRARED REGION.**<br>PLASS, G. N. KATTAWAR, G. W.<br>APPL. OPT.<br>10 ( 4 ), 738–48, 1971. |
| 61432 | **SPECIFIC HEAT OF ARGON XENON SOLID SOLUTIONS AT LOW TEMPERATURES.**<br>MANZHELII, V. G. CHAUSOV, G. P.<br>KOVALENKO, S. I.<br>FIZ. TVERD. TELA<br>12 ( 9 ), 2764–6, 1970.<br>( FOR ENGLISH TRANSLATION SEE TPRC NO. 61433 ) |
| 61433 | **SPECIFIC HEAT OF ARGON-XENON SOLID SOLUTIONS AT LOW TEMPERATURES.**<br>MANZHELII, V. G. CHAUSOV, G. P.<br>KOVALENKO, S. I.<br>SOV. PHYS. - SOLID STATE<br>12 ( 9 ), 2233–4, 1971.<br>( ENGLISH TRANSLATION OF FIZ. TVERD. TELA, 12 ( 9 ), 2764–6, 1970., FOR ORIGINAL SEE TPRC NO. 61432 ) |
| 61455 | **NUMERICAL EVALUATION OF THE ELECTROMAGNETIC PROPERTIES OF A HALF-SPACE RELATIVISTIC PLASMA.**<br>IMRE, K. OZIZMIR, E.<br>PHYS. FLUIDS<br>14 ( 3 ), 603–5, 1971. |
| 61461 | **DIFFUSION IN MIXED SOLVENTS. II. IODINE IN BINARY SOLUTIONS OF ETHANOL WITH HYDROCARBONS AND CARBON TETRACHLORIDE.**<br>NAKANISHI, K. OZASA, T. ASHITANI, K.<br>J. PHYS. CHEM.<br>75 ( 7 ), 963–7, 1971. |

| TPRC Number | Bibliographic Citation |
|---|---|
| 61478 | **APPLICATION OF THE ENVIRONMENTAL RELAXATION MODEL TO THE TEMPERATURE DEPENDENCE OF THE VISCOSITY.**<br>TWEER, H. SIMMONS, J. H. MACEDO, P. B.<br>J. CHEM. PHYS.<br>54 ( 5 ), 1952–9, 1971. |
| 61549 | **A METHOD OF DETERMINING THE VISCOSITY OF A PLASMA AT ATMOSPHERIC PRESSURE.**<br>ROTHER, W.<br>BEITR. PLASMAPHYS.<br>10 ( 6 ), 497–506, 1970. |
| 61577 | **SQUEEZE FILMS BETWEEN RUBBER AND GLASS.**<br>ROBERTS, A. D.<br>J. PHYS. D<br>4 ( 3 ), 423–32, 1971. |
| 61579 | **THE SHEAR OF THIN LIQUID FILMS.**<br>ROBERTS, A. D.<br>J. PHYS. D<br>4 ( 3 ), 433–40, 1971. |
| 61580 | **THE THERMAL CONDUCTIVITY OF MOLTEN SALTS.**<br>CORNWELL, K.<br>J. PHYS. D<br>4 ( 3 ), 441–5, 1971. |
| 61585 | **ELECTRICAL CONDUCTANCES AND IONIZATION BEHAVIOR OF SODIUM CHLORIDE IN DIOXANE – WATER SOLUTIONS AT 100 DEGREES AND PRESSURES TO 4000 BARS.**<br>YEATTS, L. B. DUNN, L. A. MARSHALL, W. L.<br>J. PHYS. CHEM.<br>75 ( 8 ), 1099–106, 1971. |
| 61603 | **INTERPRETATION OF SOLID SOLUTION HARDENING WITH VIBRATIONAL SPECTRA.**<br>PLENDL, J. N. GIELISSE, P. J. MANSUR, L. C.<br>MITRA, S. S. SMAKULA, A. TARTE, P. C.<br>APPL. OPT.<br>10 ( 5 ), 1129–33, 1971. |
| 61648 | **PARTIAL MOLAL ENTHALPIES, EXCESS PARTIAL MOLAL HEAT CAPACITIES AND STRUCTURAL EFFECTS OF TETRABUTYLAMMONIUM BROMIDE AND SODIUM TETRAPHENYLBORATE IN AQUEOUS BINARY SOLVENT MIXTURES.**<br>MOHANTY, R. K. SARMA, T. S. SUBRAMANIAN, S.<br>AHLUWALIA, J. C.<br>TRANS. FARADAY SOC.<br>67, ( PT 2 ) ( 578 ), 305–15, 1971. |
| 61671 | **ELECTRONIC SPECTRA OF TRAPPED ELECTRONS IN ORGANIC GLASSES AT 4 K.**<br>HASE, H. NODA, M. HIGASHIMURA, T.<br>J. CHEM. PHYS.<br>54 ( 7 ), 2975–8, 1971. |
| 61674 | **EXISTENCE OF MINIMA AND MAXIMA IN THE COMPOSITION DEPENDENCE OF THE THERMAL DIFFUSION FACTOR AND THERMAL CONDUCTIVITY.**<br>BIOLSI, L. MASON, E. A.<br>J. CHEM. PHYS.<br>54 ( 7 ), 3020–2, 1971. |
| 61756 | **DROP CALORIMETRIC MEASUREMENTS ON SOME CHLORIDES SULFIDES, AND BINARY MELTS.**<br>THOMPSON, W. T. FLENGAS, S. N.<br>CAN. J. CHEM.<br>49 ( 9 ), 1550–63, 1971. |
| 61760 | **THERMAL CONDUCTIVITY OF MULTICOMPONENT MIXTURES.**<br>DULNEV, G. N. ZARICHNYAK, YU. P.<br>J. ENG. PHYS.<br>12 ( 4 ), 215–9, 1967.<br>( ENGLISH TRANSLATION OF INZH. FIZ. ZH., 12 ( 4 ), 419–25, 1967; FOR ORIGINAL SEE TPRC NO. 45419 ) |
| 61837 | **ULTRASONIC INVESTIGATION OF ACETIC ACID HYDROGEN BOND FORMATION. I. ACETIC ACID – ACETONE SOLUTIONS.**<br>CORSARO, R. D. ATKINSON, G.<br>J. CHEM. PHYS.<br>54 ( 9 ), 4090–5, 1971. |
| 61866 | **THE CONDUCTANCE OF ELECTROLYTES IN ACETONE AND IN 1 – PROPANOL – ACETONE MIXTURES AT 25 DEGREES K.**<br>EVANS, D. F. THOMAS, J. NADAS, J. A.<br>MATESICH, M. A.<br>J. PHYS. CHEM.<br>75 ( 11 ), 1714–22, 1971. |
| 61867 | **TRANSPORT BEHAVIOR IN DIMETHYL SULFOXIDE. I I. VISCOSITY STUDIES.**<br>YAO, N–P BENNION, D. N.<br>J. PHYS. CHEM.<br>75 ( 11 ), 1727–34, 1971. |
| 61868 | **AIRBORNE MEASUREMENTS OF EARTH SURFACE TEMPERATURE ( OCEAN AND LAND ) IN THE 10–12 MICRONS AND 8–14 MICRONS REGIONS.**<br>WEISS, M.<br>APPL. OPT.<br>10 ( 6 ), 1280–7, 1971. |
| 61933 | **VISCOSITY OF FUSED CESIUM CHLORIDE– BARIUM DICHLORIDE SALT MIXTURES.**<br>SMIRNOV, M. V. KHOKHLOV, V. A.<br>TR. INST. ELEKTROKHIM., AKAD. NAUK S S S R, URAL. FILIAL<br>( 12 ), 45–9, 1969. |
| 61935 | **VISCOSITY OF CADMIUM IODIDE–CADMIUM CHLORIDE, LEAD BROMIDE–POTASSIUM CHLORIDE, LEAD CHLORIDE–POTASSIUM BROMIDE SYSTEM MELTS.**<br>BARSEGOV, D. G. IL'YASOV, I. I.<br>UKR. KHIM. ZH.<br>36 ( 8 ), 798–801, 1970. |
| 61937 | **SYSTEMS OF SALTS AND MIXED SOLVENTS. IX. SYSTEM CALCIUM CHLORIDE – METHANOL – WATER.**<br>EMONS, H. H. GRUNER, H. WINKLER, F.<br>Z. CHEM.<br>10 ( 11 ), 444–6, 1970. |
| 61943 | **EFFECT OF REFRACTORY ELEMENTS ON THE THERMAL AND ELECTRICAL CONDUCTIVITY OF GRAPHITE.**<br>LUTKOV, A. I. VOLGA, V. I. DYMOV, B. K.<br>DEMIN, A. V. RAKCHEEVA, V. I. PERKOVA, G. A.<br>TSVET. METAL.<br>43 ( 8 ), 48–51, 1970. |
| 62078 | **THERMAL CONDUCTIVITY OF MIXTURES OF ARGON – HELIUM, ARGON – NITROGEN, AND ARGON – NEON.**<br>PETERSON, J. N. HAHN, T. F. COMINGS, E. W.<br>A I CH E J.<br>17 ( 2 ), 289–91, 1971. |
| 62104 | **PHYSICOCHEMICAL PROPERTIES OF SOLUTIONS OF THE LITHIUM NITRITE–LANTHANUM NITRITE–WATER SYSTEM AT 20 DEGREES.**<br>BERDYUKOVA, V. A. PROTSENKO, P. I.<br>SB. NAUCH. SOOBSHCH., DAGESTAN. GOS. UNIV., KAFEDRA KHIM.<br>( 5 ), 60–2, 1969. |
| 62116 | **VISCOSITY AND ACTIVITY OF BINARY IRON, NICKEL, COBALT, AND COPPER SULFIDE COMPONENTS.**<br>BARMIN, L. N. ESIN, O. A.<br>DOBROVINSKII, I. E.<br>TERMODIN. SVOISTVA RASPLAVOV, MATER. KONF.<br>119–27, 1969. |
| 62127 | **VISCOUS FLOW AND DIELECTRIC RELAXATION IN GLYCEROL–ISOPROPYL ALCOHOL AND GLYCEROL–ISOBUTYL ALCOHOL MIXTURES.**<br>RIZK, H. A. ELANWAR, I. M.<br>Z. PHYS. CHEM. ( LEIPZIG )<br>245 ( 5–6 ), 299–307, 1970. |
| 62128 | **VISCOSITY OF AQUEOUS ELECTROLYTE SOLUTIONS IN A WIDE RANGE OF TEMPERATURES AND CONCENTRATIONS. I. APPARATUS FOR MEASURING THE VISCOSITY OF ELECTROLYTE SOLUTIONS AT UP TO 275 DEGREES.**<br>MASHOVETS, V. P. PUCHKOV, L. V.<br>SARGAEV, P. M. FEDOROV, M. K.<br>ZH. PRIKL. KHIM. ( LENINGRAD )<br>44 ( 1 ), 90–4, 1971.<br>( FOR ENGLISH TRANSLATION SEE TPRC NO. 66472 ) |
| 62129 | **SOLUBILITY AND PHYSICAL PROPERTIES OF AQUEOUS SOLUTIONS OF ADIPIC ACID.**<br>NALIVAIKO, A. S. GRANZHAN, V. A.<br>POSTNIKOV, V. A.<br>ZH. PRIKL. KHIM. ( LENINGRAD )<br>44 ( 1 ), 219–21, 1971.<br>( FOR ENGLISH TRANSLATION SEE TPRC NO. 66474 ) |
| 62151 | **PHYSICAL PROPERTIES OF WATER–PROPYLENE CARBONATE MIXTURES AT 25 C. ENVIRONMENTS RICH IN PROPYLENE CARBONATE.**<br>COURTOT–COUPEX, J. L'HER, M.<br>COMPT. REND.<br>275 C ( 3 ), 195–8, 1972. |
| 62325 | **SURFACE STATES OF HELIUM–3 ON AN ADSORBED HELIUM–3 – HELIUM–4 FILM AT LOW TEMPERATURES.**<br>BREWER, D. F. EVENSON, A. THOMSON, A. L.<br>PHYS. LETT.<br>35 A ( 4 ), 307–8, 1971. |
| 62352 | **VISCOSITY OF THE TERNARY LIQUID MIXTURE BUTYL ACETATE–BUTANOL–PROPYLENE GLYCOL.**<br>NAGESHWAR, G. D. MENE, P. S.<br>TRANS., INDIAN INST. CHEM. ENG.<br>118–20, 1968. |
| 62368 | **EFFECT OF THERMAL CONDUCTIVITY ON THE PROPAGATION OF THE ABSORPTION WAVE OF LASER RADIATION.**<br>VOLOSEVICH, P. P. LEVANOV, E. I.<br>AKAD. NAUK SSSR, DOKL.<br>194, 49–52, 1970.<br>( FOR ENGLISH TRANSLATION SEE TPRC NO. 62369 ) |
| 62369 | **EFFECT OF THERMAL CONDUCTIVITY ON THE PROPAGATION OF THE ABSORPTION WAVE OF LASER RADIATION.**<br>VOLOSEVICH, P. P. LEVANOV, E. I.<br>SOV. PHYS.–DOKL.<br>15, 814–6, 1971.<br>( ENGLISH TRANSLATION OF DOKL. AKAD. NAUK SSSR, 194, 49–52, 1970., FOR ORIGINAL SEE TPRC NO. 62368 ) |

| TPRC Number | Bibliographic Citation |
|---|---|
| 62378 | PHYSICOCHEMICAL ANALYSIS OF THE BINARY SYSTEMS OF PHOSPHORIC ACID WITH SOME KETONES.<br>KRUPATKIN, I. L. BRATUSHCHAK, A. N.<br>ZH. FIZ. KHIM.<br>45 ( 1 ), 177–, 1971.<br>( FOR ENGLISH TRANSLATION SEE TPRC NO. 62379 ) |
| 62379 | PHYSICOCHEMICAL ANALYSIS OF THE BINARY SYSTEMS OF PHOSPHORIC ACID WITH SOME KETONES.<br>KRUPATKIN, I. L. BRATUSHCHAK, A. N.<br>RUSS. J. PHYS. CHEM.<br>45 ( 1 ), 99–100, 1971.<br>( ENGLISH TRANSLATION OF ZH. FIZ. KHIM. 45 ( 1 ), 177–, 1971., FOR ORIGINAL SEE TPRC NO. 62378 ) |
| 62400 | INHOMOGENEOUS PLASMAS - ON THE REFLECTION AND TRANSMISSION OF FIELDS OF ARBITRARY POLARIZATION.<br>TAYLOR, C. D. HARRISON, C. W. JR.<br>J. APPL. PHYS.<br>42 ( 7 ), 2676–81, 1971. |
| 62404 | SCATTERED AND REFLECTED LIGHT INTENSITIES ABOVE THE ATMOSPHERE.<br>THOMPSON, B. C. WELLS, M. B.<br>APPL. OPT.<br>10 ( 7 ), 1539–49, 1971. |
| 62412 | DETERMINATION OF SOME PHYSICAL PROPERTIES OF POTASSIUM BENZOATE AND ITS AQUEOUS SOLUTIONS.<br>NAVRATIL, J. NYVLT, J.<br>CHEM. PRUM.<br>20 ( 45 ), 466–7, 1970. |
| 62419 | THERMODYNAMICS AND MECHANICAL PROPERTIES OF ANTHRACENE-CARBAZOLE ALLOYS.<br>ROBINSON, P. M. SCOTT, H. G.<br>MOL. CRYST. LIQUID CRYST.<br>11 ( 3 ), 239–55, 1970. |
| 62433 | HEAT CAPACITY OF AQUEOUS SOLUTIONS OF PERCHLORATES OF DIVALENT MANGANESE, COBALT, NICKEL, COPPER, AND ZINC.<br>LATYSHEVA, V. A. KOZHEVNIKOV, O. A.<br>TERMODIN. TERMOKHIM. KONSTANTY<br>146–52, 1970. |
| 62440 | PHYSICOCHEMICAL STUDY OF A STRONTIUM HYDROXIDE- BORIC ACID- WATER SYSTEM AT 30 DEGREES.<br>KARIBYAN, A. N. BURNAZYAN, A. S.<br>BABAYAN, G. G. TOROSYAN, E. E.<br>ARM. KHIM. ZH.<br>23 ( 8 ), 684–9, 1970. |
| 62442 | VISCOSITY MEASUREMENTS IN CONCENTRATED ELECTROLYTE SOLUTIONS ( EFFECT OF IONS ON THE STRUCTURE OF WATER ).<br>EBERT, G. WENDORFF, J.<br>BER. BUNSENGES. PHYS. CHEM.<br>75 ( 1 ), 82–8, 1971. |
| 62455 | EFFECT OF THE DIELECTRIC CONSTANT OF A MIXTURE OF SOLVENTS, ACETONE- BENZENE AND ACETONE- CARBON TETRACHLORIDE, ON THE VISCOSITY AND ELECTRICAL CONDUCTIVITY OF TETRABUTYLAMMONIUM IODIDE SOLUTIONS.<br>GORENBEIN, E. YA. KARPOVICH, A. I.<br>IZV. VYSSH. UCHEB. ZAVED., KHIM. KHIM. TEKHNOL.<br>13 ( 11 ), 1588–92, 1970. |
| 62456 | THERMAL DIFFUSIVITY OF WATER- ALCOHOL SOLUTIONS.<br>MANDZYUK, A. I. STABNIKOVA, E. V.<br>NOVINKOVA, O. E.<br>IZV. VYSSH. UCHEB. ZAVED., PISHCH. TEKHNOL.<br>( 5 ), 119–21, 1970. |
| 62463 | THERMAL CONDUCTIVITY OF MIXTURES WITH INTERPENETRATING COMPONENTS.<br>DULNEV, G. N.<br>HEAT TRANSFER- SOV. RES.<br>3 ( 2 ), 169–82, 1971. |
| 62464 | EXPERIMENTAL STUDY OF THERMAL CONDUCTIVITY OF MONATOMIC GASES AND THEIR MIXTURES.<br>SHASHKOV, A. G. MARCHENKO, YE. I.<br>HEAT TRANSFER- SOV. RES.<br>3 ( 3 ), 139–56, 1971. |
| 62466 | ENTHALPIES OF DECALINS AND OF TRANS-DECALIN AND N-PENTANE MIXTURES.<br>LENOIR, J. M. HAYWORTH, K. E. HIPKIN, H. G.<br>J. CHEM. ENG. DATA<br>16 ( 2 ), 129–33, 1971. |
| 62467 | DENSITY, VISCOSITY, AND CARBON DIOXIDE SOLUBILITY AND DIFFUSIVITY IN AQUEOUS ETHYLENE GLYCOL SOLUTIONS.<br>HAYDUK, W. MALIK, V. K.<br>J. CHEM. ENG. DATA<br>16 ( 2 ), 143–6, 1971. |
| 62470 | EXPERIMENTAL DATA AND PROCEDURES FOR PREDICTING THERMAL CONDUCTIVITY OF BINARY MIXTURES OF NONPOLAR GASES.<br>SAXENA, S. C. TONDON, P. K.<br>J. CHEM. ENG. DATA<br>16 ( 2 ), 212–20, 1971. |
| 62471 | DIELECTRIC CONSTANTS, VISCOSITIES, FUSION POINT CURVES, AND OTHER PROPERTIES OF THREE NON-AQUEOUS BINARY SYSTEMS.<br>SEARS, P. G. STOECKINGER, T. M.<br>DAWSON, L. R.<br>J. CHEM. ENG. DATA<br>16 ( 2 ), 220–2, 1971. |
| 62472 | DENSITY AND VISCOSITY OF AQUEOUS SOLUTIONS OF METHANOL AND ACETONE FROM THE FREEZING POINT TO 10 DEGREES C.<br>YERGOVICH, T. W. SWIFT, G. W. KURATA, F.<br>J. CHEM. ENG. DATA<br>16 ( 2 ), 222–6, 1971. |
| 62475 | THE POTASSIUM SULFATE-SULFURIC ACID-WATER SYSTEM AT 25 DEGREES C.<br>STOROZHENKO, V. A. SHEVCHUK, V. G.<br>ZH. NEORG. KHIM.<br>16 ( 1 ), 235–, 1971.<br>( FOR ENGLISH TRANSLATION SEE TPRC NO. 62476 ) |
| 62476 | THE POTASSIUM SULFATE-SULFURIC ACID-WATER SYSTEM AT 25 DEGREES C.<br>STOROZHENKO, V. A. SHEVCHUK, V. G.<br>RUSS. J. INORG. CHEM.<br>16 ( 1 ), 121–4, 1971.<br>( ENGLISH TRANSLATION OF ZH. NEORG. KHIM. 16 ( 1 ), 235–, 1971., FOR ORIGINAL SEE TPRC NO. 62475 ) |
| 62479 | REFLECTION AND TRANSMISSION BY SINGLE-DOMAIN CHOLESTERIC LIQUID CRYSTAL FILMS.. THEORY AND VERIFICATION.<br>BERREMAN, D. W. SCHEFFER, T. J.<br>MOL. CRYST. LIQUID CRYST.<br>11 ( 4 ), 395–405, 1970. |
| 62486 | VISCOSITY OF A SOLUTION NEAR THE CRITICAL POINT OF DISPLACEMENT.<br>KUSKOVA, N. V. MATIZEN, E. V.<br>PIS'MA ZH. EKSP. TEOR. FIZ.<br>12 ( 5 ), 255–9, 1970.<br>( FOR ENGLISH TRANSLATION SEE TPRC NO. 69014 ) |
| 62496 | MEAN EMISSIVITY OF A LUMINOUS FLAME ( SPRAY COMBUSTION OF LIQUID FUEL ).<br>KUNITOMO, T.<br>HEAT TRANSFER - JAP. RES.<br>1 ( 2 ), 57–64, 1972. |
| 62505 | REFLECTION SPECTRA OF N-NITROSODIMETHYLAMINE AND N-NITROSODIETHYLAMINE.<br>FREDHOLM, H.<br>ACTA CHEM. SCAND.<br>24 ( 9 ), 3419–20, 1970. |
| 62518 | PHYSICOCHEMICAL PROPERTIES OF SOME THREE - COMPONENT SYSTEMS.<br>SHATALOVA, V. V.<br>IZV. VYSSH. UCHEB. ZAVED., PISHCH. TEKHNOL.<br>( 5 ), 122–4, 1970. |
| 62536 | VISCOSITY OF MOLTEN LITHIUM CHLORIDE-POTASSIUM CHLORIDE MIXTURES.<br>ZUCA, S. BORCAN, R.<br>REV. ROUM. CHIM.<br>15 ( 11 ), 1681–8, 1970. |
| 62543 | PHYSICOCHEMICAL PROPERTIES OF AMMONIUM NITRITE-AMMONIUM NITRATE-WATER SYSTEM SOLUTIONS SATURATED AT 20 DEGREES.<br>PROTSENKO, P. I. UGRYUMOVA, A. A.<br>KUTSENKO, T. E.<br>ZH. PRIKL. KHIM. ( LENINGRAD )<br>44 ( 1 ), 30–4, 1971.<br>( FOR ENGLISH TRANSLATION SEE TPRC NO. 66471 ) |
| 62561 | CORRELATION OF MOLECULAR PARAMETERS FOR TRANSPORT PROPERTIES OF POLAR GASES.<br>BAE, J. H. REED, T. M., III<br>IND. ENG. CHEM., FUNDAM.<br>10 ( 2 ), 269–72, 1971. |
| 62583 | PHYSICAL PROPERTIES IN THE SYSTEM 2-AMINOETHANOL-WATER.<br>HALL, J. L. TINNEY, S. K.<br>PROC. W. VA. ACAD. SCI.<br>41, 239–41, 1969. |
| 62595 | PHYSICOCHEMICAL PROPERTIES OF THE SODIUM CHLORIDE- SODIUM CHLORATE- SODIUM PERCHLORATE- WATER SYSTEM AT 25 DEGREES.<br>LOSEVA, G. K. SEMCHENKO, D. P. ILIN, K. G.<br>TR. NOVOCHERKASSK. POLITEKH. INST.<br>( 197 ), 45–50, 1969. |
| 62600 | TEMPERATURE DEPENDENCES OF THE VISCOSITY OF AQUEOUS ELECTROLYTE SOLUTIONS.<br>PENKINA, N. V.<br>ZH. PRIKL. KHIM. ( LENINGRAD )<br>44 ( 2 ), 349–54, 1971.<br>( FOR ENGLISH TRANSLATION SEE TPRC NO. 66475 ) |

| TPRC Number | Bibliographic Citation |
|---|---|

62602 NONLINEAR REFLECTION AND TRANSMISSION OF ELECTROMAGNETIC WAVES BY MOVING PLASMAS.
CHAKRAVARTI, A. K.
I E E E TRANS. ANTENNAS PROPAGA.
A P-19, 145-7, 1971.

62620 DEPENDENCE OF TRANSPORT COEFFICIENTS ON PLASMA PARAMETERS IN THE TOKAMAK INSTALLATION.
BOBROVSKII, G. A. KUZNETSOV, E. I.
RAZUMOVA, K. A.
ZH. EKSP. TEOR. FIZ.
59 ( 4 ), 1103-14, 1970.
( FOR ENGLISH TRANSLATION SEE TPRC NO. 62621 )

62621 DEPENDENCE OF TRANSPORT COEFFICIENTS ON PLASMA PARAMETERS IN THE TOKAMAK INSTALLATION.
BOBROVSKII, G. A. KUZNETSOV, E. I.
RAZUMOVA, K. A.
SOB. PHYS. JETP
32 ( 4 ), 599-605, 1971.
( ENGLISH TRANSLATION OF ZH. EKSP. TEOR. FIZ. 59 ( 4 ), 1103-14, 1970; FOR ORIGINAL SEE TPRC NO. 62620 )

62622 ENTHALPIES OF CYCLOHEXANE AND MIXTURES OF N - PENTANE AND CYCLOHEXANE.
LENOIR, J. M. KURAVILA, G. K. HIPKIN, H. G.
J. CHEM. ENG. DATA
16 ( 3 ), 271-6, 1971.

62623 ENTHALPIES OF MIXTURES OF BENZENE AND HEXADECANE.
HAYWORTH, K. E. LENOIR, J. M. HIPKIN, H. G.
J. CHEM. ENG. DATA
16 ( 3 ), 276-9, 1971.

62624 ENTHALPIES OF BENZENE AND MIXTURES OF BENZENE WITH N-OCTANE.
LENOIR, J. M. HAYWORTH, K. E. HIPKIN, H. G.
J. CHEM. ENG. DATA
16 ( 3 ), 280-5, 1971.

62625 ENTHALPIES OF MIXTURES OF BENZENE AND CYCLOHEXANE.
LENOIR, J. M. HAYWORTH, K. E. HIPKIN, H. G.
J. CHEM. ENG. DATA
16 ( 3 ), 285-8, 1971.

62626 VISCOSITIES OF BINARY MIXTURES OF N - BUTYL ACETATE - METHYL ETHYL KETONE OR ISOAMYL ALCOHOL AT 35 DEGREES C.
THAYUMANASUNDARAM, R. RAO, P. B.
J. CHEM. ENG. DATA
16 ( 3 ), 323-4, 1971.

62627 DENSITY, VISCOSITY, AND CONDUCTIVITY OF LITHIUM TRIFLUOROMETHANESULFONATE SOLUTIONS IN DIMETHYL SULFITE.
TIEDEMANN, W. H. BENNION, D. N.
J. CHEM. ENG. DATA
16 ( 3 ), 368-70, 1971.

62671 A CRITICAL SURVEY OF SIX THERMODYNAMIC AND TRANSPORT PROPERTIES OF AIR.
GARDINER, G. K.
NAVAL WEAPONS CENTER
104PP., 1971.
( AD-722 402, NWC-TP-4997, N71-29742, AVAIL. NTIS )

62685 THERMODYNAMIC PROPERTIES OF HYDROGEN-HELIUM PLASMAS.
NELSON, H. F.
NTIS
121PP., 1971.
( NASA-CR-1861, AVAIL. NTIS )

62744 THE LITHIUM CHROMATE-UREA-WATER SYSTEM AT 25 DEGREES C.
KARNAUKHOV, A. S. LEPESHKOV, I. N.
ZAVORUEVA, G. N.
ZH. NEORG. KHIM.
16 ( 2 ), 510-, 1971.
( FOR ENGLISH TRANSLATION SEE TPRC NO. 62745 )

62745 THE LITHIUM CHROMATE-UREA-WATER SYSTEM AT 25 DEGREES C.
KARNAUKHOV, A. S. LEPESHKOV, I. N.
ZAVORUEVA, G. N.
RUSS. J. INORG. CHEM.
16 ( 2 ), 269-71, 1971.
( ENGLISH TRANSLATION OF ZH. NEORG. KHIM. 16 ( 2 ), 510-, 1971., FOR ORIGINAL SEE TPRC NO. 62744 )

62763 THERMAL EXPANSION OF SOLID SOLUTIONS IN ZIRCONIUM CARBIDE-NIOBIUM CARBIDE AND HAFNIUM CARBIDE-TANTALUM CARBIDE SYSTEMS.
BARANTSEVA, I. G. PADERNO, V. N.
TUGOPLAVKIE KARBIDY
168-9, 1970.

62768 PHYSICOCHEMICAL PROPERTIES OF A POLYPHOSPHORIC ACID-AMMONIA-WATER SYSTEM.
BUGAI, P. M. EREMENKO, R. A.
ZH. PRIKL. KHIM. ( LENINGRAD )
44 ( 3 ), 657-60, 1971.
( FOR ENGLISH TRANSLATION SEE TPRC NO. 66477 )

62769 PROPERTIES OF SOLUTIONS OF NITRITES OF SOME UNI- AND DIVALENT METALS.
PROTSENKO, P. I. IVLEVA, T. I.
AVDEEV, N. YA.
ZH. PRIKL. KHIM. ( LENINGRAD )
43 ( 12 ), 2593-6, 1970.
( FOR ENGLISH TRANSLATION SEE TPRC NO. 66469 )

62771 CONDUCTIVITY IN CONCENTRATED AQUEOUS URANYL SALT SOLUTIONS.
KALMAN, E.
Z. PHYS. CHEM. ( LEIPZIG )
246 ( 1-2 ), 141-4, 1971.

62773 VISCOSITIES OF SODIUM CHLORIDE + ZINC CHLORIDE MELTS.
BLOOM, H. WEEKS, I. A.
TRANS. FARADAY SOC.
( PT. 5 ) 67 ( 581 ), 1420-5, 1971.

62797 TRANSPIRATION COOLING OF THE CONSTRICTOR WALLS OF AN ELECTRIC HIGH - INTENSITY ARC.
HEBERLEIN, J. PFENDER, E.
J. HEAT TRANSFER, TRANS. A S M E
93 C ( 2 ), 146-54, 1971.

62816 TEMPERATURE RISE AND RADIAL PROFILES IN CARBON DIOXIDE LASERS.
LADERMAN, A. J. BYRON, S. R.
J. APPL. PHYS.
42 ( 8 ), 3138-44, 1971.

62926 SOME OBSERVATIONS ON THE VISCOSITY COEFFICIENTS OF IONS IN VARIOUS SOLVENTS.
CRISS, C. M. MASTROIANNI, M. J.
J. PHYS. CHEM.
75 ( 16 ), 2532-4, 1971.

62985 VISCOSITY OF LIQUID HELIUM-3.
BLACK, M. A. HALL, H. E. THOMPSON, K.
J. PHYS.
4 C ( 2 ), 129-42, 1971.

62991 INFRARED ABSORPTION IN BARIUM SULFATE-POTASSIUM PERMANGANATE SOLID SYSTEMS.
BYCHKOVA, G. S. DMITRIEVA, L. P.
ORG. KATAL.
32-6, 1970.

63003 CONVECTION-FREE METHOD OF MEASURING THE THERMAL CONDUCTIVITY OF FLUIDS.
HECKLE, M.
PROGR. REFRIG. SCI. TECHNOL., PROC. INT. CONGR. REFRIG., 12TH
2, 307-16, 1969.

63034 VISCOSITY AND THERMODYNAMIC CHARACTERISTICS OF THE VISCOUS FLOW OF AQUEOUS SOLUTIONS OF TETRABUTYLAMMONIUM NITRATE.
KUDRYAVTSEVA, I. V. KRUMGAL'Z, B. S.
MISHCHENKO, K. P.
ZH. PRIKL. KHIM. ( LENINGRAD )
44 ( 1 ), 201-5, 1971.
( FOR ENGLISH TRANSLATION SEE TPRC NO. 66473 )

63052 INFRARED SPECTRA AND VAPOR PRESSURE ISOTOPE EFFECT OF CRYSTALLIZED AMMONIA AND ITS DEUTERIUM DERIVATIVES.
WOLFF, H. ROLLAR, H. G. WOLFF, E.
J. CHEM. PHYS.
55 ( 3 ), 1373-8, 1971.

63113 TRACER DIFFUSION AND SHEAR VISCOSITY FOR THE SYSTEM ISOBUTYRIC ACID - WATER NEAR THE CRITICAL MIXING POINT.
ALLEGRA, J. C. STEIN, A. ALLEN, G. F.
J. CHEM. PHYS.
55 ( 4 ), 1716-20, 1971.

63114 ULTRASONIC INVESTIGATION OF ACETIC ACID HYDROGEN BOND FORMATION. II. ACETIC ACID - WATER SOLUTIONS.
CORSARO, R. D. ATKINSON, G.
J. CHEM. PHYS.
55 ( 4 ), 1971-8, 1971.

63123 VISCOSITIES, DENSITIES, AND REFRACTIVE INDICES OF BINARY LIQUID MIXTURES.
KATZ, M. LOBO, P. W. MINANO, A. S.
SOLIMO, H.
CAN. J. CHEM.
49 ( 15 ), 2605-9, 1971.

63144 SPATIALLY DAMPED TEMPERATURE OSCILLATIONS IN A HIGHLY IONIZED PLASMA COLUMN.
STEFANOV, B.
J. APPL. PHYS.
42 ( 9 ), 3540-4, 1971.

63169 APPLICATION OF THE CONGRUENCE PRINCIPLE TO VISCOSITIES OF 1- CHLOROALKANE BINARY MIXTURES.
COURSEY, B. M. HERIC, E. L.
CAN. J. CHEM.
49 ( 16 ), 2631-5, 1971.

| TPRC Number | Bibliographic Citation |
|---|---|
| 63173 | **VISCOSITY OF WATER-METHANOL SOLUTIONS OF LITHIUM PERCHLORATE, SODIUM PERCHLORATE, AND SODIUM CHLORIDE.**<br>WERBLAN, L. ROTOWSKA, A. MINC, S.<br>ELECTROCHIM. ACTA<br>16 ( 1 ), 41–59, 1971. |
| 63174 | **USE OF A MIXTURE OF MAGNESIUM AND ZINC NITRATES FOR THE CONCENTRATION OF WEAK NITRIC ACID.**<br>BARANOV, A. V. CHENTSOVA, L. I.<br>RUBTSOV, YU. I.<br>FIZ., KHIM. KHIM. TEKHNOL., SB. MATER. NAUCH.-TEKH. KONF. RAB. NAUKI PROIZVOD.<br>151–4, 1969. |
| 63183 | **SPECIFIC HEAT OF THE WATER - 1,4-DIOXANE - SODIUM IODIDE SYSTEM AT 25 DEGREES C.**<br>TUNGUSOV, V. P. SHADSKII, S. V.<br>IZV. VYSSH. UCHEB. ZAVED., KHIM. KHIM. TEKHNOL.<br>14 ( 2 ), 203–7, 1971. |
| 63200 | **VISCOSITY OF POTASSIUM AMIDE SOLUTIONS IN AMMONIA AND IN AMMONIA - TRIMETHYLAMINE MIXTURES.**<br>VAAST, C. LEPOUTRE, G.<br>J. CHIM. PHYS. PHYSICOCHIM. BIOL.<br>68 ( 3 ), 400–5, 1971. |
| 63201 | **THERMODYNAMIC STUDY OF BINARY MIXTURES. II. STUDY OF THE LIQUID-PHASE INTERACTIONS BETWEEN PYRIDINE BASES AND CARBON TETRACHLORIDE BY MEASURING THE VISCOSITY COEFFICIENTS.**<br>MEYER, M. MEYER, R. METZGER, J.<br>PENELOUX, A.<br>J. CHIM. PHYS. PHYSICOCHIM. BIOL.<br>68 ( 3 ), 413–6, 1971. |
| 63202 | **DENSITIES AND VISCOSITY COEFFICIENTS OF METAL - AMMONIA SOLUTIONS.**<br>DEMORTIER, A. LOBRY, P. LEPOUTRE, G.<br>J. CHIM. PHYS. PHYSICOCHIM. BIOL.<br>68 ( 3 ), 498–503, 1971. |
| 63204 | **THERMAL CONDUCTIVITY OF POLYATOMIC GASEOUS MIXTURES AT ELEVATED TEMPERATURES.**<br>JAISWAL, V. K.<br>J. PHYS.<br>4 A ( 3 ), 437–8, 1971. |
| 63207 | **PHYSIOCHEMICAL PROPERTIES OF POLYPHOSPHORIC ACIDS.**<br>MIKHAILIN, A. D. KLENITSKII, A. I.<br>SMIRNOVA, E. P. OGNEVA, V. K.<br>KHIM. TEKHNOL. KONDENS. FOSFATOV, TR. VSES. SOVESHCH., 2ND<br>108–14, 1970. |
| 63212 | **EFFECT OF BORON ON THE THERMAL AND ELECTRICAL CONDUCTIVITY OF GRAPHITE.**<br>LUTKOV, A. I. DYMOV, B. K. VOLGA, V. I.<br>PERKOVA, G. S. SHIPKOV, N. N.<br>KONSTR. MATER. OSN. GRAFITA<br>( 5 ), 104–11, 1970. |
| 63222 | **DETERMINATION OF TRANSPORT PROPERTIES FROM ARC EXPERIMENTS. METHODS AND RESULTS.**<br>BAUDER, W. H. MAECKER, H. H.<br>PROC. IEEE<br>59 ( 4 ), 588–92, 1971.<br>( AD-731 519, ARL-71-0123 ) |
| 63227 | **GRAPHITIZATION OF MICROPOWDERS OF SYNTHETIC DIAMONDS STUDIES FROM INFRARED ABSORPTION SPECTRA.**<br>MALOGOLOVETS, V. G. GATILOVA, E. G.<br>SIN. ALMAZY<br>2 ( 4 ), 29–32, 1970. |
| 63245 | **POTASSIUM SULFATE-MANGANESE SULFATE-WATER TERNARY SYSTEM AT 20 DEGREES.**<br>KARNAUKHOV, A. S. RUNOV, N. N.<br>FTD<br>6PP., 1971.<br>( ENGLISH TRANSLATION OF IZV. VYSSH. UCHEB. ZAVED., KHIM. KHIM. TEKHNOL., 10 ( 12 ), 1299–302, 1967., FOR ORIGINAL SEE TPRC NO. 50605 )<br>( FTD-HT-23-1349-71, AD-735638 ) |
| 63252 | **PHYSICOCHEMICAL PROPERTIES OF A SODIUM CHLORIDE- SODIUM CHLORATE- SODIUM PERCHLORATE- WATER SYSTEM AT 40 DEGREES.**<br>ILIN, K. G. LOSEVA, G. K. SEMCHENKO, D. P.<br>TR. NOVOCHERKASSK. POLITEKH. INST.<br>( 197 ), 37–44, 1969. |
| 63260 | **RAPID METHOD FOR DETERMINING THE AVERAGE SPECIFIC HEAT OF AQUEOUS SOLUTIONS OF INORGANIC COMPOUNDS.**<br>TERESHKIN, I. V. FILIPPOV, S. N.<br>ZAVOD. LAB.<br>37 ( 3 ), 313, 1971.<br>( FOR ENGLISH TRANSLATION SEE TPRC NO. 66485 ) |
| 63262 | **CALCULATION OF VISCOSITY FOR AQUEOUS SOLUTIONS OF POTASSIUM SULFATE AND SODIUM SULFATE FROM NMR DATA.**<br>GRITSUS, B. V. KUSHCHENKO, V. V.<br>MISHCHENKO, K. P.<br>ZH. PRIKL. KHIM. ( LENINGRAD )<br>43 ( 12 ), 2741–4, 1970.<br>( FOR ENGLISH TRANSLATION SEE TPRC NO. 66470 ) |
| 63267 | **VISCOSITY OF DILUTED SOLUTIONS OF ALKALI METAL HALIDES IN METHANOL.**<br>EINFELDT, J. GERDES, E.<br>Z. PHYS. CHEM. ( LEIPZIG )<br>246 ( 3–4 ), 221–33, 1971. |
| 63359 | **COMPARISON OF SHEAR AND CONDUCTIVITY RELAXATION TIMES FOR CONCENTRATED LITHIUM CHLORIDE SOLUTIONS.**<br>MOYNIHAN, C. T. BALITACTAC, N. BOONE, L.<br>LITOVITZ, T. A.<br>J. CHEM. PHYS.<br>55 ( 6 ), 3013–9, 1971. |
| 63380 | **COMPLEX FORMATION BY ZINC AND CADMIUM IODIDES WITH AMMONIUM, TETRAETHYLAMMONIUM, AND TETRABUTYL - AMMONIUM IODIDES IN NON-AQUEOUS MEDIA.**<br>GORENBEIN, E. YA. SHEVCHENKO, V. M.<br>ZH. NEORG. KHIM.<br>16 ( 3 ), 629– , 1971.<br>( FOR ENGLISH TRANSLATION SEE TPRC NO. 63381 ) |
| 63381 | **COMPLEX FORMATION BY ZINC AND CADMIUM IODIDES WITH AMMONIUM, TETRAETHYLAMMONIUM, AND TETRABUTYL - AMMONIUM IODIDES IN NON-AQUEOUS MEDIA.**<br>GORENBEIN, E. YA. SHEVCHENKO, V. M.<br>RUSS. J. INORG. CHEM.<br>16 ( 3 ), 335–7, 1971.<br>( ENGLISH TRANSLATION OF ZH. NEORG. KHIM., 16 ( 3 ), 629– , 1971., FOR ORIGINAL SEE TPRC NO. 63380 ) |
| 63382 | **THE COBALT CHLORIDE-MANGANESE CHLORIDE-WATER SYSTEM AT 25 DEGREES C.**<br>SHEVCHUK, V. G. PAVLENKO, A. I.<br>ZH. NEORG. KHIM.<br>16 ( 3 ), 809– , 1971.<br>( FOR ENGLISH TRANSLATION SEE TPRC NO. 63383 ) |
| 63383 | **THE COBALT CHLORIDE-MANGANESE CHLORIDE-WATER SYSTEM AT 25 DEGREES C.**<br>SHEVCHUK, V. G. PAVLENKO, A. I.<br>RUSS. J. INORG. CHEM.<br>16 ( 3 ), 430–2, 1971.<br>( ENGLISH TRANSLATION OF ZH. NEORG. KHIM., 16 ( 3 ), 809– , 1971., FOR ORIGINAL SEE TPRC NO. 63382 ) |
| 63392 | **HEAT CONDUCTIVITY OF AQUEOUS SOLUTIONS OF ELECTROLYTES.**<br>KERIMOV, A. M. EL'DAROV, F. G.<br>EL'DAROV, V. S.<br>NASA<br>8PP., 1971.<br>( ENGLISH TRANSLATION OF IZV. AKAD. NAUK AZERB. SSR, SER. FIZ.-TEKH. MAT. NAUK, ( 6 ), 112–6, 1969., FOR ORIGINAL SEE TPRC NO. 61083 )<br>( N71-24815, NASA-TT-F-13660 ) |
| 63398 | **THE CALCULATION OF SOLAR RADIATION RECEIVED AND ABSORBED AT THE SEA SURFACE.**<br>DISCHEL, R.<br>GEOPHYSICAL SCIENCES LAB., NEW YORK UNIV.<br>20PP., 1971.<br>( AD-723 853, GSL-TR-71-2, AVAIL. NTIS ) |
| 63414 | **ATMOSPHERIC EFFECTS ON INFRARED MULTISPECTRAL SENSING OF SEA-SURFACE TEMPERATURE FROM SPACE.**<br>ANDING, D. KAUTH, R. TURNER, R.<br>NASA AND NTIS<br>98PP., 1971.<br>( NASA-CR-1858 ) |
| 63422 | **THE DESIGN OF AN INSTRUMENT TO MEASURE SMALL CHANGES OF COMPOSITION IN GAS MIXTURES.**<br>WAKEHAM, W. W.<br>J. PHYS. SCI. INSTR.<br>4 E ( 6 ), 443–5, 1971. |
| 63424 | **ELECTRONIC EXCITATION AND THERMODYNAMIC PROPERTIES OF HIGH TEMPERATURE GASES.**<br>CAPITELLI, M. FICOCELLI V, E. MOLINARI, E.<br>Z. NATURFORSCH.<br>26 A ( 4 ), 672–83, 1971. |
| 63430 | **TRANSPORT COEFFICIENTS OF GAS MIXTURES OF ATOMIC AND MOLECULAR HYDROGEN.**<br>WISE, H.<br>CHEM. PHYS. LETT.<br>9 ( 2 ), 133–4, 1971. |
| 63531 | **THERMAL CONDUCTIVITY OF AQUEOUS SOLUTIONS OF DI AND TRIETHYLENE GLYCOLS.**<br>GAZDIEV, M. A. RASTORGUEV, YU. L.<br>ZH. FIZ. KHIM.<br>45 ( 3 ), 692–3, 1971.<br>( FOR ENGLISH TRANSLATION SEE TPRC NO. 63532 ) |
| 63532 | **THERMAL CONDUCTIVITY OF AQUEOUS SOLUTIONS OF DI- AND TRI-ETHYLENE GLYCOLS.**<br>GAZDIEV, M. A. RASTORGUEV, YU. L.<br>RUSS. J. PHYS. CHEM.<br>45 ( 3 ), 383–5, 1971.<br>( ENGLISH TRANSLATION OF ZH. FIZ. KHIM., 45 ( 3 ), 692–3, 1971., FOR ORIGINAL SEE TPRC NO. 63531 ) |

| TPRC Number | Bibliographic Citation |
|---|---|
| 63550 | **HEAT CAPACITY OF SODIUM IODIDE SOLUTIONS IN WATER, METHANOL, AND ACETONE., THEIR CONCENTRATION AND TEMPERATURE DEPENDENCIES.**<br>NOVOSELOV, N. P. MISHCHENKO, K. P.<br>ZH. OBSHCH. KHIM.<br>41 ( 2 ), 255–63, 1971.<br>( FOR ENGLISH TRANSLATION SEE TPRC NO. 67243 ) |
| 63554 | **MEASUREMENTS OF THE PRODUCT OF VISCOSITY AND NORMAL DENSITY IN AN INTERVAL ABOUT T LAMBDA FOR PURE HELIUM–4 AND FOR THREE DILUTE MIXTURES OF HELIUM–3 IN HELIUM–4.**<br>WEBELER, R. W. H. ALLEN, G.<br>LEWIS RESEARCH CENTER, CLEVELAND, OHIO<br>20PP., 1971.<br>( NASA–TN–D–6516, AVAIL. NTIS ) |
| 63558 | **DETERMINING THE THERMAL CONDUCTIVITY OF LIQUIDS BY THE INTERPOLATION METHOD.**<br>VOLKENSHTEIN, V. S. MEDVEDEV, N. N.<br>LEONTEVA, Z. M.<br>IND. LAB., USSR<br>39 ( 6 ), 964–5, 1973.<br>( ENGLISH TRANSLATION OF ZAVOD. LAB., 39 ( 6 ), 724, 1973; FOR ORIGINAL SEE TPRC NO. 72087 ) |
| 63573 | **VISCOSITY OF LIQUID MIXTURES.**<br>BLOOMFIELD, V. A. DEWAN, R. K.<br>J. PHYS. CHEM.<br>75 ( 20 ), 3113–9, 1971. |
| 63577 | **ESTRONE: ULTRAVIOLET REFLECTION SPECTRUM IN SOLID 1–TETRADECANOL.**<br>FREDHOLM, H.<br>BIOCHEM. BIOPHYS. RES. COMMUN.<br>43 ( 2 ), 285–8, 1971. |
| 63597 | **INVESTIGATION OF THE VISCOSITY OF DRY AIR AT ELEVATED PRESSURES AND TEMPERATURES USING A STEADY–FLOW CAPILLARY VISCOSIMETER.**<br>GORING, G. E. EAGAN, D. P.<br>CAN. J. CHEM. ENG.<br>49 ( 1 ), 25–31, 1971. |
| 63598 | **PREDICTION OF THE VISCOSITIES OF LIQUID MIXTURES BY A GROUP SOLUTION MODEL.**<br>RATCLIFF, G. A. KHAN, M. A.<br>CAN. J. CHEM. ENG.<br>49 ( 1 ), 125–9, 1971. |
| 63600 | **CHANGE IN THE VISCOSITY OF WATER AND SOME AQUEOUS SOLUTIONS UNDER THE ACTION OF MAGNETIC AND ELECTRIC FIELDS.**<br>BANTYSH, L. A. POPOVSKII, V. G.<br>ELEKTRON. OBRAB. MATER.<br>( 5 ), 49–51, 1970. |
| 63637 | **EXPERIMENTAL INVESTIGATION OF TRANSPORT PROPERTIES OF LOW–TEMPERATURE PLASMA BY MEANS OF ELECTRIC ARC.**<br>ASINOVSKII, E. I. KIRILLIN, A. V.<br>PAKHOMOV, E. P. SHABASHOV, V. I.<br>PROC. IEEE<br>59 ( 4 ), 592–601, 1971. |
| 63638 | **FORM–FACTOR METHOD YIELDS THERMAL CONDUCTIVITY AND RADIATED POWER OF NITROGEN ARC.**<br>PFLANZ, H. M. TER HORST, D. TH. J.<br>PROC. IEEE<br>59 ( 4 ), 601–4, 1971. |
| 63651 | **TRANSPORT PROPERTIES OF GASES AND GASEOUS MIXTURES AT HIGH TEMPERATURES.**<br>SAXENA, S. C.<br>HIGH TEMP. SCI.<br>3 ( 2 ), 168–88, 1971. |
| 63661 | **MATRIX INFRARED SPECTRUM OF OXYGEN FLUORIDE AND DETECTION OF LITHIUM OXYFLUORIDE.**<br>ANDREWS, L. RAYMOND, J. I.<br>J. CHEM. PHYS.<br>55 ( 7 ), 3078–86, 1971. |
| 63662 | **ARGON MATRIX INFRARED SPECTRUM OF THE OXYGEN CHLORIDE RADICAL.**<br>ANDREWS, L. RAYMOND, J. I.<br>J. CHEM. PHYS.<br>55 ( 7 ), 3087–94, 1971. |
| 63689 | **INFRARED SPECTRAL REFLECTANCE OF SEA WATER.**<br>HOBSON, D. E., JR. WILLIAMS, D.<br>APPL. OPT.<br>10 ( 10 ), 2372–3, 1971. |
| 63708 | **COMPOSITION DEPENDENCE OF VISCOSITY OF SOME BINARY SOLUTIONS.**<br>GHAI, R. K. DULLIEN, F. A. L.<br>CAN. J. CHEM. ENG.<br>49 ( 2 ), 260–8, 1971. |
| 63713 | **ELECTRICAL CONDUCTIVITY, VISCOSITY, AND DENSITY OF SODIUM CHLORIDE AND CESIUM CHLORIDE SOLUTION IN N–METHYLFORMAMIDE OF 5 AND 25 DEGREES.**<br>MOSTKOVA, R. I. KESSLER, YU. M.<br>SEMENOVA, V. N.<br>ELEKTROKHIMIYA<br>7 ( 5 ), 642–8, 1971.<br>( FOR ENGLISH TRANSLATION SEE TPRC NO. 67376 ) |
| 63728 | **ENTHALPY AND THERMODYNAMIC PROPERTIES OF SILICON CARBIDE AT TEMPERATURES UP TO 2900 DEGREES K.**<br>CHEKHOVSKOI, V. YA.<br>J. CHEM. THERMODYN.<br>3 ( 3 ), 289–96, 1971. |
| 63737 | **OPTICAL AND ELECTRICAL PROPERTIES OF ( ZINC, CADMIUM, MERCURY ) SULFIDE AND CADMIUM ( SULFIDE, SELENIDE ) SOLID SOLUTIONS.**<br>KAWAI, S. KIRIYAMA, R. NAKAHARA, F.<br>MEM. INST. SCI. IND. RES., OSAKA UNIV.<br>28, 101–12, 1971. |
| 63740 | **SELECTIVE REFLECTION BY CHOLESTERIC LIQUID CRYSTALS.**<br>DREHER, R. MEIER, G. SAUPE, A.<br>MOL. CRYST. LIQUID CRYST.<br>13 ( 1 ), 17–26, 1971. |
| 63745 | **EXPERIMENTAL INVESTIGATION OF THE VISCOSITY OF NITROGEN–OXYGEN AND NITROGEN–ARGON MIXTURES.**<br>HELLEMANS, J. ZINK, H.<br>PROC. INT. CRYOG. ENG. CONF., 3RD<br>498–503, 1970. |
| 63756 | **EXPERIMENTAL STUDY OF THERMAL CONDUCTIVITY OF COMPRESSED AIR AT LOW TEMPERATURES.**<br>TSEDERBERG, N. V. IVANOVA, Z. A.<br>TEPLOENERGETIKA<br>18 ( 6 ), 69–71, 1971.<br>( FOR ENGLISH TRANSLATION SEE TPRC NO. 67154 ) |
| 63767 | **PARTIAL MOLAL HEAT CAPACITIES OF SODIUM IODIDE IN WATER, METHANOL, AND ACETONE AT 0.8, 25, AND 50 DEGREES.**<br>NOVOSELOV, N. P. MISHCHENKO, K. P.<br>ZH. OBSHCH. KHIM.<br>41 ( 3 ), 489–94, 1971.<br>( FOR ENGLISH TRANSLATION SEE TPRC NO. 67244 ) |
| 63783 | **PHYSICOCHEMICAL PROPERTIES OF A WATER – ETHYL ALCOHOL – ISOAMYL ALCOHOL THREE – COMPONENT SYSTEM.**<br>KHARIN, S. E. SOROKINA, G. S.<br>SHATALOVA, V. V.<br>FERMENT. SPIRT. PROM.<br>37 ( 4 ), 16–7, 1971. |
| 63819 | **SOLVENT AND TEMPERATURE EFFECTS ON NITROGEN NUCLEAR SPIN RELAXATION IN METHYL NITRATE.**<br>SHEPPERD, C. M. SCHAEFER, T. GOODWIN, B. W.<br>T RAA, J.<br>CAN. J. CHEM.<br>49 ( 19 ), 3158–64, 1971. |
| 63834 | **VISCOSITY B–COEFFICIENTS AT 15 AND 25 DEGREES C FOR GLYCINE, BETA–ALANINE, 4–AMINO–N–BUTYRIC ACID, AND 6–AMINO–N–HEXANOIC ACID IN AQUEOUS SOLUTION.**<br>LOWE, B. M. DEVINE, W.<br>J. CHEM. SOC.<br>A ( 13 ), 2113–6, 1971. |
| 63835 | **TRACER AND MUTUAL DIFFUSIVITIES IN THE SYSTEM CHLOROFORM–CARBON TETRACHLORIDE AT 25 DEGREES.**<br>KELLY, C. M. WIRTH, G. B. ANDERSON, D. K.<br>J. PHYS. CHEM.<br>75 ( 21 ), 3293–6, 1971. |
| 63836 | **INTERMOLECULAR HYDROGEN BONDING. I. EFFECTS ON THE PHYSICAL PROPERTIES OF TETRAMETHYLUREA–WATER MIXTURES.**<br>LINDFORS, K. R. OPPERMAN, S. H.<br>GLOVER, M. E. SEESE, J. D.<br>J. PHYS. CHEM.<br>75 ( 21 ), 3313–6, 1971. |
| 63862 | **INFRARED ABSORPTION SPECTRA AND STRUCTURE OF PERCHLORYLAMIDE DERIVATIVES.**<br>KARELIN, A. I. KHARITONOV, YU. YA.<br>ROSOLOVSKII, V. YA.<br>KOLEBATEL'NYE SPEKTRY NEORG. KHIM.<br>182–218, 1971. |
| 63875 | **CALCULATION OF THE EFFECT OF THERMAL DIFFUSION ON THE THERMAL CONDUCTIVITY OF GAS MIXTURES.**<br>SHASHKOV, A. G. IVASHKEVICH, E. V.<br>ABRAMENKO, T. N.<br>PROBL. TEPLO – MASSOPERENOSA<br>15–20, 1970. |
| 63876 | **EXPERIMENTAL STUDY OF THE THERMAL CONDUCTIVITY OF MONOATOMIC GASES AND THEIR MIXTURES.**<br>SHASHKOV, A. G. MARCHENKOV, E. I.<br>PROBL. TEPLO – MASSOPERENOSA<br>297–317, 1970.<br>( FOR ENGLISH TRANSLATION SEE TPRC NO. 38291 ) |

| TPRC Number | Bibliographic Citation |
|---|---|
| 63878 | PROPERTIES OF N,N-DIMETHYLFORMAMIDE-WATER MIXTURES.<br>ONCESCU, T. JURCONI, E.<br>REV. ROUM. CHIM.<br>16 ( 7 ), 1033–8, 1971. |
| 63879 | SIMULTANEOUS VISCOUS FLOW AND DIFFUSION OF GASES THROUGH SMALL APERTURES.<br>PASTERNAK, R. A. CHRISTENSEN, M. V.<br>J. APPL. PHYS.<br>42 ( 11 ), 4183–7, 1971. |
| 63893 | USE OF THE PROPERTIES OF P–N JUNCTIONS FOR STUDYING THE THERMAL CONDUCTIVITY OF BINARY GAS MIXTURES AT LOW TEMPERATURES.<br>BULGAKOVA, L. V.<br>TEPLO–MASSOOBMEN NIZK. TEMP.<br>98–108, 1970. |
| 63894 | DETERMINATION OF THE COEFFICIENTS OF THERMAL CONDUCTIVITY AND DIFFUSIVITY OF NITRIC ACID AND NITROGEN TETROXIDE IN THE -196 TO +20 DEGREE RANGE.<br>SAN'KO, YU. P. AHASHKOV, A. G.<br>TEPLO- MASSOOBMEN NIZK. TEMP.<br>131–41, 1970. |
| 63896 | THERMOPHYSICAL PROPERTIES OF THE COMBUSTION PRODUCTS OF METHANE, REFORMED METHANE, AND KEROSINE AT HIGH TEMPERATURES.<br>POPYRIN, L. S. PSHENICHNOV, N. N.<br>ROSHCHIN, A. M. KHOKHLOV, L. K.<br>SOKOLSKII, A. G. ROGACHEV, A. P.<br>TEPLOFIZ. SVOISTVA NIZKOTEMP. PLAZMY, MATER. VSES. TEPLOFIZ. KONF. SVOISTVAM VESCHESTV VYS. TEMP., 3RD<br>69–71, 1970. |
| 63898 | APPLICABILITY OF ZDANOVSKIIS RULE FOR THE HEAT CAPACITIES AND SPECIFIC VOLUMES OF MAGNESIUM CHLORIDE- STRONTIUM CHLORIDE- WATER THREE- COMPONENT SOLUTIONS.<br>VASILEV, V. A. FEDYAINOV, N. V.<br>KARAPETYANTS, M. KH. FEDOTOVA, T. F.<br>TR. MOSK. KHIM.– TEKHNOL. INST.<br>( 67 ), 31–4, 1970. |
| 63910 | VISCOSITY OF AMMONIA AND ITS MIXTURES WITH NOBLE GASES IN AN ELECTRIC FIELD.<br>TOMMASINI, F. LEVI, A. C. SCOLES, G.<br>Z. NATURFORSCH.<br>26 A ( 6 ), 1098–100, 1971. |
| 63914 | SHEAR VISCOELASTIC RELAXATION OF A NITRATE MELT.<br>GLOVER, G. M. MATHESON, A. J.<br>TRANS. FARADAY SOC.<br>( PT. 7 ) 67 ( 583 ), 1960–70, 1971. |
| 63921 | MAGNETIC DENSIMETER–VISCOMETER.<br>HODGINS, M. G. BEAMS, J. W.<br>REV. SCI. INSTRUM.<br>42 ( 10 ), 1455–7, 1971. |
| 63930 | ENTHALPY OF CIS–2–PENTENE AND A MIXTURE WITH N–PENTANE.<br>LENOIR, J. M. REBERT, C. J. HIPKIN, H. G.<br>J. CHEM. ENG. DATA<br>16 ( 4 ), 401–4, 1971. |
| 63941 | MEASUREMENT OF ELECTRICAL AND THERMAL CONDUCTIVITY COEFFICIENTS OF AN AIR AND CARBON DIOXIDE PLASMA BY USING A STABILIZED ELECTRIC ARC.<br>ASINOVSKII, E. I. NIZOVSKII, V. L.<br>SHABASHOV, V. I.<br>TEPLOFIZ. VYS. TEMP.<br>9 ( 1 ), 37–41, 1971.<br>( FOR ENGLISH TRANSLATION SEE TPRC NO. 63942 ) |
| 63942 | MEASUREMENT OF THE COEFFICIENTS OF ELECTRICAL AND THERMAL CONDUCTIVITY OF AIR AND CARBON DIOXIDE GAS PLASMA BY MEANS OF A STABILIZED ELECTRIC ARC.<br>ASINOVSKII, E. I. NIZOVSKII, V. L.<br>SHABASHOV, V. I.<br>HIGH TEMP.<br>9 ( 1 ), 33–6, 1971.<br>( ENGLISH TRANSLATION OF TEPLOFIZ. VYS. TEMP., 9 ( 1 ), 37–41, 1971., FOR ORIGINAL SEE TPRC NO. 63941 ) |
| 63955 | VISCOSITY AND THERMAL CONDUCTIVITY OF METHANE COMBUSTION PRODUCTS. III.<br>SAMUILOV, E. V. TSITELAURI, N. N.<br>TEPLOFIZ. VYS. TEMP.<br>9 ( 1 ), 197–200, 1971.<br>( FOR ENGLISH TRANSLATION SEE TPRC NO. 63956 ) |
| 63956 | VISCOSITY AND THERMAL CONDUCTIVITY OF METHANE COMBUSTION PRODUCTS. III.<br>SAMUILOV, E. V. TSITELAURI, N. N.<br>HIGH TEMP.<br>9 ( 1 ), 173–4, 1971.<br>( ENGLISH TRANSLATION OF TEPLOFIZ. VYS. TEMP., 9 197–200, 1971., FOR ORIGINAL SEE TPRC NO. 63955 ) |
| 63967 | ELECTRON – SPIN – RESONANCE STUDY OF HIGH – CONCENTRATION MANGANESE NITRATE AQUEOUS SOLUTIONS.<br>STROTHER, E. F. FARACH, H. A.<br>POOLE, C. P., JR.<br>PHYS. REV.<br>4 A ( 5 ), 2079–87, 1971. |
| 63970 | ULTRASONIC INVESTIGATION OF THE LOWER CONSOLUTE POINT OF THE 2,6–LUTIDINE. WATER SYSTEM.<br>GUTSCHICK, V. P. PINGS, C. J.<br>J. CHEM. PHYS.<br>55 ( 8 ), 3845–50, 1971. |
| 63977 | A GENERAL STAGNATION–POINT CONVECTIVE–HEATING EQUATION FOR ARBITRARY GAS MIXTURES.<br>SUTTON, K. GRAVES, R. A., JR.<br>LANGLEY RESEARCH CENTER, HAMPTON, VA.<br>64PP., 1971.<br>( NASA–TR–R–376 ) |
| 63981 | NUMERICAL SIMULATION OF SPARK DISCHARGES IN AIR.<br>PLOOSTER, M. N.<br>PHYS. FLUIDS<br>14 ( 10 ), 2111–23, 1971. |
| 63982 | DRIFT EFFECTS ON THE BEHAVIOR OF A LOW–BETA AXISYMMETRIC TOROIDAL PLASMA.<br>BOWERS, E. WINSOR, N. K.<br>PHYS. FLUIDS<br>14 ( 10 ), 2203–16, 1971. |
| 63984 | ELECTRICAL CONDUCTIVITY AND VISCOSITY OF SODIUM IODIDE SOLUTIONS IN METHANOL AND DIOXANE MIXTURES.<br>SHKODIN, A. M. LEVITSKAYA, N. K.<br>FTD–HT–23–71–1451<br>7PP., 1971.<br>( ENGLISH TRANSLATION OF UKR. KHIM. ZH., 34 ( 4 ), 330–3, 1968; FOR ORIGINAL SEE TPRC NO. 50148 ) |
| 63986 | STUDYING DISTRIBUTIONS OF NEODYMIUM AND SAMARIUM CHLORIDES BETWEEN TRIBUTYLPHOSPHATE AND WATER.<br>MIKHAILICHENKO, A. I. KLIMENKO, M. A.<br>BOGDANOV, V. P.<br>FTD–HT–23–71–1467<br>10PP., 1971.<br>( ENGLISH TRANSLATION OF ZH. NEORG. KHIM., 14 ( 11 ), 3076–80, 1969; FOR ORIGINAL SEE TPRC NO. 56961 ) |
| 63989 | VISCOSITY OF METHANE – NITROGEN AND METHANE – NITROGEN – HYDROGEN MIXTURES AT 273–473 K AND PRESSURES TO 490.3 X 10 TO THE 5TH POWER NEWTONS PER SQUARE METER.<br>GNEZDILOV, N. E. GOLUBEV, I. F.<br>FTD<br>11PP., 1971.<br>( ENGLISH TRANSLATION OF GAZOV. PROM., 13 ( 4 ), 46–8, 1968., FOR ORIGINAL SEE TPRC NO. 50752 )<br>( FTD–MT–24–1583–71, AD–742798 ) |
| 63999 | TECHNICAL SUPPORT PACKAGE FOR TECH BRIEF 69–10055 THERMAL EXPANSION PROPERTIES OF AEROSPACE MATERIALS.<br>NATIONAL AERONAUTICS AND SPACE ADMINISTRATION<br>NATIONAL AERONAUTICS AND SPACE ADMINISTRATION, WASHINGTON, D. C.<br>174PP., 1969.<br>( PB–184 749 ) |
| 64038 | PROPAGATION OF A STRONG CIRCULARLY POLARIZED ELECTROMAGNETIC WAVE THROUGH AFTERGLOW PLASMAS NEAR THE ELECTRON CYCLOTRON FREQUENCY.<br>SHKAROFSKY, I. P.<br>CAN. J. PHYS.<br>49 ( 22 ), 2797–2824, 1971. |
| 64047 | ANALYSIS OF COMPOSITION OF GASES BY THERMAL CONDUCTIVITY. PHYSICAL PRINCIPLES.<br>KOCHEROV, L. E. SHUTOV, M. D.<br>FTD<br>15PP., 1971.<br>( ENGLISH TRANSLATION OF ENTSIKL. IZMER., KONTR. AVTOMAT., ( 11 ), 5–6, 1968; FOR ORIGINAL SEE TPRC NO. 57538 )<br>( FTD–MT–2J–1860–71, AD–742682 ) |
| 64054 | OSCILLATING TOROIDAL CRUCIBLE VISCOMETER.<br>GALLINA, V. BOSCO, M. M. OMINI, M.<br>REV. SCI. INSTRUM.<br>42 ( 11 ), 1708–9, 1971. |
| 64056 | SPECIFIC HEAT OF LIQUID HELIUM–3 – HELIUM–4 MIXTURES. NEAR THE JUNCTION OF THE LAMBDA AND PHASE – SEPARATION CURVES. I.<br>ALVESALO, T. A. BERGLUND, P. M.<br>ISLANDER, S. T. PICKETT, G. R.<br>ZIMMERMANN, W., JR.<br>PHYS. REV.<br>4 A ( 6 ), 2354–68, 1971. |
| 64086 | SHEAR VISCOSITY OF 3–METHYLPENTANE – NITROETHANE NEAR THE CRITICAL POINT.<br>STEIN, A. ALLEGRA, J. C. ALLEN, G. F.<br>J. CHEM. PHYS.<br>55 ( 9 ), 4265–9, 1971. |

| TPRC Number | Bibliographic Citation |
|---|---|
| 64090 | NON-STEADY-STATE, HOT WIRE, THERMAL CONDUCTIVITY APPARATUS.<br>DAVIS, P. S. THEEUWES, F. BEARMAN, R. J.<br>GORDON, R. P.<br>J. CHEM. PHYS.<br>55 ( 10 ), 4776-83, 1971. |
| 64103 | MEASUREMENT OF THE VISCOSITY OF THE ATOMIC HYDROGEN.<br>CHENG, D. Y. BLACKSHEAR, P. L., JR.<br>J. CHEM. PHYS.<br>56 ( 1 ), 213-23, 1972. |
| 64129 | VISCOSITY COEFFICIENTS OF A PLASMA IN A STRONG MAGNETIC FIELD - APPLICATION OF THE QUANTUM MECHANICAL UNIFIED THEORY.<br>KANEKO, S.<br>J. PHYS. SOC. JAPAN<br>31 ( 4 ), 1240-5, 1971. |
| 64137 | TRANSPORT BEHAVIOR IN DIMETHYL SULFOXIDE. III. CONDUCTANCE-VISCOSITY BEHAVIOR OF TETRA-N-AMYLAMMONIUM THIOCYANATE FROM INFINITE DILUTION TO MOLTEN SALT AT 55 DEGREES.<br>YAO, N.-P. BENNION, D. N.<br>J. PHYS. CHEM.<br>75 ( 23 ), 3586-91, 1971.<br>( AD-741804, UCLA-ENG-7211 ) |
| 64141 | TRANSPORT PROCESSES IN LOW MELTING SALTS. SILVER NITRATE-THALLIUM NITRATE SYSTEM.<br>JANZ, G. J. TOMKINS, R. P. T.<br>SIEGENTHALER, H. BALASUBRAHMANYAM, K.<br>LURIE, S. W.<br>J. PHYS. CHEM.<br>75 ( 26 ), 4025-8, 1971. |
| 64142 | MEASUREMENTS OF THE THERMAL CONDUCTIVITY OF GASES. III. DATA FOR BINARY MIXTURES OF HELIUM AND ARGON WITH HYDROGEN ISOTOPES.<br>CAUWENBERGH, H. VAN DAEL, W.<br>PHYSICA ( UTRECHT )<br>54 ( 3 ), 347-60, 1971. |
| 64146 | PARTIAL MOLAL VOLUME AND VISCOSITY OF ANILINIUM CHLORIDE IN WATER AT VARIOUS TEMPERATURES.<br>CHO, B.-R. LEE, Y.-J. KIM, K.-H.<br>DAEHAN HWAHAK HWOEJEE<br>15 ( 2 ), 85-93, 1971. |
| 64152 | SPECIFIC HEAT, DENSITY, AND COMPRESSIBILITY OF AQUEOUS SOLUTIONS OF SOME METAL PERCHLORATES.<br>KOZHEVNIKOV, O. A. MAKARENKOVA, L. P.<br>LATYSHEVA, V. A.<br>IZV. VYSSH. UCHEB. ZAVED., KHIM. KHIM. TEKHNOL.<br>14 ( 5 ), 694-7, 1971. |
| 64154 | VISCOSITY OF SOME MONO MONOVALENT SALTS IN DIOXANE WATER MIXTURES AT 35 DEGREES C.<br>DAS, P. K.<br>J. INDIAN CHEM. SOC.<br>48 ( 5 ), 490-2, 1971. |
| 64176 | A MODIFICATION OF THE MEASURING CELL FOR THE INVESTIGATION OF THE THERMAL CONDUCTIVITY OF LIQUIDS AND GASES BY THE HOT-WIRE METHOD.<br>RASTORGUEV, YU. L. GELLER, V. Z.<br>J. ENG. PHYS.<br>13 ( 1 ), 9-14, 1967.<br>( ENGLISH TRANSLATION OF INZH. FIZ. ZH., 13 ( 1 ), 16-23, 1967., FOR ORIGINAL SEE TPRC NO. 46165 ) |
| 64178 | A SIMPLE CONTINUOUS-FLOW CALORIMETER FOR MEASURING THE SPECIFIC HEAT OF GASES.<br>GRUZDEV, V. A. SHUMSKAYA, A. I.<br>J. ENG. PHYS.<br>13 ( 1 ), 19-21, 1967.<br>( ENGLISH TRANSLATION OF INZH. FIZ. ZH., 13 ( 1 ), 31-6, 1967. FOR ORIGINAL SEE TPRC NO. 45767 ) |
| 64189 | THERMAL CONDUCTIVITY OF BINARY GAS MIXTURES AT LOW TEMPERATURES.<br>SHASHKOV A. G. ABRAMENKO, T. N.<br>J. ENG. PHYS.<br>13 ( 4 ), 294-6, 1967.<br>( ENGLISH TRANSLATION OF INZH. FIZ. ZH., 13 ( 4 ), 549-53, 1967., FOR ORIGINAL SEE TPRC NO. 49226 ) |
| 64211 | VISCOSITY AND MOLECULAR INTERACTIONS IN BINARY MIXTURES OF BENZENE AND TOLUENE WITH CHLORO - AND BROMO - BENZENES.<br>NIGAM, R. K. SINGH, P. P.<br>INDIAN J. CHEM.<br>9 ( 7 ), 691-5, 1971. |
| 64236 | THERMAL CONDUCTIVITY OF AN ETHANOL - WATER SYSTEM AT 20-300 DEGREES AND PRESSURE TO 30 MEGANEWTONS PER SQUARE METER.<br>POPOV, V. N. MALOV, B. A.<br>TEPLOENERGETIKA<br>18 ( 8 ), 88-90, 1971.<br>( FOR ENGLISH TRANSLATION SEE TPRC NO. 69511 ) |
| 64245 | HEAT CAPACITY OF SOLUTIONS OF ELECTROLYTES AND THEIR MIXTURES. I. EXPERIMENTAL DETERMINATION OF THE HEAT CAPACITY OF AQUEOUS SOLUTIONS OF AMMONIUM BICARBONATE, SODIUM BICARBONATE, SODIUM CARBONATE, AMMONIA, AND LIQUIDS FOR SODA PRODUCTION AT 25 DEGREES.<br>CHERNEN'KAYA, E. I.<br>ZH. PRIKL. KHIM. ( LENINGRAD )<br>44 ( 7 ), 1543-9, 1971.<br>( FOR ENGLISH TRANSLATION SEE TPRC NO. 70814 ) |
| 64255 | CALCULATION OF A FROZEN COMPONENT OF THE EFFECTIVE COEFFICIENT OF THERMAL CONDUCTIVITY OF DENSE CHEMICALLY REACTING GASES.<br>KOTELEVSKII, YU. G. SVIRIDOVA, V. D.<br>DISSOTSIIRUYUSHCHIE GAZY TEPLONOSITELI RAB. TELA ENERG. USTANOVOK, MATER. VSES. KONF.<br>184-9, 1970. |
| 64281 | CRITICAL PHENOMENA IN THE BINARY SYSTEM CARBON DISULFIDE-METHANOL. BEHAVIOR OF SPECIFIC HEAT.<br>GAMBHIR, R. D. VISWANATHAN, B.<br>GOPAL, E. S. R.<br>PROC. NUCL. PHYS. SOLID STATE PHYS. SYMP., 14TH<br>3, 207-10, 1970. |
| 64299 | THE HEAT CONDUCTIVITY OF POLYATOMIC GASES IN MAGNETIC FIELDS.<br>HERMANS, L. J. F.<br>UNIVERSITY OF LEIDEN, HOLLAND, PH.D. THESIS<br>57PP., 1970. |
| 64301 | VISCOSITY OF GAS MIXTURES AT NORMAL PRESSURES. BINARY POLAR-NOPOLAR SYSTEMS.<br>HATTIKUDUR, U. R. THODOS, G.<br>AICHE J.<br>17 ( 5 ), 1220-5, 1971. |
| 64303 | EXCESS THERMODYNAMIC FUNCTIONS OF THE SYSTEMS. BENZENE + P- XYLENE AND BENZENE + P- DIOXAN.<br>KHAN, V. H. SUBRAHMANYAM, S. V.<br>TRANS. FARADAY SOC.<br>( PT. 8 ) 67 ( 584 ), 2282-91, 1971. |
| 64305 | DILUTE AQUEOUS SOLUTIONS OF UNSYMMETRICAL QUATERNARY AMMONIUM IODIDES. 1. VISCOSITY AND DENSITY MEASUREMENTS.<br>LOWE, B. M. RENDALL, H. M.<br>TRANS. FARADAY SOC.<br>67 ( 8 ), 2318-27, 1971. |
| 64321 | EFFECT OF A CARBON MONOXIDE IMPURITY ON THE HEAT CAPACITY OF SOLID ARGON.<br>CHAUSOV, G. P. FREIMAN, YU. A.<br>MANZHELII, V. G.<br>FIZ. TVERD. TELA<br>13 ( 5 ), 1491-3, 1971.<br>( FOR ENGLISH TRANSLATION SEE TPRC NO. 64322 ) |
| 64322 | EFFECT OF CARBON OXIDE IMPURITIES ON THE HEAT CAPACITY OF SOLID ARGON.<br>CHAUSOV, G. P. FREIMAN, YU. A.<br>MANZHELII, V. G.<br>SOV. PHYS. - SOLID STATE<br>13 ( 5 ), 1248-9, 1971.<br>( ENGLISH TRANSLATION OF FIZ. TVERD. TELA, 13 ( 5 ), 1491-3, 1971., FOR ORIGINAL SEE TPRC NO. 64321 ) |
| 64327 | THERMAL CONDUCTIVITY OF WATER-ETHANOL-ETHER TERNARY SYSTEMS.<br>KHARIN, V. M.<br>FERMENT. SPIRT. PROM.<br>37 ( 5 ), 11-2, 1971. |
| 64328 | DENSITY AND VISCOSITY OF AQUEOUS SOLUTIONS OF ZINC NITRATE AND A MIXTURE OF MAGNESIUM AND ZINC NITRATES.<br>BURYAGIN, V. I. KAREV, V. G. BARANOV, A. V.<br>FIZ., KHIM. KHIM. TEKHNOL.<br>138-42, 1969. |
| 64329 | CONDUCTIVITY AND EXPANSION.<br>ZANKER, A.<br>IND. PROCESS HEAT.<br>11 ( 8 ), 16, 18-19, 1971. |
| 64330 | MEASUREMENTS ON IONIZED NITROGEN IN JETS AND ESTIMATIONS OF PRANDTL NUMBER.<br>MYRONUK, D. J. SOO, S. L.<br>INT. J. HEAT MASS TRANSFER<br>14 ( 9 ), 1433-43, 1971. |
| 64359 | VISCOSITY OF BORIC ACID SOLUTIONS IN WATER.<br>POPOV, V. N. MOROZOVA, N. A.<br>TR. MOSK. ENERG. INST.<br>( 75 ), 75-8, 1970. |
| 64393 | HEAT CAPACITIES OF AQUEOUS AND NON - AQUEOUS SOLUTIONS OF ELECTROLYTES AND THE EXTRAPOLATION OF APPARENT VALUES TO INFINITE DILUTION.<br>NOVOSELOV, N. P. MISHCHENKO, K. P.<br>ZH. FIZ. KHIM.<br>45 ( 5 ), 1254-, 1971.<br>( FOR ENGLISH TRANSLATION SEE TPRC NO. 64394 ) |

| TPRC Number | Bibliographic Citation |
|---|---|
| 64394 | **HEAT CAPACITIES OF AQUEOUS AND NON - AQUEOUS SOLUTIONS OF ELECTROLYTES AND THE EXTRAPOLATION OF APPARENT VALUES TO INFINITE DILUTION.**<br>NOVOSELOV, N. P. MISHCHENKO, K. P.<br>RUSS. J. PHYS. CHEM.<br>45 ( 5 ), 709–11, 1971.<br>( ENGLISH TRANSLATION OF ZH. FIZ. KHIM., 45 ( 5 ), 1254–, 1971., FOR ORIGINAL SEE TPRC NO. 64393 ) |
| 64427 | **TRANSFER PHENOMENA IN REACTING GAS MIXTURES.**<br>ALEKSEEV, B. V.<br>FTD<br>37PP., 1970.<br>( ENGLISH TRANSLATION OF CHISLENNYE METODY TEOR. RAZREZH. GAZOV, 140–74, 1969; FOR ORIGINAL SEE TPRC NO. 56446 )<br>( FTD-MT-60-70 ) |
| 64428 | **THERMODYNAMIC PROPERTIES AND COMBUSTION-PRODUCT TRANSFER PROPERTIES OF HYDROCARBON FUELS AT TEMPERATURES UP TO 4000 K.**<br>SAMUILOV, E. V. ROZHDESTVENSKII, I. B.<br>VOSKRSENSKAYA, N. V. TSITELAURI, N. N.<br>OLEVINSKII, K. K. SHEVELEV, V. P.<br>FTD<br>29PP., 1968.<br>( ENGLISH TRANSLATION OF ELEC. MHD, PROC. SYMP., 4, 2249–76, 1968; FOR ORIGINAL SEE TPRC NO. 55010 ) |
| 64466 | **VISCOSITIES OF BINARY LIQUID MIXTURES OF NONELECTROLYTES.**<br>NIGAM, R. K. DHILLON, M. S.<br>J. CHEM. ENG. DATA<br>17 ( 1 ), 35–7, 1972. |
| 64521 | **THERMAL CHARACTERISTICS OF A ROCKET JET CONTAINING ALUMINA PARTICLES.**<br>CRABOL, J.<br>OFF. NAT. ETUD. RECH. AEROSP., PUBL.<br>( 133 ), 38PP., 1970. |
| 64571 | **BEHAVIOR OF AN ARC IN AN AXIAL GAS FLOW.**<br>SCHRAMM, H. H.<br>Z. ANGEW. PHYS.<br>31 ( 5–6 ), 311–9, 1971. |
| 64575 | **DENSITY AND VISCOSITY OF AN OCTANE - QUINOLINE SYSTEM IN THE MINUS 20 TO PLUS 40 DEGREES RANGE.**<br>ENNAN, A. A. SAMOILENKO, V. G.<br>CHOBOTAREV, A. N. ANIKEEV, V. A.<br>ZH. PRIKL. KHIM. ( LENINGRAD )<br>44 ( 6 ), 1427–9, 1971.<br>( FOR ENGLISH TRANSLATION SEE TPRC NO. 66479 ) |
| 64576 | **CALCULATION OF SOME PHYSICOCHEMICAL PROPERTIES OF SODIUM-POTASSIUM HYDROXIDE AND ALUMINATE SOLUTIONS.**<br>PENKINA, N. V. PUCHKOV, L. V.<br>ZH. PRIKL. KHIM. ( LENINGRAD )<br>44 ( 7 ), 1554–8, 1971.<br>( FOR ENGLISH TRANSLATION SEE TPRC NO. 69784 ) |
| 64583 | **VIBRATIONAL SPECTRA OF SILANE AND SILANE-PERDEUTERO SILANE MIXTURES IN THE CONDENSED STATES.**<br>FOURNIER, R. P. SAVOIE, R. THE, N. D.<br>BELZILE, R. CABANA, A.<br>CAN. J. CHEM.<br>50 ( 1 ), 35–42, 1972. |
| 64641 | **VISCOSITY AND THERMAL CONDUCTIVITY OF ARGON AT HIGH TEMPERATURES.**<br>KULIK, P. P.<br>TEPLOFIZ. VYS. TEMP.<br>9 ( 2 ), 431–2, 1971.<br>( FOR ENGLISH TRANSLATION SEE TPRC NO. 64642 ) |
| 64642 | **HIGH - TEMPERATURE VISCOSITY AND THERMAL CONDUCTIVITY OF ARGON.**<br>KULIK, P. P.<br>HIGH TEMP.<br>9 ( 2 ), 389–90, 1971.<br>( ENGLISH TRANSLATION OF TEPLOFIZ. VYS. TEMP., 9 ( 2 ), 431–2, 1971., FOR ORIGINAL SEE TPRC NO. 64641 ) |
| 64727 | **BEHAVIOUR OF ZIRCONIUM DISULPHATE IN AQUEOUS SOLUTION.**<br>GODNEVA, M. M. MOTOV, D. L.<br>ZH. NEORG. KHIM.<br>16 ( 7 ), 1885–, 1971.<br>( FOR ENGLISH TRANSLATION SEE TPRC NO. 64728 ) |
| 64728 | **BEHAVIOUR OF ZIRCONIUM DISULPHATE IN AQUEOUS SOLUTION.**<br>GODNEVA, M. M. MOTOV, D. L.<br>RUSS. J. INORG. CHEM.<br>16 ( 7 ), 1001–4, 1971.<br>( ENGLISH TRANSLATION OF ZH. NEORG. KHIM. 16 ( 7 ), 1885–, 1971., FOR ORIGINAL SEE TPRC NO. 64727 ) |
| 64729 | **THE MAGNESIUM CHROMATE- UREA- WATER SYSTEM AT 25 DEGREES C.**<br>SLOBODCHIKOV, A. M. KARNAUKHOV, A. S.<br>LEPESHKOV, I. N.<br>ZH. NEORG. KHIM.<br>16 ( 7 ), 1999–, 1971.<br>( FOR ENGLISH TRANSLATION SEE TPRC NO. 64730 ) |
| 64730 | **THE MAGNESIUM CHROMATE- UREA- WATER SYSTEM AT 25 DEGREES C.**<br>SLOBODCHIKOV, A. M. KARNAUKHOV, A. S.<br>LEPESHKOV, I. N.<br>RUSS. J. INORG. CHEM.<br>16 ( 7 ), 1064–6, 1971.<br>( ENGLISH TRANSLATION OF ZH. NEORG. KHIM. 16 ( 7 ), 1999–, 1971., FOR ORIGINAL SEE TPRC NO. 64729 ) |
| 64753 | **ARGON AND NITROGEN PLASMA VISCOSITY MEASUREMENTS.**<br>SCHREIBER, P. W. HUNTER, A. M., II<br>BENEDETTO, K. R.<br>PHYS. FLUIDS<br>14 ( 12 ), 2696–702, 1971.<br>( AD-739 835, ARL 72-0005 ) |
| 64761 | **HEAT CONDUCTIVITY OF POLAR- NONPOLAR GAS MIXTURES IN AN ELECTRIC FIELD.**<br>DE GROOT, J. J. VAN DEN BROEKE, J. W.<br>MARTINIUS, H. J. VAN DEN MEIJDENBERG, C. J. N.<br>BEENAKKER, J. J. M.<br>PHYSICA ( UTRECHT )<br>56 ( 3 ), 388–424, 1971. |
| 64762 | **A SECOND- ORDER ELECTRIC- FIELD EFFECT FOR THE HEAT CONDUCTIVITY.**<br>DE GROOT, J. J. VAN DEN BROEKE, J. W.<br>VAN DEN MEIJDENBERG, C. J. N. BEENAKKER, J. J. M.<br>PHYSICA ( UTRECHT )<br>56 ( 3 ), 425–8, 1971. |
| 64774 | **STARCH IN FOODS.**<br>SCHOCH, T. J.<br>DENPUN KOGYO GAKKAISHI<br>17 ( 2 ), 261–73, 1969. |
| 64788 | **SPECIFIC HEAT AND DENSITY OF BERYLLIUM CHLORIDE - WATER AND BERYLLIUM CHLORIDE - STRONTIUM CHLORIDE - WATER SOLUTIONS AT 25 DEGREES.**<br>VASILEV, V. A. FEDYAINOV, N. V.<br>KARAPETYANTS, M. KH. SERAFIMOVA, T. I.<br>ZH. FIZ. KHIM.<br>45 ( 7 ), 1867, 1971.<br>( FOR ENGLISH TRANSLATION SEE TPRC NO. 64789 ) |
| 64789 | **THE HEAT CAPACITIES AND DENSITIES OF BERYLLIUM CHLORIDE - WATER AND BERYLLIUM CHLORIDE - STRONTIUM CHLORIDE - WATER SOLUTIONS AT 25 DEGREES C.**<br>VASILEV, V. A. FEDYAINOV, N. V.<br>KARAPETYANTS, M. KH. SERAFIMOVA, T. I.<br>RUSS. J. PHYS. CHEM.<br>45 ( 7 ), 1066–7, 1971.<br>( ENGLISH TRANSLATION OF ZH. FIZ. KHIM., 45 ( 7 ), 1867, 1971., FOR ORIGINAL SEE TPRC NO. 64788 ) |
| 64799 | **TRANSFER DIFFUSION. III. KINETICS AND MECHANISM OF THE TRIIODIDE-IODIDE EXCHANGE REACTION.**<br>RUFF, I. FRIEDRICH, V. J. CSILLAG, K.<br>J. PHYS. CHEM.<br>76 ( 2 ), 162–5, 1972. |
| 64802 | **ELECTROLYTE VISCOSITIES IN ASSOCIATED SOLVENTS.**<br>BARE, J. P. SKINNER, J. F.<br>J. PHYS. CHEM.<br>76 ( 3 ), 434–41, 1972. |
| 64803 | **EFFECTIVE VISCOSITY OF DILUTE SOLUTIONS OF LIQUID HELIUM-3 IN HELIUM-4 BETWEEN 20 MK AND 1 K.**<br>BERTINAT, M. P. BETTS, D. S. BREWER, D. F.<br>BUTTERWORTH, G. J.<br>PHYS. REV. LETT.<br>28 ( 8 ), 472–4, 1972. |
| 64864 | **PHYSICOCHEMICAL ANALYSIS OF TERNARY AQUEOUS SALT SYSTEMS WITH BARIUM NITRATE AS BASIC COMPONENT.**<br>VIDENOV, N. B. KHRISTOVA, L. N.<br>IVANOVA, R. A. DIMITROV, D. I.<br>DOKL. BOLG. AKAD. NAUK<br>24 ( 6 ), 737–40, 1971. |
| 64869 | **CHANGE IN PHYSICAL PROPERTIES OF BORNITE DURING THE DISSOLUTION OF CHALCOPYRITE IN IT.**<br>SLAVSKAYA, A. I. KACHALOVSKAYA, V. M.<br>SHARYBKINA, M. A. RUDNICHENKO, V. E.<br>GEOL. RUD. MESTOROZHD.<br>13 ( 4 ), 113–15, 1971. |
| 64874 | **EFFECT OF GAMMA - QUANTUM IRRADIATION ON LINEAR EXPANSION TEMPERATURE COEFFICIENT OF THE POTASSIUM CHLORIDE - POTASSIUM BROMIDE SYSTEM SOLID SOLUTIONS.**<br>IVANKINA, M. S. YANTSEN, I. V.<br>IZV. TOMSK. POLITEKH. INST.<br>( 184 ), 48–51, 1970. |
| 64892 | **DENSITY AND VISCOSITY OF LITHIUM NITRATE SOLUTIONS AND MIXTURES OF MAGNESIUM AND LITHIUM NITRATES.**<br>POLOVNIKOVA, G. T. KAREV, V. G.<br>ZH. PRIKL. KHIM. ( LENINGRAD )<br>44 ( 8 ), 1696–700, 1971.<br>( FOR ENGLISH TRANSLATION SEE TPRC NO. 69785 ) |

| TPRC Number | Bibliographic Citation |
|---|---|

64893 **SPECIFIC HEAT OF SODIUM AND POTASSIUM ALUMINATE SOLUTIONS CONTAINING CARBONATE ION AT 25–300 DEGREES.**
PENKINA, N. V. PUCHKOV, L. V.
BARANOVA, T. A.
ZH. PRIKL. KHIM. ( LENINGRAD )
44 ( 8 ), 1890–2, 1971.
( FOR ENGLISH TRANSLATION SEE TPRC NO. 69787 )

64901 **THEORY OF TRANSPORT PROPERTIES OF GASES.**
MASON, E. A. MONCHICK, L.
NASA AND NTIS
16PP., 1962.
( IMP–NASA–31 )

64913 **THE POTASSIUM MOLYBDATE–POTASSIUM THIOCYANATE SYSTEM AT 25 DEGREES C.**
KAROV, Z. G. LEPESHKOV, I. N.
SEMENOVA, S. B.
ZH. NEORG. KHIM.
16 ( 8 ), 2273–, 1971.
( FOR ENGLISH TRANSLATION SEE TPRC NO. 64914 )

64914 **THE POTASSIUM MOLYBDATE–POTASSIUM THIOCYANATE SYSTEM AT 25 DEGREES C.**
KAROV, Z. G. LEPESHKOV, I. N.
SEMENOVA, S. B.
RUSS. J. INORG. CHEM.
16 ( 8 ), 1212–5, 1971.
( ENGLISH TRANSLATION OF ZH. NEORG. KHIM. 16 ( 8 ), 2273–, 1971., FOR ORIGINAL SEE TPRC NO. 64913 )

64917 **THE POTASSIUM CARBONATE–POTASSIUM MOLYBDATE–WATER SYSTEM AT 25 DEGREES C.**
KAROV, Z. G. ASHKHOTOV, V. G.
SEMENOVA, S. B.
ZH. NEORG. KHIM.
16 ( 9 ), 2580–, 1971.
( FOR ENGLISH TRANSLATION SEE TPRC NO. 64918 )

64918 **THE POTASSIUM CARBONATE–POTASSIUM MOLYBDATE–WATER SYSTEM AT 25 DEGREES C.**
KAROV, Z. G. ASHKHOTOV, V. G.
SEMENOVA, S. B.
RUSS. J. INORG. CHEM.
16 ( 9 ), 1376–8, 1971.
( ENGLISH TRANSLATION OF ZH. NEORG. KHIM. 16 ( 9 ), 2580–, 1971., FOR ORIGINAL SEE TPRC NO. 64917 )

64929 **SPECIFIC HEATS OF MELTS IN IRON - SULFUR, COBALT - SULFUR, AND NICKEL - SULFUR BINARY SYSTEMS.**
VAISBURD, S. E. ZEDINA, I. N.
ZH. FIZ. KHIM.
45 ( 8 ), 2038–40, 1971.
( FOR ENGLISH TRANSLATION SEE TPRC NO. 64930 )

64930 **HEAT CAPACITY OF IRON - SULPHUR, COBALT - SULPHUR, AND NICKEL - SULPHUR BINARY MELTS.**
VAISBURD, S. E. ZEDINA, I. N.
RUSS. J. PHYS. CHEM.
45 ( 8 ), 1156–7, 1971.
( ENGLISH TRANSLATION OF ZH. FIZ. KHIM., 45 ( 8 ), 2038–40, 1971., FOR ORIGINAL SEE TPRC NO. 64929 )

64935 **ELECTRICAL CONDUCTIVITY IN RELATION WITH DISPERSION DEGRESS OF LYOPHILIC COLLOIDS. XI. 0.2 SODIUM STEARATE AQUEOUS SOLUTIONS WITH ADDITIONS OF O–CRESOL AND VARIOUS HYDROCARBONS.**
RADU, M.
REV. ROUMAINE CHIM.
15, 767–73, 1970.

64937 **DEVELOPMENT OF A SATELLITE MICROWAVE RADIOMETER TO SENSE THE SURFACE TEMPERATURE OF THE WORLD OCEANS.**
HIDY, G. M. HALL, W. F. HARDY, W. N.
HO, W. W. JONES, A. C. LOVE, A. W.
VAN MELLE, M. J. WANG, H. H. WHEELER, A. E.
NTIS
266PP., 1972.
( NASA–CR–1960 )

64975 **TRANSPORT PROPERTIES OF MIXED PLASMAS.. HELIUM–NITROGEN, ARGON–NITROGEN AND XENON–NITROGEN PLASMAS AT ONE ATMOSPHERE, BETWEEN 5000 AND 35000 K.**
CAPITELLI, M. FICOCELLI, E.
BARI UNIV., ITALY
55PP., 1970.
( N71–23082, AVAIL. NTIS )

65003 **VISCOSITIES OF TETRAALKYLAMMONIUM CHLORIDE SOLUTIONS IN ISOPROPYL ALCOHOL–WATER MIXTURES AT 30 DEGREES.**
CHO, B.–R. LEE, Y.–J. LEE, I.
DAEHAN HWAHAK HWOEJEE
15 ( 3 ), 159–63, 1971.

65016 **BASIC THERMOPHYSICAL PROPERTIES OF BINARY MIXTURES OF HYDROGEN AND OXYGEN WITH WATER VAPOR.**
MALASHENKO, V. I. MOROZENKOV, YU. M.
STAN'KOV, V. KH.
IZV. AKAD. NAUK SSSR, ENERG. TRANSP.
( 4 ), 162–8, 1971.

| TPRC Number | Bibliographic Citation |
|---|---|

65018 **PHYSICOCHEMICAL ANALYSIS OF A HYDRAZINE HYDRATE–PHENOL BINARY SYSTEM.**
KRUPATKIN, I. L. BALINA, T. K.
IZV. VYSSH. UCHEB. ZAVED., KHIM. KHIM. TEKHNOL.
14 ( 8 ), 1161–4, 1971.

65020 **MEASUREMENTS OF CLOUD EMISSIVITY IN THE 8–13 MICRON WAVEBAND.**
ALLEN, J. R.
JOURNAL OF APPL. METEOROL.
10, 260–5, 1971.

65038 **EMISSIVITY OF HIGH LAYER CLOUDS BY COMBINED LIDAR AND RADIOMETRIC TECHNIQUES.**
PLATT, C. M. R.
ROY. METERO. SOC., QUART. J.
97, 322–5, 1971.

65044 **COMPARATIVE STUDY, BY INFRARED SPECTROSCOPY, OF THE ROTATIONAL MOTION OF SMALL MOLECULES IN LIQUID SOLUTION AND SOLID SOLUTION IN CARBON TETRACHLORIDE.**
DEVAURE, J. MARTIN, J. J. PERROT, M.
PHAM–VAN–HUONG LASCOMBE, J.
SPECTROCHIM. ACTA
27 A ( 10 ), 2283–5, 1971.

65051 **PHYSICOCHEMICAL ANALYSIS OF TWO - COMPONENT SYSTEMS. DIMETHYLFORMAMIDE - METHANOL AND ACETONE - ACETIC ACID.**
KURMAEVA, A. I. FEDOROVA, I. G.
BARABANOV, V. P.
TR. KAZAN. KHIM. - TEKHNOL. INST.
40 ( 1 ), 228–31, 1969.

65066 **N M R STUDY OF MOLECULAR ASSOCIATION OF ANILINE AND QUINOLINE IN LIQUID MIXTURES.**
MODAK, S. G. KHARE, P. L. MANDE, C.
INDIAN J. PURE APPL. PHYS.
9 ( 9 ), 735–7, 1971.

65083 **THE EXAMINATION OF APPARENT TERRAIN REFLECTANCES.**
SLAGTER, P. C. DEUTEKOM, M.
PHYSICS LAB., RVO–TNO, THE HAGUE, NETHERLANDS
30PP., 1970.
( N71–31052, PHL–1970–25, TDCK–57503, AVAIL. NTIS )

65091 **THEORY OF THE ACTIVITY OF SUCROSE IN AQUEOUS SOLUTION.**
HEITZ, F.
SUCR. BELGE
90 ( 8 ), 383–96., ( 9 ), 441–51, 1971.

65094 **NEWTONIAN LIMIT FOR VISCOSITY MEASUREMENTS OF KMER PHOTORESIST SOLUTIONS.**
YARBROUGH, D. W. SHIRES, E. A. METTE, H. L.
DDC
6PP., 1972.
( ECOM–3526, AD–737596, N72–24512 )

65111 **SPECIFIC HEAT AND COEXISTENCE CURVE NEAR THE CRITICAL POINT OF THE CARBON DISULPHIDE - METHANOL SYSTEM.**
GAMBHIR, R. D. VISWANATHAN, B.
GOPAL, E. S. R.
INDIAN J. PURE APPL. PHYS.
9 ( 10 ), 787–92, 1971.

65118 **HEAT- FLOW MEASUREMENTS IN BAFFIN BAY AND THE LABRADOR SEA.**
PYE, G. D. HYNDMAN, R. D.
J. GEOPHYS. RES.
77 ( 5 ), 938–44, 1972.

65122 **THE THERMAL CONDUCTIVITY OF BINARY MIXTURES OF POLAR GASES.**
SHASHKOV, A. G. ABRAMENKO, T. N.
J. ENG. PHYS.
14 ( 2 ), 171–5, 1968.
( ENGLISH TRANSLATION OF INZH. FIZ. ZH. 14 ( 2 ), 322–8, 1968., FOR ORIGINAL SEE TPRC NO. 49920 )

65123 **DETERMINATION OF THE THERMOPHYSICAL CHARACTERISTICS OF FREE–FLOWING MATERIALS AND THICK LAYERS OF VARIOUS SUBSTANCES.**
MEDVEDEV, N. N.
J. ENG. PHYS.
14 ( 2 ), 176–9, 1968.
( ENGLISH TRANSLATION OF INZH. FIZ. ZH. 14 ( 2 ), 329–33, 1968., FOR ORIGINAL SEE TPRC NO. 49504 )

65173 **HEAT CAPACITY OF SOLUTIONS BY FLOW MICROCALORIMETRY.**
PICKER, P. LEDUC, P. A. PHILIP, P. R.
DESNOYERS, J. E.
J. CHEM. THERMODYN.
3 ( 5 ), 631–42, 1971.

65181 **SPECIFIC HEAT OF ADSORBED FILMS OF HELIUM–3, HELIUM–4, AND ISOTOPIC HELIUM MIXTURES NEAR MONOLAYER COVERAGES.**
BREWER, D. F. EVENSON, A. THOMSON, A. L.
J. LOW TEMP. PHYS.
3 ( 6 ), 603–17, 1970.

| TPRC Number | Bibliographic Citation |
|---|---|

65187 **HEAT CONDUCTIVITY OF THE NONPOLAR-POLAR GAS MIXTURES ARGON-CHLOROMETHANE AND XENON-CHLOROMETHANE.**
BHATTACHARYYA, S. K. BANDOPADHYAY, P. K.
BARUA, A. K.
J. PHYS.
4 A ( 6 ), 944-51, 1971.

65197 **LOGARITHMIC SINGULARITY OF VISCOSITY NEAR THE CRITICAL POINT.**
BALLARO, S. MIGLIARDO, P. WANDERLINGH, F.
LETT. NUOVO CIMENTO SOC. ITAL. FIS.
1 ( 4 ), 129-32, 1971.

65213 **VISCOSITY OF MIXTURES OF BENZENE AND TERT-BUTYL ALCOHOL AT VARIOUS TEMPERATURES.**
OTIN, S. GRACIA, M. GUTIERREZ-LOSA, C.
REV. ACAD. CIENC. EXACTAS, FIS.-QUIM. NATUR. ZARAGOZA
26 ( 2 ), 407-12, 1971.

65228 **RESEARCH STUDY AND EXPERIMENTAL PROGRAM TO DETERMINE THE TRANSPORT PROPERTIES OF HIGH TEMPERATURE AND HIGH DENSITY GASES.**
CAREY, C. BRADSHAW, J. CARNEVALE, E. H.
NTIS
81PP., 1971.
( AD-729 784, AEDC-TR-71-191 )

65230 **ABSORPTION OF SOUND NEAR CRITICAL STATES.**
PINGS, C. J. GUTSCHICK, V. P.
CHEM. ENG. PROGR. SYMPOSIUM SERIES
67 ( 109 ), 13-7, 1971.
( AD-730 136, AFOSR-TR-71-2404 )

65266 **HOT WIRE METHOD FOR MEASURING HEAT CAPACITY OF GASES AT LOW PRESSURES.**
GOVINDARAJAN, K. GOPAL, E. S. R.
J. INDIAN INST. SCI.
53 ( 1 ), 21-7, 1971.

65285 **CALORIMETRIC STUDY OF AN ARGON-NITROGEN SOLID SOLUTION.**
CHAUSOV, G. P. MANZHELII, V. G.
SB. NAUCH. TR., FIZ.-TEKH. INST. NIZK. TEMP., AKAD. NAUK UKR. SSR
( 5 ), 97-105, 1969.

65293 **VISCOSITY, DENSITY, AND ELECTRICAL CONDUCTIVITY OF NICKEL SULFATE SOLUTIONS AT HIGH TEMPERATURES.**
FEDOTOV, N. V. MAKSIMOVA, I. N.
ZH. PRIKL. KHIM. ( LENINGRAD )
14 ( 9 ), 1986-9, 1971.
( FOR ENGLISH TRANSLATION SEE TPRC NO. 69789 )

65294 **PHYSICOCHEMICAL PROPERTIES OF SULFURIC ACID IN THE 0 TO 45 DEGREES RANGE.**
GERZHBERG, YU. I. KRUMGAL'Z, B. S.
RYABIKOVA, V. M. TRABER, D. G. KOZLOV, D. A.
TREPALIN, A. I.
ZH. PRIKL. KHIM. ( LENINGRAD )
44 ( 10 ), 2321-2, 1971.
( FOR ENGLISH TRANSLATION SEE TPRC NO. 66480 )

65303 **EQUATION FOR THE ENTHALPY AND HEAT CAPACITY OF SILICON CARBIDE IN THE 273.15-2900 DEGREES K RANGE.**
CHEKHOVSKOI, V. YA.
TEPLOFIZ. VYS. TEMP.
9 ( 4 ), 751-4, 1971.
( FOR ENGLISH TRANSLATION SEE TPRC NO. 65304 )

65304 **EQUATION FOR THE ENTHALPY AND SPECIFIC HEAT OF SILICON CARBIDE IN THE TEMPERATURE RANGE 273.15-2900 DEGREES K.**
CHEKHOVSKOI, V. YA.
HIGH TEMP.
9 ( 4 ), 682-5, 1971.
( ENGLISH TRANSLATION OF TEPLOFIZ. VYS. TEMP., 9 ( 4 ), 751-4, 1971., FOR ORIGIANL SEE TPRC NO. 65303 )

65317 **THE SODIUM PERCHLORATE-HEXAETHYLENE TETRAMINE-WATER SYSTEM AT 25 DEGREES C.**
SALNIKOVA, L. N. KARNAUKHOV, A. S.
LEPESHKOV, I. N.
ZH. NEORG. KHIM.
16 ( 10 ), 2840- , 1971.
( FOR ENGLISH TRANSLATION SEE TPRC NO. 65318 )

65318 **THE SODIUM PERCHLORATE-HEXAETHYLENE TETRAMINE-WATER SYSTEM AT 25 DEGREES C.**
SALNIKOVA, L. N. KARNAUKHOV, A. S.
LEPESHKOV, I. N.
RUSS. J. INORGAN. CHEM.
16 ( 10 ), 1511-3, 1971.
( ENGLISH TRANSLATION OF ZH. NEORG. KHIM. 16 ( 10 ), 2840- , 1971., FOR ORIGINAL SEE TPRC NO. 65317 )

65328 **SPECIFIC HEATS OF SOLID HYDROGEN, SOLID DEUTERIUM, AND SOLID MIXTURES OF HYDROGEN AND DEUTERIUM.**
ROBERTS, R. J. DAUNT, J. G.
J. LOW TEMP. PHYS.
6 ( 1-2 ), 97-129, 1972.

65329 **INTERPRETATION OF THE T-SQUARED LAW IN HEAT HELIUM SUBMONOLAYERS.**
BLANDIN, A. TOULOUSE, G.
PHYS. LETT.
38 A ( 6 ), 383-4, 1972.

65336 **THERMAL CONDUCTIVITY OF SOLUTIONS.**
RASTORGUEV, YU. L. GANIEV, YU. A.
J. ENG. PHYS.
14 ( 4 ), 372-7, 1968.
( ENGLISH TRANSLATION OF INZH. - FIZ. ZH. 14 ( 4 ), 689-97, 1968., FOR ORIGINAL SEE TPRC NO. 50756 )

65339 **LITHIUM + METHYLAMINE SOLUTIONS. PART 2. - VISCOSITIES.**
YAMAMOTO, M. NAKAMURA, Y. SHIMOJI, M.
J. CHEM. SOC., FARADAY TRANS.
( PT. 1 ) 68 ( 589 ), 135-9, 1972.

65349 **SPECIFIC HEATS AND DENSITIES OF AQUEOUS SOLUTIONS CONTAINING CHLORIDES OF LITHIUM AND BERYLLIUM SUBGROUP METAL STUDIED BY COMPARATIVE CALCULATION METHODS.**
KARAPETYANTS, M. KH. VASILEV, V. A.
FEDYAINOV, N. V.
ZH. FIZ. KHIM.
45 ( 9 ), 2378, 1971.
( FOR ENGLISH TRANSLATION SEE TPRC NO. 65350 )

65350 **SPECIFIC HEATS AND DENSITIES OF AQUEOUS SOLUTIONS CONTAINING LITHIUM AND BERYLLIUM SUBGROUP METAL CHLORIDES BY METHODS OF COMPARATIVE CALCULATION.**
KARAPETYANTS, M. KH. VASILEV, V. A.
FEDYAINOV, N. V.
RUSS. J. PHYS. CHEM.
45 ( 9 ), 1349-50, 1971.
( ENGLISH TRANSLATION OF ZH. FIZ. KHIM. 45 ( 9 ), 2378, 1971., FOR ORIGINAL SEE TPRC NO. 65349 )

65357 **EFFECTS OF HIGH-PRESSURE HYDROGEN CHLORIDE ON TRANSPORT PROPERTIES OF THE MOLTEN ( 46 MOL PERCENT ) POTASSIUM CHLORIDE - ( 54 MOL PERCENT ) ZINC CHLORIDE EUTECTIC SYSTEM.**
HUBBLE, B. R. COPELAND, J. L.
J. PHYS. CHEM.
76 ( 6 ), 904-8, 1972.

65359 **AIRBORNE INFRARED RADIANCE MEASUREMENTS ( 10- TO 12- MICRON WAVELENGTH ) OFF TROPICAL EAST- COAST AUSTRALIA.**
PLATT, C. M. R.
J. GEOPHYS. RES.
77 ( 9 ), 1597-1609, 1972.

65365 **ORDER VERSUS TEMPERATURE IN CHOLESTERIC LIQUID CRYSTALS FROM REFLECTANCE SPECTRA.**
BERREMAN, D. W. SCHEFFER, T. J.
PHYS. REV.
5 A ( 3 ), 1397-403, 1972.

65366 **STIMULATED CONCENTRATION SCATTERING IN THE BINARY-GAS MIXTURES XENON - HELIUM AND SULFUR HEXAFLUORIDE - HELIUM**
LOWDERMILK, W. H. BLOEMBERGEN, N.
PHYS. REV.
5 A ( 3 ), 1423-43, 1972.

65367 **AN EXTENDED LAW OF CORRESPONDING STATES FOR THE EQUILIBRIUM AND TRANSPORT PROPERTIES OF THE NOBLE GASES.**
KESTIN, J. RO, S. T. WAKEHAM, W.
PHYSICA ( UTRECHT )
58 ( 2 ), 165-211, 1972.

65399 **INFRARED REFRACTIVE INDEX OF ATMOSPHERIC AEROSOL SUBSTANCES.**
VOLZ, F. E.
APPL. OPT.
11 ( 4 ), 755-9, 1972.
( AD-744397, AFCRL-720300 )

65400 **THERMAL EXPANSION MEASUREMENTS ON FOUR OPTICAL MATERIALS FROM ROOM TEMPERATURE TO 10 K**
BROWDER, J. S. BALLARD, S. S.
APPL. OPT.
11 ( 4 ), 841-3, 1972.

65414 **THERMAL AND ELASTIC PROPERTIES OF THALLIUM-DOPED SODIUM IODIDE SINGLE CRYSTALS.**
BERGER, L. I. BAKRADZE, R. V.
BALANEVSKAYA, A. E. KUTSIKOVICH, M. B.
IZV. AKAD. NAUK SSSR, NEORG. MATER.
7 ( 10 ), 1851-2, 1971.
( FOR ENGLISH TRANSLATION SEE TPRC NO. 65415 )

65415 **THERMAL AND ELASTIC PROPERTIES OF THALLIUM-DOPED SODIUM IODIDE SINGLE CRYSTALS.**
BERGER, L. I. BAKRADZE, R. V.
BALANEVSKAYA, A. E. KUTSIKOVICH, M. B.
INORG. MAT.
7 ( 10 ), 1651-2, 1971.
( ENGLISH TRANSLATION OF IZV. AKAD. NAUK SSSR, NEORG. MATER. 7 ( 10 ), 1851-2, 1971., FOR ORIGINAL SEE TPRC NO. 65414 )

| TPRC Number | Bibliographic Citation |
|---|---|
| 65423 | **FLUOROALCOHOLS. 14. DENSITIES, REFRACTIVE INDEXES, VISCOSITIES, AND ISOTHERMAL VAPOR - LIQUID EQUILIBRIUMS OF HEXAFLUOROACETONE - WATER MIXTURES.**<br>MURTO, J. KIVINEN, A. LUNDSTROM, G.<br>ACTA CHEM. SCAND.<br>25 ( 7 ), 2451-6, 1971. |
| 65429 | **THERMAL CONDUCTIVITY OF WATER - ETHANOL - ALDEHYDE SOLUTIONS.**<br>KHARIN, V. M. YASYUNAS, B. V.<br>IZV. VYSSH. UCHEB. ZAVED., PISHCH. TEKHNOL.<br>( 4 ), 74-6, 1971. |
| 65443 | **THERMAL CONDUCTIVITY OF DISPERSED SYSTEMS.**<br>TANAEVA, S. A. DOMOROD, L. S.<br>TEPLO - MASSOOBMEN NIZK. TEMP.<br>74-81, 1970. |
| 65456 | **THERMOPHYSICAL PROPERTIES OF BORIC ACID SOLUTIONS IN WATER.**<br>POPOV, V. N.<br>TR. VSES. NAUCH. - TEKH. KONF. TERMODIN., SB. DOKL. SEKTS. TERMODIN. FAZOVYKH PEREKHODOV, POTOKA NEOBRATIMYKH PROTSESSOV TEPLOFIZ. SVOISTVA VESHCHESTV, 3RD<br>371-9, 1970. |
| 65461 | **EXPERIMENTAL INVESTIGATION OF THE CHARACTERISTICS OF THE SPECIFIC HEAT AT THE CRITICAL SOLUTION POINT OF A BINARY MIXTURE.**<br>ANISIMOV, M. A. VORONEL, A. V.<br>OVODOVA, T. M.<br>ZH. EKSP. TEOR. FIZ.<br>61 ( 3 ), 1092-100, 1971.<br>( FOR ENGLISH TRANSLATION SEE TPRC NO. 68704 ) |
| 65462 | **VISCOSITY AND DENSITY OF LITHIUM CHLORIDE, LITHIUM BROMIDE, AND LITHIUM IODIDE AQUEOUS SOLUTIONS AT AVERAGE AND LOW TEMPERATURES.**<br>MASHOVETS, V. P. BARON, N. M.<br>SHCHERBA, M. U.<br>ZH. PRIKL. KHIM. ( LENINGRAD )<br>44 ( 9 ), 1981-6, 1971.<br>( FOR ENGLISH TRANSLATION SEE TPRC NO. 69788 ) |
| 65495 | **VISCOSITY OF THREE-COMPONENT WATER - ETHANOL - METHANOL SOLUTIONS.**<br>KHARIN, S. E. KOKHANOVSKII, N. I.<br>SOROKINA, G. S.<br>FTD-HC-23-33-72<br>5PP., 1972.<br>( ENGLISH TRANSLATION OF IZV. VYSSH. UCHEB. ZAVED., KHIM. KHIM. TEKHNOL., 12 ( 9 ), 1219-22, 1969; FOR ORIGINAL SEE TPRC NO. 56638 ) |
| 65496 | **VISCOSITY AND PROGRESSIVE DIFFUSION IN AQUEOUS KETONE SOLUTIONS.**<br>BAZUNOVA - YAROSHEVSKAYA, L. G.<br>MORACHEVSKII, A. G. NERONOV, YU. I.<br>SHAPIL, L. G<br>FTD-HC-20-39-72<br>5PP., 1972.<br>( ENGLISH TRANSLATION OF KHIM. TERMODIN. RASTVOROV ( 2 ), 73-7, 1968; FOR ORIGINAL SEE TPRC NO. 56795 ) |
| 65502 | **PHYSICOCHEMICAL STUDY OF THE FERROUS CHLORIDE - ZINC CHLORIDE - WATER SYSTEM.**<br>SHCHEDRINA, A. P. ZMIEVA, L. V.<br>KRASNOVA, L. I.<br>FTD-HC-23-60-72<br>5PP., 1972.<br>( ENGLISH TRANSLATION OF IZV. VYSSH. UCHEB. ZAVED., KHIM. KHIM TEKHNOL., 12 ( 6 ), 714-7, 1969., FOR ORIGINAL SEE TPRC NO. 55597 ) |
| 65514 | **VISCOSITY OF CALCIUM CHLORIDE SOLUTIONS.**<br>EZROKHI, L. L.<br>FTD-HC-23-228-72<br>8PP., 1972.<br>( ENGLISH TRANSLATION OF TR. VSES. NAUCH.-ISSLED. INST. GALURGII, ( 52 ), 3-9, 1967; FOR ORIGINAL SEE TPRC NO. 53325 )<br>( AD-744 152 ) |
| 65518 | **NITROGEN TETROXIDE VISCOSITY IN A DENSE GASEOUS STATE.**<br>MAKSIMOV B. G. MISHINA, L. V.<br>SEREBRYANYI, G. Z. DUNETS, V. M.<br>FTD-HT-23-1280-71<br>5PP., 1972.<br>( ENGLISH TRANSLATION OF DOKL. AKAD. NAUK BELORUSS. SSR, 13 ( 8 ), 691-3, 1969; FOR ORIGINAL SEE TPRC NO. 56170 )<br>( AD-745 424 ) |
| 65532 | **PHYSICOCHEMICAL PROPERTIES OF THREE-COMPONENT SOLUTIONS OF WATER, ETHANOL, AND ISOBUTANOL ( HOMOGENEOUS REGION ).**<br>KHARIN, S. E. KNIGA, A. A. SOROKINA, G. S.<br>FTD-HC-23-2142-71<br>6PP., 1972.<br>( ENGLISH TRANSLATION OF FERMENT. SPIRT. PROM., 35 ( 7 ), 9-12, 1969., FOR ORIGINAL SEE TPRC NO. 56229 ) |
| 65541 | **HYDROMATIC EQUATIONS FOR ANISOTROPIC PLASMAS INCLUDING TRANSPORT EFFECTS.**<br>CHODURA, R. POHL, F.<br>INSTITUT FUER PLASMAPHYS. G.M.B.H., GARCHING, WEST GERMANY<br>37PP., 1970.<br>( IPP-1/112 ) |
| 65550 | **VISCOSITY OF ANILINE AND CYCLOHEXANE MIXTURES IN THE CRITICAL REGION.**<br>YANG, C. C.<br>UNIV. CINCINNATI, PH.D. THESIS<br>93PP., 1970.<br>( UM 71-1521, AVAIL. UNIV. MICROFILMS ) |
| 65551 | **EXPERIMENTAL STUDY OF THE TEMPERATURE DEPENDENCE IN THE FLOW EQUATIONS FOR POROUS MEDIA AS APPLIED TO HIGH TEMPERATURE GAS VISCOSITY MEASUREMENTS.**<br>PULKRABEK, W. W.<br>UNIV. MINNESOTA, PH.D. THESIS<br>217PP., 1970.<br>( UM 71-8203, AVAIL. UNIV. MICROFILMS ) |
| 65564 | **HEAT CAPCAITY OF LIQUID HELIUM-4 AND HELIUM-( 3 ) - HELIUM-( 4 ) MIXTURES AT THE LAMBDA TRANSITION.**<br>GASPARINI, F. M.<br>UNIV. MINNESOTA, PH.D. THESIS<br>166PP., 1970.<br>( UM 71-18,726, AVAIL. UNIV. MICROFILMS ) |
| 65568 | **DENSE FLUID VISCOSITY AND A CORRELATION BASED ON THE SQUARE-WELL POTENTIAL FUNCTION FOR ETHANE AND THREE MIXTURES OF METHANE AND ETHANE.**<br>WALKER, J. D.<br>UNIV. KANSAS, PH.D. THESIS<br>175PP., 1970.<br>( UM 70-25,420, AVAIL. UNIV. MICROFILMS ) |
| 65575 | **ELECTRICAL CONDUCTANCE AND IONIZATION BEHAVIOR OF SODIUM CHLORIDE IN DIOXANE - WATER SOLUTIONS AT 300 DEGREES AND PRESSURES TO 4000 BARS.**<br>YEATTS, L. B. MARSHALL, W. L.<br>J. PHYS. CHEM.<br>76 ( 7 ), 1053-62, 1972. |
| 65593 | **VISCOSITY OF ETHYL ALCOHOL AQUEOUS SOLUTIONS AT HIGH PRESSURES IN THE 30-200 DEGREE RANGE.**<br>ABASZADE, A. AGAEV, N. A. KERIMOV, A. M.<br>ZH. FIZ. KHIM.<br>45 ( 10 ), 2672-3, 1971.<br>( FOR ENGLISH TRANSLATION SEE TPRC NO. 65594 ) |
| 65594 | **VISCOSITY OF AQUEOUS ETHANOL UNDER HIGH PRESSURES AT 30-200 DEGREES C.**<br>ABASZADE, A. AGAEV, N. A. KERIMOV, A. M.<br>RUSS. J. PHYS. CHEM.<br>45 ( 10 ), 1517, 1971.<br>( ENGLISH TRANSLATION OF ZH. FIZ. KHIM. 45 ( 10 ), 2672-3, 1971., FOR ORIGINAL SEE TPRC NO. 65593 ) |
| 65599 | **EXPERIMENTAL INVESTIGATION OF THE TEMPERATURE AND CONCENTRATION DEPENDENCE OF THE VISCOSITY OF BINARY MIXTURES OF OCTANE WITH ETHANOL AND WITH PROPANOL.**<br>MAMEDOV, A. A. KERIMOV, A. M.<br>SHAKHALIEV, YA.<br>ZH. FIZ. KHIM.<br>45 ( 10 ), 2696- , 1971.<br>( FOR ENGLISH TRANSLATION SEE TPRC NO. 65600 ) |
| 65600 | **EXPERIMENTAL INVESTIGATION OF THE TEMPERATURE AND CONCENTRATION DEPENDENCE OF THE VISCOSITY OF BINARY MIXTURES OF OCTANE WITH ETHANOL AND WITH PROPANOL.**<br>MAMEDOV, A. A. KERIMOV, A. M.<br>SHAKHALIEV, YA.<br>RUSS. J. PHYS. CHEM.<br>45 ( 10 ), 1532-3, 1971.<br>( ENGLISH TRANSLATION OF ZH. FIZ. KHIM. 45 ( 10 ), 2696- , 1971., FOR ORIGINAL SEE TPRC NO. 65599 ) |
| 65613 | **THE VISCOSITY DEPENDENCE OF A PUTATIVE DIFFUSION-LIMITED REACTION.**<br>CERJAN, C. BARNETT, R. E.<br>J. PHYS. CHEM.<br>76 ( 8 ), 1192-5, 1972. |
| 65620 | **THERMAL EXPANSION OF POWDERED METAL FLUORIDES.**<br>MURAT, M. CHATELUT, F. BARDOT, C.<br>BULL. SOC. CHIM. FR.<br>( 9 ), 3101-5, 1971. |
| 65623 | **METHOD FOR CALCULATING THE RELATIVE MOLAR RESPONSE OF A GAS-CHROMATOGRAPH DETECTOR.**<br>MONFORT, J. P.<br>CHIM. ANAL. ( PARIS )<br>53 ( 10 ), 646-51, 1971. |
| 65632 | **PLASMA DIFFUSION IN TOROIDAL SYSTEMS.**<br>KADOMTSEV, B. B. POGUTSE, O. P.<br>ZH. EKSP. TEOR. FIZ.<br>58 ( 5 ), 1675-82, 1970.<br>( FOR ENGLISH TRANSLATION SEE TPRC NO. 65633 ) |

| TPRC Number | Bibliographic Citation |
|---|---|
| 65633 | **PLASMA DIFFUSION IN TOROIDAL SYSTEMS.**<br>KADOMTSEV, B. B. POGUTSE, O. P.<br>SOVIET PHYSICS JETP<br>31 ( 5 ), 898–901, 1970.<br>( ENGLISH TRANSLATION OF ZH. EKSP. TEOR. FIZ., 58 ( 5 ), 1675–82, 1970., FOR ORIGINAL SEE TPRC NO. 65632 ) |
| 65677 | **ROTATIONAL HEAT CAPACITY OF BETA – NITROGEN.**<br>SMITH, D.<br>CHEM. PHYS. LETT.<br>11 ( 4 ), 405–6, 1971. |
| 65686 | **DENSITY AND VISCOSITY OF MIXED AQUEOUS SOLUTIONS OF PHOSPHORIC ACID AND SOME SALTS.**<br>OREKHOV, I. I. VLASOVA, T. L.<br>CHEBOTARENKO, N. M.<br>IZV. VYSSH. UCHEB. ZAVED., KHIM. KHIM. TEKHNOL.<br>14 ( 9 ), 1302–5, 1971. |
| 65697 | **VISCOSITY OF AQUEOUS POTASSIUM SILICATE SOLUTIONS.**<br>TATOMYR, YA. T. KORDIYAK, KH. S. KUZ, V. I.<br>KOLLOID. ZH.<br>33 ( 6 ), 905–8, 1971.<br>( FOR ENGLISH TRANSLATION SEE TPRC NO. 69339 ) |
| 65727 | **THE DETERMINATION OF THE PRANDTL NUMBER AND THERMAL CONDUCTIVITY OF A REACTING GAS.**<br>ANDOW, P. K.<br>UNIVERSITY OF DELAWARE, M. S. THESIS,<br>75PP., 1971. |
| 65730 | **CHANGE OF VISCOSITY AND ITS FUNCTION IN THE ALKALI–ACID–WATER SYSTEMS.**<br>ZDANOVSKII, A. B. IVANOVA, F. I.<br>ISSLED. ELEKTROKHIM. TEKHNOL. ELEKTROKHIM. PROIZVOD.<br>57–64, 1969. |
| 65734 | **VISCOSITY OF SUGAR SOLUTIONS DURING THE ADDITION OF SODIUM CHLORIDE.**<br>SAMSONOVA, A. I. TSISIN, A. D.<br>TKACHENKO, I. P.<br>KHLEBOPEK. KONDITER. PROM.<br>15 ( 10 ), 15–6, 1971. |
| 65743 | **CRITICAL PHENOMENA IN THE SPECIFIC HEAT OF THE BINARY LIQUID MIXTURE METHANOL–HEPTANE.**<br>VISWANATHAN, B. GOVINDARAJAN, K.<br>GOPAL, E. S. R.<br>PROC. NUCL. PHYS. SOLID STATE PHYS. SYMP., 15TH<br>3, 293–8, 1971. |
| 65756 | **PHYSICOCHEMICAL PROPERTIES OF ELECTROLYTES IN BUTANOL AND ISOBUTANOL.**<br>PIRTSKHALAVA, N. I. KETILADZE, D. D.<br>SOOBSHCH. AKAD. NAUK GRUZ. SSR<br>64 ( 1 ), 65–8, 1971. |
| 65758 | **SHOCK STRUCTURED BY THERMAL CONDUCTIVITY OF ATOMS IN AN IONIZED GAS MIXTURE.**<br>RACZELOWSKI, S. F.<br>BROWN UNIVERSITY, M. S. THESIS,<br>31PP., 1971. |
| 65766 | **PHYSICAL PROPERTIES OF N–VINYLPYRROLIDONE, ALPHA–PYRROLIDONE, AND SOLUTIONS OF POTASSIUM PYRROLIDONE IN ALPHA–PYRROLIDONE.**<br>MYASNIKOVA, L. F. BUSHINSKII, V. I.<br>SUBBOTINA, L. A. VAISMAN, I. L.<br>FREIDLIN, G. N.<br>ZH. PRIKL. KHIM. ( LENINGRAD )<br>44 ( 8 ), 1849–53, 1971.<br>( FOR ENGLISH TRANSLATION SEE TPRC NO. 69786 ) |
| 65773 | **EXPERIMENTAL HEAT CAPACITIES OF CARBON DIOXIDE – METHANE MIXTURES AT ELEVATED PRESSURES.**<br>BISHNOI, P. R. ROBINSON, D. B.<br>CAN. J. CHEM. ENG.<br>49 ( 5 ), 657–62, 1971. |
| 65789 | **PROPERTIES OF A SOLUTION OF METHANOL AND LITHIUM BROMIDE STUDIED AS WORKING SUBSTANCES OF ABSORPTION REFRIGERATION MACHINES.**<br>GROSMAN, E. R.<br>KHOLOD. TEKH. TEKHNOL.<br>( 11 ), 84–6, 1971. |
| 65821 | **HEAT CAPACITY OF AQUEOUS ELECTROLYTE SOLUTIONS.**<br>SYRNIKOV, YU. P.<br>ZH. STRUKT. KHIM.<br>12 ( 6 ), 1082–4, 1971.<br>( FOR ENGLISH TRANSLATION SEE TPRC NO. 67354 ) |
| 65832 | **THE IN SITU FORMATION OF GERMANIUM THIN FILMS.**<br>KIVLIGHN, H.<br>GRUMMAN AEROSPACE CORP., RESEARCH DEPT., BETHPAGE, N. Y.<br>18PP., 1971.<br>( AD–734 745, AVAIL. NTIS ) |
| 65835 | **SOLUTIONS OF COPPER( I ) THIOCYANATE IN MOLTEN TETRA–N–PENTYLAMMONIUM THIOCYANATE.**<br>KOWALSKI, M. A. HARRINGTON, G. W.<br>ANAL. CHEM.<br>44 ( 3 ), 479–83, 1972. |
| 65840 | **STRENGTH OF INTERACTIONS IN BINARY MIXTURES OF METHYLENE CHLORIDE WITH BENZENE, TOLUENE, AND XYLENES, CHLOROFORM WITH ACETONE, ETHER, AND DIOXANE, BENZENE WITH ACETONE, N – PENTANE, AND N – HEXANE, AND CYCLOHEXANE WITH N – PENTANE.**<br>NIGAM, R. K. MAHL, B. S.<br>INDIAN J. CHEM.<br>9 ( 11 ), 1255–8, 1971. |
| 65842 | **PROPERTIES OF DIPHENYL–ETHER CALORIMETERS.**<br>DAVIES, J. V. PRITCHARD, H. O.<br>J. CHEM. THERMODYN.<br>4 ( 1 ), 9–22, 1972. |
| 65879 | **THERMAL CONDUCTIVITY OF MOLTEN ALKALI NITRATES AND HALIDES.**<br>MC DONALD, J.<br>UNIV. MINNESOTA, PH.D. THESIS<br>128PP., 1969.<br>( UM 70–15,774, AVAIL. UNIV. MICROFIILMS ) |
| 65883 | **DIFFUSION COEFFICIENT AND VISCOSITY COEFFICIENT OF THE SYSTEM 3–METHYLPENTANE–NITROETHANE IN THE VICINITY OF ITS CRITICAL SOLUTION TEMPERATURE.**<br>LEISTER, H. M.<br>TEMPLE UNIV., PH.D. THESIS<br>111PP., 1970.<br>( UM 70–19,755, AVAIL. UNIV. MICROFILMS ) |
| 65884 | **THERMAL CONDUCTIVITIES OF SULFUR DIOXIDE AND SULFURYL FLUORIDE AND BINARY MIXTURES CONTAINING THESE GASES.**<br>CHANG, K.–C.<br>ST. LOUIS UNIV., PH.D. THESIS<br>147PP., 1969.<br>( UM 70–20,375, AVAIL. UNIV. MICROFILMS ) |
| 65885 | **STUDY OF THE CONDUCTIVITY, VISCOSITY, SPECTROSCOPIC, AND THERMODYNAMIC PROPERTIES OF CONCENTRATED AQUEOUS SOLUTIONS OF TETRAALKYLAMMONIUM BROMIDES AS A FUNCTION OF TEMPERATURE.**<br>LO SURDO, A.<br>SYRACUSE UNIV., PH.D. THESIS<br>292PP., 1970.<br>( UM 70–24094, AVAIL. UNIV. MICROFILMS ) |
| 65899 | **SIMULTANEOUS MEASUREMENT OF THE SURFACE TENSION AND VISCOSITY OF LIQUIDS.**<br>BEGLYAROV, E. M. SHUMAKOV, G. YA.<br>ZAVOD. LAB.<br>37 ( 12 ), 1483–4, 1971.<br>( FOR ENGLISH TRANSLATION SEE TPRC NO. 66488 ) |
| 65912 | **SHEAR RATE DEPENDENCE OF THE THERMAL CONDUCTIVITY IN NON – NEWTONIAN LIQUIDS.**<br>KHABAKHPASHEVA, E. M. CHIRKOV, YU. S.<br>KOSTYLEV, YU. V.<br>J. ENG. PHYS.<br>14 ( 5 ), 486–7, 1968.<br>( ENGLISH TRANSLATION OF INZH. FIZ. ZH. 14 ( 5 ), 918–20, 1968., FOR ORIGINAL SEE TPRC NO. 50758 ) |
| 65913 | **THERMAL CONDUCTIVITY OF GAS MIXTURES REACTING AT A FINITE RATE.**<br>BAKALIN, YU. I. VERBA, M. I. PORTNOV, V. D.<br>J. ENG. PHYS.<br>14 ( 5 ), 491, 1968.<br>( ENGLISH TRANSLATION OF INZH. FIZ. ZH. 14 ( 5 ), 925–6, 1968., FOR ORIGINAL SEE TPRC NO. 51001 ) |
| 65942 | **COMPARISON OF THE OPTICAL PROPERTIES OF SEA WATER WITH THOSE OF PURE WATER.**<br>HOBSON, D. E., JR.<br>KANSAS STATE UNIVERSITY, M. S. THESIS,<br>38PP., 1971. |
| 65945 | **VISCOSITY OF BINARY LIQUID MIXTURES.**<br>COLEMAN, R. L.<br>LAMAR STATE COLLEGE OF TECHNOLOGY, M. S. THESIS,<br>225PP., 1971. |
| 65971 | **EFFECT OF XENON AND KRYPTON IMPURITIES ON THE HEAT CAPACITY OF SOLID METHANE AT LOW TEMPERATURES.**<br>MANZHELII, V. G. CHAUSOV, G. P.<br>FREIMAN, YU. A.<br>FIZ. TVERD. TELA ( LENINGRAD )<br>13 ( 11 ), 3441–2, 1971.<br>( FOR ENGLISH TRANSLATION SEE TPRC NO. 65972 ) |
| 65972 | **EFFECT OF XENON AND KRYPTON IMPURITIES ON THE SPECIFIC HEAT OF SOLID METHANE AT LOW TEMPERATURES.**<br>MANZHELII, V. G. CHAUSOV, G. P.<br>FREIMAN, YU. A.<br>SOV. PHYS.– SOLID STATE<br>13 ( 11 ), 2902–4, 1972.<br>( ENGLISH TRANSLATION OF FIZ. TVERD. TELA ( LENINGRAD ) 13 ( 11 ), 3441–2, 1971., FOR ORIGINAL SEE TPRC NO. 65971 ) |
| 65975 | **PREDICTION OF GAS THERMAL CONDUCTIVITY OF HALOGENATED METHANES.**<br>TAUSCHER, W. A.<br>KAELTETECH.–KLIM.<br>24 ( 3 ), 67–8, 1972. |

| TPRC Number | Bibliographic Citation |
|---|---|
| 66009 | THERMAL CONDUCTIVITY OF TWO-PHASE SYSTEMS.<br>DESHPANDE, P. B. COUPER, J. R.<br>J. HEAT TRANSFER<br>94 C ( 2 ), 249-51, 1972. |
| 66013 | HEAT FLOW, BATHYMETRY, AND THE MID- ATLANTIC RIDGE AT 43 DEGREES N.<br>KASAMEYER, P. W. VON HERZEN, R. P.<br>SIMMONS, G.<br>J. GEOPHYS. RES.<br>77 ( 14 ), 2535-42, 1972. |
| 66023 | ELASTIC BEHAVIOR OF POLAR SOLUTES AT MICROWAVE FREQUENCIES.<br>MAKHIJA, I. J. DAWAR, A. L.<br>INDIAN J. PURE APPL. PHYS.<br>9 ( 12 ), 1051-3, 1971. |
| 66024 | APPLICATION OF THE PROBE METHOD IN THE STUDY OF THERMOPHYSICAL PROPERTIES OF BIOLOGICAL COMPOUNDS.<br>TUTOVA, E. G. NOVICHENOK, L. N.<br>KONONENKO, V. D.<br>HEAT TRANSFER - SOV. RES.<br>4 ( 2 ), 50-4, 1972. |
| 66031 | LABORATORY SIMULATION OF DIFFUSE REFLECTIVITY FROM A CLOUDY PLANETARY ATMOSPHERE.<br>MARGOLIS, J. S. MC CLEESE, D. J. HUNT, G. E.<br>APPL. OPT.<br>11 ( 5 ), 1212-6, 1972. |
| 66037 | EXPERIMENTAL INVESTIGATION OF THE THERMAL CONDUCTIVITY OF LIQUID SATURATED HYDROCARBONS ( HEXANE - HEPTADECANE ) AND NORMAL ALCOHOLS ( BUTANOL - OCTADECANOL ).<br>MUKHAMEDZYANOV, G. KH. USMANOV, A. G.<br>FTD-MT-72-190<br>29PP., 1972.<br>( ENGLISH TRANSLATION OF HEAT AND MASS TRANSFER, 7, 135-45, 1968; ENERGGIIA, MOSCOW: LENINGRAD; FOR ORIGINAL SEE TPRC NO. 48088 ) |
| 66038 | VISCOSITY OF SOLUTIONS OF A SODIUM, MAGNESIUM, CHLORINE AND SULFATE IONS - WATER SYSTEM.<br>EZROKHI, L. L.<br>FTD-MT-72-0225<br>41PP., 1972.<br>( ENGLISH TRANSLATION OF TR. VSES. NAUCH.-ISSLED. INST. GALURGII, ( 52 ), 20-36, 1967., FOR ORIGINAL SEE TPRC NO. 53326 ) |
| 66039 | PHYSICAL-CHEMICAL PROPERTIES OF PROPIONIC ACID - WATER AND VALERIC ACID - WATER SYSTEMS.<br>KOVALEVA, A. G.<br>FTD-MT-72-0231<br>16PP., 1972.<br>( ENGLISH TRANSLATION OF UKR. KHIM. ZH., 34 ( 9 ), 871-6, 1968., FOR ORIGINAL SEE TPRC NO. 53364 ) |
| 66056 | ENERGY OF ACTIVATION AND RELATIVE TEMPERATURE COEFFICIENTS OF THE VISCOSITY AND ELECTRICAL CONDUCTIVITY OF BINARY LIQUID SYSTEMS.<br>FIALKOV, YU. YA. TSENDROVSKAYA, V. A.<br>FTD-MT-72-229<br>20PP., 1972.<br>( ENGLISH TRANSLATION OF UKR. KHIM. ZH., 35 ( 10 ), 1029-35, 1969., FOR ORIGINAL SEE TPRC NO. 56923 ) |
| 66063 | ACOUSTIC AND SOME PHYSICOCHEMICAL PROPERTIES OF PROPYL ALCOHOL-WATER BINARY MIXTURES ALONG THE SATURATION LINE.<br>VINITSKII, E. S. ZAKURENOV, V. M.<br>FTD-HC-23-2107-71<br>6PP., 1972.<br>( ENGLISH TRANSLATION OF PRIMEN. UL'TRAAKUST. ISSLED. VESHCHESTVA, ( 23 ), 25-30, 1967., FOR ORIGINAL SEE TPRC NO. 55527 ) |
| 66074 | COLLISON INTEGRALS FOR THE COMPONENTS OF DISSOCIATED PLANETARY ATMOSPHERES.<br>KALININ, A. P. LEONAS, V. B.<br>SERMYAGIN, A. V.<br>TEPLOFIZ. VYS. TEMP.<br>9 ( 5 ), 1066-8, 1971.<br>( FOR ENGLISH TRANSLATION SEE TPRC NO. 66075 ) |
| 66075 | COLLISON INTEGRALS FOR THE COMPONENTS OF DISSOCIATED PLANETARY ATMOSPHERES.<br>KALININ, A. P. LEONAS, V. B.<br>SERMYAGIN, A. V.<br>HIGH TEMPERATURE<br>9 ( 5 ), 971-3, 1971.<br>( ENGLISH TRANSLATION OF TEPLOFIZ. VYS. TEMP. 9 ( 5 ), 1066-8, 1971., FOR ORIGINAL SEE TPRC NO. 66074 ) |
| 66080 | LAMBDA-POINT MEASUREMENTS OF VISCOSITY TIMES DENSITY FOR PURE HELIUM-4 AND FOR THREE HELIUM-3 - HELIUM-4 MIXTURES. HELIUM(3) - HELIUM(4) MIXTURES.<br>WEBLER, R. W. H. ALLEN, G.<br>PHYS. REV.<br>5 A ( 4 ), 1820-7, 1972. |
| 66088 | INFRARED ABSORPTIONS OF INTERSTITIAL HYDROGEN ATOMS IN SOLID ARGON AND KRYPTON.<br>BONDYBEY, V. E. PIMENTEL, G. C.<br>J. CHEM. PHYS.<br>56 ( 8 ), 3832-6, 1972. |
| 66089 | VISCOSITY OF THE BINARY GASEOUS MIXTURE HELIUM - NITROGEN.<br>KESTIN, J. RO, S. T. WAKEHAM, W. A.<br>J. CHEM. PHYS.<br>56 ( 8 ), 4036-42, 1972.<br>( AD-74861, AROD-8117-6-E ) |
| 66090 | VISCOSITY OF THE BINARY GASEOUS MIXTURE NEON - KRYPTON.<br>KESTIN, J. RO, S. T. WAKEHAM, W. A.<br>J. CHEM. PHYS.<br>56 ( 8 ), 4086-91, 1972.<br>( AD-748787, AROD-8117-8-E ) |
| 66094 | VISCOSITY OF MOLTEN ZINC CHLORIDE AND SUPERCRITICAL BEHAVIOR IN ITS BINARY SOLUTIONS.<br>EASTEAL, A. J. ANGELL, C. A.<br>J. CHEM. PHYS.<br>56 ( 8 ), 4231-3, 1972. |
| 66095 | DIELECTRIC RELAXATION OF RIGID MOLECULES IN SUPERCOOLED DECALIN.<br>JOHARI, G. P. SMYTH, C. P.<br>J. CHEM. PHYS.<br>56 ( 9 ), 4411-18, 1972. |
| 66112 | THERMAL CONDUCTIVITY OF PHASE-SEPARATED SOLID HELIUM(3) - HELIUM(4) MIXTURES.<br>BURGESS, A. E. CROOKS, M. J.<br>PHYS. LETT.<br>39 A ( 3 ), 183-5, 1972. |
| 66115 | LITHIUM CHLORIDE - BORIC ACID - WATER SYSTEM AT 25 DEGREES C.<br>GODE, H. KLAVINA, L.<br>ZH. NEORG. KHIM.<br>16 ( 12 ), 3390-1, 1971.<br>( FOR ENGLISH TRANSLATION SEE TPRC NO. 66116 ) |
| 66116 | THE LITHIUM CHLORIDE - BORIC ACID - WATER SYSTEM AT 25 DEGREES C.<br>GODE, H. KLAVINA, L.<br>RUSS. J. INORG. CHEM.<br>16 ( 12 ), 1794-5, 1971.<br>( ENGLISH TRANSLATION OF ZH. NEORG. KHIM. 16 ( 12 ), 3390-1, 1971., FOR ORIGINAL SEE TPRC NO. 66115 ) |
| 66117 | INTERACTION BETWEEN THIOCARBAMIDE AND ZINC NITRATE IN SATURATED AQUEOUS SOLUTIONS AT 25 DEGREES C.<br>PROTSENKO, P. I. GLININA, A. G.<br>PROTSENKO, G. P.<br>ZH. NEORG. KHIM.<br>16 ( 12 ), 3305- , 1971.<br>( FOR ENGLISH TRANSLATION SEE TPRC NO. 66118 ) |
| 66118 | INTERACTION BETWEEN THIOCARBAMIDE AND ZINC NITRATE IN SATURATED AQUEOUS SOLUTIONS AT 25 DEGREES C.<br>PROTSENKO, P. I. GLININA, A. G.<br>PROTSENKO, G. P.<br>RUSS. J. INORG. CHEM.<br>16 ( 12 ), 1748-52, 1971.<br>( ENGLISH TRANSLATION OF ZH. NEORG. KHIM. 16 ( 12 ), 3305- , 1971., FOR ORIGINAL SEE TPRC NO. 66117 ) |
| 66119 | A PHYSICOCHEMICAL STUDY OF THE ZIRCONIUM OXIDE CHLORIDE - COPPER( II ) CHLORIDE-HYDROGEN CHLORIDE - WATER SYSTEM AT 25 DEGREES C.<br>BELYAEV, I. N. LOBAS, L. M.<br>EFIMUSHKINA, L. I. DVOSKINA, V. N.<br>ZH. NEORG. KHIM.<br>16 ( 12 ), 3392- , 1971.<br>( FOR ENGLISH TRANSLATION SEE TPRC NO. 66120 ) |
| 66120 | A PHYSICOCHEMICAL STUDY OF THE ZIRCONIUM OXIDE CHLORIDE - COPPER( II ) CHLORIDE-HYDROGEN CHLORIDE - WATER SYSTEM AT 25 DEGREES C.<br>BELYAEV, I. N. LOBAS, L. M.<br>EFIMUSHKINA, L. I. DVOSKINA, V. N.<br>RUSS. J. INORG. CHEM.<br>16 ( 12 ), 1795-7, 1971.<br>( ENGLISH TRANSLATION OF ZH. NEORG. KHIM. 16 ( 12 ), 3392- , 1971., FOR ORIGINAL SEE TPRC NO. 66119 ) |
| 66132 | DENSITIES, EXCESS VOLUMES, SURFACE TENSIONS, VISCOSITIES, AND DIELECTRIC CONSTANTS OF THE SYSTEMS: METHANOL - CYCLOHEXANE, ACETONE - METHANOL, ACETONE - CYCLOHEXANE, AND METHANOL - CYCLOHEXANE - ACETONE.<br>CAMPBELL, A. N. ANAND, S. C.<br>CAN. J. CHEM.<br>50 ( 8 ), 1109-14, 1972. |
| 66136 | A SIMPLIFIED PROCESS VISCOMETER FOR NON-NEWTONIAN FLUIDS.<br>SCHEVE, J. L.<br>IOWA STATE UNIVERSITY, M. S. THESIS,<br>97PP., 1971. |

| TPRC Number | Bibliographic Citation |
|---|---|

66156 **PARTIAL MOLAL HEAT CAPACITIES OF SELECTED ELECTROLYTES IN WATER AND METHANOL AT VARIOUS TEMPERATURES.**
MASTROIANNI, M. J.
UNIV. MIAMI, PH.D. THESIS
122PP., 1971.
( UM 71-19858, AVAIL. UNIV. MICROFILMS )

66171 **INFRARED REFLECTANCE SPECTRA OF AQUEOUS SOLUTIONS OF SOME STRONG ELECTROLYTES.**
RHINE, P. E.
KANSAS STATE UNIV., MANHATTAN, KANS., PH. D. THESIS,
104PP., 1971.
( AVAIL NTIS, UM 71-26, 619, AVAIL. UNIV. MICROFILMS )

66197 **IMPROVED METHOD FOR MEASURING THE THERMAL CONDUCTIVITY OF ELECTROLYTE SOLUTIONS.**
PRUDNIKOV, N. K.
FTD
7PP., 1972.
( ENGLISH TRANSLATION OF IZV. VYSSH. UCHEB. ZAVED., ENERG., 13 ( 1 ), 122-6, 1970; FOR ORIGINAL SEE TPRC NO. 59465 )
( FTD-HC-23-43-72, AD-747810 )

66199 **THE VISCOSITY OF AQUEOUS DIETHYLENE GLYCOL SOLUTIONS.**
KLIMENKO, A. P. KOSTYUK, V. I.
PETRUSHENKO, A. A. KRAVCHENKO, M. I.
FTD
5PP., 1972.
( ENGLISH TRANSLATION OF GAZOV. PROM., 12 ( 11 ), 47-8, 1967; FOR ORIGINAL SEE TPRC NO. 49914 )
( FTD-HC-23-61-72, AD-748746 )

66200 **TEMPERATURE DEPENDENCE OF THE VISCOSITY OF MIXTURES OF RAREFIED GASES AT LOW TEMPERATURES.**
ULYBIN, S. A.
FTD
7PP., 1972.
( ENGLISH TRANSLATION OF GAZOV. PROM., 13 ( 12 ), 36-9, 1968; FOR ORIGINAL SEE TPRC NO. 54413 )
( FTD-HC-23-68-72, AD-748747 )

66212 **SOME METHODS FOR CALCULATING THE DENSITY AND HEAT CAPACITY OF MIXED ELECTROLYTE SOLUTIONS.**
MIKULIN, G. I.
FTD
47PP., 1972.
( ENGLISH TRANSLATION OF VOP. FIZ. KHIM. RASTVOROV ELEKTROLIT., 401-16, 168; FOR ORIGINAL SEE TPRC NO. 51848 )
( FTD-MT-ST-72-762 )

66217 **TRANSPORT COEFFICIENTS OF HIGH-PRESSURE ARGON IN A MAGNETIC FIELD.**
DEVOTO, R. S.
DEPT. AERONAUT. ASTRONAUT., STANFORD UNIV., STANFORD, CALIF.
121PP., 1971.
( AD-728 678, ARL-71-0075, AVAIL. NTIS )

66238 **TEMPERATURE DEPENDENCE OF VISCOSITY AND CONDUCTIVITY OF CONCENTRATED AQUEOUS CALCIUM NITRATE SOLUTIONS.**
AMBRUS, J. H. MOYNIHAN, C. T. MACEDO, P. B.
J. ELECTROCHEM. SOC.
119 ( 2 ), 192-8, 1972.

66242 **LONGITUDINAL THERMAL CONDUCTIVITY AND ENERGY RELAXATION OF ELECTRONS IN AN AFTERGLOW PLASMA IN A MAGNETIC FIELD.**
LIVENTSEVA, I. P. JLLIN'SKII, A. P.
PHENOMENA IONIZ. GASES, CONTRIB. PAP. INT. CONF., 10
20PP., 1971.

66262 **MEASUREMENT OF THE TEMPERATURE OF A WALL-STABILIZED ELECTRIC ARC IN ARGON.**
ZHEENBAEV, ZH. ZH. NESMACHNAYA, I. P.
POLOVIKOV, A. I. ENGELSHT, V. S.
KHIM. FIZ. NIZKOTEMP. PLAZMY, TR. MEZHVUZ. KONF., 1ST
297-301, 1971.

66267 **THERMAL PROPERTIES OF HIGHLY NONEQUILIBRIUM PLASMAS.**
YEN, J. T.
PROGR. HEAT MASS TRANSFER
2, 451-8, 1969.

66274 **DENSITY, VISCOSITY, AND ELECTRIC CONDUCTIVITY OF AQUEOUS SOLUTIONS OF NORMAL SODIUM MOLYBDATE.**
KAROV, Z. G. KUKULNEVA, E. I.
SEMENOVA, S. B.
SB. STUD. NAUCH. RAB., KABARDINO-BALKAR. GOS. UNIV.
( 5 ), 126-30, 1970.

66284 **THERMODYNAMIC AND TRANSPORT PROPERTIES OF CHEMICALLY REACTING GASEOUS SYSTEMS. PART II.**
KRASIN, A.
NAUKA I TEKHNIKA: MINSK, BELORUSS, SSR
237PP., 1971.
( FOR ENGLISH TRANSLATION SEE TPRC NO. 67539 )

| TPRC Number | Bibliographic Citation |
|---|---|

66290 **THERMAL CONDUCTIVITY OF ETHANOL-WATER SOLUTIONS IN THE VAPOR PHASE.**
TSEDERBERG, N. V. POPOV, V. N. MALOV, B. A.
TR. MOSK. ENERG. INST.
( 2 ) 96-101, 1971.

66303 **PHYSICOCHEMICAL PROPERTIES OF AQUEOUS SOLUTIONS OF SODIUM AND POTASSIUM PYROPHOSPHATES.**
GALINKER, V. S. BYKOVA, M. I.
ZH. PRIKL. KHIM. ( LENINGRAD )
44 ( 12 ), 2741-4, 1971.
( FOR ENGLISH TRANSLATION SEE TPRC NO. 69790 )

66322 **THE THERMAL CONDUCTIVITY OF BINARY MIXTURES OF METHYL NITRATE VAPOR WITH HELIUM, NEON, ARGON, AND NITROGEN.**
PARKINSON, C. GRAY, P.
COMBUSTION FLAME
17, 175-8, 1971.

66342 **EFFECTIVE ATMOSPHERIC EMMISSIVITY UNDER CLEAR SKIES.**
STALEY, D. O. JURICA, G. M.
J. APPL. METEOROL.
11 ( 2 ), 349-56, 1972.

66348 **CHANGE IN THE VISCOSITY OF GAMMA-ALUMINA ENRICHED FLUXES FOR MELTING ALUMINUM AND ALUMINUM ALLOYS.**
BOTON, M. BAKURDZHIEV, P.
RUDODOBIV MET.
25 ( 7 ), 54-7, 1970.

66352 **VISCOSITY-PRESSURE BEHAVIOR OF GAS-LIQUID SOLUTIONS.**
KUSS, E. GOLLY, H.
BER. BUNSENGES. PHYS. CHEM.
76 ( 2 ), 131-8, 1972.

66353 **DIELECTRIC RELAXATION MECHANISM AND MUTUAL VISCOSITY IN SOME AROMATIC COMPOUNDS.**
MEHROTRA, N. K. PANT, T. C.
SAXENA, M. C.
BULL. CHEM. SOC. JAP.
45 ( 2 ), 356-8, 1972.

66354 **THE THERMAL CONDUCTIVITY OF BINARY MIXTURES OF METHYL NITRATE VAPOR WITH HELIUM, NEON, ARGON, AND NITROGEN.**
PARKINSON, C. GRAY, P.
COMBUST. FLAME
17, 175-8, 1971.

66356 **INFLUENCE OF CARBONATION ON THE CONDUCTIVITY AND VISCOSITY OF CONCENTRATED POTASSIUM HYDROXIDE SOLUTIONS.**
NITABAH, M. LUCESOLI, D. DEGOBERT, P.
ELECTROCHIM. ACTA
17 ( 2 ), 327-31, 1972.

66358 **DIELECTRIC RELAXATION AND MUTUAL VISCOSITY OF SOME SUBSTITUTED BENZALDEHYDES.**
MEHROTRA, N. K.
INDIAN J. PHYS.
45 ( 2 ), 83-7, 1971.

66363 **EFFECT OF VISCOSITY ON CALCIUM DIPOSITIVE IONS A GLYCEROL - WATER SOLUTION - KU2 CATION EXCHANGER SYSTEM UNDER STATIC CONDITIONS.**
TUICHIEV, I. S. RIZAEV, N. U. YUSIPOV, M. M.
GABRIELYAN, N. A.
IZV. VYSSH. UCHEB. ZAVED., KHIM. KHIM. TEKHNOL.
14 ( 12 ), 1825-9, 1971.

66364 **VISCOSITY OF POTASSIUM HEXAFLUOROZIRCONATE AND ITS MELTS.**
KURMAEV, R. KH. BELKIN, G. I. KOLESOV, I. M.
IZV. VYSSH. UCHEB. ZAVED., TSVET. MET.
14 ( 6 ), 79-80, 1971.

66367 **THERMAL CONDUCTIVITY OF SOLID-LIQUID SUSPENSIONS.**
SAEGUSA, T. WAKAO, N. IIDA, Y.
HASHIGUCHI, K.
KAGAKU KOGAKU
36 ( 2 ), 193-7, 1972.

66368 **RELATIVE DENSITY AND VISCOSITY OF SOLUTIONS IN THE UREA-AMMONIUM DIHYDROGEN PHOSPHATE-AMMONIUM MONOHYDROGEN PHOSPHATE-WATER SYSTEM.**
GRUNCHAROV, IV. N. IVANOV, D. G.
KHIM. IND.( SOFIA )
43 ( 7 ), 304-7, 1971.

66385 **VISCOSITY OF SOME TETRAALKYLAMMONIUM IODIDES IN GLYCEROL.**
GOPAL, R. HUSAIN, M. M. SINGH, P.
Z. PHYS. CHEM. ( FRANKFURT AM MAIN )
76 ( 3-4 ), 216-20, 1971.

66405 **CRITICAL VISCOSITY AND DIFFUSION IN THE BINARY - LIQUID PHASE TRANSITION.**
PERL, R. FERRELL, R. A.
PHYS. REV. LETT.
29 ( 1 ), 51-4, 1972.

66414 **WALL EFFECTS IN FALLING - BALL AND FALLING - CYLINDER VISCOMETERS.**
VENKATRAMAN, N.
UNIVERSITY OF TULSA, M. S. THESIS
71PP., 1971.

| TPRC Number | Bibliographic Citation |
|---|---|
| 66417 | **VISCOSITY MINIMUM OF SYSTEMS DURING THE COMPLEXING OF ALKALI METAL IODIDES WITH ZINC AND CADMIUM IODIDES IN NON-AQUEOUS MEDIA.**<br>GORENBEIN, E. YA. SHEVCHENKO, V. M.<br>TROFIMCHUK, V. M.<br>ZH. FIZ. KHIM.<br>45 ( 12 ), 3076–80, 1971.<br>( FOR ENGLISH TRANSLATION SEE TPRC NO. 66418 ) |
| 66418 | **VISCOSITY MINIMA ON FORMATION OF COMPLEXES BY ALKALI METAL IODIDES WITH ZINC AND CADMIUM IODIDES IN NON- AQUEOUS MEDIA.**<br>GORENBEIN, E. YA. SHEVCHENKO, V. M.<br>TROFIMCHUK, A. K.<br>RUSS. J. PHYS. CHEM.<br>45 ( 12 ), 1743–6, 1971.<br>( ENGLISH TRANSLATION OF ZH. FIZ. KHIM. 45 ( 12 ), 3076–80, 1971., FOR ORIGINAL SEE TPRC NO. 66417 ) |
| 66423 | **AN EQUILIBRIUM THEORY OF THE DILUTE SOLUTIONS OF HELIUM-3 IN SUPERFLUID HELIUM-4 AT VERY LOW TEMPERATURES.**<br>DISATNIK, Y. BRUCKER, H.<br>J. LOW TEMP. PHYS.<br>7 ( 5–6 ), 491–523, 1972. |
| 66431 | **ELECTRICAL CONDUCTANCE IN SUPERCRITICAL ISOBUTYRIC ACID - WATER SOLUTIONS.**<br>GAMMELL, P. M. ANGELL, C. A.<br>PHYS. LETT.<br>40 A ( 1 ), 49–51, 1972. |
| 66455 | **STATE OF TITANIUM( IV ) AND ITS CHANGES DURING THERMAL HYDROLYSIS IN SULFURIC ACID SOLUTIONS.**<br>DOMATOV, YU. D.<br>J. APPL. CHEM., USSR<br>42 ( 8 ), 1627–32, 1969.<br>( ENGLISH TRANSLATION OF ZH. PRIKL. KHIM. ( LENINGRAD ), 42 ( 8 ), 1725–32, 1969., FOR ORIGINAL SEE TPRC NO. 55892 ) |
| 66456 | **SPECIFIC HEATS OF SOLUTIONS IN THE SYSTEM POTASSIUM OXIDE–ALUMINUM OXIDE–WATER IN THE 25–340 DEGREES C TEMPERATURE RANGE.**<br>MASHOVETS, V. P. PUCHKOV, L. V.<br>BARANOVA, T. A. LAPIDUS, M. E.<br>J. APPL. CHEM., USSR<br>43 ( 1 ), 178–81, 1970.<br>( ENGLISH TRANSLATION OF ZH. PRIKL. KHIM., 43 ( 1 ), 183–6, 1970., FOR ORIGINAL SEE TPRC NO. 58349 ) |
| 66457 | **SPECIFIC HEATS OF AQUEOUS POTASSIUM HYDROXIDE SOLUTIONS AT TEMPERATURES UP TO 400 DEGREES C.**<br>PUCHKOV, L. V. BARANOVA, T. A.<br>LAPIDUS, M. E.<br>J. APPL. CHEM., USSR<br>43 ( 2 ), 463–5, 1970.<br>( ENGLISH TRANSLATION OF ZH. PRIKL. KHIM. ( LENINGRAD ), 43 ( 2 ), 455–7, 1970., FOR ORIGINAL SEE TPRC NO. 58553 ) |
| 66458 | **VISCOSITY OF AQUEOUS POTASSIUM HYDROXIDE SOLUTIONS AT TEMPERATURES UP TO 90 DEGREES C.**<br>PUCHKOV, L. V. SARGAEV, P. M. FEDOROV, M. K.<br>J. APPL. CHEM., USSR<br>43 ( 3 ), 679–82, 1970.<br>( ENGLISH TRANSLATION OF ZH. PRIKL. KHIM. ( LENINGRAD ), 43 ( 3 ), 677–80, 1970., FOR ORIGINAL SEE TPRC NO. 59317 ) |
| 66459 | **DENSITY, VISCOSITY, AND SURFACE TENSION DATA ON CERTAIN BINARY SYSTEMS.**<br>KONOBEEV, B. I. LYAPIN, V. V.<br>J. APPL. CHEM., USSR<br>43 ( 4 ), 806–13, 1970.<br>( ENGLISH TRANSLATION OF ZH. PRIKL. KHIM. ( LENINGRAD ), 43 ( 4 ), 803–11, 1970; FOR ORIGINAL SEE TPRC NO. 59007 ) |
| 66462 | **VISCOSITY AND SURFACE TENSION OF CYMENES, THEIR HYDROPEROXIDES, AND DECOMPOSITION PRODUCTS.**<br>SHUB, G. A. ORLOVA, O. S.<br>KHARLAMPOVICH, G. D.<br>J. APPL. CHEM., USSR<br>43 ( 5 ), 1184–6, 1970.<br>( ENGLISH TRANSLATION OF ZH. PRIKL. KHIM. ( LENINGRAD ), 43 ( 5 ), 1175–7, 1970., FOR ORIGINAL SEE TPRC NO. 60921 ) |
| 66463 | **DEPENDENCE OF THE DIFFUSION COEFFICIENT, VISCOSITY, AND DENSITY IN THE CYCLOHEXANE–CYCLOHEXANOL SYSTEM ON TEMPERATURE AND CONCENTRATION.**<br>RAVDEL, A. A. PORAI–KOSHITS, A. B.<br>PENKINA, N. V. OLEVSKII, V. M. IVANOV, V. A.<br>APPL. CHEM., USSR<br>43 ( 6 ), 1410–2, 1970.<br>( ENGLISH TRANSLATION OF ZH. PRIKL. KHIM. ( LENINGRAD ), 43 ( 6 ), 1401–2, 1970., FOR ORIGINAL SEE TPRC NO. 60922 ) |
| 66465 | **PHYSICOCHEMICAL ANALYSIS OF THE SYSTEM N-METHYL-ALPHA-PYRROLIDONE-METHANOL.**<br>GRANZHAN, V. A. KIRILLOVA, O. G.<br>J. APPL. CHEM., USSR<br>43 ( 8 ), 1898–900, 1970.<br>( ENGLISH TRANSLATION OF ZH. PRIKL. KHIM. ( LENINGRAD ), 43 ( 8 ), 1875–7, 1970., FOR ORIGINAL SEE TPRC NO. 60923 ) |
| 66469 | **PROPERTIES OF SOLUTIONS OF NITRITES OF CERTAIN UNI- AND BIVALENT METALS.**<br>PROTSENKO, P. I. IVLEVA, T. I.<br>AVDEEV, N. YA.<br>J. APPL. CHEM., USSR<br>43 ( 12 ), 2635–8, 1970.<br>( ENGLISH TRANSLATION OF ZH. PRIKL. KHIM. ( LENINGRAD ), 43 ( 12 ), 2593–6, 1970., FOR ORIGINAL SEE TPRC NO. 62769 ) |
| 66470 | **CALCULATION OF VISCOSITIES OF AQUEOUS POTASSIUM SULFATE AND SODIUM SULFATE SOLUTIONS FROM NMR DATA.**<br>GRITSUS, B. V. KUSHCHENKO, V. V.<br>MISHCHENKO, K. P.<br>J. APPL. CHEM., USSR<br>43 ( 12 ), 2784–7, 1970.<br>( ENGLISH TRANSLATION OF ZH. PRIKL. KHIM. ( LENINGRAD ), 43 ( 12 ), 2741–4, 1970., FOR ORIGINAL SEE TPRC NO. 63262 ) |
| 66471 | **PHYSICOCHEMICAL PROPERTIES OF SOLUTIONS IN THE SYSTEM AMMONIUM NITRITE–AMMONIUM NITRATE–WATER SATURATED AT 20 DEGREES C.**<br>PROTSENKO, P. I. UGRYUMOVA, A. A.<br>KUTSENKO, T. E.<br>J. APPL. CHEM., USSR<br>44 ( 1 ), 28–31, 1971.<br>( ENGLISH TRANSLATION OF ZH. PRIKL. KHIM. ( LENINGRAD ), 44 ( 1 ), 30–4, 1971., FOR ORIGINAL SEE TPRC NO. 62543 ) |
| 66472 | **APPARATUS FOR MEASURING VISCOSITIES OF ELECTROLYTE SOLUTIONS AT TEMPERATURES UP TO 275 DEGREES C.**<br>MASHOVETS, V. P. PUCHKOV, L. V.<br>SARGAEV, P. M. FEDOROV, M. K.<br>J. APPL. CHEM., USSR<br>44 ( 1 ), 85–8, 1971.<br>( ENGLISH TRANSLATION OF ZH. PRIKL. KHIM. ( LENINGRAD ), 44 ( 1 ), 90–4, 1971., FOR ORIGINAL SEE TPRC NO. 62128 ) |
| 66473 | **VICOSITY AND THERMODYNAMIC CHARACTERISTICS OF VISCOUS FLOW OF AQUEOUS TETRABUTYLAMMONIUM NITRATE SOLUTIONS.**<br>KUDRYAVTSEVA, I. V. KRUMGAL'Z, B. S.<br>MISHCHENKO, K. P.<br>J. APPL. CHEM., USSR<br>44 ( 1 ), 195–8, 1971.<br>( ENGLISH TRANSLATION OF ZH. PRIKL. KHIM. ( LENINGRAD ), 44 ( 1 ), 201–5, 1971., FOR ORIGINAL SEE TPRC NO. 63034 ) |
| 66474 | **SOLUBILITY OF ADIPIC ACID AND PHYSICAL PROPERTIES OF ITS AQUEOUS SOLUTIONS.**<br>NALIVAIKO, A. S. GRANZHAN, V. A.<br>POSTNIKOV, V. A.<br>J. APPL. CHEM., USSR<br>44 ( 1 ), 216–7, 1971.<br>( ENGLISH TRANSLATION OF ZH. PRIKL. KHIM. ( LENINGRAD ), 44 ( 1 ), 219–21, 1971., FOR ORIGINAL SEE TPRC NO. 62129 ) |
| 66475 | **TEMPERATURE DEPENDENCE OF VISCOSITY OF AQUEOUS ELECTROLYTE SOLUTIONS.**<br>PENKINA, N. V.<br>J. APPL. CHEM., USSR<br>44 ( 2 ), 343–7, 1971.<br>( ENGLISH TRANSLATION OF ZH. PRIKL. KHIM. ( LENINGRAD ), 44 ( 2 ), 349–54, 1971; FOR ORIGINAL SEE TPRC NO. 62600 ) |
| 66477 | **CERTAIN PHYSICOCHEMICAL PROPERTIES OF THE SYSTEM POLYPHOSPHORIC ACID–AMMONIA–WATER.**<br>BUGAI, P. M. EREMENKO, R. A.<br>J. APPL. CHEM., USSR<br>44 ( 3 ), 662–4, 1971.<br>( ENGLISH TRANSLATION OF ZH. PRIKL. KHIM. ( LENINGRAD ), 44 ( 3 ), 657–60, 1971., FOR ORIGINAL SEE TPRC NO. 62768 ) |
| 66479 | **DENSITY AND VISCOSITY OF THE OCTANE–QUINOLINE SYSTEM IN THE TEMPERATURE RANGE FROM MINUS 20 TO PLUS 40 DEGREES C.**<br>ENNAN, A. A. SAMOILENKO, V. G.<br>CHOBOTAREV, A. N. ANIKEEV, V. A.<br>J. APPL. CHEM., USSR<br>44 ( 6 ), 1449–51, 1971.<br>( ENGLISH TRANSLATION OF ZH. PRIKL. KHIM. ( LENINGRAD ), 44 ( 6 ), 1427–9, 1971., FOR ORIGINAL SEE TPRC NO. 64575 ) |

| TPRC Number | Bibliographic Citation |
|---|---|
| 66480 | **SOME PHYSIOCHEMICAL PROPERTIES OF SULFURIC ACID IN THE TEMPERATURE RANGE FROM 0 TO MINUS 45 DEGREES C.**<br>GERZHBERG, YU. I. KRUMGAL'Z, B. S.<br>RYABIKOVA, V. M. TRABER, D. G. KOZLOV, D. A.<br>TREPALIN, A. I.<br>J. APPL. CHEM., USSR<br>44 ( 10 ), 2376–7, 1971.<br>( ENGLISH TRANSLATION OF ZH. PRIKL. KHIM. ( LENINGRAD ), 44 ( 10 ), 2321–2, 1971., FOR ORIGINAL SEE TPRC NO. 65294 ) |
| 66485 | **RAPID METHOD FOR THE DETERMINATION OF THE AVERAGE SPECIFIC HEAT OF AQUEOUS SOLUTIONS OF INORGANIC COMPOUNDS.**<br>TERESHKIN, I. V. FILIPPOV, S. N.<br>IND. LAB., ( USSR )<br>37 ( 3 ), 399–400, 1971.<br>( ENGLISH TRANSLATION OF ZAVOD. LAB., 37 ( 3 ), 313, 1971., FOR ORIGINAL SEE TPRC NO. 63260 ) |
| 66488 | **SIMULTANEOUS MEASUREMENTS OF SURFACE TENSION AND VISCOSITY IN LIQUIDS ( EXCHANGE OF EXPERIENCE ).**<br>BEGLYAROV, E. M. SHUMAKOV, G. YA.<br>IND. LAB., ( USSR )<br>37 ( 12 ), 1911–2, 1971.<br>( ENGLISH TRANSLATION OF ZAVOD. LAB., 37 ( 12 ), 1483–4, 1971., FOR ORIGINAL SEE TPRC NO. 65899 ) |
| 66503 | **HETEROGENEOUS EQUILIBRIUMS IN TERNARY AQUEOUS SYSTEMS OF CALCIUM AND SODIUM PERCHLORATES AND CHLORIDES AT 40 DEGREES.**<br>IVANOV, S. A.<br>FTD–HT–23–822–72<br>16PP., 1972.<br>( ENGLISH TRANSLATION OF UCH. ZAP., YAROSLAV. GOS. PEDAGOG. INST., ( 66 ), 82–95, 1969; FOR ORIGINAL SEE TPRC NO. 56919 ) |
| 66505 | **VISCOSITY AND ELECTRICAL CONDUCTIVITY OF ALCOHOLIC SOLUTIONS OF LITHIUM CHLORIDE.**<br>GOLIK, A. Z. ALLANAZAROV, G. CHOLPAI, P. F.<br>FTD<br>5PP., 1972.<br>( ENGLISH TRANSLATION OF IZV. AKAD. NAUK TURKM. SSR, SER. FIZ.–TEKH., KHIM. GEOL. NAUK, ( 4 ), 112–4, 1969; FOR ORIGINAL SEE TPRC NO. 55656 )<br>( FTD–HT–23–1340–71, AD–748015 ) |
| 66506 | **MEASUREMENT OF THE THERMAL CONDUCTIVITY AND THE VELOCITY OF SOUND OF AQUEOUS SALT SOLUTIONS AS A FUNCTION OF THE TEMPERATURE.**<br>MULLER, W. T.<br>ZURICH EIDG. TECH. HOCHSCHULE, PH. D. THESIS<br>98PP., 1971. |
| 66508 | **THE REACTIONS OF ELECTRONS IN GLYCEROL.**<br>KAJIWARA, T. THOMAS, J. K.<br>J. PHYS. CHEM.<br>76 ( 12 ), 1700–6, 1972. |
| 66511 | **ON THE VISCOSITY OF CONCENTRATED AQUEOUS SOLUTIONS OF TETRAALKYLAMMONIUM BROMIDES.**<br>EAGLAND, D. PILLING, G.<br>J. PHYS. CHEM.<br>76 ( 13 ), 1902–6, 1972. |
| 66526 | **CONDUCTIVITY AND VISCOSITY OF THE TETRABUTYL-AMMONIUM BROMIDE- CARBON TETRACHLORIDE SYSTEM.**<br>DENNING, K. F. PLAMBECK, J. A.<br>CAN. J. CHEM.<br>50 ( 10 ), 1600–3, 1972. |
| 66530 | **VISCOSITY–DEPENDENT RATE OF ION RECOMBINATION FOLLOWING PULSE IRRADIATION.**<br>SAWAI, T. HAMILL, W. H.<br>J. CHEM. PHYS.<br>56 ( 11 ), 5524–7, 1972. |
| 66533 | **VISCOSITY OF THE BINARY GASEOUS MIXTURES HELIUM–NEON AND HELIUM–NITROGEN IN THE TEMPERATURE RANGE 25–700 DEGREES C.**<br>KESTIN, J. RO, S. T. WAKEHAM, W. A.<br>J. CHEM. PHYS.<br>56 ( 12 ), 5837–42, 1972. |
| 66538 | **TRACER DIFFUSION AND SHEAR VISCOSITY FOR THE SYSTEM 2,6–LUTIDINE – WATER NEAR THE LOWER CRITICAL POINT.**<br>STEIN, A. DAVIDSON, S. J. ALLEGRA, J. C.<br>ALLEN, G. F.<br>J. CHEM. PHYS.<br>56 ( 12 ), 6164–8, 1972. |
| 66539 | **LIGHT SCATTERING AND SHEAR VISCOSITY STUDIES OF THE BINARY SYSTEM 2,6–LUTIDINE–WATER IN THE CRITICAL REGION.**<br>GULARI, E. COLLINGS, A. F. SCHMIDT, R. L.<br>PINGS, C. J.<br>J. CHEM. PHYS.<br>56 ( 12 ), 6169–79, 1972.<br>( ERRATA: 57 ( 12 ), 5623, 1972 )<br>( AD–748 380, AFOSR–TR–72–1735 ) |
| 66540 | **THERMAL CONDUCTIVITY AND RESONANT MULTIPOLE INTERACTIONS.**<br>NYELAND, C. MASON, E. A. MONCHICK, L.<br>J. CHEM. PHYS.<br>56 ( 12 ), 6180–92, 1972. |
| 66544 | **CLASSICIAL MECHANICS CALCULATIONS OF THE TRANSPORT PROPERTIES OF GASEOUS MIXTURES WITH INTERNAL DEGREES OF FREEDOM.**<br>VAN DE REE, J. SCHOLTES, T.<br>J. CHEM. PHYS.<br>57 ( 1 ), 122–31, 1972. |
| 66545 | **VISCOSITY AND THERMAL CONDUCTIVITY OF MODERATELY DENSE GAS MIXTURES.**<br>WAKEHAM, W. A. KESTIN, J. MASON, E. A.<br>SANDLER, S. I.<br>J. CHEM. PHYS.<br>57 ( 1 ), 295–301, 1972. |
| 66548 | **THERMAL CONDUCTIVITY OF DIETHYL ETHER – METHYL CHLORIDE MIXTURES.**<br>MANNA, A. SRIVASTAVA, B. N.<br>J. CHEM. PHYS.<br>57 ( 2 ), 1009–11, 1972. |
| 66552 | **CALCULATION OF INFRARED FLUX AT THE SEA SURFACE.**<br>CHOW, M.–D.<br>GEOPHYSICAL SCIENCES LAB., NEW YORK UNIV., N.Y.<br>49PP., 1971.<br>( AD–722 036, N71–28455, GSL–TR–71–1, AVAIL. NTIS ) |
| 66556 | **PREDICTION OF THE ENTHALPY OF SATURATED PARAFFIN BINARY HYDROCARBON MIXTURES.**<br>GHORMLEY, E. L. LENOIR, J. M.<br>CAN. J. CHEM. ENG.<br>50 ( 1 ), 89–94, 1972. |
| 66557 | **NOMOGRAPH FOR THE CALCULATION OF THE VISCOSITY OF IMMISCIBLE LIQUID MIXTURES.**<br>ZANKER, A.<br>CHEM. ENG. ( LONDON )<br>( 259 ), 121–2, 1972. |
| 66570 | **CYLINDRICAL DEVICE FOR MEASURING THE THERMAL CONDUCTIVITY OF HYDROCARBON MIXTURES.**<br>NURBERDYEV, A. A. NAZIEV, YA. M.<br>AKHMEDOV, A. K. ABASOV, A. A.<br>IZV. AKAD. NAUK TURKM. SSR, SER. FIZ.–TEKH., KHIM.<br>( 6 ), 98–100, 1971. |
| 66581 | **THERMODYNAMIC PROPERTIES OF HYDROGEN – HELIUM PLASMAS.**<br>NELSON, H. F.<br>J. SPACECR. ROCKETS<br>9 ( 3 ), 177–81, 1972. |
| 66597 | **THERMAL CONDUCTIVITY OF HYDROGEN–METHANE, HYDROGEN–NITROGEN, AND HYDROGEN–CARBON MONOXIDE MIXTURES.**<br>MAMONOV, YU. V. GOLUBEV, I. F.<br>TR. NAUCH.–ISSLED. PROEKT. INST. AZOTN. PROM. PROD. ORG. SIN.<br>( 8 ), 112–23, 1971. |
| 66602 | **DETERMINATION OF THE AVERAGE HEAT CAPACITIES OF SALIFEROUS ROCKS AND BRINES.**<br>YUSOVA, YU. I. ESELEV, I. M.<br>KARATYGIN, E. P.<br>ZH. PRIKL. KHIM. ( LENINGRAD )<br>45 ( 2 ), 427–8, 1972.<br>( FOR ENGLISH TRANSLATION SEE TPRC NO. 69793 ) |
| 66641 | **LAYERS OF HIGH THERMAL CONDUCIVITY IN THE NORTH ATLANTIC.**<br>KASAMEYER, P. W. HERZEN, R. P. V.<br>SIMMONS, G.<br>J. GEOPHYS. RES.<br>77 ( 17 ), 3162–7, 1972. |
| 66642 | **HEAT CONDUCTION IN A TURBULENT MAGNETIC FIELD, WITH APPLICATION TO SOLAR- WIND ELECTRONS.**<br>HOLLWEG, J. V. JOKIPII, J. R.<br>J. GEOPHYS. RES.<br>77 ( 19 ), 3311–6, 1972. |
| 66644 | **A PHYSIOCOCHEMICAL STUDY OF THE IONOGENIC SYSTEMS FORMED BY THIOCYANATES AND ISOTHIOCYANATES WITH 1–NAPHTHYLAMINE.**<br>MUZYCHENKO, V. A. MISKIDZH'YAN, S. P.<br>ZH. FIZ. KHIM.<br>45 ( 7 ), 1864, 1971.<br>( FOR ENGLISH TRANSLATION SEE TPRC NO. 66645 ) |
| 66645 | **A PHYSIOCOCHEMICAL STUDY OF THE IONOGENIC SYSTEMS FORMED BY THIOCYANATES AND ISOTHIOCYANATES WITH 1–NAPHTHYLAMINE.**<br>MUZYCHENKO, V. A. MISKIDZHYAN, S. P.<br>RUSS. J. PHYS. CHEM.<br>45 ( 7 ), 1064–5, 1971.<br>( ENGLISH TRANSLATION OF ZH. FIZ. KHIM., 45 ( 7 ), 1864, 1971., FOR ORIGINAL SEE TPRC NO. 66644 ) |

| TPRC Number | Bibliographic Citation |
|---|---|
| 66662 | **ELECTRICAL CONDUCTANCES AND IONIZATION BEHAVIOR OF SODIUM CHLORIDE IN DIOXANE - WATER MIXTURES AT 50 DEGREES.**<br>LEONG, T. H. DUNN, L. A.<br>J. PHYS. CHEM.<br>76 ( 16 ), 2294-8, 1972. |
| 66692 | **ELECTRONIC BAND STRUCTURE AND OPTICAL PROPERTIES OF CARBON-SILICON CARBIDE, BORON PHOSPHIDE, AND BORON NITRIDE.**<br>HEMSTREET, L. A. JR. FONG, C. Y.<br>PHYS. REV.<br>6 B ( 4 ), 1464-80, 1972. |
| 66729 | **DETERMINATION OF THE CONTRIBUTION OF THE MAGNETIC INTERACTION TO THE SPECIFIC HEAT OF PARAMAGNETIC SALTS BY A MAGNETO-OPTICAL METHOD.**<br>HAGENE, B.<br>COMPT. REND.<br>274 B ( 25 ), 1375-8, 1972. |
| 66757 | **OCEAN COLOR DETERMINATION THROUGH A SCATTERING ATMOSPHERE.**<br>CURRAN, R. J.<br>APPL. OPT.<br>11 ( 8 ), 1857-66, 1972. |
| 66773 | **DENSITIES AND VISCOSITIES OF 2 - BUTYNE AND 2,4 - HEXADIYNE IN SEVERAL SOLVENTS.**<br>PARKER, R. G. JONAS, J.<br>J. CHEM. ENG. DATA<br>17 ( 3 ), 300-2, 1972. |
| 66774 | **MEASURED ENTHALPIES FOR MIXTURES OF BENZENE WITH N - PENTANE.**<br>LENOIR, J. M. HIPKIN, H. G.<br>J. CHEM. ENG. DATA<br>17 ( 3 ), 319-23, 1972. |
| 66775 | **VISCOSITIES OF SOME ORGANIC GLASSES USED AS TRAPPING MATRICES. III.**<br>HUTZLER, J. S. COLTON, R. J.<br>CAMPBELL LING, A.<br>J. CHEM. ENG. DATA<br>17 ( 3 ), 324-7, 1972. |
| 66776 | **VISCOSITY STUDIES OF CERTAIN ALKYL PHOSPHATES AND THEIR AQUEOUS MIXTURES.**<br>THAU - ALEXANDROWICZ, G. MICHAELI, G.<br>J. CHEM. ENG. DATA<br>17 ( 3 ), 339-41, 1972. |
| 66777 | **SPECIFIC GRAVITY, VISCOSITY, AND SOLUBILITY FOR AQUEOUS NICKEL SULFATE SOLUTIONS.**<br>PHILLIPS, V. R.<br>J. CHEM. ENG. DATA<br>17 ( 3 ), 357-60, 1972. |
| 66806 | **NMR IN NONDILUTE SOLID HELIUM( 3 ) - HELIUM( 4 ) MIXTURES.**<br>GUYER, R. A.<br>PHYS. REV.<br>5 A ( 6 ), 2541-4, 1972. |
| 66829 | **TRANSMISSION AND REFLECTION OF AN ELECTROMAGNETIC WAVE INCIDENT NORMALLY ON A PLASMA HALF- SPACE.**<br>MASON, D. P.<br>PHYSICA ( UTRECHT )<br>60 ( 1 ), 41-62, 1972. |
| 66853 | **DIFFUSION AND VISCOSITY IN WATER-ALCOHOL MIXTURES. BEHAVIOR OF THE THALLIUM( I ) ION.**<br>HURTADO, G. G. VIDAL-ABARCA, J. B. SERNA, A.<br>AN. QUIM.<br>67 ( 12 ), 1161-7, 1971. |
| 66854 | **VISCOSITY OF BINARY LIQUID MIXTURES. IV. ACETONITRILE-ALCOHOL SYSTEMS.**<br>MATO, F. COCA, J.<br>AN. QUIM.<br>68 ( 1 ), 17-24, 1972. |
| 66855 | **MECHANISM OF ION MIGRATION IN THE MIXTURES OF MONOHYDRIC AND POLYHYDRIC ALCOHOLS AND WATER.**<br>ERDEY-GRUZ, T. KUGLER, E. MAJTHENYI, L.<br>ANN. UNIV. SCI. BUDAPEST. ROLANDO EOTVOS NOMINATAE, SECT. CHIM.<br>12, 13-33, 1971. |
| 66860 | **HEAT CAPACITY OF THE BINARY LIQUID MIXTURE TETRAFLUOROMETHANE-TRIFLUOROMETHANE NEAR THE UPPER CONSOLUTE POINT.**<br>COPE, A. F. G. REAMER, H. H. PINGS, C. J.<br>BER. BUNSENGES. PHYS. CHEM.<br>76 ( 3-4 ), 318-20, 1972.<br>( AFOSR-TR-72-1365, AD-746 439 ) |
| 66881 | **VISCOSITY OF WATER-ETHANOL-ALEHYDE THREE-COMPONENT SOLUTIONS.**<br>KHARIN, S. E. KNIGA, A. A. SOROKINA, G. S.<br>FERMENT. SPIRT. PROM.<br>( 1 ), 10-1, 1972. |
| 66889 | **SCHMIDT AND PRANDTL NUMBERS OF BINARY GAS MIXTURES.**<br>ROETZEL, W. SCHABER, A.<br>FORSCH. INGENIEURW.<br>38 ( 1 ), 25-30, 1972. |
| 66917 | **PHYSICOCHEMICAL ANALYSIS OF HYDRAZINE HYDRATE-PHENOL BINARY SYSTEMS. 2. PHYSICOCHEMICAL ANALYSIS OF SYSTEMS OF HYDRAZINE HYDRATE WITH M- AND P-CRESOLS.**<br>KRUPATKIN, I. L. BALINA, T. K.<br>IZV. VYSSH. UCHEB. ZAVED., KHIM. KHIM. TEKHNOL.<br>15 ( 1 ), 50-4, 1972. |
| 66919 | **VISCOSITY OF FUSED SALTS SATURATED WITH CHLORINE GAS.**<br>KURMAEV, R. KH. UGLEV, N. P.<br>IZV. VYSSH. UCHEB. ZAVED., TSVET. MET.<br>15 ( 1 ), 47-9, 1972. |
| 66920 | **VISCOSITY OF METHANOL + BENZENE MIXTURES.**<br>MEDANI, M. S.<br>J. APPL. CHEM. BIOTECHNOL.<br>22 ( 3 ), 293-302, 1972. |
| 66926 | **HEAT CAPACITIES OF TETRAALKYLAMMONIUM BROMIDES IN WATER AND IN ANHYDROUS METHANOL AT VARIOUS TEMPERATURES.**<br>MASTROIANNI, M. J. CRISS, C. M.<br>J. CHEM. THERMODYN.<br>4 ( 2 ), 321-30, 1972. |
| 66967 | **PHYSIOCHEMICAL ANALYSIS OF THE BARIUM NITRATE-ALUMINUM NITRATE-WATER TERNARY SYSTEM AT 25.**<br>VIDENOV, N. B. KHRISTOVA, L. N.<br>IVANOVA, R. A. DIMITROV, D. I.<br>KHIM. IND. ( SOFIA )<br>43 ( 10 ), 435-6, 1971. |
| 67004 | **EFFECTS OF POTASSIUM CHLORIDE AND SODIUM CHLORIDE ADDITIONS ON CERTAIN PHYSICAL PROPERTIES OF CESIUM CHLORIDE.**<br>FULLAM, H. T.<br>MATER. RES. BULL.<br>7 ( 4 ), 289-300, 1972. |
| 67063 | **DIFFUSION IN MIXED SOLVENTS. III. IODINE IN NONPOLAR SOLUTIONS. EVIDENCE FOR CHARGE - TRANSFER INTERACTION.**<br>NAKANISHI, K. KITAJIMA, S.<br>J. PHYS. CHEM.<br>76 ( 17 ), 2470-4, 1972. |
| 67064 | **TRACER DIFFUSION AND VISCOSITY STUDY AT 25 DEGREES IN BINARY AND TERNARY LIQUID SYSTEMS.**<br>AOYAGI, K. ALBRIGHT, J. G.<br>J. PHYS. CHEM.<br>76 ( 18 ), 2572-7, 1972. |
| 67090 | **CALCULATION OF THE THERMAL CONDUCTIVITY OF GASES AND GAS MIXTURES.**<br>BOGOMOL'NYI, A. M. SOBOLEVA, G. N.<br>LEBEDEVA, E. E.<br>TEOR. OSN. KHIM. TEKHNOL.<br>6 ( 2 ), 276-7, 1972. |
| 67115 | **THERMAL CONDUCTIVITY OF ETHYLENE-NITROGEN AND ETHYLENE-HYDROGEN GAS MIXTURES AT VARIOUS TEMPERATURES AND PRESSURES.**<br>MAMONOV, YU. V. GOLUBEV, I. F.<br>TR. NAUCH.-ISSLED. PROEKT. INST. AZOTN. PROM. PROD. ORG. SIN.<br>( 8 ), 124-36, 1971. |
| 67134 | **PHYSIOCHEMICAL STUDIES OF SOME NONAQUEOUS BINARY LIQUID SYSTEMS. PYRIDINE AND PROPIONIC ACID.**<br>SUNDARAM, K. M. S.<br>Z. PHYS. CHEM. ( LEIPZIG )<br>249 ( 3-4 ), 285-8, 1972. |
| 67139 | **PHYSICOCHEMICAL PROPERTIES OF AQUEOUS ORGANIC SYSTEMS. II. DENSITY AND VISCOSITY OF DIMETHYFORMAMIDE - WATER AND WATER - UROTROPINE SYSTEMS IN THE MINUS 15 TO 40 DEGREE RANGE.**<br>ENNAN, A. A. CHOBOTAREV, A. N.<br>ANIKEEV, V. A. KORNELLI, M. E.<br>YUR'EVA, E. I. KATS, B. M.<br>ZH. PRIKL. KHIM. ( LENINGRAD )<br>45 ( 3 ), 622-5, 1972.<br>( FOR ENGLISH TRANSLATION SEE TPRC NO. 69794 ) |
| 67140 | **SALTS OF POLYNITROALKANES IN SOLUTION. I. DENSITY AND VISCOSITY OF AQUEOUS SOLUTIONS OF THE SODIUM SALT OF NITROFORM.**<br>PENKINA, N. V. ODOKIENKO, S. S.<br>SELIVANOV, V. F. GIDASPOV, B. V.<br>ZH. PRIKL. KHIM. ( LENINGRAD )<br>45 ( 3 ), 667-8, 1972.<br>( FOR ENGLISH TRANSLATION SEE TPRC NO. 69795 ) |
| 67143 | **EMISSIVITY OF CARBON DIOXIDE PLASMA IN THE VISIBLE SPECTRAL REGION.**<br>KISELEVSKII, L. I. SOLOV'YANCHIK, D. A.<br>FESHCHENKO, Z. P.<br>ZH. PRIKL. SPEKTROSK.<br>16 ( 2 ), 234-8, 1972.<br>( FOR ENGLISH TRANSLATION SEE TPRC NO. 76925 ) |

| TPRC Number | Bibliographic Citation |
|---|---|
| 67154 | **EXPERIMENTAL STUDY OF THE THERMAL CONDUCTIVITY OF COMPRESSED AIR AT LOW TEMPERATURES.**<br>TSEDERBERG, N. V. IVANOVA, Z. A.<br>THERM. ENG., USSR<br>18 ( 6 ), 100-2, 1971.<br>( ENGLISH TRANSLATION OF TEPLOENERGETIKA, 18 ( 6 ), 69-71, 1971., FOR ORIGINAL SEE TPRC NO. 63756 ) |
| 67157 | **THERMAL CONDUCTIVITIES OF GASEOUS MIXTURES CONTAINING HYDROCARBONS. PART 1. N - BUTANE, ISO - BUTANE, ISO-PENTANE AND NEO-PENTANE, AND THEIR BINARY MIXTURES WITH ARGON.**<br>PARKINSON, C. GRAY, P.<br>J. CHEM. SOC., FARADAY TRANS.<br>68 ( 6 ), ( PT. 1 ), 1065-76, 1972. |
| 67158 | **THERMAL CONDUCTIVITES OF GASEOUS MIXTURES CONTAINING HYDROCARBONS. PART 2. 1- CYCLOPROPANE, PROPENE, BUT-1-ENE AND TRANS - BUT-2-ENE, AND THEIR BINARY MIXTURES WITH ARGON.**<br>PARKINSON, C. MUKHOPADHYAY, P. GRAY, P.<br>J. CHEM. SOC., FARADAY TRANS.<br>68 ( 6 ), ( PT. 1 ), 1077-85, 1972. |
| 67237 | **PHYSICOCHEMICAL PROPERTIES OF SATURATED SOLUTIONS OF THE LITHIUM NITRITE-SODIUM NITRITE-WATER SYSTEM.**<br>RAZUMOVSKAYA, O. N. PROTSENKO, P. I.<br>IVLEVA, T. I.<br>RUSS. J. INORG. CHEM.<br>14 ( 2 ), 292-4, 1969.<br>( ENGLISH TRANSLATION OF ZH. NEORG. KHIM., 14 ( 2 ), 563-6, 1969., FOR ORIGINAL SEE TPRC NO. 53690 ) |
| 67238 | **EXTREME ON CURVES FOR THE RELATION OF RETENTION VALUES TO THE COMPOSITIONS OF BINARY STATIONARY PHASES.**<br>SHEVCHUK, I. M. GRANZHAN, V. A.<br>SAKHAROV, V. M. KOVALENKO, N. V.<br>J. GEN. CHEM., USSR<br>39 ( 12 ), 2576-8, 1969.<br>( ENGLISH TRANSLATION OF ZH. OBSHCH. KHIM., 39 ( 12 ), 2638-41, 1969., FOR ORIGINAL SEE TPRC NO. 58481 ) |
| 67239 | **REACTIONS OF DIMETHYL SULFOXIDE AND DIMETHYL ACETAMIDE WITH ACETIC ACID IN THE LIQUID PHASE.**<br>KHAZOVA, I. V. CHESNOKOV, V. F.<br>J. GEN. CHEM., USSR<br>40 ( 1 ), 6-9, 1970.<br>( ENGLISH TRANSLATION OF ZH. OBSHCH. KHIM., 40 ( 1 ), 7-11, 1970., FOR ORIGINAL SEE TPRC NO. 58348 ) |
| 67240 | **SOME PHYSICAL PROPERTIES OF SOLUTIONS OF THE PHENOL - CHLOROACETIC ACID SYSTEM.**<br>PANCHENKO, M. F.<br>J. GEN. CHEM., USSR<br>40 ( 2 ), 239-41, 1970.<br>( ENGLISH TRANSLATION OF ZH. OBSHCH. KHIM., 40 ( 2 ), 266-70, 1970., FOR ORIGINAL SEE TPRC NO. 58552 ) |
| 67243 | **SPECIFIC HEATS OF SOLUTIONS OF SODIUM IODIDE IN WATER, METHANOL, AND ACETONE, AND THEIR CONCENTRATION AND TEMPERATURE DEPENDENCES.**<br>NOVOSELOV, N. P. MISHCHENKO, K. P.<br>J. GEN. CHEM., USSR<br>41 ( 2 ), 250-7, 1971.<br>( ENGLISH TRANSLATION OF ZH. OBSHCH. KHIM., 41 ( 2 ), 255-63, 1971., FOR ORIGINAL SEE TPRC NO. 63550 ) |
| 67244 | **PARTIAL MOLAL HEAT CAPACITIES OF SODIUM IODIDE IN WATER, METHANOL, AND ACETONE AT 0.8, 25, AND 50 DEGREES.**<br>NOVOSELOV, N. P. MISHCHENKO, K. P.<br>J. GEN. CHEM., USSR<br>41 ( 3 ), 485-9, 1971.<br>( ENGLISH TRANSLATION OF ZH. OBSHCH. KHIM., 41 ( 3 ), 489-94, 1971., FOR ORIGINAL SEE TPRC NO. 63767 ) |
| 67255 | **SOME PHYSICO - CHEMICAL PROPERTIES OF SOLUTIONS IN SYSTEM SODIUM OXIDE-POTASSIUM OXIDE-ALUMINUM OXIDE-WATER.**<br>MASHOVETS, V. P. KUROCHKINA, V. V.<br>PENKINA, N. V. PUCHKOV, L. V. FEDOROV, M. K.<br>SOV. J. NON - FERROUS METALS<br>41 ( 7 ), 67-8, 1968.<br>( ENGLISH TRANSLATION OF TSVET. METAL., 41 ( 7 ), 49-51, 1968., FOR ORIGINAL SEE TPRC NO. 51196 ) |
| 67279 | **MEASUREMENT OF NITROGEN PLASMA TRANSPORT PROPERTIES.**<br>SCHREIBER, P. W. HUNTER, A. M., II<br>BENEDETTO, K. R.<br>AIAA J.<br>10 ( 5 ), 670-4, 1972.<br>( AD-748559, ARL-72-0097 ) |
| 67344 | **VISCOSITY OF TWO-COMPONENT LIQUID MIXTURES.**<br>KOZLOWSKI, Z.<br>SOC. SCI. LODZ., ACTA CHIM.<br>16, 17-24, 1971. |
| 67354 | **HEAT CAPACITY OF AQUEOUS ELECTROLYTE SOLUTIONS.**<br>SYRNIKOV, YU. P.<br>J. STRUCT. CHEM., USSR<br>12 ( 6 ), 994-5, 1971.<br>( ENGLISH TRANSLATION OF ZH. STRUKT. KHIM., 12 ( 6 ), 1082-4, 1971., FOR ORIGINAL SEE TPRC NO. 65821 ) |
| 67368 | **INFRARED SPECTRA OF LITHIUM ALUMINUM FLUORIDE AND SODIUM ALUMINUM FLUORIDE.**<br>SNESON, A. CYVIN, S. J. CYVIN, B. N.<br>J. PHYS. CHEM.<br>75 ( 17 ), 2609-15, 1971.<br>( AD-735 200, AFOSR-TR-72-0065 ) |
| 67376 | **CONDUCTIVITY, VISCOSITY, AND DENSITY OF SODIUM CHLORIDE AND CESIUM CHLORIDE SOLUTIONS IN N-METHYLFORMAMIDE AT 5 AND 25 DEGREES.**<br>MOSTKOVA, R. I. KESSLER, YU. M.<br>SEMENOVA, V. N.<br>SOV. ELECTROCHEM.<br>7 ( 5 ), 620-5, 1971.<br>( ENGLISH TRANSLATION OF ELEKTROKHIMIYA, 7 ( 5 ), 642-8, 1971., FOR ORIGINAL SEE TPRC NO. 63713 ) |
| 67419 | **INHOMOGENEOUS MODELS OF THE ATMOSPHERE OF JUPITER.**<br>AXEL, L.<br>ASTROPHYS. J.<br>173 ( 2 ), 451-68, 1972.<br>( AD-742 156, AFOSR-TR-72-1137 ) |
| 67433 | **OPTICAL CONSTANTS IN THE INFRARED FOR AQUEOUS SOLUTIONS OF SODIUM CHLORIDE.**<br>QUERRY, M. R. WARING, R. C. HOLLAND, W. E.<br>HALE, G. M. NIJM, W.<br>J. OPT. SOC. AMER.<br>62 ( 7 ), 849-55, 1972. |
| 67500 | **PHYSICAL PROPERTIES OF SEA WATER SOLUTIONS. VISCOSITY.**<br>ISDALE, J. D. SPENCE, C. M. TUDHOPE, J. S.<br>DESALINATION<br>10, 319-28, 1972. |
| 67501 | **TRANSPORT PROCESSES NEAR THE CRITICAL POINT OF GASES AND BINARY LIQUIDS IN THE HYDRODYNAMIC REGIME.**<br>SENGERS, J. V.<br>BER. BUNSENGES. PHYS. CHEM.<br>76 ( 3-4 ), 234-49, 1972. |
| 67514 | **EFFECT OF IONIC VISCOSITY ON THE STABILITY OF A FINITE PRESSURE PLASMA.**<br>TSYPIN, V. S.<br>ZH. TEKH. FIZ.<br>42 ( 4 ), 727-37, 1972.<br>( FOR ENGLISH TRANSLATION SEE TPRC NO. 69670 ) |
| 67539 | **THERMODYNAMIC AND TRANSPORT PROPERTIES OF CHEMICALLY REACTING GASEOUS SYSTEMS. PART II.**<br>KRASIN, A.<br>FOREIGN TECHNOLOGY DIVISION<br>220PP., 1971.<br>( ENGLISH TRANSLATION OF TERMODIN. PERENOSNYE SVOISTVA KHIM. REAGIRUYUSHCHIKH GAZOV. SIST., 2, 237PP., 1971; FOR ORIGINAL SEE TPRC NO. 66284 )<br>( FTD-HC-23-0768-72 AD 756 597 ) |
| 67551 | **ALBEDO OF THE SEA SURFACE.**<br>PAYNE, R. E.<br>J. ATMOS. SCI.<br>29 ( 5 ), 959-70, 1972.<br>( AD-750 027, WHOI-72-66, AVAIL. NTIS ) |
| 67567 | **ANOMALOUS BEHAVIOR OF SOLUTIONS OF SIMPLE AROMATIC COMPOUNDS IN STRAIGHT CHAIN HYDROCARBONS AND MINERAL OILS. III. FURTHER VISCOSITY MEASUREMENTS IN SOME BINARY SYSTEMS, AND DEVELOPMENT OF A FORMULA FOR PREDICTING THE FREE ENERGY OF MIXING FOR AROMATIC COMPOUNDS.**<br>QURESHI, T. M. QURASHI, M. M.<br>PAK. J. SCI. IND. RES.<br>14 ( 4-5 ), 311-22, 1971. |
| 67576 | **VISCOSITIES OF BINARY SOLUTIONS OF NONELECTROLYTES.**<br>RAMAN, G. K. NAIDU, P. R.<br>PROC. INDIAN ACAD. SCI., SECT. A<br>75 ( 2 ), 68-72, 1972. |
| 67595 | **COMPOSITIONAL AND TEMPERATURE DEPENDENCY OF EMISSION SPECTRA OF EUROPIUM TRIPOSITIVE IONS IN GLASSES.**<br>KOMIYAMA, T. UENO, T.<br>YOGYO KYOKAI SHI<br>80 ( 6 ), 227-32, 1972. |
| 67600 | **FLOW BEHAVIOR OF COMPENSATED CHOLESTERIC LIQUID CRYSTAL MIXTURES.**<br>POLLMANN, P.<br>Z. NATURFORSCH.<br>27 A ( 4 ), 719-20, 1972. |
| 67624 | **THERMAL CONDUCTIVITY OF BINARY GAS MIXTURES.**<br>DUL'NEV, G. N. ZARICHNYAK, YU. P.<br>J. ENG. PHYS., USSR<br>15 ( 4 ), 972-6, 1968.<br>( ENGLISH TRANSLATION OF INZH. -FIZ. ZHUR., 15 ( 4 ), 669-77, 1968; FOR ORIGINAL SEE T53433 ) |
| 67640 | **CONDUCTANCES, DENSITIES, AND VISCOSITIES OF SOLUTIONS OF SODIUM NITRATE IN WATER AND IN DIOXANE-WATER, AT 25 DEGREES C.**<br>KARTZMARK, E. M.<br>CAN. J. CHEM.<br>50 ( 17 ), 2845-50, 1972. |

| TPRC Number | Bibliographic Citation |
|---|---|
| 67668 | THERMAL CONDUCTIVITY OF SOLUTIONS COMPOSED OF WATER, ETHANOL, AND CRUDE ALCOHOL IMPURITIES.<br>KHARIN, V. M. MILYAKOVA, A. A.<br>FERMENT. SPIRT. PROM.<br>( 4 ), 13–5, 1972. |
| 67680 | THERMAL CONDUCTIVITY OF BINARY GAS MIXTURES.<br>SHASHKOV, A. G. KAMCHATOV, F. P.<br>INZH.–FIZ. ZH.<br>22 ( 5 ), 843–9, 1972.<br>( FOR ENGLISH TRANSLATION SEE TPRC NO. 75005 ) |
| 67681 | DYNAMIC MEASUREMENT OF THE THERMAL CONDUCTIVITY OF GASES AT HIGH TEMPERATURE.<br>MUSTAFAEV, R. A.<br>INZH.–FIZ. ZH.<br>22 ( 5 ), 850–3, 1972.<br>( FOR ENGLISH TRANSLATION SEE TPRC NO. 75006 ) |
| 67696 | VACUUM ADIABATIC CALORIMETER FOR SPECIFIC HEAT STUDIES IN LIQUID MIXTURES, ESPECIALLY IN THEIR CRITICAL REGION.<br>GOPAL, E. S. R. GAMBHIR, R. D.<br>GOVINDARAJAN, K. VISWANATHAN, B.<br>J. INDIAN. INST. SCI.<br>53 ( 4 ), 355–67, 1971. |
| 67710 | THERMODYNAMIC STUDIES OF SALT–CONTAINING ETHANOL–WATER MIXTURES.<br>DECKER, U. F. RUDAKOFF, G.<br>NEUE BERGBAUTECH.<br>2 ( 3 ), 216–19, 1972. |
| 67725 | PAVLOVS LAW ON THE LINEARITY OF CHEMICOTECHNICAL FUNCTIONS IN APPLICATION TO THE CALCULATION OF THERMOPHYSICAL GAS PARAMETERS ON A COMPUTER.<br>ZIGMUND, F. F. GALIMULLIN, R. G.<br>TR. KAZAN. KHIM.–TEKHNOL. INST.<br>( 45 ), 208–12, 1970. |
| 67731 | AQUEOUS SYSTEM WITH NICKEL AND SODIUM PERCHLORATES STUDIED AT 25 DEGREES.<br>KARNAUKHOV, A. S. TARAKANOV, V, F.<br>UCH. ZAP. YAROSLAV. GOS. PEDAGOG. INST.<br>( 79 ), 28–31 1970. |
| 67732 | MAGNESIUM CHROMATE–MAGNESIUM NITRATE–WATER SYSTEM AT 25 DEGREES.<br>LEPESHKOV, I. N. SLOBODCHIKOV, A. M.<br>UCH. ZAP. YAROSLAV. GOS. PEDAGOG. INST.<br>( 79 ), 76–80 1970. |
| 67733 | CALCIUM PERCHLORATE – MAGNESIUM PERCHLORATE – WATER SYSTEM AT 25 DEGREES.<br>RUNOV, N. N. ZAKHAROVA, V. P.<br>UCH. ZAP. YAROSLAV. GOS. PEDAGOG. INST.<br>( 79 ), 120–4 1970. |
| 67734 | TERNARY AQUEOUS SYSTEM WITH BARIUM AND AMMONIUM NITRATES AT 25 DEGREES.<br>KARNAUKHOV, A. S. BITOKOV, V. T.<br>BOGACHEV, A. V<br>UCH. ZAP. YAROSLAV. GOS. PEDAGOG. INST.<br>( 79 ), 161–6, 1970. |
| 67737 | SIMPLIFIED EXPRESSIONS FOR THE CALCULATION OF THE CONTRIBUTION OF THE HEAVY COMPONENTS TO THE TRANSPORT COEFFICIENTS OF PARTIALLY IONIZED GASES.<br>CAPITELLI, M.<br>Z. NATURFORSCH.<br>27 A ( 5 ), 809–12, 1972. |
| 67745 | PARAMETERS INFLUENCING THE THEORETICAL CALCULATION OF THE THERMAL CONDUCTIVITY OF A NITROGEN PLASMA.<br>MOURGUES, W. FAUCHAIS, P.<br>COMPT. REND.<br>275 B ( 7 ), 251–4, 1972. |
| 67751 | PHYSICAL PROPERTIES OF MATERIALS BASED ON SILICON NITRIDE.<br>ANDREEVA, T. V. KAZAKOV, V. K.<br>DIELEKTRIKI<br>( 1 ), 77–81, 1971. |
| 67762 | TRANSPORT PROPERTIES OF LIQUID MIXTURES ( N–PARAFFINS ) AND THE THEOREM OF CORRESPONDING STATES.<br>DOAN, M. H. BRUNET, J.<br>IND. ENG. CHEM., FUNDAM.<br>11 ( 3 ), 356–61, 1972. |
| 67807 | PARAMETERS OF GAS JETS FOR THE ATOMIZATION OF MELTS.<br>GOLDAEV, I. P. MOTORNENKO, A. P.<br>SHEVCHENKO, A. P. LASTIVNYAK, YU. A.<br>SAMOLETOSTR. TEKH. VOZDUSH. FLOTA<br>( 25 ) 15–21, 1971. |
| 67818 | DETERMINATION OF THE ION HYDRATION NUMBER.<br>AMIRKHANOV, A. KH.<br>TR. UFIM. AVIATS. INST.<br>( 28 ), 110–15, 1971. |
| 67822 | HEAT CAPACITY AND ENTHALPY OF PHOSPHORIC ACID.<br>WAKEFIELD, Z. T. LUFF, B. B. REED, R. B.<br>J. CHEM. ENG. DATA<br>17 ( 4 ), 420–3, 1972. |
| 67823 | DENSITIES, VISCOSITIES, AND VAPOR PRESSURES OF COPPER ( II ) CHLORIDE SOLUTIONS IN HYDROCHLORIC ACID.<br>GRITZNER, G. PHILLIPS, R. F.<br>J. CHEM. ENG. DATA<br>17 ( 4 ), 425–8, 1972. |
| 67824 | THERMODYNAMIC AND PHYSICAL PROPERTIES OF AMMONIA – LITHIUM THIOCYANATE SYSTEM.<br>MACRISS, R. A. PUNWANI, D. RUSH, W. F.<br>BIERMANN, W. J.<br>J. CHEM. ENG. DATA<br>17 ( 4 ), 443–6, 1972. |
| 67825 | SOLUBILITY AND DIFFUSIVITY OF NITROUS OXIDE IN EQUIMOLAR POTASSIUM CARBONATE – POTASSIUM BICARBONATE SOLUTIONS AT 25 DEGREES C AND 1 ATM.<br>JOOSTEN, G. E. H. DANCKWERTS, P. V.<br>J. CHEM. ENG. DATA<br>17 ( 4 ), 452–4, 1972. |
| 67826 | ENTHALPIES OF TERNARY SYSTEM PENTANE – CYCLOHEXANE BENZENE.<br>LENOIR, J. M. BAL, M. HIPKIN, H. G.<br>J. CHEM. ENG. DATA<br>17 ( 4 ), 461–4, 1972. |
| 67827 | ENTHALPY OF TWO TERNARY MIXTURES OF BENZENE – OCTANE TETRALIN.<br>LENOIR, J. M. HIPKIN, H. G.<br>J. CHEM. ENG. DATA<br>17 ( 4 ), 476–8, 1972. |
| 67845 | VISCOSITY OF ETHYLBENZENE – ETHYL ALCOHOL AND ETHYLBENZENE – PROPYL ALCOHOL SYSTEMS DEPENDENT ON TEMPERATURE AND CONCENTRATION.<br>MAMEDOV, A. A. SHIKHALIEV, YA. A.<br>IZV. VYSSH. UCHEB. ZAVED., NEFT GAZ<br>15 ( 3 ), 63–6, 1972. |
| 67941 | EFFECT OF HEAVY WATER ON PROPERTIES OF CORN AND POTATO STARCH.<br>KRETOVICH, V. I. ZOTOVA, N. N.<br>BUSAREVA, N. N.<br>PRIKL. BIOKHIM. MIKROBIOL.<br>251–3, 1970. |
| 67965 | VISCOSITY OF MIXTURES OF BENZENE AND METHANOL.<br>IRVING, H. M. N. H. SIMPSON, R. B.<br>J. INORG. NUCL. CHEM.<br>34 ( 7 ), 2241–7, 1972. |
| 67978 | SPECIFIC HEAT OF ION EXCHANGE MEMBRANE BRINE AND ITS CONCENTRATES.<br>MOTOYAMA, M. KADOTA, M. OKA, S.<br>NIPPON KAISUI GAKKAI–SHI<br>25 ( 136 ), 264–9, 1972. |
| 68014 | THE VARIATION OF LATTICE PARAMETER OF URANIUM CARBIDE–ZIRCONIUM CARBIDE SOLID SOLUTIONS WITH TEMPERATURE AND COMPOSITION.<br>BOWMAN, A. L. KRIKORIAN, N. H. NERESON, N. G.<br>AIP CONF. PROC.<br>( 3 ), 119–30, 1972. |
| 68030 | VISCOSITY OF AQUEOUS SOLUTIONS OF ELECTROLYTES. III. CALCULATION OF THE VISCOSITY OF AQUEOUS SOLUTIONS OF ELECTROLYTES.<br>PENKINA, N. V.<br>ZH. PRIKL. KHIM. ( LENINGRAD )<br>45 ( 4 ), 880, 1972.<br>( FOR ENGLISH TRANSLATION SEE TPRC NO. 70603 ) |
| 68031 | VISCOSITY OF AQUEOUS SOLUTIONS OF ELECTROLYTES. V. NEGATIVE VISCOSITY OF AQUEOUS SOLUTIONS OF ELECTROLYTES.<br>PENKINA, N. V. KHRAMTSOVA, T. S.<br>ZH. PRIKL. KHIM. ( LENINGRAD )<br>45 ( 4 ), 881, 1972.<br>( FOR ENGLISH TRANSLATION SEE TPRC NO. 70604 ) |
| 68039 | BRILLOUIN SCATTERING IN AQUEOUS GLYCEROL SOLUTIONS.<br>BELL, C. E.<br>NAVAL ORDNANCE LAB., WHITE OAK, MD.<br>103PP., 1972.<br>( AD–739 683, NOLTR–72–40, AVAIL. NTIS ) |
| 68056 | ION PAIR FORMATION IN SODIUM NITRATE – DEUTERIUM OXIDE SOLUTIONS. RAMAN AND INFRARED SPECTRA, PARTIAL MOLAL VOLUMES, CONDUCTANCE, AND VISCOSITY.<br>RIDDELL, J. D. LOCKWOOD, D. J. IRISH, D. E.<br>CAN. J. CHEM.<br>50 ( 18 ), 2951–62, 1972. |
| 68072 | THERMAL CONDUCTIVITY IN STELLAR ATMOSPHERES. I. WITHOUT MAGNETIC FIELD.<br>ULMSCHNEIDER, P.<br>ASTRON. ASTROPHYS.<br>4 ( 1 ), 144–51, 1970. |
| 68099 | VISCOSITY OF THE BINARY GASEOUS MIXTURES OF NITROGEN WITH ARGON AND KRYPTON.<br>HELLEMANS, J. M. KESTIN, J. RO, S. T.<br>J. CHEM. PHYS.<br>57 ( 9 ), 4038–42, 1972. |

| TPRC Number | Bibliographic Citation |
|---|---|
| 68151 | **DISSOLUTION OF METALLIC SODIUM IN AQUEOUS ELECTROLYTE SOLUTIONS.**<br>MISKINOVA, T. A. SEMIRENKO, L. G.<br>NADEINSKII, B. P.<br>ZH. FIZ. KHIM.<br>46 ( 4 ), 1000– , 1972.<br>( FOR ENGLISH TRANSLATION SEE TPRC NO. 68152 ) |
| 68152 | **DISSOLUTION OF METALLIC SODIUM IN AQUEOUS ELECTROLYTE SOLUTIONS.**<br>MISKINOVA, T. A. SEMIRENKO, L. G.<br>NADEINSKII, B. P.<br>RUSS. J. PHYS. CHEM.<br>46 ( 4 ), 579–81, 1972.<br>( ENGLISH TRANSLATION OF ZH. FIZ. KHIM., 46 ( 4 ), 1000– , 1972., FOR ORIGINAL SEE TPRC NO. 68151 ) |
| 68163 | **COMPLEX FORMATION BY COBALT CHLORIDE WITH PYRIDINE IN MIXED SOLUTIONS.**<br>MUFTAKHOV, A. G. ALIMOVA, G. A.<br>RAKHIMOV, KH. R.<br>ZH. NEORG. KHIM.<br>17 ( 2 ), 429– , 1972.<br>( FOR ENGLISH TRANSLATION SEE TPRC NO. 68164 ) |
| 68164 | **COMPLEX FORMATION BY COBALT CHLORIDE WITH PYRIDINE IN MIXED SOLUTIONS.**<br>MUFTAKHOV, A. G. ALIMOVA, G. A.<br>RAKHIMOV, KH. R.<br>RUSS. J. INORG. CHEM.<br>17 ( 2 ), 224–6, 1972.<br>( ENGLISH TRANSLATION OF ZH. NEORG. KHIM., 17 ( 2 ), 429– , 1972., FOR ORIGINAL SEE TPRC NO. 68163 ) |
| 68165 | **THE SODIUM MOLYBDATE–SODIUM THIOCYANTE–WATER SYSTEM AT 25 DEGREES C.**<br>KAROV, Z. G. LEPESHKOV, I. N.<br>KUKULIEVA, E. I.<br>ZH. NEORG. KHIM.<br>17 ( 2 ), 509– , 1972.<br>( FOR ENGLISH TRANSLATION SEE TPRC NO. 68166 ) |
| 68166 | **THE SODIUM MOLYBDATE–SODIUM THIOCYANATE–WATER SYSTEM AT 25 DEGREES C.**<br>KAROV, Z. G. LEPESHKOV, I. N.<br>KUKULIEVA, E. I.<br>RUSS. J. INORG. CHEM.<br>17 ( 2 ), 267–9, 1972.<br>( ENGLISH TRANSLATION OF ZH. NEORG. KHIM., 17 ( 2 ), 509– , 1972., FOR ORIGINAL SEE TPRC NO. 68165 ) |
| 68167 | **PROPERTIES OF SOLUTIONS OF THE POTASSIUM NITRATE – NICKEL NITRATE – WATER AND CESIUM NITRATE – NICKEL NITRATE – WATER SYSTEMS AT 25 C.**<br>ANDREEVA, T. A. POZHARSKAYA, S. S.<br>ZH. NEORG. KHIM.<br>17 ( 2 ), 517– , 1972.<br>( FOR ENGLISH TRANSLATION SEE TPRC NO. 68168 ) |
| 68168 | **PROPERTIES OF SOLUTIONS OF THE POTASSIUM NITRATE – NICKEL NITRATE – WATER AND CESIUM NITRATE – NICKEL NITRATE – WATER SYSTEMS AT 25 C.**<br>ANDREEVA, T. A. POZHARSKAYA, S. S.<br>RUSS. J. INORG. CHEM.<br>17 ( 2 ), 271–3, 1972.<br>( ENGLISH TRANSLATION OF ZH. NEORG. KHIM., 17 ( 2 ), 517– , 1972., FOR ORIGINAL SEE TPRC NO. 68167 ) |
| 68169 | **PHYSICOCHEMICAL PROPERTIES OF THE LITHIUM NITRATE – NICKEL NITRATE – WATER AND SODIUM NITRATE – NICKEL NITRATE – WATER SYSTEMS AT 25 DEGREES.**<br>ANDREEVA, T. A. POZHARSKAYA, S. S.<br>GOLOVANOVA, T. G.<br>ZH. NEORG. KHIM.<br>17 ( 2 ), 522–6, 1972.<br>( FOR ENGLISH TRANSLATION SEE TPRC NO. 68170 ) |
| 68170 | **PHYSICOCHEMICAL PROPERTIES OF THE LITHIUM NITRATE – NICKEL NITRATE–WATER AND SODIUM NITRATE–NICKEL NITRATE–WATER SYSTEMS AT 25 C.**<br>ANDREEVA, T. A. POZHARSKAYA, S. S.<br>GOLOVANOVA, T. G.<br>RUSS. J. INORG. CHEM.<br>17 ( 2 ), 273–5, 1972.<br>( ENGLISH TRANSLATION OF ZH. NEORG. KHIM., 17 ( 2 ), 522–6, 1972., FOR ORIGINAL SEE TPRC NO. 68169 ) |
| 68175 | **MOLTEN SALTS: VOLUME3. NITRATES, NITRITES, AND MIXTURES. ELECTRICAL CONDUCTANCE, DENSITY, VISCOSITY, AND SURFACE TENSION DATA.**<br>JANZ, G. J. KREBS, U. SIEGENTHALER, H. F.<br>TOMKINS, R. P. T.<br>J. PHYS. CHEM. REF. DATA<br>1 ( 3 ), 581–746, 1972. |
| 68186 | **A CORRELATION OF THE VISCOSITIES OF ELEVEN GASES.**<br>WATSON, J. T. R.<br>NATIONAL ENGINEERING LAB., GLASGOW<br>70PP., 1972.<br>( NEL – REPORT–516, AVAIL. NEL ) |
| 68203 | **VISCOSITY OF CONCENTRATED SOLUTIONS.**<br>FURUSE, H.<br>JAP. J. APPL. PHYS.<br>11 ( 10 ), 1537–41, 1972. |
| 68213 | **THE AMMONIUM NITRATE – SODIUM NITRATE – WATER SYSTEM.**<br>BERGMAN, A. G. SHULYAK, L. F.<br>ZH. NEORG. KHIM.<br>17 ( 4 ), 1141– , 1972.<br>( FOR ENGLISH TRANSLATION SEE TPRC NO. 68214 ) |
| 68214 | **THE AMMONIUM NITRATE – SODIUM NITRATE – WATER SYSTEM.**<br>BERGMAN, A. G. SHULYAK, L. F.<br>RUSS. J. INORG. CHEM.<br>17 ( 4 ), 593–6, 1972.<br>( ENGLISH TRANSLATION OF ZH. NEORG. KHIM., 17 ( 4 ), 1141– , 1972., FOR ORIGINAL SEE TPRC NO. 68213 ) |
| 68218 | **VISCOSITY AND CONDUCTIVITY OF POTASSIUM CHLORIDE AND POTASSIUM IODIDE IN GLYCINE AND UREA SOLUTIONS.**<br>PHANG, S.<br>AUST. J. CHEM.<br>25 ( 7 ), 1575–8, 1972. |
| 68223 | **NEW METHOD FOR DETERMINATION OF SURFACE TENSION OF VISCOUS LIQUIDS.**<br>ELMAS, M.<br>CHEM. ENG. SCI.<br>27 ( 1 ), 163–5, 1972. |
| 68226 | **MEASUREMENT OF THE THERMAL CONDUCTIVITY OF PROPELLANTS BY A PROBE METHOD.**<br>ZARKO, V. E. BUFETOV, N. S.<br>ZLATOMREZHEVA, N. I. KHLEVNOI, S. S.<br>FIZ. GORENIYA VZRYVA<br>7 ( 4 ), 611–13, 1971. |
| 68236 | **LOW–TEMPERATURE FLUORINE CHEMISTRY.**<br>MAMANTOV, G. CLARKE, M. R.<br>DEPT. OF CHEMISTRY, TENNESSEE UNIV., KNOXVILLE, TENN.<br>44PP., 1971.<br>( AD–727 059, AFRPL–TR–71–79, AVAIL. NTIS ) |
| 68266 | **VISCOSITY OF SOLUTIONS OF POTASSIUM CHLORIDE, SODIUM CHLORIDE, POTASSIUM BROMIDE, AND SODIUM BROMIDE IN DIOXANE–WATER MIXTURES AT 30 AND 40 DEGREES AND THE ACTIVATION ENERGIES AND ENTROPIES OF VISCOUS FLOW.**<br>DAS, B. SINGH, K. DAS, P. K.<br>J. INDIAN CHEM. SOC.<br>49 ( 6 ), 561–4, 1972. |
| 68305 | **ENERGY BALANCE IN DEVICES OF THE TOKAMAK TYPE CONSIDERING CHARGE EXCHANGE.**<br>DNESTROVSKII, YU. N. KOSTOMAROV, D. P.<br>PAVLOVA, N. L.<br>TRANSTEK ASSOCIATES, READING, MASS.<br>52PP., 1971.<br>( N72–23733, MATT–TRANS–107, AVAIL. NTIS ) |
| 68340 | **MICHELSON–TYPE FOURIER SPECTROMETER FOR THE FAR INFRARED.**<br>SAKAI, K.<br>APPL. OPT.<br>11 ( 12 ), 2894–901, 1972. |
| 68367 | **NMR STUDY OF THE EFFECT OF FREE RADICALS AND PARAMAGNETIC IONS ON RELAXATION TIMES IN VISCOUS SYSTEMS.**<br>MODAK, S. G. KHARE, P. L. MANDE, C.<br>INDIAN J. PURE APPL. PHYS.<br>10 ( 4 ), 293–5, 1972. |
| 68390 | **VISCOSITY–PRESSURE BEHAVIOR OF BINARY MIXTURES WITH EXTREME VISCOSITY VALUES.**<br>KUSS, E.<br>RHEOL. ACTA<br>11 ( 1 ), 13–8, 1972. |
| 68391 | **VISCOSITY MEASUREMENTS ON VAPOR–GAS MIXTURES. THE METHANOL–METHANE SYSTEM AT ATMOSPHERIC PRESSURE.**<br>MEERLENDER, G. ABDEL AZIZ, N. E.<br>RHEOL. ACTA<br>11 ( 1 ), 19–28, 1972. |
| 68398 | **THERMODYNAMIC PROPERTIES OF THE BINARY SYSTEM DIMETHYL SULFOXIDE–P–XYLENE.**<br>NISSEMA, A. KOSKENNISKA, L.<br>SUOM. KEMISTILEHTI<br>45 B ( 7–8 ), 203–6, 1972. |
| 68399 | **THERMODYNAMIC PROPERTIES OF THE BINARY SYSTEM DIMETHYL SULFOXIDE–HEXAFLUOROBENZENE.**<br>NISSEMA, A. KUVAJA, A.<br>SUOM. KEMISTILCHTI<br>45 B ( 7–8 ), 206–11, 1972. |
| 68415 | **DYNAMICAL BEHAVIOR OF HELIUM( 3 ) – HELIUM( 4 ) MIXTURES NEAR THE TRICRITICAL POINT.**<br>KAWASAKI, K. GUNTON, J. D.<br>PHYS. REV. LETT.<br>29 ( 25 ), 1661–4, 1972. |
| 68453 | **DAMPING OF ADIABATIC PERTURBATIONS IN AN EXPANDING UNIVERSE.**<br>CHIBISOV, G. V.<br>ASTRON. ZH.<br>49 ( 1 ), 74–84, 1972.<br>( FOR ENGLISH TRANSLATION SEE TPRC NO. 69683 ) |

| TPRC Number | Bibliographic Citation |
|---|---|
| 68474 | **A NUMERICAL METHOD FOR THE THERMODYNAMIC PROPERTIES OF MONATOMIC IONIZED GASES.**<br>SCHMIDT, W.<br>INT. NUMERICAL METHODS ENG.<br>4, 39–47, 1972. |
| 68489 | **TRANSMISSION AND REFLECTION OF A WAVE OBLIQUELY INCIDENT ON A NONUNIFORM MAGNETIZED PLASMA.**<br>PERATT, A. L. KUEHL, H. H.<br>RADIO SCI.<br>7, 309–23, 1972. |
| 68510 | **THE LOW TEMPERATURE THERMODYNAMIC PROPERTIES OF SUPERFLUID SOLUTIONS OF HELIUM 3 IN HELIUM 4.**<br>EBNER, C. EDWARDS, D. O.<br>PHYS. LETTERS<br>2 C ( 2 ), 77–154, 1970. |
| 68527 | **PROPERTIES OF ALKALI METAL CHLORIDE SOLUTIONS.**<br>KUZNETSOV, S. I. VAULIN, L. V.<br>YASHINA, G. M.<br>IZV. VYSSH. UCHEB. ZAVED., TSVET. MET.<br>15 ( 3 ), 104–10, 1972. |
| 68529 | **MEASURED ENTHALPY CHANGES FOR A SIMULATED NATURAL GAS MIXTURE AND THEIR PREDICTION USING A MODIFIED BWR ( BENEDICT-WEBB-RUBIN ) EQUATION.**<br>SOOD, S. K. HASELDEN, G. G.<br>AICHE J.<br>18 ( 5 ), 999–1004, 1972. |
| 68536 | **ESTIMATING SPECIFIC HEAT OF LIQUID MIXTURES.**<br>DIMOPLON, W., JR.<br>CHEM. ENG. ( N. Y. )<br>79 ( 22 ), 64–6, 1972. |
| 68545 | **THERMODYNAMIC AND PHYSICAL BEHAVIOR OF BINARY MIXTURES INVOLVING SULFOLANE. II. VISCOSITY, DIELECTRIC CONSTANT, SOLID + LIQUID PHASE DIAGRAM OF MIXTURES OF BENZENE + SULFOLANE.**<br>JANNELLI, L. SACCO, A.<br>J. CHEM. THERMODYN.<br>4 ( 5 ), 715–22, 1972. |
| 68551 | **COMPARATIVE TESTING OF VISCOSITY EQUATIONS RECOMMENDED FOR CONCENTRATED AQUEOUS SOLUTIONS.**<br>MOULIK, S. P.<br>J. INDIAN CHEM. SOC.<br>49 ( 5 ), 483–8, 1972. |
| 68595 | **PHYSICAL PROPERTIES OF FOAMED PLASTICS AND THEIR DEPENDENCE ON STRUCTURE.**<br>BAXTER, S. JONES, T. T.<br>PLAST. POLYM.<br>40 ( 146 ), 69–76, 1972. |
| 68604 | **STRUCTURAL CHEMISTRY OF VITREOUS FLUORBERYLLATES.**<br>PETROVSKII, G. T.<br>STEKLOOBRAZNOE SOSTOYANIE, TR. VSES. SOVESHCH., 5TH<br>76–80, 1971. |
| 68627 | **DIFFUSE REFLECTION SPECTRA FOR STUDY AND CONTROL IN THE PRODUCTION OF ZINC-CADMIUM SULFIDE LUMINOPHORS.**<br>KRONGAUZ, V. G. PIVNEVA, S. P.<br>PAN'KOVA, L. T.<br>ZH. PRIKL. SPEKTROSK.<br>17 ( 1 ), 146–9, 1972.<br>( FOR ENGLISH TRANSLATION SEE TPRC NO. 77027 ) |
| 68630 | **SUPERHIGH-FREQUENCY HEATING OF PLASMA AND LONGITUDINAL ELECTRONIC THERMAL CONDUCTIVITY IN A MAGNETIC FIELD.**<br>ZHILINSKII, A. P. LIVENTSEVA, I. F.<br>SAFRONOV, YU. M.<br>ZH. TEKH. FIZ.<br>42 ( 7 ), 1404–12, 1972.<br>( FOR ENGLISH TRANSLATION SEE TPRC NO. 69671 ) |
| 68663 | **PHASE STATES AND PHYSICOCHEMICAL PROPERTIES OF THE HYDROCHLORIC ACID - WATER - HEPTYL ALCOHOL SYSTEM.**<br>IONIN, M. V. SHANINA, P. I.<br>ZH. NEORG. KHIM.<br>17 ( 5 ), 1444–9, 1972.<br>( FOR ENGLISH TRANSLATION SEE TPRC NO. 68664 ) |
| 68664 | **PHASE STATES AND PHYSICOCHEMICAL PROPERTIES OF THE HYDROCHLORIC ACID-WATER-HEPTYL ALCOHOL SYSTEM.**<br>IONIN, M. V. SHANINA, P. I.<br>RUSS. J. INORG. CHEM.<br>17 ( 5 ), 747–50, 1972.<br>( ENGLISH TRANSLATION OF ZH. NEORG. KHIM., 17 ( 5 ), 1444–9, 1972., FOR ORIGINAL SEE TPRC NO. 68663 ) |
| 68665 | **HEAT CAPACITY NEAR THE CONSOLUTE POINT IN SOLID METHANE-ARGON.**<br>GREER, S. C. SENGERS, J. M. H. L.<br>FURUKAWA, G. T.<br>J. CHEM. PHYS.<br>57 ( 12 ), 5052–8, 1972. |
| 68667 | **DIELECTRIC RELAXATION AND COMPLEX FORMATION IN DILUTE SOLUTIONS OF CHLOROFORM AND BROMOFORM IN NONPOLAR SOLVENTS.**<br>BOULE, P.<br>J. CHEM. PHYS.<br>57 ( 12 ), 5285–91, 1972. |

| TPRC Number | Bibliographic Citation |
|---|---|
| 68678 | **VISCOSITY OF A NITROGEN AND AIR PLASMA.**<br>PAKHOMOV, E. P. SAUTKIN, V. V.<br>TEPLOFIZ. VYS. TEMP.<br>10 ( 2 ), 221–6, 1972.<br>( FOR ENGLISH TRANSLATION SEE TPRC NO. 68679 ) |
| 68679 | **VISCOSITY OF NITROGEN AND AIR PLASMA.**<br>PAKHOMOV, E. P. SAUTKIN, V. V.<br>HIGH TEMP.<br>10 ( 2 ), 196–200, 1972.<br>( ENGLISH TRANSLATION OF TEPLOFIZ. VYS. TEMP., 10 ( 2 ), 221–6, 1972., FOR ORIGINAL SEE TPRC NO. 68678 ) |
| 68704 | **EXPERIMENTAL INVESTIGATION OF THE SINGULARITY OF SPECIFIC HEAT AT THE CRITICAL STRATIFICATION POINT OF A BINARY MIXTURE.**<br>ANISIMOV, M. A. VORONEL, A. V.<br>ORODOVA, T. M.<br>SOV. PHYS. JETP<br>34 ( 3 ), 583–7, 1972.<br>( ENGLISH TRANSLATION OF ZH. EKSP. TEOR. FIZ., 61 ( 3 ), 1092–100, 1971., FOR ORIGINAL SEE TPRC NO. 65461 ) |
| 68707 | **INCREASING THE PROTECTIVE PROPERTIES OF AQUEOUS SOLUTIONS OF SODIUM NITRITE BY THICKENING THEM WITH XYLITOL.**<br>ROZENFEL'D, I. L. PERSIANTSEVA, V. P.<br>KUZNETSOV, YU. I.<br>PROT. METALS, ( USSR )<br>6 ( 5 ), 526–9, 1970.<br>( ENGLISH TRANSLATION OF ZASHCH. METAL., 6 ( 5 ), 577–81, 1970., FOR ORIGINAL SEE TPRC NO. 60759 ) |
| 68719 | **GAS HEATING EFFECTS IN THE CONSTRICTION OF A HIGH-PRESSURE GLOW DISCHARGE COLUMN.**<br>ROGOFF, G. L.<br>PHYS. FLUIDS<br>15 ( 11 ), 1931–40, 1972. |
| 68736 | **DROP CALORIMETRIC DETERMINATION OF ENTHALPY CONTENT MIXTURES OF POTASSIUM, RUBIDIUM, CESIUM, AMMONIUM AND TETRAMETHYLAMMONIUM IODIDES WITH SILVER IODIDE.**<br>JOHNSON, K. E. SIME, S. J. DUDLEY, J.<br>J. CHEM. SOC., FARADAY TRANS.<br>68 ( PT. 11 ), 2015–20, 1972. |
| 68738 | **THERMAL CONDUCTIVITIES OF SOLID - LIQUID SUSPENSIONS.**<br>SAEGUSA, T. HASHIGUCHI, K. IIDA, Y.<br>WAKAO, N.<br>HEAT TRANSFER - JAP. RES.<br>1 ( 3 ), 25–9, 1972. |
| 68796 | **VISCOSITIES AND MOLAR POLARIZATION IN THE TERNARY SYSTEM DIMETHYLSULFOXIDE-CYCLOHEXANE-BENZENE.**<br>NISSEMA, A.<br>SUOM. KEMISTILEHTI<br>45 B ( 9 ), 274–6, 1972. |
| 68806 | **VISCOSITY OF GASES AND GAS MIXTURES OF HIGH-BOILING SUBSTANCES AT ATMOSPHERIC PRESSURE AND DIFFERENT TEMPERATURES.**<br>GOLUBEV, I. F. KOVARSKAYA, G. G.<br>TR. NAUCH.-ISSLED. PROEKT. INST. AZOTN. PROM. PROD. ORG. SIN.<br>( 8 ), 54–66, 1971. |
| 68819 | **THE SILICON NITRIDE - SILICON CARBIDE COMPOSITE SYSTEM: EFFECT OF MICROSTRUCTURE ON STRENGTH.**<br>LANGE, F. F.<br>WESTINGHOUSE RESEARCH LABS, PITTSBURGH, PA.<br>51PP., 1972.<br>( AD–743 510, 72–9D4–KERAM–R1, AVAIL. NTIS ) |
| 68837 | **CARBON RESEARCH.**<br>BOKROS, J. C. AKINS, R. J. KAAE, J. L.<br>KOYAMA, K. PRICE, R. J. SCHOEN, F. J.<br>GULF ELECTRONIC SYSTEMS, SAN DIEGO, CALIF.<br>83PP., 1972.<br>( AD–748 223, GULF–EL–A12250, AVAIL. NTIS ) |
| 68848 | **TRANSPORT PROPERTIES IN HIGH POWER ARCS.**<br>HAECKER, H. H.<br>TECHNISCHE UNIVERSITAET, MUNICH ( WEST GERMANY )<br>71PP., 1972.<br>( AD–744 518, ARL–720070, AVAIL. NTIS ) |
| 68858 | **APPROPRIATE HARDENING MECHANISMS IN ALKALI HALIDE MATERIALS FOR HIGH POWER 10.6 MU WINDOWS.**<br>POSEN, H. ARMINGTON, A. BRUCE, J.<br>AIR FORCE CAMBRIDGE RESEARCH LABS., L. G. HANSCOM FIELD, BEDFORD, MASS.<br>17PP., 1972.<br>( AD–749 867, AFCL–72–0434, AVAIL. NTIS ) |
| 68888 | **PHYSICAL PROPERTIES OF GAS MIXTURES CONTAINING HYDROGEN, METHANE, WATER, CARBON MONOXIDE AND CARBON DIOXIDE.**<br>HOHLEIN, B.<br>KERNFORSCHUNGSANLAGE JULICH<br>67PP., 1972.<br>( REPORT NO. JUL–847–RB ) |

| TPRC Number | Bibliographic Citation |
|---|---|
| 68890 | **THERMAL ENERGY TRANSPORT IN THE SOLAR WIND.**<br>MONTGOMERY, M. D.<br>COSMIC PLASMA PHYSICS: PROCEEDINGS OF THE CONFERENCE, FRASCATI, ITALY, SEPTEMBER, 20–24, 1971.<br>61–72PP., 1972.<br>( A72-33901 16-30 ) |
| 68908 | **TEMPERATURE DEPENDANCE OF THE ELECTRON THERMAL CONDUCTIVITY IN RARE GAS PLASMAS.**<br>VUKOVIC, S. M.   MARKOVIC, P. D.<br>YUGOSLAV SYMP. ON PHY. IONIZ. GASES<br>185–7, 1972. |
| 68939 | **REQUIREMENTS FOR PRODUCTION OF GRAINED CONFECTIONERY.**<br>CAKEBREAD, S. H.<br>MFG. CONFECT.<br>52 ( 3 ), 39–40,42,44,46, 1972. |
| 68940 | **CONTRIBUTION TO THE STUDY OF THE SPECIFIC HEAT OF ACTINIDE COMPOUNDS.**<br>CHRISTIAN, A.<br>C. E. N., COMMIS. ENERG. AT., FONTENAY-AUX-ROSES, FRANCE<br>91PP., 1972.<br>( CEA-R-4266, AVAIL. NTIS ) |
| 68954 | **REVERSIBILITY OF MAIZE CORN PROTEIN DENATURATION.**<br>BRATERSKII, F. D.<br>IZV. VYSSH. UCHEB. ZAVED. PISHCHEVAYA TEKHNOL.<br>( 5 ), 24–8, 1971. |
| 68955 | **SUGAR SUBSTITUTES IN CONFECTIONERY PRODUCTION.**<br>KAMMERER, F. X.<br>KAKAO ZUCKER<br>24 ( 5 ), 184–6, 188–90, 1972. |
| 68957 | **CHANGES IN VISCOSITY OF SOLUTIONS OF STARCHES OBTAINED FROM PEA, BEAN AND WHEAT SEEDS AT DIFFERENT STAGES OF DEVELOPMENT.**<br>VESEL'TSOVA, L. M.<br>PRIKL. BIOKHIM. MIKROBIOL.<br>5 ( 2 ), 199–202, 1969. |
| 68958 | **STUDY OF THE FRACTIONAL COMPOSITION OF RYE MUCOIDS, USING DEAE-SEPHADEX.**<br>L'VOVA, L. S.   GOLENKOV, V. F.<br>PRIEZZHEVA, L. G.<br>PRIKL. BIOKHIM. MIKROBIOL.<br>6 ( 2 ), 133–7, 1970. |
| 68959 | **COMPOSITION OF MUCOUS SUBSTANCES ISOLATED FROM RYE GRAIN AT DIFFERENT STAGES OF RIPENING.**<br>PRIEZZHEVA, L. G.   GOLENKOV, V. F.<br>PRIKL. BIOKHIM. MIKROBIOL.<br>6 ( 5 ), 574–7, 1970. |
| 68960 | **VISCOSITY OF CASEINATE SOLUTIONS FROM THE TECHNOLOGICAL VIEW POINT.**<br>DOLEZALEK, J.   BREZINA, P.   PETERKOVA, L.<br>PRUM. POTRAVIN<br>23 ( 5 ), 156–9, 1972. |
| 68964 | **VARIETAL DETERMINATION OF QUALITY CHARACTERISTICS IN RYE. RESULTS OF STUDY OF FOREIGN RYE VARIETIES PROPAGATED IN THE GERMAN FEDERAL REPUBLIC.**<br>SEIBEL, W.   DREWS, E.   REIMERS, H.<br>Z. PFLANZENZUCH.<br>66 ( 2 ), 130–50, 1971. |
| 68974 | **KRAMERS-KRONIG ANALYSIS OF RELATIVE REFLECTANCE SPECTRA MEASURED AT AN OBLIQUE ANGLE.**<br>HALE, G. M.   HOLLAND, W. E.   QUERRY, M. R.<br>APPL. OPT.<br>12 ( 1 ), 48–51, 1973. |
| 68977 | **VARIABLE BANDWIDTH TRANSMISSION FILTER FOR THE VACUUM ULTRAVIOLET.. NONSTOICHIOMETRIC LANTHANUM CERIUM FLUORIDE.**<br>ELIAS, L. R.   FLACH, R.   YEN, W. M.<br>APPL. OPT.<br>12 ( 1 ), 138–9, 1973. |
| 68979 | **A STUDY OF RADIOMETRIC EMISSION FROM A ROUGH SEA SURFACE.**<br>STOGRYN, A.<br>AEROJET-GENERAL CORP., EL MONTE, CALIF.<br>80PP., 1972.<br>N72-29116,   NASA-CR-2088,   AVAIL. NTIS |
| 68987 | **SCATTERING OF MICROWAVES BY A STRATIFIED OVER-DENSE PLASMA AT HIGH COLLISION FREQUENCIES.**<br>KLEIN, M. M.<br>J. APPL. PHYS.<br>44 ( 1 ), 91–5, 1973. |
| 69014 | **VISCOSITY OF SOLUTION NEAR THE CRITICAL MIXING POINT.**<br>KUSKOVA, N. V.   MATIZEN, E. V.<br>SOV. PHYS., JETP<br>31 ( 1 ), 174–7, 1970.<br>( ENGLISH TRANSLATION OF PIS'MA ZH. EKSP. TEOR. FIZ., 12 ( 5 ), 255–9, 1970., FOR ORIGINAL SEE TPRC NO. 62486 ) |
| 69016 | **INFRARED SPECTRA OF MATRIX ISOLATED BORIC ACID AND METABORIC ACID MONAMER AND TIMER.**<br>MACK, J. L.   WILMOT, G. B.<br>PROC. MEETING INTERAGENCY CHEM. ROCKET PROPULSION GROUP THERMOCHEM., 6TH, CALIFORNIA<br>1, 59–70, 1968. |
| 69020 | **PROPERTIES OF SOLUTIONS OF ALKALINE EARTH METAL COMPLEXONATES.**<br>VOROB'EV, G. I.   RYKOVA, G. A.<br>SHTERNINA, E. B.<br>ZH. NEORG. KHIM.<br>17 ( 7 ), 1833–7, 1972.<br>( FOR ENGLISH TRANSLATION SEE TPRC NO. 69021 ) |
| 69021 | **SOME PROPERTIES OF SOLUTIONS OF ALKALINE EARTH METAL COMPLEXONATES.**<br>VOROB'EV, G. I.   RYKOVA, G. A.<br>SHTERNINA, E. B.<br>RUSS. J. INORG. CHEM.<br>17 ( 7 ), 950–2, 1972.<br>( ENGLISH TRANSLATION OF ZH. NEORG. KHIM., 17 ( 7 ), 1833–7, 1972., FOR ORIGINAL SEE TPRC NO. 69020 ) |
| 69022 | **NICKEL PERCHLORATE-UREA-WATER SYSTEM AT 25 DEGREES C.**<br>KARNAUKHOV, A. S.   LEPESHKOV, I. N.<br>TARAKANOV, V. F.<br>ZH. NEORG. KHIM.<br>17 ( 7 ), 2027– , 1972.<br>( FOR ENGLISH TRANSLATION SEE TPRC NO. 69023 ) |
| 69023 | **NICKEL PERCHLORATE-UREA-WATER SYSTEM AT 25 DEGREES C.**<br>KARNAUKHOV, A. S.   LEPESHKOV, I. N.<br>TARAKANOV, V. F.<br>RUSS. J. INORG. CHEM.<br>17 ( 7 ), 1051–3, 1972.<br>( ENGLISH TRANSLATION OF ZH. NEORG. KHIM., 17 ( 7 ), 2027– , 1972., FOR ORIGINAL SEE TPRC NO. 69022 ) |
| 69028 | **MATRIX REACTIONS OF LITHIUM ATOMS WITH NITROUS OXIDE. INFRARED SPECTRA OF LITHIUM OXIDES.**<br>SPIKER, R. C., JR.   ANDREWS, L.<br>J. CHEM. PHYS.<br>58 ( 2 ), 702–12, 1972. |
| 69031 | **VISCOSITY DATA OF BINARY MIXTURES IN THE SYSTEM METHYL ISOBUTYLKETONE-N-BUTANOL.**<br>DAKSHINAMURTY, P.   RAO, K. V.   RAO, P. V.<br>CHIRANJIVI, C.<br>J. CHEM. ENG. DATA<br>18 ( 1 ), 39–41, 1973. |
| 69032 | **HEATS OF FUSION AND HEAT CAPACITIES OF LITHIUM CHLORIDE-POTASSIUM CHLORIDE EUTECTIC AND POTASSIUM NITRATE.**<br>CLARK, R. P.<br>J. CHEM. ENG. DATA<br>18 ( 1 ), 67–70, 1973. |
| 69064 | **THE SPECTRAL ALBEDO OF WATER CLOUDS IN THE 1-TO 6-MICRON RANGE.**<br>MALKOVA, V. S.   ROMANOVA, L. M.<br>IZV. AKAD. NAUK SSSR, FIZ. ATMOSFERY I OKEANA<br>( 8 ), 763–767, 1972. |
| 69065 | **VISCOSITY OF BINARY LIQUID MIXTURES. V. SYSTEMS CONSISTING OF ACETONE AND P-DIOXANE WITH CHLOROMETHANES.**<br>MATO, F.   COCA, J.<br>AN. QUIM.<br>68 ( 7–8 ), 1043–51, 1972. |
| 69076 | **SPECIFIC HEATS OF AQUEOUS SODIUM HYDROXIDE SOLUTIONS UP TO 400.**<br>PUCHKOV, L. V.   BARANOVA, T. A.<br>MATVEEVA, R. P.<br>FIZ. KHIM. RASTVOROV<br>86–90, 1972. |
| 69077 | **SPECIFIC HEATS AND THERMODYNAMIC SOLUTION FUNCTIONS IN THE LITHIUM HYDROXIDE-WATER SYSTEM.**<br>PUCHKOV, L. V.   MATVEEVA, R. P.<br>FIZ. KHIM. RASTVOROV<br>95–8, 1972. |
| 69078 | **PROPERTIES OF NITROGEN-CONTAINING SOLVENTS AND THEIR BINARY MIXTURES. II. PROPERTIES OF THE BUTYLAMINE-WATER SYSTEM.**<br>ZHIGUNOV, I. S.   AMPELOGOVA, N. I.<br>FIZ. KHIM. RASTVOROV<br>181–5, 1972. |
| 69085 | **ON MERCURY CLOUDS IN THE ATMOSPHERE OF VENUS.**<br>POTTER, J. F.<br>ICARUS<br>17, 79–87, 1972. |
| 69124 | **EVALUATION OF THE SHEAR VISCOSITY FOR A BINARY DENSE MIXTURE, WITH THE SINGLE-LIQUID MODEL.**<br>COLDEA, S.<br>STUD. CERCET. FIZ.<br>24 ( 5 ), 643–6, 1972. |

| TPRC Number | Bibliographic Citation |
|---|---|
| 69136 | **THERMAL CONDUCTIVITY OF TERNARY SYSTEMS IN THE PRODUCTION OF ALCOHOLS.**<br>KHARIN, V. M. JASIUNAS, B.<br>TR. VORONEZH. TEKHNOL. INST.<br>19 ( 2 ), 59–62, 1971. |
| 69219 | **VISCOSITY STUDIES OF AQUEOUS SOLUTIONS OF ALCOHOLS, UREAS, AND AMIDES.**<br>HERSKOVITS, T. T. KELLY, T. M.<br>J. PHYS. CHEM.<br>77 ( 3 ), 381–8, 1973. |
| 69230 | **SPECIFIC HEAT OF LIQUID HELIUM( 3 ) – HELIUM( 4 ) MIXTURES NEAR THE JUNCTION OF THE LAMBDA AND PHASE-SEPARATION CURVES. II.**<br>ISLANDER, S. T. ZIMMERMANN, W., JR.<br>PHYS. REV.<br>7 A ( 1 ), 188–97, 1973. |
| 69252 | **TRANSPORT EQUATIONS FOR DILUTE PLASMA IN A MAGNETIC FIELD.**<br>PRICE, J. C.<br>PHYS. FLUIDS<br>16 ( 1 ), 90–4, 1973. |
| 69256 | **EXPERIMENTAL HEAT CAPACITIES OF NITROGEN-CARBON DIOXIDE MIXTURES AT ELEVATED PRESSURES.**<br>BISHNOI, P. R. HAMALIUK, G. P.<br>ROBINSON, D. B.<br>CAN. J. CHEM. ENG.<br>50 ( 5 ), 677–9, 1972. |
| 69270 | **VAPOR PRESSURE, VISCOSITY, DENSITY, AND ELECTRIC CONDUCTIVITY OF ELECTROLYTE SOLUTIONS AT HIGH CONCENTRATIONS AND TEMPERATURES.**<br>MAKSIMOVA, I. N.<br>FIZ. KHIM. RASTVOROV<br>98–104, 1972. |
| 69271 | **VISCOSITY AND ENERGY OF THE INTERMOLECULAR INTERACTION OF LITHIUM CHLORIDE SOLUTIONS IN METHANOL.**<br>ORISHCHENKO, A. V.<br>FIZ. KHIM. RASTVOROV<br>135–9, 1972. |
| 69283 | **DETERMINING THE THERMAL CONDUCTIVITY OF REFRIGERANTS, ANTIFREEZE OILS, AND THEIR MIXTURES.**<br>HEIDE, R.<br>LUFT-KAELTETECH.<br>8 ( 3 ), 128–33, 1972. |
| 69324 | **EFFECT OF MODIFIED THERMAL CONDUCTIVITY ON THE TEMPERATURE DISTRIBUTION IN THE PROTONOSPHERE.**<br>MAYR, H. G. FONTHEIM, E. G.<br>MAHAJAN, K. K.<br>GODDARD SPACE FLIGHT CENTER, GREENBELT, MD.<br>13PP., 1972.<br>( NASA-TM-X-65900, N72-25339, X-621-72-107, AVAIL. NTIS ) |
| 69326 | **HUMIDITY MEASUREMENT BY MEANS OF INFRARED REFLECTION.**<br>HOFFMAN, K.<br>CHEM.-ING.-TECH.<br>35 ( 1 ), 55–62, 1963.<br>( FOR ENGLISH TRANSLATION SEE TPRC NO. 69327 ) |
| 69327 | **HUMIDITY MEASUREMENT BY MEANS OF INFRARED REFLECTION.**<br>HOFFMAN, K.<br>SCIENTIFIC TRANSLATION SERVICE, SANTA BARBARA, CALIF.<br>27PP., 1972.<br>( ENGLISH TRANSLATION OF CHEM.-ING.-TECH., 35 ( 1 ), 55–62, 1963; FOR ORIGINAL SEE TPRC NO. 69326 )<br>( NASA-TT-F-14333, N72-27640, AVAIL. NTIS ) |
| 69328 | **THE NASA EARTH RESOURCES SPECTRAL INFORMATION SYSTEM: A DATA COMPILATION.**<br>LEEMAN, V. EARING, D. VINCENT, R. K.<br>LADD, S.<br>WILLOW RUN LABS., MICHIGAN UNIV., ANN ARBOR<br>574PP., 1971.<br>( NASA-CR-115757, N72-28366, WRL-31650-T AVAIL. NTIS ) |
| 69329 | **THERMAL CONDUCTIVITY AND RESONANT MULTIPOLE INTERACTIONS.**<br>NYELAND, C. MASON, E. A. MONCHICK, L.<br>CHEMISTRY LAB., COPENHAGEN UNIV. ( DENMARK )<br>53PP., 1971.<br>( N72-25639, NASA-CR-126729, AVAIL. NTIS ) |
| 69339 | **VISCOSITY OF AQUEOUS POTASSIUM SILICATE SOLUTIONS.**<br>TATOMYR, YA. T. KORDIYAK, KH. S. KUZ, V. I.<br>COLLOID J., USSR<br>33 ( 6 ), 757–9, 1971.<br>( ENGLISH TRANSLATION OF KOLL. ZH., 33 ( 6 ), 905–8, 1971., FOR ORIGINAL SEE TPRC NO. 65697 ) |
| 69340 | **THE NASA EARTH RESOURCES SPECTRAL INFORMATION SYSTEM: A DATA COMPILATION, FIRST SUPPLEMENT.**<br>LEEMAN, V.<br>INFRARED AND OPTICS LAB., MICHIGAN UNIV.<br>159PP., 1972.<br>( N72-30345, NASA-CR-115756, WRL-31650-69-SUPPL-1 AVAIL. NTIS ) |
| 69342 | **PROPERTIES OF SEAWATER AND ITS CONCENTRATES AND RELATED SOLUTIONS AT TEMPERATURES UP TO 400 F.**<br>BROMLEY, L. A.<br>CALIFORNIA UNIV., BERKELEY<br>109PP., 1972.<br>( N72-28404, PB-208064, INT-OSW-RDPR-72-747, AVAIL. NTIS ) |
| 69362 | **HEAT CAPACITIES OF THE SODIUM CARBONATE-WATER AND SODIUM SULFATE-WATER SYSTEMS.**<br>GINZBURG, D. M. KOCHKALDA, V. E.<br>ZH. FIZ. KHIM.<br>46 ( 8 ), 2142–3, 1972.<br>( FOR ENGLISH TRANSLATION SEE TPRC NO. 69363 ) |
| 69363 | **HEAT CAPACITY OF THE SODIUM CARBONATE-WATER AND SODIUM SULPHATE-WATER SYSTEMS.**<br>GINZBURG, D. M. KOCHKALDA, V. E.<br>RUSS. J. PHYS. CHEM.<br>46 ( 8 ), 1226–7, 1972.<br>( ENGLISH TRANSLATION OF ZH. FIZ. KHIM., 46 ( 8 ), 2142–3, 1972., FOR ORIGINAL SEE TPRC NO. 69362 ) |
| 69364 | **SPECIFIC HEAT OF SOLUTIONS OF POTASSIUM CHLORIDE IN WATER AND 30 PERCENT AQUEOUS SOLUTION OF ISOPROPYL ALCOHOL IN THE LESS THAN 0.01 M RANGE OF ELECTROLYTE CONCENTRATION AT 25 DEGREES.**<br>KLOPOV, V. I. KOLKER, A. M. KRESTOV, G. A.<br>ZH. FIZ. KHIM.<br>46 ( 8 ), 2155–6, 1972.<br>( FOR ENGLISH TRANSLATION SEE TPRC NO. 69365 ) |
| 69365 | **SPECIFIC HEATS OF POTASSIUM CHLORIDE SOLUTIONS BELOW 0.01 MOLAL IN WATER AND IN 30 PERCENT AQUEOUS ISOPROPYL ALCOHOL AT 25 C.**<br>KLOPOV, V. I. KOLKER, A. M. KRESTOV, G. A.<br>RUSS. J. PHYS. CHEM.<br>46 ( 8 ), 1235, 1972.<br>( ENGLISH TRANSLATION OF ZH. FIZ. KHIM., 46 ( 8 ), 2155–6, 1972., FOR ORIGINAL SEE TPRC NO. 69364 ) |
| 69372 | **RECIPROCAL EFFECT OF NITRITE AND NITRATE IONS IN SOLUTION.**<br>PLEKHOTKIN, V. F. BOBROVSKAYA, L. P.<br>ZH. FIZ. KHIM.<br>46 ( 9 ), 2403–5, 1972.<br>( FOR ENGLISH TRANSLATION SEE TPRC NO. 69373 ) |
| 69373 | **INTERACTION OF NITRITE AND NITRATE IONS IN SOLUTION.**<br>PLEKHOTKIN, V. F. BOBROVSKAYA, L. P.<br>RUSS. J. PHYS. CHEM.<br>46 ( 9 ), 1375–7, 1972.<br>( ENGLISH TRANSLATION OF ZH. FIZ. KHIM., 46 ( 9 ), 2403–5, 1972., FOR ORIGINAL SEE TPRC NO. 69372 ) |
| 69474 | **RELATION BETWEEN THE THERMAL CONDUCTIVITY OF BINARY SOLUTION AND THEIR CONCENTRATION.**<br>LOSENICKY, Z.<br>POTRAVIN. CHLADICI TECH.<br>3 ( 5 ), 156–7, 1972. |
| 69502 | **THE KINETIC THEORY OF HEAT CONDUCTION IN GAS MIXTURES AT LOW TEMPERATURES.**<br>SHASHKOV, A. G. ABRAMENKO, T. N.<br>J. ENG. PHYS., USSR<br>16 ( 6 ), 689–96, 1969.<br>( ENGLISH TRANSLATION OF INZH. FIZ. ZH., 16 ( 6 ), 1009–18, 1969., FOR ORIGINAL SEE TPRC NO. 55757 ) |
| 69503 | **THERMAL CONDUCTIVITY OF GAS MIXTURES AT LOW TEMPERATURES. II.**<br>SHASHKOV, A. G. ABRAMENKO, T. N.<br>J. ENG. PHYS., USSR<br>17 ( 1 ), 833–7, 1969.<br>( ENGLISH TRANSLATION OF INZH. FIZ. ZH., 17 ( 1 ), 65–71, 1969., FOR ORIGINAL SEE TPRC NO. 55851 ) |
| 69511 | **THERMAL CONDUCTIVITY OF THE ETHANOL-WATER SYSTEM IN THE TEMPERATURE RANGE 20–300 DEGREES C AND AT PRESSURES UP TO 30 MEGANEWTONS PER SQUARE METER.**<br>POPOV, V. N. MALOV, B. A.<br>THERM. ENG., USSR<br>18 ( 8 ), 125–9, 1971.<br>( ENGLISH TRANSLATION OF TEPLOENERGETIKA, 18 ( 8 ), 88–90, 1971; FOR ORIGINAL SEE TPRC NO. 64236 ) |
| 69535 | **VISCOSITY AND STRUCTURE OF POTASSIUM IODIDE SOLUTIONS IN ETHYLENE GLYCOL.**<br>ARTEMCHENKO, A. I.<br>FIZ. KHIM. RASTVOROV<br>128–34, 1972. |
| 69560 | **PHYSICOCHEMICAL PROPERTIES AND EXTRACTION CHARACTERISTICS OF A DIMETHYLFORMAMIDE-DIETHYLENE GLYCOL MIXED SOLVENT.**<br>PROKOPETS, M. M. ZELIZNYI, A. M.<br>KUNTII, L. I<br>KHIM. TEKHNOL. TOPL. MASEL<br>17 ( 11 ), 31–4, 1972.<br>( FOR ENGLISH TRANSLATION SEE TPRC NO. 72680 ) |

| TPRC Number | Bibliographic Citation |
|---|---|

69566 THERMOPHYSICAL PROPERTIES OF PETROLEUMS AND WATER-PETROLEUM EMULSIONS IN THE TATAR ASSR.
SATTAROV, U. G. FARRAKHOV, E. Z.
GAZIMOV, M. G.
NEFTEPROM. DELO
( 8 ), 28-31, 1972.

69601 THERMAL CONDUCTIVITY OF AIR AT ATMOSPHERIC PRESSURE.
IRVING, J. B. JAMIESON, D. T. PAGET, D. S.
TRANS. INST. CHEM. ENG.
51 ( 1 ), 10-3, 1972.

69609 ELECTRICAL CONDUCTIVITY OF SODIUM IODIDE AND CESIUM IODIDE SOLUTIONS IN METHANOL AND DECANOL MIXTURES.
SERGEEVA, I. A. LISACHENKO, T. F.
ZAMAZII, G. P.
VISN. KHARKIV. UNIV., KHIM.
73 ( 2 ), 22-5, 1971.

69611 TEMPERATURE DEPENDENCE OF THE VISCOSITY OF ETHANOL-DIOXANE AND ACETONITRILE-DIOXANE BINARY SYSTEMS.
LEVITSKAYA, N. K. TKACHENKO, L. I.
SHKODIN, A. M.
VISN. KHARKIV. UNIV., KHIM.
73 ( 2 ), 27-9, 1971.

69615 EXPERIMENTAL DETERMINATION OF THE THERMAL CONDUCTIVITY OF AQUEOUS SOLUTIONS OF SALTS AND AMMONIA AT 25 AND 50 DEGREES.
CHERNEN'KAYA, E. I. VERNIGORA, G. A.
ZH. PRIKL. KHIM. ( LENINGRAD )
45 ( 8 ), 1704-7, 1972.
( FOR ENGLISH TRANSLATION SEE TPRC NO. 70815 )

69620 TRANSPORT PROPERTIES OF GASES AND BINARY LIQUIDS NEAR THE CRITICAL POINT.
SENGERS, J. V.
NATL. AERONAUTICS AND SPACE ADMIN.
74PP., 1972.
NASA-CR-WIIW, AVAIL. NTIS N72-32938

69650 PHYSICOCHEMICAL ANALYSIS OF BINARY LIQUID SYSTEMS COMPRISING PHOSPHORIC ACID WITH BUTYL AND PENTYL ACETATES.
KRUPATKIN, I. L. BRATUSHCHAK, A. N.
ZH. FIZ. KHIM.
46 ( 10 ), 2674- , 1972.
( FOR ENGLISH TRANSLATION SEE TPRC NO. 69651 )

69651 PHYSICOCHEMICAL ANALYSIS OF BINARY LIQUID SYSTEMS COMPRISING PHOSPHORIC ACID WITH BUTYL AND PENTYL ACETATES.
KRUPATKIN, I. L. BRATUSHCHAK, A. N.
RUSS. J. PHYS. CHEM.
46 ( 10 ), 1532-3, 1972.
( ENGLISH TRANSLATION OF ZH. FIZ. KHIM., 46 ( 10 ), 2674- , 1972., FOR ORIGINAL SEE TPRC NO. 69650 )

69652 PHYSICOCHEMICAL ANALYSIS OF BINARY LIQUID SYSTEMS COMPRISING PHOSPHORIC ACID AND ETHERS.
KRUPATKIN, I. L. BRATUCHSHAK, A. M.
ZH. FIZ. KHIM.
46 ( 10 ), 2676- , 1972.
( FOR ENGLISH TRANSLATION SEE TPRC NO. 69653 )

69653 PHYSICOCHEMICAL ANALYSIS OF BINARY LIQUID SYSTEMS COMPRISING PHOSPHORIC ACID AND ETHERS.
KRUPATKIN, I. L. BRATUCHSHAK, A. M.
RUSS. J. PHYS. CHEM.
46 ( 10 ), 1533-4, 1972.
( ENGLISH TRANSLATION OF ZH. FIZ. KHIM., 46 ( 10 ), 2676- , 1972., FOR ORIGINAL SEE TPRC NO. 69652 )

69654 ENTHALPIES AND SPECIFIC HEATS IN THE POTASSIUM CARBONATE-WATER SYSTEM.
GINZBURG, D. M. KOCHKALDA, V. E.
ZH. FIZ. KHIM.
46 ( 10 ), 2677- , 1972.
( FOR ENGLISH TRANSLATION SEE TPRC NO. 69655 )

69655 ENTHALPIES AND SPECIFIC HEATS IN THE POTASSIUM CARBONATE-WATER SYSTEM.
GINZBURG, D. M. KOCHKALDA, V. E.
RUSS. J. PHYS. CHEM.
46 ( 10 ), 1535, 1972.
( ENGLISH TRANSLATION OF ZH. FIZ. KHIM., 46 ( 10 ), 2677- , 1972., FOR ORIGINAL SEE TPRC NO. 69654 )

69670 EFFECT OF ION VISCOSITY ON THE STABILITY OF A FINITE-PRESSURE PLASMA.
TSYPIN, V. S.
SOV. PHYS.-TECH. PHYS.
17 ( 4 ), 573-80, 1972.
( ENGLISH TRANSLATION OF ZH. TEKH. FIZ., 42 ( 4 ), 727-37, 1972., FOR ORIGINAL SEE TPRC NO. 67514 )

69671 MICROWAVE HEATING OF A PLASMA AND LONGITUDINAL ELECTRONIC THERMAL CONDUCTIVITY IN A MAGNETIC FIELD.
ZHILINSKII, A. P. LIVENTSEVA, I. F.
SAFRONOV, YU. M.
SOV. PHYS.-TECH. PHYS.
17 ( 7 ), 1116-22, 1973.
( ENGLISH TRANSLATION OF ZH. TEKH. FIZ., 42 ( 7 ), 1404-12, 1972., FOR ORIGINAL SEE TPRC NO. 68630 )

69683 DAMPING OF ADIABATIC PERTURBATIONS IN AN EXPANDING UNIVERSE.
CHIBISOV, G. V.
SOV. ASTRON.-AJ
16 ( 1 ), 56-63, 1972.
( ENGLISH TRANSLATION OF ASTRON. ZH., 49 ( 1 ), 74-84, 1972., FOR ORIGINAL SEE TPRC NO. 68453 )

69692 DEVELOPMENT OF THE VARIABLE-GAP TECHNIQUE FOR MEASURING THE THERMAL CONDUCTIVITY OF FLUORIDE SALT MIXTURES.
COOKE, J. W.
OAK RIDGE NATIONAL LAB., OAK RIDGE, TENN.
114PP., 1973.
( REPORT ORNL-4831, AVAIL. DEP. NTIS )

69701 INFRARED SPECTRA OF CARBON MONOXIDE TRAPPED IN SOLID ARGON. DOUBLE-DOPING EXPERIMENTS WITH WATER, AMMONIA, AND MOLECULAR NITROGEN.
DUBOST, H. ABOUAF-MARGUIN, L.
CHEM. PHYS. LETT.
17 ( 2 ), 269-73, 1972.

69713 THERMAL CONDUCTIVITY OF AQUEOUS ELECTROLYTE SOLUTIONS.
KERIMOV, A. M. FATULLAEV, F. G.
IZV. AKAD. NAUK AZERB. SSR, SER. FIZ.-TEKH. MAT. NAUK
( 1 ), 108-12, 1972.

69716 HEATS OF SUCROSE DISSOLUTION AND THE HEAT CAPACITY OF AQUEOUS SUGAR SOLUTIONS.
ZELTSER, YA. V. MALYAROV, B. M.
IZV. VYSSH. UCHEB. ZAVED., PISHCH. TEKHNOL.
( 4 ), 49-52, 1972.

69723 VISCOSITY OF AQUEOUS SOLUTIONS OF ALKALI AND TETRAALKYLAMMONIUM HALIDES AT 25.
DESNOYERS, J. E. PERRON, G.
J. SOLUTION CHEM.
1 ( 3 ), 199-212, 1972.

69724 ACTIVITIES AND VISCOSITIES OF AQUEOUS BILE ACID SALT-LECITHIN SOLUTIONS.
CUSSLER, E. L. DUNCAN, C. L.
J. SOLUTION CHEM.
1 ( 3 ), 269-77, 1972.

69758 MEASUREMENT OF THE SHEAR VISCOSITY COEFFICIENT IN THE SEC-BUTANOL-TERT-BUTANOL SYSTEM IN A WIDE TEMPERATURE RANGE.
SPERKACH, V. S.
VESTN. MOSK. UNIV., KHIM.
13 ( 5 ), 602-3, 1972.

69760 SALTS OF POLYNITROALKANES IN SOLUTIONS. II. DENSITY AND VISCOSITY OF AQUEOUS SOLUTIONS OF SODIUM NITROETHANE.
PENKINA, N. V. ODOKIENKO, S. S.
SELIVANOV, V. F. GIDASPOV, B. V.
ZH. PRIKL. KHIM. ( LENINGRAD )
45 ( 10 ), 2340-1, 1972.
( FOR ENGLISH TRANSLATION SEE TPRC NO. 70681 )

69764 KINETIC THEORY FOR MIXTURES OF DILUTE GASES OF LINEAR ROTATING MOLECULES. II. SENFTLEBEN-BEENAKKER EFFECT OF THE HEAT CONDUCTIVITY.
KOEHLER, W. E. RAUM, H. H.
Z. NATURFORSCH.
27 A ( 10 ), 1383-93, 1972.

69767 TRANSPORT PROPTERTIES OF FLUIDS NEAR CRITICAL POINTS.
SENGERS, J. V.
CRITICAL PHENOMENA
445-507, 1971.

69784 CALCULATION OF CERTAIN PHYSICOCHEMICAL PROPERTIES OF SODIUM-POTASSIUM HYDROXIDE AND ALUMINATE SOLUTIONS.
PENKINA, N. V. PUCHKOV, L. V.
J. APPL. CHEM., USSR
44 ( 7 ), 1571-5, 1971.
( ENGLISH TRANSLATION OF ZH. PRIKL. KHIM., 44 ( 7 ), 1554-8, 1971., FOR ORIGINAL SEE TPRC NO. 64576 )

69785 DENSITY AND VISCOSITY AND AQUEOUS SOLUTIONS OF LITHIUM NITRATE AND MIXED MAGNESIUM AND LITHIUM NITRATES.
POLOVNIKOVA, G. T. KAREV, V. G.
J. APPL. CHEM., USSR
44 ( 8 ), 1720-3, 1971.
( ENGLISH TRANSLATION OF ZH. PRIKL. KHIM., 44 ( 8 ), 1696-1700, 1971; FOR ORIGINAL SEE T64892 )

69786 PHYSICAL PROPERTIES OF N-VINYLPYRROLIDONE, ALPHA-PYRROLIDONE, AND POTASSIUM PYRROLIDONE SOLUTIONS IN ALPHA-PYRROLIDONE.
MYASNIKOVA, L. F. BUSHINSKII, V. I.
SUBBOTINA, L. A. VAISMAN, I. L.
FREIDLIN, G. N.
J. APPL. CHEM., USSR
44 ( 8 ), 1867-70, 1971.
( ENGLISH TRANSLATION OF ZH. PRIKL. KHIM., 44 ( 8 ), 1849-53, 1971; FOR ORIGINAL SEE TPRC NO. 65766 )

| TPRC Number | Bibliographic Citation |
|---|---|
| 69787 | **SPECIFIC HEATS OF SODIUM AND POTASSIUM ALUMINATE SOLUTIONS CONTAINING THE CARBONATE - ION AT 25-300 DEGREES.**<br>PENKINA, N. V. PUCHKOV, L. V.<br>BARANOVA, T. A.<br>J. APPL. CHEM., USSR<br>44 ( 8 ), 1913-5, 1971.<br>( ENGLISH TRANSLATION OF ZH. PRIKL. KHIM., 44 ( 8 ), 1890-2, 1971., FOR ORIGINAL SEE TPRC NO. 64893 ) |
| 69788 | **VISCOSITY AND DENSITY OF LITHIUM CHLORIDE, LITHIUM BROMIDE, AND LITHIUM IODIDE AQUEOUS SOLUTIONS AT AVERAGE AND LOW TEMPERATURES.**<br>MASHOVETS, V. P. BARON, N. M.<br>SHCHERBA, M. U.<br>J. APPL. CHEM., USSR<br>44 ( 9 ), 2014-8, 1971.<br>( ENGLISH TRANSLATION OF ZH. PRIKL. KHIM., 44 ( 9 ), 1981-6, 1971., FOR ORIGINAL SEE TPRC NO. 65462 ) |
| 69789 | **VISCOSITY, DENSITY, AND CONDUCTIVITY OF NICKEL SULFATE SOLUTIONS AT ELEVATED TEMPERATURES.**<br>FEDOTOV, N. V. MAKSIMOVA, I. N.<br>J. APPL. CHEM., USSR<br>44 ( 9 ), 2019-22, 1971.<br>( ENGLISH TRANSLATION OF ZH. PRIKL. KHIM., 44 ( 9 ), 1986-9, 1971., FOR ORIGINAL SEE TPRC NO. 65293 ) |
| 69790 | **SOME PHYSICOCHEMICAL PROPERTIES OF AQUEOUS SODIUM AND POTASSIUM PYROPHOSPHATE SOLUTIONS.**<br>GALINKER, V. S. BYKOVA, M. I.<br>J. APPL. CHEM., USSR<br>44 ( 12 ), 2827-9, 1971.<br>( ENGLISH TRANSLATION OF ZH. PRIKL. KHIM., 44 ( 12 ), 2741-4, 1971., FOR ORIGINAL SEE TPRC NO. 66303 ) |
| 69793 | **DETERMINATION OF THE AVERAGE SPECIFIC HEATS OF SALT ROCKS AND BRINES.**<br>YUSOVA, YU. I. ESELEV, I. M.<br>KARATYGIN, E. P.<br>J. APPL. CHEM., USSR<br>45 ( 2 ), 425-6, 1972.<br>( ENGLISH TRANSLATION OF ZH. PRIKL. KHIM., 45 ( 2 ), 427-8, 1972., FOR ORIGINAL SEE TPRC NO. 66602 ) |
| 69794 | **DENSITY AND VISCOSITY IN THE SYSTEMS DIMETHYLFORMAMIDE-WATER AND WATER-UROTROPIN IN THE TEMPERATURE RANGE FROM MINUS 15 TO PLUS 40 C.**<br>ENNAN, A. A. CHOBOTAREV, A. N.<br>ANIKEEV, V. A. KORNELLI, M. E.<br>YUR'EVA, E. I. KATS, B. M.<br>J. APPL. CHEM., USSR<br>45 ( 3 ), 627-30, 1972.<br>( ENGLISH TRANSLATION OF ZH. PRIKL. KHIM., 45 ( 3 ), 622-6, 1972., FOR ORIGINAL SEE TPRC NO. 67139 ) |
| 69795 | **VISCOSITY AND DENSITY OF AQUEOUS SOLUTIONS OF THE SODIUM SALT OF NITROFORM.**<br>PENKINA, N. V. ODOKIENKO, S. S.<br>SELIVANOV, V. F. GIDASPOV, B. V.<br>J. APPL. CHEM., USSR<br>45 ( 3 ), 676-7, 1972.<br>( ENGLISH TRANSLATION OF ZH. PRIKL. KHIM., 45 ( 3 ), 667-8, 1972., FOR ORIGINAL SEE TPRC NO. 67140 ) |
| 69807 | **CLASSICAL THERMAL CONDUCTIVITY OF A STATIONARY TOROIDAL PLASMA.**<br>MASCHKE, E. K.<br>PLASMA PHYS.<br>14, 141-51, 1972. |
| 69808 | **VISCOSITY AND THERMAL CONDUCTIVITY EFFECTS ON THE CONTAINMENT OF A TOROIDALLY CONFINED RESISTIVE PLASMA WITH FLOW.**<br>GRIMM, R. C. JOHNSON, J. L.<br>PLASMA PHYS.<br>14, 617-34, 1972. |
| 69811 | **VISCOSITY OF GASES IN METRIC UNITS.**<br>WATSON, J. T. R.<br>HER MAJESTY'S STATIONERY OFFICE, EDINBURGH<br>48PP., 1972. |
| 69824 | **MONOCHROMATIC EMISSIVITY OF THE TITANIUM CARBIDE-ZIRCONIUM CARBIDE SYSTEM.**<br>KIRILLOV, V. S. SAFONOV, V. I.<br>TEPLOFIZ. VYS. TEMP.<br>10 ( 4 ), 902-5, 1972.<br>( FOR ENGLISH TRANSLATION SEE TPRC NO. 69825 ) |
| 69825 | **MONOCHROMATIC EMITTANCE OF THE TITANIUM CARBIDE-ZIRCONIUM CARBIDE SYSTEM.**<br>KIRILLOV, V. S. SAFONOV, V. I.<br>HIGH TEMP.<br>10 ( 4 ), 814-6, 1972.<br>( ENGLISH TRANSLATION OF TEPLOFIZ. VYS. TEMP., 10 ( 4 ), 902-5, 1972., FOR ORIGINAL SEE TPRC NO. 69824 ) |
| 69830 | **VISCOSITY, DENSITY, AND ELECTRICAL CONDUCTIVITY IN THE LANTHANUM NITRATE-MAGNESIUM NITRATE-WATER SYSTEM AT 25 DEGREES.**<br>GUZHAVINA, E. I. NOSOVA, N. F.<br>YAKIMOV, M. A.<br>ZH. NEORG. KHIM.<br>17 ( 11 ), 3072-4, 1972.<br>( FOR ENGLISH TRANSLATION SEE TPRC NO. 69831 ) |
| 69831 | **VISCOSITY, DENSITY, AND SPECIFIC CONDUCTANCE OF THE LANTHANUM NITRATE-MAGNESIUM NITRATE-WATER SYSTEM AT 25 C.**<br>GUZHAVINA, E. I. NOSOVA, N. F.<br>YAKIMOV, M. A.<br>RUSS. J. INORG. CHEM.<br>17 ( 11 ), 1615-6, 1972.<br>( ENGLISH TRANSLATION OF ZH. NEORG. KHIM., 17 ( 11 ), 3072-4, 1972., FOR ORIGINAL SEE TPRC NO. 69830 ) |
| 69832 | **A PHYSICOCHEMICAL STUDY OF THE ZIRCONIUM OXIDE CHLORIDE-CADMIUM CHLORIDE-WATER SYSTEM AT 25 C.**<br>BELYAEV, I. N. LOBAS, L. M.<br>EFIMUSHKINA, L. I.<br>ZH. NEORG. KHIM.<br>17 ( 11 ), 3075- , 1972.<br>( FOR ENGLISH TRANSLATION SEE TPRC NO. 69833 ) |
| 69833 | **A PHYSICOCHEMICAL STUDY OF THE ZIRCONIUM OXIDE CHLORIDE-CADMIUM CHLORIDE-WATER SYSTEM AT 25 C.**<br>BELYAEV, I. N. LOBAS, L. M.<br>EFIMUSHKINA, L. I.<br>RUSS. J. INORG. CHEM.<br>17 ( 11 ), 1617-9, 1972.<br>( ENGLISH TRANSLATION OF ZH. NEORG. KHIM., 17 ( 11 ), 3075- , 1972., FOR ORIGINAL SEE TPRC NO. 69832 ) |
| 69851 | **PHYSICO-CHEMICAL PROPERTIES AND THERMODYNAMICS OF ACTIVATION OF THE SYSTEMS SODIUM HYDROXIDE - WATER AND SODIUM HYDROXIDE - WATER - ALUMINA.**<br>RIEDL, I.<br>BAUXITE, ALUMINA, ALUM., PROC. INT. SYMP. ICSOBA, 2ND<br>3, 77-88, 1971. |
| 69860 | **ELOCITY AND ABSORPTION OF ULTRASOUND IN ACETONE-CYCLOHEXANE BINARY MIXTURES.**<br>TURABAEV, F.<br>DOKL. AKAD. NAUK TADZH. SSR<br>15 ( 10 ), 20-3, 1972. |
| 69886 | **SHEAR VISCOSITY OF A BINARY MIXTURE OF SIMPLE LIQUIDS.**<br>BRUK-LEVINSON, E. T. VIKHRENKO, V. S.<br>NEMTSOV, V. B. ROTT, L. A.<br>SOV. PHYS. J., USSR<br>13 ( 2 ), 194-8, 1970.<br>( ENGLISH TRANSLATION OF IZV. VYSSH. UCHEB. ZAVED., FIZ., 13 ( 2 ), 70-5, 1970., FOR ORIGINAL SEE TPRC NO. 58511 ) |
| 69908 | **HEAT CAPACITY OF LIQUID CARBON DISULFIDE + ACETONITRILE, ESPECIALLY NEAR THE UPPER CRITICAL SOLUTION TEMPERATURE.**<br>GOVINDARAJAN, K. VISWANATHAN, B.<br>GOPAL, E. S. R.<br>J. CHEM. THERMODYN.<br>5 ( 1 ), 73-81, 1973. |
| 69943 | **OPTICAL AND PHOTOELECTRICAL PROPERTIES AND BAND STRUCTURE OF SINGLE CRYSTALS OF SOLID SOLUTIONS OF THE SYSTEM ( CADMIUM SULFIDE ) - ( INDIUM SULFIDE ).**<br>RADAUTSAN, S. I. SYRBU, N. N.<br>TEZLEVAN, V. E. SHERBAN, K. F.<br>STRUMBAN, E. E.<br>PHYS. STATUS SOLIDI<br>15 A ( 1 ), 295-302, 1973. |
| 69961 | **VISCOSITY OF NON-NEWTONIAN LIQUIDS.**<br>AVANESYAN, V. G.<br>PROM. ARM.<br>( 3 ), 19-23, 1971. |
| 69996 | **ELECTRICAL CONDUCTIVITY AND VISCOSITY OF HIGHLY CONCENTRATED SOLUTIONS OF IRON CHLORIDES.**<br>SHVETSOV, A. N. MELKOVA, I. M.<br>ZH. PRIKL. KHIM. ( LENINGRAD )<br>45 ( 11 ), 2551-3, 1972.<br>( FOR ENGLISH TRANSLATION SEE TPRC NO. 70683 ) |
| 70000 | **SPECIFIC HEAT OF SOLUTIONS OF ELECTROLYTES AND THEIR MIXTURES. III. EXPERIMENTAL DETERMINATION OF SPECIFIC HEATS OF SODIUM CARBONATE, SODIUM BICARBONATE, AND AMMONIUM BICARBONATE AQUEOUS SOLUTIONS AND SODA PRODUCTION LIQUIDS AT 35 AND 50 DEGREES.**<br>CHERNEN'KAYA, E. I. BRATASH, E. G.<br>ZH. PRIKL. KHIM. ( LENINGRAD )<br>45 ( 10 ), 2218-24, 1972.<br>( FOR ENGLISH TRANSLATION SEE TPRC NO. 70682 ) |
| 70014 | **PRINCIPAL THERMOPHYSICAL PROPERTIES OF BINARY MIXTURES OF HYDROGEN AND OXYGEN WITH WATER VAPOR.**<br>MALASHENKO, V. I. MOROZENKOV, YU. M.<br>STAN'KOV, V. KH.<br>HEAT TRANSFER-SOV. RES.<br>5 ( 1 ), 37-45, 1973. |

| TPRC Number | Bibliographic Citation |
|---|---|

70027 **RAMJET ENGINE FUEL INFECTION STUDIES.**
HOJNACKI, J. T.
AIR FORCE AERO PROPULSION LAB., WPAFB, OHIO
42PP., 1972.
( AD-754 852, AFAPL-TR-72-76, AVAIL. NTIS )

70041 **EFFECT OF THE DIELECTRIC CONSTANT AND NATURE OF METHANOL-CARBON TETRACHLORIDE AND METHANOL-BENZENE MIXED SOLVENTS ON THE VISCOSITY AND ELECTRICAL CONDUCTIVITY OF TETRABUTYLAMMONIUM BROMIDE.**
GORENBEIN, E. YA. KARPOVICH, A. I.
IZV. VYSSH. UCHEB. ZAVED., KHIM. KHIM. TEKHNOL.
15 ( 12 ), 1808-12, 1972.

70048 **PHYSICOCHEMICAL INVESTIGATIONS OF AQUEOUS TERNARY ELECTROLYTE SOLUTIONS. VII. SOLUBILITY OF MAGNESIUM CHLORIDE AND THE VISCOSITY AND DENSITY OF THE MAGNESIUM CHLORIDE-HYDROCHLORIC ACID-WATER SYSTEM.**
BERECZ, E. BADER, I.
MAGY. KEM. FOLY.
79 ( 1 ), 31-45, 1973.

70101 **VISCOSITY OF MELTS OF THE CALCIUM NITRATE - LITHIUM NITRATE, CALCIUM NITRATE - SODIUM NITRATE, AND CALCIUM NITRATE - POTASSIUM NITRATE SYSTEMS.**
SMOTRAKOV, V. G. POPOVSKAYA, N. P.
TERESHCHENKO, V. A.
ZH. PRIKL. KHIM. ( LENINGRAD )
45 ( 12 ), 2627-30, 1972.
( FOR ENGLISH TRANSLATION SEE TPRC NO. 71431 )

70102 **DENSITY, VISCOSITY, AND ELECTRICAL CONDUCTIVITY OF MONOAMMONIUM PHOSPHATE AQUEOUS SOLUTIONS.**
SARBAEV, A. N. POLYAKOV, E. V.
RUCHKOVA, A. KH.
ZH. PRIKL. KHIM. ( LENINGRAD )
45 ( 12 ), 2749-51, 1972.
( FOR ENGLISH TRANSLATION SEE T71433 )

70107 **ULTRASONIC STUDY OF VOLUME VISCOSITY AND STRUCTURE OF BINARY MIXTURES OF WATER AND HEAVY WATER.**
KOR, S. K. SINGH, B. K. TANDON, U. S.
INDIAN J. PURE APPL. PHYS.
10 ( 10 ), 742-3, 1972.

70125 **ENGINEERING PROPERTY DATA ON ROCKET PROPELLANTS.**
CONSTANTINE, M. T. YOUEL, K. J.
HAMILTON, J. RODRIGUEZ, S. E.
LECCE, J. V. QUAGLINE, J.
AIR FORCE ROCKET PROPULSION LAB., EDWARDS AIR FORCE BASE, CALIF.
53PP., 1966.
( AD-376 814, AFRPL-TR-22-289, AVAIL. AFRPL )

70128 **FINAL REPORT, ENGINEERING PROPERTIES OF ROCKET PROPELLANTS.**
CONSTANTINE, M. T.
ROCKETDYNE RESEARCH DEPT., CANOGA PARK, CALIF.
132PP., 1966.
( AD-373 662, AFRPL-TR-66-122, AVAIL. DDC )

70129 **QUARTERLY PROGRESS REPORT, ENGINEERING PROPERTIES OF ROCKET PROPELLANTS, 1 OCT. 1965-DEC. 1965.**
CONSTANTINE, M. T. ROCK, A. H.
HAMILTON, J. V. RODRIGUEZ, S. E.
LECCE, J. V. QUAGLINO, J. SERIC, M. J.
ROCKETDYNE RESEARCH DEPT., CANOGA PARK, CALIF.
23PP., 1966.
( AD-368 978, R-6248-3, AVAIL. DDC )

70136 **THERMODYNAMIC PROPERTIES OF A LEAN NATURAL GAS AT CRYOGENIC CONDITIONS.**
GREGORY, D. P. DJORDJEVICH, L.
KAO, R. ANDERSON, G. L.
INSTITUTE OF GAS TECHNOLOGY ( AMERICAN GAS ASSOCIATION, INC. )
83PP., 1971.
( IGT-TR-11 AVAIL. AMERICAN GAS ASSO., INC. )

70137 **HANDBOOK OF SPECIFIC HEAT, VISCOSITY, AND THERMAL CONDUCTIVITY FOR DCK PROPELLANT GAS IN THE ROCKET CHAMBER, THROAT AND EXIT, ( U. ).**
CHAO, G. T. Y.
ALLEGANY BALLISTICS LAB., CUMBERLAND, MD.
77PP., 1964.
( AD-348843, ABL/X-112 )

70160 **THERMAL DIFFUSIVITY MEASURING TECHNIQUE FOR HAZARDOUS MATERIALS.**
MURRAY, R. C. COOPER, T. E.
J. APPL. PHYS.
44 ( 4 ), 1420-5, 1973.

70167 **SPECIFIC HEAT CAPACITY AT CONSTANT PRESSURE OF WATER SUBSTANCE.. ICE, WATER AND STEAM, SOLID LIQUID AND GASEOUS HEAVY WATER, AND SEA WATER.**
ENGINEERING SCIENCES DATA UNIT.
ENGINEERING SCIENCES DATA, LONDON ( ENGLAND )
36PP., 1968.
( N73-10957, ESDU-68008, AVAIL. ESDU )

70168 **DYNAMIC VISCOSITY OF WATER SUBSTANCE.. WATER AND STEAM, LIQUID AND GASEOUS HEAVY WATER, AND SEA WATER.**
ENGINEERING SCIENCES DATA UNIT.
ENGINEERING SCIENCES DATA, LONDON ( ENGLAND )
45PP., 1968.
( N73-10303, ESDU-68009, AVAIL. ESDU )

70169 **THERMAL CONDUCTIVITY OF LIQUID HALOGENATED ALIPHATIC HYDROCARBONS INCLUDING COMMERCIAL REFRIGERANTS.**
ENGINEERING SCIENCES DATA UNIT.
ENGINEERING SCIENCES DATA, LONDON ( ENGLAND )
52PP., 1971.
( N73-10956, ESDU-68024, AVAIL. ESDU )

70184 **TRANSPORT PROPERTIES AND SECOND VIRIAL COEFFICIENT OF NONPOLAR-POLAR GAS MIXTURES.**
RAKSHIT, A. B. BHATTACHARYYA, S. K.
BARUA, A. K.
J. PHYS., B, AT. MOL. PHYS.
5 ( 10 ), 1950-8, 1972.

70212 **VISCOSITY FUNCTIONS OF BINARY SOLUTIONS OF NITROPARAFFINS IN CARBON TETRACHLORIDE.**
MUSULIN, B.
CHEM. SCR.
3 ( 2 ), 91-3, 1973.

70272 **RELATION OF TRANSFER COEFFICIENTS OF GAS MIXTURES TO COMPONENT CONCENTRATION.**
SHASHKOV, A. G. ABRAMENKO, T. N.
TEPLO-MASSOPERENOS, DOKL. VSES. SOVESHCH.
7, 4TH, 109-18, 1972.

70280 **VISCOSITY OF POTASSIUM NITRATE - DIVALENT METAL NITRATE SYSTEM MELTS ( METAL IS CALCIUM, STRONTIUM, OR BARIUM ).**
PROTSENKO, P. I. SMOTRAKOV, V. G.
POPOVSKAYA, N. P.
ZH. PRIKL. KHIM. ( LENINGRAD )
45 ( 12 ), 2745-7, 1972.
( FOR ENGLISH TRANSLATION SEE TPRC NO. 71432 )

70286 **EFFECT OF ION VISCOSITY AND THERMAL CONDUCTIVITY ON THE DRIFT INSTABILITY IN AN INHOMOGENEOUS HIGH-PRESSURE COLLISIONAL PLASMA.**
NEMOV, V. V.
NUCL. FUSION
12, 643-8, 1972.

70289 **ELECTRICAL AND THERMAL CONDUCTIVITIES OF A RELATIVISTIC DEGENERATE PLASMA.**
KELLY, D. C.
ASTROPHYS. J.
( 179 ), ( PT. 1 ), 599-606, 1973.

70290 **HEAT CONDUCTIVITY, PLASMA INSTABILITIES, AND RADIO STAR SCINTILLATIONS IN THE SOLAR WIND.**
PERKINS, F.
ASTROPHYSICAL JOURNAL
179, ( PT. 1 ), 637-642, 1973.
( AD 761 709 AFOSR-73-1015 )

70292 **THERMAL CONDUCTIVITY IN SOLAR MAGNETO PLASMAS.**
YUN, H. S. WYLLER, A. A.
SOL. PHYS.
27, 44-60, 1972.

70303 **COMPOSITION OF THE VENUS CLOUD TOPS IN LIGHT OF RECENT SPECTROSCOPIC DATA.**
LEWIS, J. S.
ASTROPHYS. J.
171 ( PT. 2 ), L75-9, 1972.

70322 **PROPERTIES OF SILVER-CONTAINING SULFIDE GLASSES.**
KAWAMOTO, Y. NAGURA, N.
TSUCHIHASHI, S.
J. AMER. CERAM. SOC.
56 ( 5 ), 289-90, 1973.

70328 **PYROGRAPHALLOY RESEARCH AND DEVELOPMENT.**
PAPPIS, J. BRODERICK, R. CAPRIULO, A.
HAGEN, L.
RAYTHEON CO., WALTHAM, MASS.
59PP., 1963.
( AD-350 705, S618, AVAIL. DDC )

70372 **HIGH-TEMPERATURE THERMAL LATTICE EXPANSION OF MIXED CARBONITRIDE AND URANIUM CARBONITRIDE.**
IHARA, S. SUZUKI, M. AKIMOTO, Y.
J. NUCL. MATER.
39 ( 3 ), 311-4, 1971.

70392 **THERMOPHYSICAL PROPERTIES OF HIGH TEMPERATURE INSULATORS BASED ON NONMETALLIC HIGH MELTING COMPOUNDS.**
SAMSONOV, G. V. ANDREEVA, T. V.
DUBOVIK, T. V.
HIGH TEMP.-HIGH PRESSURES
4 ( 5 ), 537-9, 1972.

70408 **VISCOSIMETRIC AND DIELECTRIC MEASUREMENTS ON HEXAMETHYLPHOSPHOROTRIAMIDE FREE OF PEROXIDE.**
GAL, J. Y. MOLITON-BOUCHETOUT, C.
BULL. SOC. CHIM. FR.
( 2 ) ( PT. 1 ), 464-6, 1973.

| TPRC Number | Bibliographic Citation |
|---|---|
| 70463 | **THERMAL DIFFUSIVITY AND CONDUCTIVITY OF URANIUM CARBIDE UP TO 2200 K.**<br>DE CONINCK, R. VAN LIERDE, W. GIJS, A.<br>J. NUCL. MATER.<br>46 ( 2 ), 213–16, 1973. |
| 70475 | **RADIATIVE TRANSFER THROUGH CARBON ABLATION LAYERS.**<br>NELSON, H. F.<br>J. QUANT. SPECTROSC. RADIAT. TRANSFER<br>13 ( 5 ), 427–45, 1973. |
| 70476 | **THERMODYNAMIC PROPERTIES AND VISCOSITIES OF AQUEOUS 1,1 PRIME-DIMETHYL-4,4 PRIME-DIPYRIDINIUM DICHLORIDE SOLUTIONS AT 25.**<br>PERRON, G. DESNOYERS, J. E.<br>J. SOLUTION CHEM.<br>1 ( 6 ), 537–44, 1972. |
| 70492 | **ANALYSIS OF PHYSICOCHEMICAL CHARACTERISTICS AND ACTIVATION THERMODYNAMICAL FUNCTIONS OF THE SODIUM HYDROXIDE - WATER AND SODIUM HYDROXIDE - WATER - ALUMINUM OXIDE SYSTEMS.**<br>RIEDL, I.<br>NEHEZIP. MUSZ. EGYET., MISKOLC, IDEGENNYELVU KOZLEM.<br>32, 217–31, 1972. |
| 70515 | **REFLECTION OF LIGHT AND THICKNESS OF THE SURFACE LAYER NEAR THE CRITICAL POINT.**<br>PSHENITSYN, V. I. RUSANOV, A. I.<br>POVERKH. YAVLENIYA ZHIDK. ZHIDK. RASTVORAKH<br>( 1 ), 150–1, 1972. |
| 70522 | **COMPARISON OF DIFFERENTIAL THERMAL ANALYSIS AND CALORIMETRY MEASUREMENTS FOR THE SYSTEM LITHIUM CHLORIDE-POTASSIUM CHLORIDE-CALCIUM CHROMATE.**<br>CLARK, R. P.<br>THERM. ANAL., PROC. INT. CONF., 3RD<br>2, 713–24, 1972.<br>( EDITED BY WIEDEMANN, HANS G. BIRKHAEUSER: BASEL, SWITZ. ) |
| 70524 | **DETERMINATION OF VISCOSITY IN WATERS WITH HIGH TOTAL MINERAL CONTENT FROM CHEMICAL COMPOSITION.**<br>MOROZOV, V. P.<br>RAZVED. OKHR. NEDR<br>( 12 ), 45–7, 1972. |
| 70583 | **NATURE OF THE NEGATIVE VISCOSITY OF AQUEOUS AND NONAQUEOUS ELECTROLYTE SOLUTIONS.**<br>ORISHCHENKO, A. V.<br>ZH. FIZ. KHIM.<br>47 ( 2 ), 477, 1973.<br>( FOR ENGLISH TRANSLATION SEE TPRC NO. 72340 ) |
| 70588 | **SPECIFIC HEAT OF AQUEOUS SOLUTIONS OF SODIUM AND POTASSIUM NITRATES IN THE 25–340 DEGREES C RANGE.**<br>PUCHKOV, L. V. MATVEEVA, R. D.<br>BARANOVA, T. L.<br>ZH. PRIKL. KHIM. ( LENINGRAD )<br>46 ( 2 ), 443–5, 1973.<br>( FOR ENGLISH TRANSLATION SEE TPRC NO. 72553 ) |
| 70589 | **THERMAL CONDUCTIVITY OF THE ELECTRON COMPONENT OF PLASMA ACROSS A MAGNETIC FIELD.**<br>ZHILINSKII, A. P. LIVENTSEVA, I. F.<br>ZH. TEKH. FIZ.<br>42 ( 12 ), 2616–18, 1972.<br>( FOR ENGLISH TRANSLATION SEE TPRC NO. 72302 ) |
| 70593 | **EMISSIVITY OF HIGH TEMPERATURE AIR. INTERIM REPORT.**<br>KECK, J. KIVEL, B. WENTINK, T., JR.<br>AVCO EVERETT RESEARCH LAB., EVERETT, MASS.<br>115PP., 1956. |
| 70603 | **CALCULATION OF VISCOSITIES OF AQUEOUS ELECTROLYTE SOLUTIONS.**<br>PENKINA, N. V.<br>J. APPL. CHEM., USSR<br>45 ( 4 ), 897, 1972.<br>( ENGLISH TRANSLATION OF ZH. PRIKL. KHIM., 45 ( 4 ), 880, 1972; FOR ORIGINAL SEE TPRC NO. 68030 ) |
| 70604 | **NEGATIVE VISCOSITY OF AQUEOUS ELECTROLYTE SOLUTIONS.**<br>PENKINA, N. V. KHRAMTSOVA, T. S.<br>J. APPL. CHEM., USSR<br>45 ( 4 ), 898, 1972.<br>( ENGLISH TRANSLATION OF ZH. PRIKL. KHIM., 45 ( 4 ), 881, 1972., FOR ORIGINAL SEE TPRC NO. 68031 ) |
| 70621 | **CALCULATION OF THE TRANSFER COEFFICIENTS OF PLANETARY ATMOSPHERES FORMED BY MIXTURES OF CARBON DIOXIDE AND NITROGEN.**<br>ZDUNKEVICH, M. D. LEONAS, V. B.<br>TEPLOFIZ. VYS. TEMP.<br>10 ( 5 ), 1110–2, 1972.<br>( FOR ENGLISH TRANSLATION SEE TPRC NO. 70622 ) |
| 70622 | **CALCULATION OF THE TRANSFER COEFFICIENTS OF PLANETARY ATMOSPHERES FORMED BY MIXTURES OF CARBON DIOXIDE AND NITROGEN.**<br>ZDUNKEVICH, M. D. LEONAS, V. B.<br>HIGH TEMP.<br>10 ( 5 ), 994–6, 1972.<br>( ENGLISH TRANSLATION OF TEPLOFIZ. VYS. TEMP., 10 ( 5 ), 1110–2, 1972., FOR ORIGINAL SEE TPRC NO. 70621 ) |

| TPRC Number | Bibliographic Citation |
|---|---|
| 70660 | **ENTHALPY AND HEAT CAPACITY OF STEAM-HELIUM AND STEAM-OXYGEN MIXTURES.**<br>LUKER, J. A. GNIEWEK, T. MOSIER, S. A.<br>CHEM. ENG. DATA SER.<br>3 ( 1 ), 11–7, 1958. |
| 70681 | **SALTS OF POLYNITROALKANES IN SOLUTIONS. II. DENSITY AND VISCOSITY OF AQUEOUS SOLUTIONS OF SODIUM NITROETHANE.**<br>PENKINA, N. V. ODOKIENKO, S. S.<br>SELIVANOV, V. F. GIDASPOV, B. V.<br>J. APPL. CHEM., USSR<br>45 ( 10 ), 2455, 1972.<br>( ENGLISH TRANSLATION OF ZH. PRIKL. KHIM., 45 ( 10 ), 2340–1, 1972; FOR ORIGINAL SEE TPRC NO. 69760 ) |
| 70682 | **SPECIFIC HEAT OF SOLUTIONS OF ELECTROLYTES AND THEIR MIXTURES. III. EXPERIMENTAL DETERMINATION OF SPECIFIC HEATS OF SODIUM CARBONATE, SODIUM BICARBONATE, AND AMMONIUM BICARBONATE AQUEOUS SOLUTIONS AND SODA PRODUCTION LIQUIDS AT 35 AND 50 DEGREES.**<br>CHERNEN'KAYA, E. I. BRATASH, E. G.<br>J. APPL. CHEM., USSR<br>45 ( 10 ), 2325–9, 1972.<br>( ENGLISH TRANSLATION OF ZH. PRIKL. KHIM. ( LENINGRAD ), 45 ( 10 ), 2218–24, 1972; FOR ORIGINAL SEE TPRC NO. 70000 ) |
| 70683 | **ELECTRICAL CONDUCTIVITY AND VISCOSITY OF HIGHLY CONCENTRATED SOLUTIONS OF IRON CHLORIDES.**<br>SHVETSOV, A. N. MELKOVA, I. M.<br>J. APPL. CHEM., USSR<br>45 ( 11 ), 2670–2, 1972.<br>( ENGLISH TRANSLATION OF ZH. PRIKL. KHIM., 45 ( 11 ), 2551–3, 1972; FOR ORIGINAL SEE TPRC NO. 69996 ) |
| 70689 | **HEAT CAPACITY OF POLYPHOSPHORIC ACIDS AS A FUNCTION OF STRUCTURE.**<br>SMIRNOVA, E. P. MATVEICHUK, N. L.<br>MIKHAILIN, A. D. ILLARIONOV, V. V.<br>BULL. ACAD. SCI. USSR, CHEM. SER.<br>189 ( 2 ), 916–8, 1969.<br>( ENGLISH TRANSLATION OF DOKL. AKAD. NAUK SSSR, 189 ( 2 ), 344–6, 1969; FOR ORIGINAL SEE TPRC NO. 57532 ) |
| 70692 | **SPECIFIC HEAT OF SOME GLASSY MIXTURES RADIO-POLYMERIZING AT LOW TEMPERATURE.**<br>DWORKIN, A.<br>J. CHIM. PHYS. PHYSICOCHIM. BIOL.<br>67 ( 1 ), 329–34, 1970. |
| 70706 | **MATRIX OPERATOR THEORY OF RADIATIVE TRANSFER. 2: SCATTERING FROM MARITIME HAZE.**<br>KATTAWAR, G. W. PLASS, G. N.<br>CATCHINGS, F. E.<br>APPL. OPT.<br>12 ( 5 ), 1071–84, 1973. |
| 70729 | **THE SOLUTION OF A SOLID BODY AT ITS HORIZONTAL SURFACE, GOVERNED BY FREE CONVECTIVE CELLULAR MOTION.**<br>HRMA, P. BELOHOUBEK, I.<br>INT. J. HEAT MASS TRANSFER<br>16 ( 7 ), 1397–1406, 1973. |
| 70750 | **CONCERNING THE MEASUREMENT OF THE GAS FLOW TEMPERATURE IN SOOT GENERATORS BY MEANS OF A RADIATION PYROMETER.**<br>MALYSHEV, P. A.<br>HEAT TRANSFER-SOV. RES.<br>5 ( 2 ), 90–1, 1973. |
| 70783 | **IRRADIANCE REFLECTIVITY OF A FLAT OCEAN AS A FUNCTION OF ITS OPTICAL PROPERTIES.**<br>GORDON, H. R. BROWN, O. B.<br>APPL. OPT.<br>12 ( 7 ), 1549–51, 1973. |
| 70788 | **EXCESS HEAT CAPACITY OF ATHERMAL ASSOCIATED MIXTURES OF THE MECKE-KEMPTER TYPE.**<br>TRESZCZANOWICZ, A.<br>BULL. ACAD. POL. SCI., SER. SCI. CHIM.<br>21 ( 3 ), 197–202, 1973. |
| 70794 | **VISCOSITY OF GASEOUS MIXTURES.**<br>KESTIN, J.<br>BROWN UNIV., PROVIDENCE, R. I.<br>23PP., 1973.<br>( AD–759 198, AROD–8117:12–E, AVAIL. NTIS ) |
| 70797 | **DIELECTRIC CONSTANT AND VISCOSITY OF PRESSURIZED HELIUM( 3 ) - HELIUM( 4 ) SOLUTIONS NEAR THE POINT.**<br>LAI, C. M. KITCHENS, T. A.<br>BROOKHAVEN NATL. LAB., UPTON, N. Y.<br>11PP., 1972.<br>( BNL-17122, CONF-720833-8, AVAIL. NTIS ) |

| TPRC Number | Bibliographic Citation |
|---|---|

70814 **EXPERIMENTAL DETERMINATION OF THE SPECIFIC HEATS OF AQUEOUS SOLUTIONS OF AMMONIUM BICARBONATE, SODIUM BICARBONATE, SODIUM CARBONATE, AMMONIA, AND LIQUIDS FOR SODA PRODUCTION AT 25 C.**
CHERNEN'KAYA, E. I.
J. APPL. CHEM., USSR
44 ( 7 ), 1562–6, 1971.
( ENGLISH TRANSLATION OF ZH. PRIKL. KHIM., 44 ( 7 ), 1543–9, 1971; FOR ORIGINAL SEE TPRC NO. 64245 )

70815 **EXPERIMENTAL DETERMINATION OF THERMAL CONDUCTIVITIES OF AQUEOUS SOLUTIONS OF SALTS AND AMMONIA AT 25 AND 50 DEGREES.**
CHERNEN'KAYA, E. I. VERNIGORA, G. A.
J. APPL. CHEM., USSR
45 ( 8 ), 1779–82, 1972.
( ENGLISH TRANSLATION OF ZH. PRIKL. KHIM., 45 ( 8 ), 1704–7, 1972; FOR ORIGINAL SEE TPRC NO. 69615 )

70859 **VISCOSITY OF SOLUTIONS OF AMMONIUM BICARBONATE, SODIUM BICARBONATE, AMMONIUM CHLORIDE, AND AMMONIA IN THE 20–100 DEGREES RANGE. IV.**
RASHKOVSKAYA, E. A. BRATASH, E. G.
ZH. PRIKL. KHIM. ( LENINGRAD )
46 ( 2 ), 273–8, 1973.
( FOR ENGLISH TRANSLATION SEE TPRC NO. 71138 )

70878 **HYDROLYSIS OF QUADRIVALENT URANIUM.**
GLEVOB, V. A. KLYGIN, A. E. SMIRNOVA, I. D.
KOLYADA, N. S.
ZH. NEORG. KHIM.
17 ( 12 ), 3312–, 1972.
( FOR ENGLISH TRANSLATION SEE TPRC NO. 70879 )

70879 **HYDROLYSIS OF QUADRIVALENT URANIUM.**
GLEVOB, V. A. KLYGIN, A. E. SMIRNOVA, I. D.
KOLYADA, N. S.
RUSS. J. INORG. KHIM.
17 ( 12 ), 1740–2, 1972.
( ENGLISH TRANSLATION OF ZH. NEORG. KHIM., 17 ( 12 ), 3312–, 1972; FOR ORIGINAL SEE TPRC NO. 70878 )

70901 **DEPOLARIZED RAYLEIGH SCATTERING AND ORIENTATIONAL RELAXATION OF MOLECULES IN SOLUTION. I. BENZENE, TOLUENE, AND PARA-XYLENE.**
ALMS, G. R. BAUER, D. R. BRAUMAN, J. I.
PECORA, R.
J. CHEM. PHYS.
58 ( 12 ), 2270–8, 1973.

70917 **HEAT CAPACITIES OF AQUEOUS SODIUM CHLORIDE, POTASSIUM CHLORIDE, MAGNESIUM CHLORIDE, MAGNESIUM SULFATE, AND SODIUM SULFATE SOLUTIONS BETWEEN 80 AND 200 DEGREES.**
LIKKE, S. BROMLEY, L A.
J. CHEM. ENG. DATA
18 ( 2 ), 189–95, 1973.

70957 **SPECIFIC HEAT MEASUREMENTS OF THE SOLID HYDROGENS AND THEIR MIXTURES.**
ROBERTS, R. J.
STEVENS INST. TECHNOL., HOBOKEN, N. J., PH. D. THESIS
133PP., 1971.
( UM 72–1630, AVAIL. UNIV. MICROFILMS )

70973 **APPARENT AND PARTIAL MOLAL HEAT CAPACITIES OF SOME AQUEOUS RARE EARTH NITRATES AND PERCHLORATES FROM TENTH MOLAL TO SATURATION AT 25 DEGREES.**
BAKER, J. L.
IOWA STATE UNIV, PH.D. THESIS
131PP., 1971.
( UM 72–12,535, AVAIL. UNIV. MICROFILMS )

71002 **CORRECTED THERMAL CONDUCTIVITY OF LIQUIDS BY THE A.C.-HEATED WIRE METHOD ( DOWTHERM A AND ISOPAR M ).**
SIEGEL, J. R. BONILLA, C. F.
PROC. SIXTH SYMP. THERMOPHYSICAL PROPERTIES, AMER. SOC. MECHANICAL ENGINEERS, NEW YORK, N. Y.
86–7, 1973.

71033 **A GENERALIZATION OF COMPOSITION AND MEAN THERMAL CONDUCTIVITY VALUES FOR NONEQUILIBRIUM MIXTURES OF DINITROPE TETROXIDE WITH NITROGEN DIOXIDE IN EQUILIBRIUM WITH NITRIC OXIDE AND OXYGEN.**
MISHINA, L. V. SEREBRYANYY, G. Z.
IZV. AKAD. NAUK BELORUS. SSR, SER. FIZ. ENERG. NAUK
( 2 ), 85–90, 1972.

71034 **EXPERIMENTAL STUDY OF ISOBARIC THERMAL CAPACITY OF DISSOCIATING NITROGEN TETROXIDE AT SUPERCRITICAL PRESSURES.**
NESTERENKO, V. B. ILYUKHIN, YU. D.
IZV. AKAD. NAUK BELORUS. SSR, SER. FIZ. ENERG. NAUK
( 3 ), 89–94, 1972.

71050 **VISCOSITIES OF ANILINE CHLORIDE SOLUTIONS IN ISOPROPANOL-WATER MIXTURES.**
CHO, B.-R. LEE, Y.-J. LEE, I.
DAEHAN HWAHAK HWOEJEE
17 ( 1 ), 10–14, 1973.

71082 **VISCOSITY OF CITRIC ACID SOLUTIONS.**
AVERBUKH, D. A. METKIN, V. P.
MAKSMENYUK, M. N.
KHLEBOPEK. KONDITER. PROM.
( 3 ), 21–3, 1973.

71090 **VISCOSITY OF OXYGEN AND OF SOME OF ITS MIXTURES WITH OTHER GASES.**
HELLEMANS, J. M. KESTIN, J. RO, S. T.
PHYSICA ( UTRECHT )
65 ( 2 ), 362–75, 1973.

71112 **THERMODYNAMIC PROPERTIES OF GASEOUS CARBON DIOXIDE OF TECHNICAL PURITY.**
ALTUNIN, V. V. KUZNETSOV, D. O.
BONDARENKO, V. F.
TEPLOENERGETIKA
( 4 ), 61–3, 1973.
( FOR ENGLISH TRANSLATION SEE TPRC NO. 74224 )

71119 **SPECIAL BENEFICATION METHODS. 1. PHYSICAL-CHEMICAL CHARACTERISTICS OF MAGNETIC SALTS AND LIQUIDS.**
BUNIN, G. M.
TSENTROBEZH.-GRAVITATSIONNOE OBOGASHCH. UGLEI
109–15, 1972.

71132 **THE INFRARED MATRIX ISOLATION SPECTRA AND NORMAL COORDINATE ANALYSIS OF LITHIUM BERYLLIUM FLUORIDES. ( FROM THERMODYNAMIC PROPERTIES OF ROCKET COMBUSTION PRODUCTS BY THE MATRIX ISOLATION TECHNIQUE ).**
CYVIN, S. J. SNELSON, A.
IIT RESEARCH INSTITUTE, CHICAGO, ILLINOIS
14–24A, 1973.
( IITRI–C6238–11 )

71138 **VISCOSITY OF SOLUTIONS OF AMMONIUM BICARBONATE, SODIUM BICARBONATE, AMMONIUM CHLORIDE, AND AMMONIA IN THE 20–100 DEGREE RANGE. IV.**
RASHKOVSKAYA, E. A. BRATASH, E. G.
J. APPL. CHEM., USSR
46 ( 2 ), 288–92, 1973.
( ENGLISH TRANSLATION OF ZH. PRIKL. KHIM. 46 ( 2 ), 273–8, 1973; FOR ORIGINAL SEE TPRC NO. 70859 )

71151 **SHEAR DEPENDENCE OF THE VISCOSITIES OF BOTH NITROETHANE - 3-METHYLPENTANE AND POLYSTYRENE - CYCLOHEXANE MIXTURES IN THE CRITICAL REGION.**
TSAI, B. C. MYERS, T. M. MC INTYRE, D.
J. CHEM. PHYS.
58 ( 10 ), 4582–5, 1973.

71182 **TRANSPORT COEFFICIENTS OF IONIZED ARGON.**
DEVOTO, R. S.
PHYS. FLUIDS
16 ( 5 ), 616–23, 1973.

71210 **A VISCOMETER FOR STUDYING STARCH SLURRY BEHAVIOR DURING COOKING.**
VOISEY, P. W. MURRAY, R. KEIGHTLY, G.
CAN. INST. FOOD TECHNOL. J.
5 ( 3 ), 129–133, 1972.

71214 **DIFFERENCES IN THE PROPERTIES OF STARCH OF BURSTING MAIZE.**
SOROCHINSKAYA, E. N. BABICHENKO, L. V.
IZV. VYSSH. UCHEB. ZAVED. PISHCH. TEKHNOL.
( 1 ), 61–64, 1970.

71225 **MIXTURES OF LIQUID SUGAR AND STARCH HYDROLYSATES AND THEIR USE.**
KRUGER, C. STEINMETZER, W.
ZUCKER
25 ( 11 ), 356–60, ( 12 ), 396–401, 1972.

71230 **THE SYSTEM POTASSIUM SULFATE-CESIUM SULFATE.**
LEVIN, E. M. BENEDICT, J. T.
SCIARELLO, J. P. MONSOUR, S.
J. AMER. CERAM. SOC.
56 ( 8 ), 427–30, 1973.

71315 **DIFFUSION AND VISCOSITY IN AQUEOUS MEDIA. BEHAVIOR OF CALCIUM( II ) AND CERIUM( III ) IONS AND THEIR COMPLEXES WITH EDTA AND DCTA.**
HURTADO, G. G. MATEO, P. L. SERNA, A.
VIDAL-ABARCA, J. B.
AN. QUIM.
69 ( 3 ), 295–303, 1973.

71356 **EFFECT OF LONGITUDINAL THERMAL CONDUCTIVITY ON THE OHMIC HEATING OF PLASMA IN THE TUMAN-1 APPARATUS.**
KAGANSKII, M. G. KALMUKOV, S. G.
SHAKHOVETS, K. G.
ZH. TEKH. FIZ.
43 ( 2 ), 302–9, 1973.
( FOR ENGLISH TRANSLATION SEE TPRC NO. 73481 )

71359 **EFFECT OF MICROSTRUCTURE ON STRENGTH OF SILICON NITRIDE - SILICON CARBIDE.**
LANGE, F. F.
J. AM. CERAM. SOC.
56 ( 9 ), 445–50, 1973.

71361 **NEW TECHNIQUES FOR FAR-INFRARED FILTERS.**
ARMSTRONG, K. R. LOW, F. J.
APPL. OPT.
12 ( 9 ), 2007–9, 1973.

| TPRC Number | Bibliographic Citation |
|---|---|
| 71386 | **VISCOSITY OF GAS MIXTURES AND AN EXTENSION OF THE LAW OF CORRESPONDING STATES FOR SIMPLE GASES.**<br>RO, S. T.<br>BROWN UNIV., PROVIDENCE, R. I., PH.D. THESIS<br>191PP., 1972.<br>( UM 73-2325, AVAIL. UNIV. MICROFILMS ) |
| 71388 | **EFFECT OF PRESSURE ON THE CONSTANT-PRESSURE HEAT CAPACITY OF THE TRIFLUOROMETHANE-TETRAFLUOROMETHANE AND THE NITROGEN-TRIFLUOROMETHANE-TETRAFLUOROMETHANE SYSTEMS.**<br>RODRIGUEZ, F. A. M.<br>LEHIGH UNIVERSITY, PH.D. THESIS<br>163PP., 1972.<br>( UM 73-4278, AVAIL. UNIV. MICROFILMS ) |
| 71401 | **THERMAL CONDUCTIVITY, ELECTRICAL RESISTIVITY, AND THERMOELECTRIC POWER OF THORIUM NITRIDE-URANIUM NITRIDE ALLOYS.**<br>WEAVER, S. C.<br>UNIV. TENNESSEE, KNOXVILLE, TENN., PH.D. THESIS<br>217PP. 1972.<br>( UM 72-21,368, AVAIL. UNIV. MICROFILMS ) |
| 71404 | **TRANSPORT COEFFICIENTS OF MULTICOMPONENT POLAR GAS MIXTURES.**<br>NASER, P.<br>ST. LOUIS UNIV., PH.D. THESIS<br>125PP., 1971.<br>( UM 72-23,980, AVAIL. UNIV. MICROFILMS ) |
| 71409 | **VISCOSITY OF METHANE - HYDROGEN SYSTEM FROM MINUS 100 TO 0 DEGREES AT PRESSURES UP TO 500 ATMOSPHERES.**<br>CHUANG, S. Y.<br>RICE UNIVERSITY, PH.D. THESIS<br>221PP., 1972.<br>( UM 72-26,400, AVAIL. UNIV. MICROFILMS ) |
| 71413 | **SQUARE-PLANAR TETRACYANO D(8) COMPLEXES OF NICKEL( II ), PALLADIUM( II ), AND PLATINUM( II ). POLARIZED SINGLE-CRYSTAL REFLECTION SPECTRA AND THE ELECTRONIC STRUCTURE IN SOLUTION AND SOLID STATE.**<br>MUSSELMAN, R. L.<br>NEW MEXICO STATE UNIV., PH.D. THESIS<br>187PP., 1972.<br>( UM 72-28,316, AVAIL. UNIV. MICROFILMS ) |
| 71431 | **VISCOSITY OF MELTS OF THE CALCIUM NITRATE - LITHIUM NITRATE, CALCIUM NITRATE - SODIUM NITRATE, AND CALCIUM NITRATE - POTASSIUM NITRATE SYSTEMS.**<br>SMOTRAKOV, V. G. POPOVSKAYA, N. P.<br>TERESHCHENKO, V. A.<br>J. APPL. CHEM., USSR<br>45 ( 12 ), 2761-4, 1972.<br>( ENGLISH TRANSLATION OF ZH. PRIKL. KHIM., 45 ( 12 ), 2627-30, 1972; FOR ORIGINAL SEE TPRC NO. 70101 ) |
| 71432 | **VISCOSITIES OF MELTS IN THE SYSTEMS POTASSIUM NITRATE ALKALINE EARTH METAL NITRATE.**<br>PROTSENKO, P. I. SMOTRAKOV, V. G.<br>POPOVSKAYA, N. P.<br>J. APPL. CHEM., USSR<br>45 ( 12 ), 2880-2, 1972.<br>( ENGLISH TRANSLATION OF ZH. PRIKL. KHIM., 45 ( 12 ), 2745-7, 1972; FOR ORIGINAL SEE TPRC NO. 70280 ) |
| 71433 | **DENSITY, VISCOSITY AND ELECTRICAL CONDUCTIVITY OF AQUEOUS MONOAMMONIUM PHOSPHATE SOLUTIONS.**<br>SARBAEV, A. N. POLYAKOV, E. V.<br>RUCHKOVA, A. KH.<br>J. APPL. CHEM., USSR<br>45 ( 12 ), 2885-7, 1972.<br>( ENGLISH TRANSLATION OF ZH. PRIKL. KHIM., 45 ( 12 ), 2749-51, 1972; FOR ORIGINAL SEE TPRC NO. 70102 ) |
| 71442 | **VISCOSITY BEHAVIOR OF SOLUTIONS OF SODIUM TETRAPHENYLBORON AND ITS GLYME COMPLEXES IN ETHEREAL SOLVENTS.**<br>SMID, J. GROTENS, A. M.<br>J. PHYS. CHEM.<br>77 ( 20 ), 2377-82, 1973. |
| 71477 | **APPARATUS FOR SPECIFIC HEAT MEASUREMENTS BETWEEN 1.5 AND 310 K.**<br>DANAN, J. CONTE, R. R. DE NOVION, C. H.<br>BULL. INFORM. SCI. TECH., COMMIS. ENERG. AT. ( FR. )<br>( 180 ), 29-32, 1973. |
| 71555 | **HEAT CAPACITY OF SEA WATER SOLUTIONS FROM 5 TO 35 AND 0.5 TO 22 PERCENT CHLORINITY.**<br>MILLERO, F. J. PERRON, G. DESNOYERS, J. E.<br>J. GEOPHYS. RES.<br>78 ( 21 ), 4499-507, 1973. |
| 71577 | **DENSITY AND VISCOSITY OF ALUMINUM AND POTASSIUM SULFATE SOLUTIONS.**<br>LIKHODED, A. D. ZAPOL'SKII, A. K.<br>SAZHIN, V. S.<br>KHIM. TEKHNOL. ( KIEV )<br>( 2 ), 55-7, 1973. |
| 71590 | **RECENT RESULTS IN THE INVESTIGATIONS OF THE STRUCTURE OF MURIATIC ACID-METAL CHLORIDE SOLUTIONS.**<br>BERECZ, E.<br>NEHEZIP. MUSZ. EGYET., MISKOLC, IDEGENNYELVU KOZLEM.<br>32, 131-43, 1972. |
| 71594 | **HEAT CONDUCTION IN BINARY GAS MIXTURES BETWEEN CONCENTRIC CYLINDERS.**<br>YEH, B. T. FROHN, A.<br>PHYS. FLUIDS<br>16 ( 6 ), 801-5, 1973. |
| 71641 | **THERMAL CONDUCTIVITY AND SPECIFIC ELECTRICAL RESISTANCE OF SOME SILVER-BASED CONTACT CERMET COMPOSITIONS.**<br>DEMIDCHENKO, V. I. ELEMA, V. A.<br>MELASHENKO, I. P. SERYKH, G. M.<br>ELEKTROTEKHNIKA<br>( 10 ), 44-6, 1972. |
| 71643 | **THERMODYNAMIC AND TRANSPORT PROPERTIES OF AIR AND CARBON OXIDE.**<br>PIRIE, M. A. M.<br>CENTRAL ELECTRICITY GENERATING BOARD, BERKELEY, ( ENGLAND )<br>34PP., 1971.<br>( RD/B/M-1955, AVAIL. NTIS ) |
| 71669 | **VISCOSITY OF AQUEOUS SOLUTIONS OF ELECTROLYTES IN A WIDE TEMPERATURE AND CONCENTRATION RANGE. II. VISCOSITY OF LITHIUM, SODIUM, AND POTASSIUM HYDROXIDE SOLUTIONS AT UP TO 275C.**<br>MASHOVETS, V. P. PUCHKOV, L. V.<br>SARGAEV, P. M. FEDOROV, M. K.<br>ZH. PRIKL. KHIM. ( LENINGRAD )<br>46 ( 5 ), 992-6, 1973.<br>( FOR ENGLISH TRANSLATION SEE TPRC NO. 72555 ) |
| 71672 | **SOLUBILITY, DENSITY, AND VISCOSITY IN THE POLYPHOSPHORIC ACID-POTASSIUM SULFATE-WATER SYSTEM.**<br>BUGAI, P. M. EREMENKO, R. A.<br>ZH. PRIKL. KHIM. ( LENINGRAD )<br>46 ( 6 ), 1372-5, 1973.<br>( FOR ENGLISH TRANSLATION SEE TPRC NO. 72559 ) |
| 71765 | **THERMODYNAMIC PROPERTIES OF LIQUID HELIUM( 3 ) - HELIUM( 4 ) MIXTURES NEAR THE TRICRITICAL POINT DERIVED FROM SPECIFIC HEAT MEASUREMENTS. FROM PROC. INT. CONF. LOW TEMP. PHYS., BOULDER, COLORADO.**<br>ISLANDER, S. T. ZIMMERMANN, W., JR.<br>TECHNICAL UNIV. OF HELSINKI, FINLAND<br>10PP., 1971.<br>( COO-1569-90, AVAIL. NTIS ) |
| 71767 | **MEASURING THE TRANSMITTANCE OF ATMOSPHERIC DUST IN POTASSIUM BROMIDE PELLETS USING A CARBON DIOXIDE LASER.**<br>WHITE, K. O. LINDBERG, J. D.<br>APPL. OPT.<br>12 ( 11 ), 2544-5, 1973. |
| 71776 | **CALCULATED LIQUID PHASE THERMODYNAMIC PROPERTIES AND LIQUID-VAPOR EQUILIBRIA FOR FLUORINE - OXYGEN ( FLOX ) MIXTURES.**<br>PARRISH, W. R. HIZA, M. J.<br>NATIONAL BUREAU OF STANDARDS, BOULDER, COLORADO<br>27PP., 1973.<br>( NBSIR-73-338, AVAIL. NTIS ) |
| 71785 | **VISCOSITY B-COEFFICIENTS OF THE ANIONIC SPECIES OF BENZOIC ACID AND ITS MONOSUBSTITUTED DERIVATIVES.**<br>YASUDA, M. MIZUTANI, K.<br>BULL. CHEM. SOC. JAP.<br>46 ( 7 ), 2226-7, 1973. |
| 71794 | **CRITICAL BEHAVIOR OF SPECIFIC HEAT AND COEXISTENCE CURVE IN THE BINARY LIQUID SYSTEM METHANOL + HEPTANE.**<br>VISWANATHAN, B. GOVINDARAJAN, K.<br>GOPAL, E. S. R.<br>INDIAN J. PURE APPL. PHYS.<br>11 ( 3 ), 157-62, 1973. |
| 71801 | **USE OF EQUATIONS RELATING THE SPECIFIC HEAT CAPACITY OR DENSITY OF ELECTROLYTE SOLUTIONS TO CONCENTRATION.**<br>VASILEV, V. A. FEDYAINOV, N. V.<br>KARAPET'YANTS, M. KH. KARPACHEV, A I.<br>IZV. VYSSH. UCHEB. ZAVED., KHIM. KHIM. TEKHNOL.<br>16 ( 5 ), 689-93, 1973. |
| 71809 | **THERMAL CONDUCTIVITY OF THE SYSTEM NITROGEN TETROXIDE YIELDS 2 NITROGEN DIOXIDE WITH DILUENTS.**<br>BANDYOPADHYAY, P. K.<br>J. PHYS.<br>6 D ( 10 ), 1283-6, 1973. |
| 71842 | **PHYSICAL PROPERTIES OF DIMETHYL SULFOXIDE-PROPYLENE CARBONATE MIXTURES AT 25 DEGREES.**<br>COURTOT-COUPEX, J. MADEC, C.<br>C. R. ACAD. SCI.<br>277 ( 1 ), 15-18, 1973. |
| 71887 | **SPECIFIC HEATS AND DENSITIES OF SODIUM CHLORATE AND SODIUM PERCHLORATE AQUEOUS SOLUTIONS AT 25, 40, AND 60 DEGREES.**<br>LOSEVA, G. K. SEMCHENKO, D. P.<br>ZH. FIZ. KHIM.<br>47 ( 6 ), 1621, 1973.<br>( FOR ENGLISH TRANSLATION SEE T73461 ) |

*TPRC Number* *Bibliographic Citation*

71888 THERMAL CONDUCTIVITY OF SOLUTIONS OF ELECTROLYTES AND THEIR MIXTURES. II. EXPERIMENTAL AND CALCULATED DETERMINATION OF THE THERMAL CONDUCTIVITY OF SOLUTIONS FOR AMMONIA SODA PRODUCTION.
CHERNEKAYA, E. I. VERNIGORA, G. A.
ZH. PRIKL. KHIM. ( LENINGRAD )
46 ( 6 ), 1224–7, 1973.
( FOR ENGLISH TRANSLATION SEE TPRC NO. 72558 )

71943 COMPUTATION OF THE HEAT CONDUCTIVITY OF BINARY GAS MIXTURES.
DULNEV, G. N. ZARICHNYAK, YU. P.
MURATOVA, B. L.
J. ENG. PHYS.
18 ( 5 ), 589–93, 1970.
( ENGLISH TRANSLATION OF INZH.–FIZ. ZH., 18 ( 5 ), 849–55, 1970; FOR ORIGINAL SEE TPRC NO. 59019 )

71949 VISCOSITY OF THE BINARY GAS MIXTURES ARGON – METHANE AND ARGON – AMMONIA.
RAKSHIT, A. B. ROY, C. S. BARUA, A. K.
J. CHEM. PHYS.
59 ( 7 ), 3633–8, 1973.

71983 CALORIMETRIC STUDIES OF SILVER BROMIDE–COPPER ( I ) BROMIDE SOLID SOLUTIONS.
BREMER, F. NOELTING, J.
BER. BUNSENGES. PHYS. CHEM.
77 ( 6 ), 398–402, 1973.

71995 VISCOSITY OF SALT SOLUTIONS SIMULATING NATURAL SOLUTIONS.
EPIFANOVA, S. S. KAPYRINA, YU. D.
KAPTSOVA, T. N. KORCHEMSKAYA, K. M.
MAKAROV, A. V. MAKSAREVA, T. S.
MAKSIMOVA, M. P. FOKEEV, V. M.
IZV. VYSSH. UCHEB. ZAVED., GEOL. RAZVED.
16 ( 5 ), 154–6, 1973.

72020 VISCOSITY OF GASES AND GAS MIXTURES.
GOLUBEV, I. F. PETROV, V. A.
TR. GOS. NAUCH. ISSLED. PROEKT. INST. AZOT. PROM.
( 1 ), 7–30, 1952.

72026 SIMPLE CALCULATION OF THE DIFFUSE REFLECTANCE OF THE OCEAN.
GORDON, H. R.
APPL. OPT.
12 ( 12 ), 2803–4, 1973.

72040 COMPARISON OF THE GROUP SOLUTION MODEL AND THE PRINCIPLE OF CONGRUENCE FOR THE PREDICTION OF LIQUID MIXTURE VISCOSITIES.
WEDLAKE, G. D. RATCLIFF, G. A.
CAN. J. CHEM. ENG.
51 ( 4 ), 511–14, 1973.

72052 VISCOSITY, FREEZING POINT, VAPOR–LIQUID EQUILIBRIUMS, AND OTHER PROPERTIES OF AQUEOUS–TETRAHYDROFURAN SOLUTIONS.
HAYDUK, W. LAUDIE, H. SMITH, O. H.
J. CHEM. ENG. DATA
18 ( 4 ), 373–6, 1973.

72087 DETERMINATION OF THE THERMAL CONDUCTIVITY OF LIQUIDS BY AN INTERPOLATION METHOD.
VOLKENSHTEIN, V. S. MEDVEDEV, N. N.
LEONTEVA, Z. M.
ZAVOD. LAB.
39 ( 6 ), 724, 1973.
( FOR ENGLISH TRANSLATION SEE TPRC NO. 63558 )

72144 VISCOSITY OF GAS MIXTURES AND POLAR–NONPOLAR MOLECULAR INTERACTIONS.
ROY, C. S. RAKSHIT, A. B. BARUA, A. K.
PHYSICA ( UTRECHT )
66 ( 3 ), 442–54, 1973.

72152 RADIATIVE PROPERTIES OF TERRESTRIAL CLOUDS AT VISIBLE AND INFRARED THERMAL WINDOW WAVELENGTHS.
HUNT, G. E.
QUART. J. ROY. METEOROL. SOC.
99 ( 420 ), 346–69, 1973.

72161 THERMAL CONDUCTIVITY OF ORTHOPHOSPHORIC ACID.
BLANKSHTEIN, V. A. LYUBCHENKO, T. V.
AZIEV, R. G. VOL'FKOVICH, S. I.
VESTN. MOSK. UNIV., KHIM.
14 ( 3 ), 365–7, 1973.

72165 THERMAL CONDUCTIVITY OF SODIUM CARBONATE, POTASSIUM CARBONATE, POTASSIUM CHLORIDE, AND POTASSIUM SULFATE SOLUTIONS AT DIFFERENT TEMPERATURES.
CHERNEN'KAYA, E. I. VERNIGORA, G. A.
ZH. PRIKL. KHIM. ( LENINGRAD )
46 ( 7 ), 1603–5, 1973.
( FOR ENGLISH TRANSLATION SEE TPRC NO. 73281 )

72172 THERMAL CONDUCTIVITY OF MIXTURES WITH INTERPENETRATING COMPONENTS.
DULNEV, G. N.
J. ENG. PHYS., USSR
19 ( 3 ), 1195–1206, 1970.
( ENGLISH TRANSLATION OF INZH.–FIZ. ZH., 19 ( 3 ), 562–77, 1970; FOR ORIGINAL SEE TPRC NO. 60734 )

72185 VISCOSITY OF CELLULAR PROTOPLASM.
KEITH, A. D. SNIPES, W.
SCIENCE
183, 666–8, 1974.

72217 PRESSURE AND CONCENTRATION EFFECTS ON THE MOLECULAR REORIENTATION IN WATER–DIOXANE MIXTURES.
LEE, Y. JONAS, J.
J. CHEM. PHYS.
59 ( 9 ), 4845–54, 1973.

72220 DEPOLARIZED RAYLEIGH SCATTERING AND ORIENTATIONAL RELAXATION OF MOLECULES IN SOLUTION. II. CHLOROFORM AND NITROBENZENE.
ALMS, G. R. BAUER, D. R.
BRAUMAN, J. I. PECORA, R.
J. CHEM. PHYS.
59 ( 10 ), 5310–20, 1973.

72221 DEPOLARIZED RAYLEIGH SCATTERING AND ORIENTATIONAL RELAXATION OF MOLECULES IN SOLUTION. III. CARBOXYLIC ACIDS.
ALMS, G. R. BAUER, D. R.
BRAUMAN, J. I. PECORA, R.
J. CHEM. PHYS.
59 ( 10 ), 5321–8, 1973.

72239 ENTHALPY AND SPECIFIC HEAT OF ZIRCONIUM CARBIDE–BASED MATERIALS.
SHEINDLIN, A. E. BELEVICH, I. S.
KOZHEVNIKOV, I. G.
TEPLOFIZ. VYS. TEMP.
11 ( 3 ), 666–8, 1973.
( FOR ENGLISH TRANSLATION SEE TPRC NO. 72240 )

72240 INVESTIGATION OF THE ENTHALPY AND SPECIFIC HEAT OF MATERIALS BASED ON ZIRCONIUM CARBIDE.
SHEINDLIN, A. E. BELEVICH, I. S.
KOZHEVNIKOV, I. G.
HIGH TEMP.
11 ( 3 ), 599–601, 1974.
( ENGLISH TRANSLATION OF TEPLOFIZ. VYS. TEMP., 11 ( 3 ), 666–8, 1973; FOR ORIGINAL SEE TPRC NO. 72239 )

72250 MEASUREMENT OF THE THERMAL CONDUCTIVITY OF A HYDROGEN PLASMA IN A STABILIZED ELECTRIC ARC.
NIZOVSKII, V. L. SHABASHOV, V. I.
TEPLOFIZ. VYS. TEMP.
11 ( 2 ), 251–5, 1973.
( FOR ENGLISH TRANSLATION SEE TPRC NO. 72251 )

72251 MEASUREMENT OF THE THERMAL CONDUCTIVITY OF A HYDROGEN PLASMA IN A STABILIZED ELECTRIC ARC.
NIZOVSKII, V. L. SHABASHOV, V. I.
HIGH TEMP.
11 ( 2 ), 217–21, 1973.
( ENGLISH TRANSLATION OF TEPLOFIZ. VYS. TEMP., 11 ( 2 ), 251–5, 1973; FOR ORIGINAL SEE TPRC NO. 72250 )

72280 HYDROSTATIC LIMITS IN LIQUIDS AND SOLIDS TO 100 KBAR.
PIERMARINI, G. J. BLOCK, S. BARNETT, J. D.
J. APPL. PHYS.
44 ( 12 ), 5377–82, 1973.

72302 TRANSVERSE ELECTRON THERMAL CONDUCTIVITY FOR A PLASMA IN A MAGNETIC FIELD.
ZHILINSKII, A. P. LIVENTSEVA, I. F.
SOV. PHYS. TECH. PHYS.
17 ( 12 ), 2034–6, 1973.
( ENGLISH TRANSLATION OF ZH. TEKH. FIZ., 42 ( 12 ), 2616–8, 1972; FOR ORIGINAL SEE TPRC NO. 70589 )

72308 EXCESS HEAT CAPACITIES OF DILUTE SOLUTIONS OF N–HEXANOL IN N–ALKANES.
KALINOWSKA, B. WOYCICKA, M.
BULL. ACAD. POL. SCI., SER. SCI. CHIM.
21 ( 11 ), 845–8, 1973.

72340 THE NATURE OF THE NEGATIVE VISCOSITY OF AQUEOUS AND NON–AQUEOUS ELECTROLYTE SOLUTIONS.
ORISHCHENKO, A. V.
RUSS. J. PHYS. CHEM.
47 ( 2 ), 272–3, 1973.
( ENGLISH TRANSLATION OF ZH. FIZ. KHIM., 47 ( 2 ), 477, 1973; FOR ORIGINAL SEE TPRC NO. 70583 )

72364 ENGINEERING PROPERTY DATA ON ROCKET PROPELLANTS.
CONSTANTINE, M. T. YOUEL, K. J.
GERHAUSER, J. UNTERBERG, W. RODRIGUEZ, S. E.
ROCKETDYNE, A DIVISION OF NORTH AMERICAN AVIATION, INC., CANOGA PARK, CALIF.
66PP., 1967.
( AD–378 858, AFRPL–TR–67–26, AVAIL. DDC )

72365 ENTHALPIES OF FUSION OF LITHIUM TETRAFLUOROBERYLLATE, LITHIUM TRIFLUOROBERYLLATE, AND SODIUM TETRAFLUOROBERYLLATE AND THE HEAT CAPACITIES OF THE LIQUID MIXTURES.
HOLM, J. L. HOLM, B. J. GRONVOLD, F.
ACTA CHEM. SCAND.
27 ( 6 ), 2035–42, 1973.

| TPRC Number | Bibliographic Citation |
|---|---|
| 72366 | TRANSPORT PROPERTIES OF GASES AND BINARY LIQUIDS NEAR THE CRITICAL STATE.<br>SENGERS, J. V.<br>AIP ( AMER. INST. PHYS. ) CONF. PROC.<br>( 11 ), 229-77, 1973. |
| 72373 | APPARENT MOLAL HEAT CAPACITIES AND VOLUMES OF TETRABUTYLAMMONIUM CARBOXYLATES AND RELATED SOLUTES IN WATER AT 25 DEGREES.<br>LEDUC, P. A. DESNOYERS, J. E.<br>CAN. J. CHEM.<br>51 ( 17 ), 2993-8, 1973. |
| 72399 | PROPERTIES OF AQUEOUS THORIUM NITRATE SOLUTIONS. 1. DENSITIES, VISCOSITIES, CONDUCTIVITIES, P H, SOLUBILITY, AND ACTIVITIES AT FREEZING POINT.<br>APELBLAT, A. AZOULAY, D. SAHR, A.<br>J. CHEM. SOC., FARADAY TRANS.<br>69 1 ( 9 ), 1618-23, 1973. |
| 72406 | EFFECT OF ADDITION OF SODIUM SALTS TO SALICYLIC ACID-CETRIMIDE SYSTEM.<br>WAN, L. S. C.<br>J. PHARM. SCI.<br>62 ( 10 ), 1724-6, 1973. |
| 72409 | LIDAR AND RADIOMETRIC OBSERVATIONS OF CIRRUS CLOUDS.<br>PLATT, C. M. R.<br>J. ATMOS. SCI.<br>30, 1191-1204, 1973. |
| 72430 | ENTHALPY OF DILUTION AND HEAT CAPACITY OF AQUEOUS SOLUTIONS OF TETRABUTYLAMMONIUM BUTYRATE AS A FUNCTION OF TEMPERATURE.<br>LEVINE, A. S. LINDENBAUM, S.<br>J. SOLUTION CHEM.<br>2 ( 5 ), 445-55, 1973. |
| 72445 | VISCOSITY-TEMPERATURE PROPERTIES OF MIXTURES OF ESTERS.<br>SIZOV, N. I. KUKOVITSKII, M. M.<br>BIKKULOV, A. Z. SHARAFUTDINOVA, S. M.<br>NEFTEPERERAB. NEFTEKHIM. ( MOSCOW )<br>( 9 ), 19-22, 1973. |
| 72492 | ISOBARIC HEAT CAPACITY OF ELECTROLYTES.<br>EGOROV, B. N. KILESSO, V. S.<br>SLEPCHENKO, V. I. TRUNOVA, T. K.<br>TEPLOFIZ. SVOISTVA ZHIDK., MATER. VSES. TEPLOFIZ. KONF., 4TH<br>89-90, 1973.<br>( EDITED BY Z. I. GELLER, NAUKA: MOSCOW, USSR ) |
| 72505 | ELECTRICAL CONDUCTIVITY, VISCOSITY, AND DENSITY OF ALKALI METAL ACETATE AQUEOUS SOLUTIONS.<br>BAKEEV, M. . ANDAMASOV, R. S.<br>TR. KHIM.-MET. INST. AKAD. NAUK KAZ. SSR<br>( 17 ), 131-7 1972. |
| 72511 | SOLUBILITY STUDIES IN THE SODIUM NITRATE-ACETONE-WATER AND POTASSIUM NITRATE-ACETONE-WATER SYSTEMS.<br>HOPPE, H. WINKLER, F.<br>Z. CHEM.<br>13 ( 8 ), 304-5, 1973. |
| 72514 | CONDUCTANCE AND VISCOSITY OF STRONG ELECTROLYTES NEAR THE PHASE TRANSITION OF N-METHYLACETAMIDE.<br>RASTOGI, P. P.<br>Z. PHYS. CHEM. ( FRANKFUT AM MAIN )<br>85 ( 1-4 ), 1-4, 1973. |
| 72523 | AIR TRANSPORT COEFFICIENTS AT 3000-25,000 K AND 0.1, 1,10, AND 100 ATMOSPHERES.<br>SOKOLOVA, I. A.<br>ZH. PRIKL. MEKH. TEKH. FIZ.<br>( 2 ), 80-90, 1973.<br>( FOR ENGLISH TRANSLATION SEE T74346 ) |
| 72529 | VISCOSITY OF BINARY LITHIUM PERCHLORATE-SODIUM PERCHLORATE AND LITHIUM PERCHLORATE-POTASSIUM PERCHLORATE MELTS.<br>FARMKOVSKAYA, A. A. BROVKINA, I. A.<br>ZH. FIZ. KHIM.<br>47 ( 6 ), 1611, 1973.<br>( FOR ENGLISH TRANSLATION SEE TPRC NO. 73455 ) |
| 72530 | DIVISION OF THE COEFFICIENT OF VISCOSITY B OF THE JONES-DOLE EQUATION INTO IONIC COMPONENTS.<br>KRUMGAL'Z, B. S.<br>ZH. FIZ. KHIM.<br>47 ( 7 ), 1691-4, 1973.<br>( FOR ENGLISH TRANSLATION SEE TPRC NO. 74817 ) |
| 72532 | ENTHALPIES AND SPECIFIC HEATS IN THE POTASSIUM HYDROXIDE-WATER SYSTEM.<br>GINZBURG, D. M. KOCHKALDA, V. E.<br>ZH. FIZ. KHIM.<br>47 ( 8 ), 2145-6, 1973.<br>( FOR ENGLISH TRANSLATION SEE TPRC NO. 74820 ) |
| 72533 | POTASSIUM CHLORIDE-LEAD CHLORIDE SYSTEM STUDIED BY MEASURING THERMAL CONDUCTIVITY.<br>POLYAKOV, P. V. GIL'DEBRANDT, E. M.<br>KOTEL'NIKOVA, G. A. MOZHAEVA, V. N.<br>ZH. FIZ. KHIM.<br>47 ( 8 ), 2155, 1973.<br>( FOR ENGLISH TRANSLATION SEE TPRC NO. 74822 ) |
| 72537 | VISCOSITY OF A MIXTURE OF NORMAL AND HEAVY WATER AT HIGH PRESSURES IN THE 75-225 DEGREE RANGE.<br>AGAEV, N. A. ABAS-ZADE, A. KERIMOV, A. M.<br>ZH. FIZ. KHIM.<br>47 ( 8 ), 2168, 1973.<br>( FOR ENGLISH TRANSLATION SEE TPRC NO. 74825 ) |
| 72548 | REFLECTION AND TRANSMISSION OF ELECTROMAGNETIC WAVES UNIAXIAL PLASMA SLAB.<br>KALLURI, D. SHRIVASTAVA, R. K.<br>IEEE TRANSACTIONS ON ANTENNAS AND PROPAGATION<br>AP-21, 63-70, 1973. |
| 72550 | THERMAL RADIATION BY COMBUSTION GASES.<br>EDWARDS, D. K. BALAKRISHNAN, A.<br>INT. J. HEAT MASS TRANSFER<br>16, 25-40, 1973. |
| 72551 | STRUCTURE OF JUPITER AND SATURN.<br>HUBBARD, W. B.<br>SPACE SCIENCE REVIEWS<br>599-662PP., 1973. |
| 72552 | LABORATORY DETRRMINATIONS OF WATER SURFACE EMISSIVITY.<br>ROBINSON, P. J.<br>J APPL. METEOROL.<br>11, 1391-3, 1972. |
| 72553 | SPECIFIC HEATS OF AQUEOUS SOLUTIONS OF SODIUM AND POTASSIUM NITRATES AT TEMPERATURES IN THE RANGE 25-340 C.<br>PUCHKOV, L. V. MATVEEVA, R. P.<br>BARANOVA, T. L.<br>J. APPL. CHEM., USSR<br>46 ( 2 ), 460-2, 1973.<br>( ENGLISH TRANSLATION OF ZH. PRIKL. KHIM., 46 ( 2 ), 443-5, 1973; FOR ORIGINAL SEE TPRC NO. 70588 ) |
| 72555 | VISCOSITIES OF LITHIUM, SODIUM, AND POTASSIUM HYDROXIDE SOLUTIONS AT TEMPERATURES UP TO 275 C.<br>MASHOVETS, V. P. PUCHKOV, L. V.<br>SARGAEV, P. M. FEDOROV, M. K.<br>J. APPL. CHEM., USSR<br>46 ( 5 ), 1055-8, 1973.<br>( ENGLISH TRANSLATION OF ZH. PRIKL. KHIM., 46 ( 5 ), 992-6, 1973; FOR ORIGINAL SEE TPRC NO. 71669 ) |
| 72558 | EXPERIMENTAL DETERMINATION AND CALCULATION OF THERMAL CONDUCTIVITY OF LIQUORS OF THE AMMONIA-SODA PROCESS.<br>CHERNENKAYA, E. I. VERNIGORA, G. A.<br>J. APPL. CHEM., USSR<br>46 ( 6 ), 1306-9, 1973.<br>( ENGLISH TRANSLATION OF ZH. PRIKL. KHIM., 46 ( 6 ), 1224-7, 1973; FOR ORIGINAL SEE TPRC NO. 71888 ) |
| 72559 | SOLUBILITY, DENSITY, AND VISCOSITY IN THE SYSTEM POLYPHOSPHORIC ACIDS - POTASSIUM SULFATE - WATER.<br>BUGAI, P. M. EREMENKO, R. A.<br>J. APPL. CHEM., USSR<br>46 ( 6 ), 1467-9, 1973.<br>( ENGLISH TRANSLATION OF ZH. PRIKL. KHIM., 46 ( 6 ), 1372-5, 1973; FOR ORIGINAL SEE TPRC NO. 71672 ) |
| 72596 | INFRARED OPTICAL CONSTANTS OF AQUEOUS SOLUTIONS OF ELECTROLYTES. THE ALKALI HALIDES.<br>RHINE, P. WILLIAMS, D. HALE, G. M.<br>QUERRY, M. R.<br>J. PHYS. CHEM.<br>78 ( 3 ), 238-46, 1974. |
| 72680 | PHYSICOCHEMICAL PROPERTIES AND EXTRACTION CHARACTERISTICS OF THE DIMETHYLFORMAMIDE-DIETHYLENE GLYCOL MIXED SOLVENT.<br>PROKOPETS, M. M. ZELIZNYI, A. M.<br>KUNTII, L. I.<br>CHEM. TECHNOL. FUELS OILS, USSR<br>17 ( 11 ), 844-7, 1972.<br>( ENGLISH TRANSLATION OF KHIM. TEKHNOL. TOPL. MASEL, 17 ( 11 ), 31-4, 1972; FOR ORIGINAL SEE TPRC NO. 69560 ) |
| 72751 | SPECIFIC HEAT OF A SYNTHETIC ZEOLITE AND THE HEAT OF FUSION OF ITS ABSORBED WATER.<br>HALY, A. R.<br>J. PHYS. CHEM. SOLIDS<br>33 ( 1 ), 129-37, 1972. |
| 72861 | CALCULATION OF REAL GAS PROPERTIES FROM TABLES OF TRIGONOMETRIC FUNCTIONS.<br>KHAYRUTDINOV, K. A.<br>INZH. FIZ. ZH.<br>24 ( 4 ), 735-9, 1973.<br>( FOR ENGLISH TRANSLATION SEE TPRC NO. 72862 ) |

| TPRC Number | Bibliographic Citation |
|---|---|

72862 **CALCULATION OF REAL GAS PROPERTIES FROM TABLES OF TRIGONOMETRIC FUNCTIONS.**
KHAYRUTDINOV, K. A.
FOREIGN TECHNOLOGY DIV., WRIGHT PATTERSON AIR FORCE BASE, OHIO
8PP., 1973.
( ENGLISH TRANSLATION OF INZH. FIZ. ZH., 24 ( 4 ), 735–9, 1973; FOR ORIGINAL SEE TPRC NO. 72861 )
( FTD–HT–23–922–74, AVAIL. DDC )

72868 **MEASUREMENT OF MUTUAL DIFFUSION COEFFICIENTS AND THERMAL DIFFUSIVITIES BY QUASIELASTIC LIGHT SCATTERING.**
GULARI, E. BROWN, R. J. PINGS, C. J.
AICHE J.
19 ( 6 ), 1196–204, 1973.

72899 **LOW–TEMPERATURE HEAT CAPACITY OF HELIUM FILMS ADSORBED ON COPPER.**
HICKERNELL, D. C. MC LEAN, E. O.
VILCHES, O. E.
J. LOW TEMP. PHYS.
13 ( 3–4 ), 241–8, 1973.

72903 **ARGON MATRIX RAMAN SPECTRA OF DICHLORINE MONOXIDE AND ITS PHOTOLYSIS PRODUCTS MONOCHLORINE MONOXIDE AND CHLORINE–MONOCHLORINE MONOXIDE. INFRARED MATRIX SPECTRA OF MONOCHLORINE MONOXIDE AND DIMERIC MONOCHLORINE MONOXIDE.**
CHI, F. K. ANDREWS, L.
J. PHYS. CHEM.
77 ( 26 ), 3062, 1973.

72914 **ELECTRICAL CONDUCTIVITY AND VISCOSITY OF THE MOLTEN LEAD( II ) CHLORIDE–ZINC CHLORIDE BINARY SYSTEM.**
UMETSU, Y. ISHII, Y.
NIPPON KINZOKU GAKKAISHI
37 ( 9 ), 997–1004, 1973.

72941 **THERMAL CONDUCTIVITY OF SOME BINARY AND TERNARY LIQUID SOLUTIONS.**
GYLMANOV, A. A. BASHIROV, M. YA.
TEPLOFIZ. SVOISTVA ZHIDK., MATER. VSES. TEPLOFIZ. KONF., 4TH
117–20, 1973.
( EDITED BY Z. I. GELLER, NAUKA: MOSCOW, USSR )

72954 **SPECIFIC HEAT OF AQUEOUS ELECTROLYTE SOLUTIONS.**
LATYSHEVA, V. A.
USP. KHIM.
42 ( 10 ), 1757–87, 1973.

72955 **THERMAL CONDUCTIVITY OF POLYATOMIC GASES AND THEIR MIXTURES.**
SHASHKOV, A. G. KAMCHATOV, F. P.
VESTSI AKAD. NAVUK BELARUS. SSR, SER. FIZ.–ENERG. NAVUK
( 3 ), 61–6, 1973.

72958 **VISCOSITY BEHAVIOR OF AQUEOUS SOLUTIONS OF AMINES.**
PATEL, R. I. PATIL, K. M. KAULGUD, M. V.
Z. PHYS. CHEM. ( FRANKFURT AM MAIN )
86 ( 1–2 ), 67–74, 1973.

72963 **DENSITY, VISCOSITY, AND ELECTRICAL CONDUCTIVITY OF SEA WATER AT UP TO 300–350.**
MASHOVETS, V. P. PUCHKOV, L. V.
SMAEV, V. N. FEDOROV, M. K. FEDOTOV, N. V.
ZH. PRIKL. KHIM. ( LENINGRAD )
46 ( 8 ), 1865–8, 1973.
( FOR ENGLISH TRANSLATION SEE TPRC NO. 74678 )

72964 **SATURATED VAPOR PRESSURE AND HEAT CAPACITY OF SEA WATER AT 25–350.**
MASHOVETS, V. P. MATVEEVA, R. P.
PUCHKOV, L. V. FEDOROV, M. K.
ZH. PRIKL. KHIM. ( LENINGRAD )
46 ( 9 ), 1954–6, 1973.
( FOR ENGLISH TRANSLATION SEE TPRC NO. 74677 )

72969 **VISCOSITY OF SEA WATER SOLUTIONS.**
CHEN, S. F. CHAN, R. C.
READ, S. M. BROMLEY, L. A.
DESALINATION
13 ( 1 ), 37–51, 1973.

72971 **GENERAL METHOD OF CALCULATING THE COEFFICIENTS OF DYNAMIC VISCOSITY OF GASEOUS MIXTURES OF HYDROCARBONS AND NATURAL GASES AT ATMOSPHERIC PRESSURE.**
BOLOTIN, N. K. KOLOMIETS, A. YA.
LABINOV, S. D. OVCHARENKO, A. G.
GAZOV. PROM.
( 10 ), 46–9, 1973.

72974 **MEASUREMENT OF THERMAL CONDUCTIVITY OF ARGON–NITROGEN GAS MIXTURES AT HIGH TEMPERATURES.**
JORDAN, D. L. SWIFT, J. D.
INT. J. ELECTRON.
35 ( 5 ), 595–608, 1973.

72975 **VISCOSITY OF ELECTROLYTIC SOLUTIONS IN DIOXANE–WATER MIXTURES.**
FUOSS, R. M. REYNOLDS, A. F.
ISR. J. CHEM.
11 ( 2–3 ), 201–10, 1973.

73002 **CALCULATION OF ENTROPIES AND HEAT CAPACITIES OF MONOATOMIC IONS IN AN AQUEOUS SOLUTION IN THE 0–100 RANGE.**
KRESTOV, G. A. KOBENIN, V. A.
TEOR. EKSP. KHIM.
9 ( 5 ), 680–7, 1973.

73009 **SPECIFIC HEATS AND SPECIFIC VOLUMES OF ISOPIESTIC AQUEOUS SOLUTIONS OF BERYLLIUM–SUBGROUP METAL CHLORIDES.**
VASILEV, Y. A. FEDYAINOV, N. V.
KURENKOV, V. V.
ZH. FIZ. KHIM.
47 ( 1 ), 2799–803, 1973.
( FOR ENGLISH TRANSLATION SEE TPRC NO. 75855 )

73015 **MEASUREMENT OF VISCOSITY OF SOLUTION OF HELIUM–3 IN HELIUM–4 AT SUPERLOW TEMPERATURES.**
BORISOV, N. S. NEGANOV, B. S. FELLER, G.
JOINT INST. FOR NUCLEAR RESEARCH, DUBNA, USSR
13PP., 1973.
( JINR–P8–7009, N–73–33086, AVAIL. NTIS )

73019 **COLLISION INTEGRALS OF CARBON–OXYGEN ATOMS IN DIFFERENT ELECTRONIC STATES.**
CAPITELLI, M. FICOCELLI, E., VTH
J. PHYS., AT. MOL. PHYS.
6 B ( 9 ), 1819–23, 1973.

73022 **ISCTHERMS OF SOLUBILITY, DENSITY, AND VISCOSITY OF SATURATED SOLUTIONS IN MAGNESIUM CHLORIDE–MAGNESIUM IODIDE–WATER AND CALCIUM CHLORIDE–CALCIUM IODIDE–WATER SYSTEMS AT 25 DEGREES.**
ZUL'FUGARLY, D. I. SHTANCHAEVA, Z. M.
PEREL'MAN, F. M. DOLININA, R. M.
AZERB. KHIM. ZH.
( 1 ), 84–90, 1973.

73061 **THE PROPERTIES OF CONCENTRATED ELECTROLYTE SOLUTIONS. III. VISCOSITY OF AQUEOUS MAGNESIUM CHLORIDE, NITRATE, AND SULPHATE SOLUTIONS.**
ROMAN'KOV, YU. I. KOMAROVA, T. A.
ZH. FIZ. KHIM.
47 ( 10 ), 2712– , 1973.
( FOR ENGLISH TRANSLATION SEE TPRC NO. 75844 )

73063 **SPECIFIC HEATS OF SODIUM IODIDE AND CESIUM IODIDE SOLUTIONS IN DIMETHYL SULFOXIDE AT 25.**
RYABCHENKO, O. I. MISHCHENKO, K. P.
NOVOSELOV, N. P. KLYUEVA, M. L.
ZH. OBSHCH. KHIM.
43 ( 9 ), 1886–90, 1973.
( FOR ENGLISH TRANSLATION SEE TPRC NO. 74679 )

73064 **ADIABATIC CALORIMETER FOR MEASURING HEAT CAPACITIES OF LIQUID SUBSTANCES IN THE 0–100 DEGREE RANGE.**
PETROV, G. I. PUCHKOV, L. V.
ZH. PRIKL. KHIM. ( LENINGRAD )
46 ( 10 ), 2233–6, 1973.
( FOR ENGLISH TRANSLATION SEE T75430 )

73089 **REFLECTIVITY AND CURVATURE OF HYPERSONIC SHOCK WAVES AT 1 ATMOSPHERE.**
SIMPSON, D. SMY, R. R.
J. PHYS., D, APPL. PHYS.
4 ( 10 ), 1487–92, 1971.

73150 **EFFECT OF MODIFIED THERMAL CONDUCTIVITY ON THE TEMPERATURE DISTRIBUTION IN THE PROTONOSPHERE.**
MAYR, H. G. FONTHEIM, E. G. MAHAJAN, K. K.
ANN. GEOPHYSIQUE
29, 21–24, 1973.

73157 **THERMAL CONDUCTIVITY OF A GASEOUS PLASMA.**
MAJUMDAR, S. K. ADHIKARI, D. LAHIRI, A.
PLASMA PHYS.
15, 1259–1261, 1973.

73174 **APPLICABILITY OF THE GRUNBERG–NISSAN EQUATION FOR THE VISCOSITY OF LIQUID MIXTURES.**
PALIT, S. R. BHATTACHARYYA, S.
INDIAN J. CHEM.
11 ( 9 ), 953–4, 1973.

73180 **INVESTIGATION OF INTERMOLECULAR ASSOCIATION OF 4–SEC–BUTYL–2–( ALPHA–METHYLBENZYL ) PHENOL BY USE OF DIELECTRIC AND ISOPIESTIC MEASUREMENTS.**
RODDY, J. W. COLEMAN, C. F.
J. INORG. NUCL. CHEM.
35 ( 12 ), 4271–8, 1973.

73190 **THERMAL CONDUCTIVITIES OF ETHENE AND ITS FLUOROCARBON ANALOGUES AND OF THEIR BINARY MIXTURES WITH ARGON.**
GRAY, P. PARKINSON, C.
J. CHEM. SOC., FARADAY TRANS.
70 I 560–71, 1974.

73278 **TRANSPORT PROPERTIES IN GASES ( COMPARISON BETWEEN THEORY AND EXPERIMENT ).**
KESTIN, J. MASON, E. A.
AIP, CONF. PROC.
( 11 ), 137–92, 1973.

| TPRC Number | Bibliographic Citation |
|---|---|

73279 **TRANSPORT PROPERTIES IN IONIZED GASES.**
SANDLER, S. I.
AIP ( AMER. INST. PHYS. ), CONF. PROC.
( 11 ), 203–22, 1973.

73281 **THERMAL CONDUCTIVITIES OF SOLUTIONS OF SODIUM CARBONATE, POTASSIUM CARBONATE, POTASSIUM CHLORIDE, AND POTASSIUM SULFATE AT VARIOUS TEMPERATURES.**
CHERNEN'KAYA, E. I. VERNIGORA, G. A.
J. APPL. CHEM., USSR
46 ( 7 ), 1707–9, 1973.
( ENGLISH TRANSLATION OF ZH. PRIKL. KHIM., 46 ( 7 ), 1603–5, 1973; FOR ORIGINAL SEE TPRC NO. 72165 )

73285 **ELECTRICAL CONDUCTIVITY AND VISCOSITY OF SUPERCRITICAL ISOBUTYRIC ACID + WATER SOLUTIONS.**
GAMMELL, P. M. ANGELL, C. A.
J. CHEM. PHYS.
60 ( 2 ), 584–90, 1974.

73290 **MEASUREMENT OF THERMAL CONDUCTIVITY OF SOLUTION OF HELIUM-3 HELIUM-4 AT EXTREME LOW TEMPERATURES.**
BORISOV, N. S. NEGANOV, B. S. FELLER, G.
JOINT INST. FOR NUCLEAR RESEARCH, DUBNA, USSR
10PP., 1973.

73327 **EFFECT OF ELECTRON-ELECTRON COLLISIONS ON THE THERMAL CONDUCTIVITY OF A DENSE PLASMA.**
OSADA, J. BONILHA, J. R. M. FERREIRA, A. C.
CAPELATO, H. V.
REV. BRASIL. FIS.
2 ( 2 ), 95–9, 1972.

73328 **A DYNAMICAL MODEL OF THE CORONA.**
BROWNE, S. L. BESSEY, R. J.
SOL. PHYS.
31, 351–375, 1973.

73337 **THEORY OF A POWERFUL ELECTRICAL DISCHARGE IN XENON WITH A PREDOMINANCE OF RADIATION.**
LITVINOV, I. I.
TEPLOFIZ. VYS. TEMP.
11 ( 4 ), 695–705, 1973.
( FOR ENGLISH TRANSLATION SEE TPRC NO. 73338 )

73338 **THEORY OF A POWERFUL ELECTRICAL DISCHARGE IN XENON WITH A PREDOMINANCE OF RADIATION.**
LITVINOV, I. I.
HIGH TEMP.
11 ( 4 ), 628–36, 1973.
( ENGLISH TRANSLATION OF TEPLOFIZ. VYS. TEMP., 11 ( 4 ), 695–705, 1973; FOR ORIGINAL SEE T73337 )

73345 **EMISSIVITY OF GRAPHITE- AND BARIUM OXIDE-BASE MATRIX MATERIAL.**
LEVINOV, B. M. BARYKIN, B. M.
SPIRIDONOV, E. G. REKOV, A. I.
TEPLOFIZ. VYS. TEMP.
11 ( 4 ), 772–5, 1973.
( FOR ENGLISH TRANSLATION SEE TPRC NO. 73346 )

73346 **EMISSIVITY OF A MATRIX MATERIAL BASED ON GRAPHITE AND BARIUM OXIDE.**
LEVINOV, B. M. BARYKIN, B. M.
SPIRIDONOV, E. G. REKOV, A. I.
HIGH TEMP.
11 ( 4 ), 693–5, 1973.
( ENGLISH TRANSLATION OF TEPLOFIZ. VYS. TEMP., 11 ( 4 ), 772–5, 1973; FOR ORIGINAL SEE TPRC NO. 73345 )

73367 **THERMAL PROPERTIES.**
HARRIS, M.
HANDBOOK OF TEXTILE FIBERS
167–82, 1954.

73370 **REFINING THE ALGORITHM FOR A ZONAL CALCULATION OF HEAT TRANSFER IN AN OPEN-HEARTH FURNACE.**
ZHURAVLEV, YU. A. LISIENKO, V. G.
KITAEV, B. I.
INZH. FIZ. ZH.
21 ( 5 ), 829–35, 1971.
( FOR ENGLISH TRANSLATION SEE TPRC NO. 73371 )

73371 **REFINING THE ALGORITHM FOR A ZONAL CALCULATION OF HEAT TRANSFER IN AN OPEN-HEARTH FURNACE.**
ZHURAVLEV, YU. A. LISIENKO, V. G.
KITAEV, B. I.
J. ENG. PHYS.
21 ( 5 ), 1375–9, 1971.
( ENGLISH TRANSLATION OF INZH. FIZ. ZH., 21 ( 5 ), 829–35, 1971; FOR ORIGINAL SEE TPRC NO. 73370 )

73373 **INTERACTION OF ELECTROMAGNETIC WAVES WITH INTERPENETRATING DIELECTRIC MEDIA WITH DIFFERENT VELOCITIES.**
STOLYAROV, S. N.
ZH. TEKHN. FIZ.
39 ( 8 ), 1388–96, 1969.
( FOR ENGLISH TRANSLATION SEE TPRC NO. 73374 )

73374 **INTERACTION OF ELECTROMAGNETIC WAVES WITH INTERPENETRATING DIELECTRIC MEDIA WITH DIFFERENT VELOCITIES.**
STOLYAROV, S. N.
SOV. PHYS.-TECH. PHYS.
14 ( 8 ), 1044–9, 1970.
( ENGLISH TRANSLATION OF ZH. TEKHN. FIZ., 39 ( 8 ), 1388–96, 1969; FOR ORIGINAL SEE TPRC NO. 73373 )

73415 **THERMODYNAMIC PROPERTIES OF ORGANIC COMPOUNDS IN AQUEOUS SOLUTIONS. APPARENT MOLAL HEAT CAPACITIES OF AMINES AND ETHERS.**
CABANI, S. CONTI, G. MARTINELLI, A.
MATTEOLI, E.
J. CHEM. SOC., FARADAY TRANS.
69 1 ( 12 ), 2212–23, 1973.

73455 **VISCOSITIES OF LITHIUM PERCHLORATE-SODIUM PERCHLORATE AND LITHIUM PERCHLORATE-POTASSIUM PERCHLORATE BINARY MELTS.**
FARMAKOVSKAYA, A. A. BROVKINA, I. A.
RUSS. J. PHYS. CHEM.
47 ( 6 ), 915–6, 1973.
( ENGLISH TRANSLATION OF ZH. FIZ. KHIM., 47 ( 6 ), 1611, 1973; FOR ORIGINAL SEE TPRC NO. 72529 )

73461 **HEAT CAPACITIES AND DENSITIES OF AQUEOUS SODIUM CHLORATE AND PERCHLORATE SOLUTIONS AT 25 DEGREES, 40 DEGREES, AND 60 DEGREES C.**
LOSEVA, G. K. SEMCHENKO, D. P.
RUSS. J. PHYS. CHEM.
47 ( 6 ), 922, 1973.
( ENGLISH TRANSLATION OF ZH. FIZ. KHIM., 47 ( 6 ), 1621, 1973; FOR ORIGINAL SEE TPRC NO. 71887 )

73463 **PULSED DISCHARGE IN A DIELECTRIC CHAMBER.**
ZEMSKOV, A. I. PRUT, V. V. KHRABROV, V. A.
ZH. TEKHN. FIZ.
42 ( 2 ), 358–63, 1972.
( FOR ENGLISH TRANSLATION SEE TPRC NO. 73464 )

73464 **PULSED DISCHARGE IN A DIELECTRIC CHAMBER.**
ZEMSKOV, A. I. PRUT, V. V. KHRABROV, V. A.
SOV. PHYS. TECH. PHYS.
17 ( 2 ), 285–9, 1972.
( ENGLISH TRANSLATION OF ZH. TEKHN. FIZ., 42 ( 2 ), 358–63, 1972; FOR ORIGINAL SEE TPRC NO. 73463 )

73467 **KINETIC THEORY OF ELECTROMAGNETIC WAVE REFLECTION FROM A MOVING INHOMOGENEOUS PLASMA SHEATH.**
STEPANOV, N. S. SOROKIN, YU. M.
ZH. TEKHN. FIZ.
42 ( 3 ), 578–83, 1972.
( FOR ENGLISH TRANSLATION SEE TPRC NO. 73468 )

73468 **KINETIC THEORY OF ELECTROMAGNETIC WAVE REFLECTION FROM A MOVING INHOMOGENEOUS PLASMA SHEATH.**
STEPANOV, N. S. SOROKIN, YU. M.
SOV. PHYS. TECH. PHYS.
17 ( 3 ), 456–60, 1972.
( ENGLISH TRANSLATION OF ZH. TEKHN. FIZ., 42 ( 3 ), 578–83, 1972; FOR ORIGINAL SEE TPRC NO. 73467 )

73469 **REFLECTION OF ELECTROMAGNETIC WAVES FROM A MOVING PLASMA.**
MIROSHNICHENKO, V. I.
ZH. TEKH. FIZ.
43 ( 3 ), 467–74, 1973.
( FOR ENGLISH TRANSLATION SEE TPRC NO. 73470 )

73470 **REFLECTION OF ELECTROMAGNETIC WAVES FROM A MOVING PLASMA.**
MIROSHNICHENKO, V. I.
SOV. PHYS. TECH. PHYS.
18 ( 3 ), 299–303, 1973.
( ENGLISH TRANSLATION OF ZH. TEKH. FIZ., 43 ( 3 ), 467–74, 1973; FOR ORIGINAL SEE TPRC NO. 73469 )

73473 **INCOHERENT REFLECTION OF ELECTROMAGNETIC WAVES FROM A PLASMA LAYER.**
GRECHKO, L. G. KOTSARENKO, N. YA.
FEDORCHENKO, A. M.
ZH. TEKHN. FIZ.
42 ( 9 ), 1832–8, 1972.
( FOR ENGLISH TRANSLATION SEE TPRC NO. 73474 )

73474 **INCOHERENT REFLECTION OF ELECTROMAGNETIC WAVES FROM A PLASMA LAYER.**
GRECHKO, L. G. KOTSARENKO, N. YA.
FEDORCHENKO, A. M.
SOV. PHYS. TECH. PHYS.
17 ( 9 ), 1469–73, 1973.
( ENGLISH TRANSLATION OF ZH. TEKHN. FIZ. 42 ( 9 ), 1832–8, 1972; FOR ORIGINAL SEE TPRC NO. 73473 )

73479 **VISCOUS FLOW OF AN INHOMOGENEOUS PLASMA IN A COAXIAL STRUCTURE WITH A THERMIONIC CATHODE.**
NEVOLIN, V. K.
ZH. TEKH. FIZ.
42 ( 12 ), 2596–9, 1972.
( FOR ENGLISH TRANSLATION SEE TPRC NO. 73480 )

| TPRC Number | Bibliographic Citation |
|---|---|

73480 **VISCOUS FLOW OF AN INHOMOGENEOUS PLASMA IN A COAXIAL STRUCTURE WITH A THERMIONIC CATHODE.**
NEVOLIN, V. K.
SOV. PHYS. TECH. PHYS.
17 ( 12 ), 2014–6, 1973.
( ENGLISH TRANSLATION OF ZH. TEKH. FIZ., 42 ( 12 ), 2596–9, 1972; FOR ORIGINAL SEE TPRC NO. 73479 )

73481 **EFFECT OF LONGITUDINAL THERMAL CONDUCTIVITY ON OHMIC HEATING IN TUMAN–1.**
KAGANSKII, M. G. KALMUKOV, S. G.
SHAKHOVETS, K. G.
SOV. PHYS. TECH. PHYS.
18 ( 2 ), 197–201, 1973.
( ENGLISH TRANSLATION OF ZH. TEKH. FIZ., 43 ( 2 ), 302–9, 1973; FOR ORIGINAL SEE TPRC NO. 71356 )

73509 **OPTICAL PROPERTIES OF THE RARE EARTH CHALCOGENIDE GLASSES.**
LOZAC'H, A. M. BARNIER, S.
GUITTARD, M. BESANCON, P. FLAHAUT, J.
ANN. CHIM. ( PARIS )
9 ( 1 ), 127–32, 1974.

73526 **EXPERIMENTAL HEAT CAPACITIES OF NITROGEN–HYDROGEN SULFIDE MIXTURES AT ELEVATED PRESSURES.**
HAMALIUK, G. P. BISHNOI, P. R.
ROBINSON, D. B.
J. CHEM. ENG. DATA
19 ( 1 ), 78–81, 1974.

73559 **KINETIC RESONANCE IN A BINARY MIXTURE OF MOLECULAR GASES.**
BORMAN, V. D. NIKOLAEV, B. I. TROYAN, V. I.
ZH. EKSP. TEOR. FIZ.
61 ( 1 ), 208–21, 1971.
( FOR ENGLISH TRANSLATION SEE TPRC NO. 73560 )

73560 **KINETIC RESONANCE IN A BINARY MIXTURE OF MOLECULAR GASES.**
BORMAN, V. D. NIKOLAEV, B. I. TROYAN, V. I.
SOV. PHYS. JETP
34 ( 1 ), 108–13, 1972.
( ENGLISH TRANSLATION OF ZH. EKSP. TEOR. FIZ., 61 ( 1 ), 208–21, 1971; FOR ORIGINAL SEE TPRC NO. 73559 )

73602 **THE VISCOSITY OF LIQUIDS.**
DUNSTAN, A. E. THOLE, F. B.
LONGMANS, GREEN AND CO., PATERNOSTER ROW, LONDON
91PP., 1914.

73605 **SHEAR VISCOSITY OF NITROETHANE–3–METHYLPENTANE IN THE CRITICAL REGION.**
TSAI, B. C. MC INTYRE, D.
J. CHEM. PHYS.
60 ( 3 ), 937–45, 1974.

73619 **DIFFUSION COEFFICIENT AND INTERACTION POTENTIAL OF THE HYDROGEN ATOM MOLECULE SYSTEM.**
TANG, K. T. WEI, P. S. P.
J. CHEM. PHYS.
60 ( 6 ), 2454–9, 1974.

73627 **THERMAL COAGULATION OF F CENTERS IN POTASSIUM CHLORIDE CRYSTALS DOPED WITH ALKALI METALS.**
GOLUBTSOV, V. V. GOLDENBERG, A. B.
OPT. SPEKTROSK.
34 ( 3 ), 510–4, 1973.
( FOR ENGLISH TRANSLATION SEE TPRC NO. 73628 )

73628 **THERMAL COAGULATION OF F CENTERS IN POTASSIUM CHLORIDE CRYSTALS DOPED WITH ALKALI METALS.**
GOLUBTSOV, V. V. GOLDENBERG, A. B.
OPT. SPECTROS., USSR
34 ( 3 ), 291–3, 1973.
( ENGLISH TRANSLATION OF OPT. SPEKTROSK., 34 ( 3 ), 510–4, 1973; FOR ORIGINAL SEE TPRC NO. 73627 )

73694 **THE VISCOSITY OF MOIST AIR.**
STUDNIKOV, E. L.
INZH. FIZ. ZH.
19 ( 2 ), 338–40, 1970.
( FOR ENGLISH TRANSLATION SEE TPRC NO. 73695 )

73695 **THE VISCOSITY OF MOIST AIR.**
STUDNIKOV, E. L.
J. ENG. PHYS.
19 ( 2 ), 1036–7, 1970.
( ENGLISH TRANSLATION OF INZH. FIZ. ZH., 19 ( 2 ), 338–40, 1970; FOR ORIGINAL SEE TPRC NO. 73694 )

73706 **MIXING OF HOMOGENEOUS STREAMS OF DISSOCIATING AIR.**
ARTYUKH, L. YU. ZAKARIN, E. A.
INZH. FIZ. ZH.
20 ( 2 ), 246–54, 1971.
( FSOR ENGLISH TRANSLATION SEE TPRC NO. 73707 )

73707 **MIXING OF HOMOGENEOUS STREAMS OF DISSOCIATING AIR.**
ARTYUKH, L. YU. ZAKARIN, E. A.
J. ENG. PHYS.
20 ( 2 ), 175–81, 1971.
( ENGLISH TRANSLATION OF INZH. FIZ. ZH., 20 ( 2 ), 246–54, 1971; FOR ORIGINAL SEE TPRC NO. 73706 )

| TPRC Number | Bibliographic Citation |
|---|---|

73714 **USE OF PROBE METHOD FOR STUDY OF THERMOPHYSICAL PROPERTIES OF BIOLOGICAL MASSES.**
TUTOVA, E. G. NOVICHENOK, L. N.
KONONENKO, V. D.
INZH. FIZ. ZH.
20 ( 3 ), 505–9, 1971.
( FOR ENGLISH TRANSLATION SEE TPRC NO. 73715 )

73715 **USE OF PROBE METHOD FOR STUDY OF THERMOPHYSICAL PROPERTIES OF BIOLOGICAL MASSES.**
TUTOVA, E. G. NOVICHENOK, L. N.
KONONENKO, V. D.
J. ENG. PHYS.
20 ( 3 ), 368–71, 1971.
( ENGLISH TRANSLATION OF INZH. FIZ. ZH., 20 ( 3 ), 505–9, 1971; FOR ORIGINAL SEE TPRC NO. 73714 )

73718 **THERMAL CONDUCTIVITY DETERMINATION FOR NONMETALLIC MEDIA BY THE CYLINDRICAL PROBE METHOD.**
GOLOVICHER, L. E. SAN'KO, YU. P.
SHASHKOV, A. G.
INZH. FIZ. ZH.
20 ( 4 ), 647–53, 1971.
( FOR ENGLISH TRANSLATION SEE TPRC NO. 73719 )

73719 **THERMAL CONDUCTIVITY DETERMINATION FOR NONMETALLIC MEDIA BY THE CYLINDRICAL PROBE METHOD.**
GOLOVICHER, L. E. SAN'KO, YU. P.
SHASHKOV, A. G.
J. ENG. PHYS.
20 ( 4 ), 466–71, 1971.
( ENGLISH TRANSLATION OF INZH. FIZ. ZH., 20 ( 4 ), 647–53, 1971; FOR ORIGINAL SEE TPRC NO. 73718 )

73771 **CALCULATION OF THE RADIATION OF AN ISOTHERMAL GAS – DUST MEDIUM.**
POPOV, YU. A.
INZH. FIZ. ZH.
17 ( 3 ), 512–9, 1969.
( FOR ENGLISH TRANSLATION SEE T73772 )

73772 **CALCULATION OF THE RADIATION OF AN ISOTHERMAL GAS – DUST MEDIUM.**
POPOV, YU. A.
J. ENG. PHYS.
17 ( 3 ), 1142–8, 1969.
( ENGLISH TRANSLATION OF INZH. FIZ. ZH., 17 ( 3 ), 512–9, 1969; FOR ORIGINAL SEE TPRC NO. 73771 )

73805 **THERMODYNAMIC PROPERTIES OF URANIUM HEXAFLUORIDE MEASURED WITH A BALLISTIC PISTON COMPRESSOR.**
STERRITT, D. E. LALOS, G. T.
SCHNEIDER, R. T.
DEPT. OF NUCLEAR ENGINEERING SCIENCES, FLORIDA UNIVERSITY, GAINESVILLE
79PP., 1973.
( N73–24936, NASA–CR–132051, AVAIL. NTIS )

73814 **EXACT EQUATIONS FOR THE TRANSPORT COEFFICIENTS OF A FIVE–COMPONENT–MIXTURE AS A MODEL GAS OF DISSOCIATED AIR.**
STRAUB, D.
KARLSRUHE UNIVERSITY, WEST GERMANY, PH.D. THESIS
93PP., 1971.
( N73–30678, DLR–FB–72–34, AVAIL. NTIS )

73821 **EFFECT OF FIBRINOGEN CONCENTRATION ON THE RHEODYNAMICAL BEHAVIOR OF HUMAN BLOOD.**
HELD, K.
DEUTSCHE LUFT– UND RAUMFAHRT, WEST GERMANY
35PP., 1973.
( N73–33046, DLR–FB–73–54, AVAIL. NTIS )

73823 **PARTICULATE COMPOSITE MATERIALS TECHNOLOGY.**
RILEY, R. E. TAUB, J. M.
LOS ALAMOS SCIENTIFIC LAB., N. MEXICO
9PP., 1973.
( N73–33505, LA–5173, AVAIL. NTIS )

73829 **THEORETICAL STUDIES IN SUNSPOTS AND ASSOCIATED STRUCTURES.**
WYLLER, A. A. YUN, H. S.
BARTOL RESEARCH FOUNDATION, FRANKLIN INST., SWARTHMORE, PA.
51PP., 1972.
( AD–752 230, AFCRL–72–0466, N73–17872, AVAIL. DDC )

73832 **PULSED NEON LASER BATHYMETRIC STUDIES USING SIMULATED DELAWARE BAY WATERS.**
HICKMAN, G. D. HOGG, J. E. GHOVANLOU, A. H.
SPARCOM INC., ALEXANDRIA, VA.
86PP., 1972.
( AD–752 486, AVAIL. DDC )

73833 **THE THERMAL CONDUCTIVITY OF PURE WATER AND STANDARD SEA WATER AS A FUNCTION OF PRESSURE AND TEMPERATURE. PART III. STANDARD SEA WATER.**
CASTELLI, V. J. STANLEY, E. M.
NAVAL SHIP RES. AND DEVELOPMENT CENTER, ANNAPOLIS, MARYLAND
14PP., 1972.
( AD–752 491, NSRDC–3566–PT. 3, AVAIL. DDC )

| TPRC Number | Bibliographic Citation |
|---|---|
| 73874 | **GENERALISED GRAPH OF THE DEPENDENCE OF THE SPECIFIC HEAT AT CONSTANT PRESSURE OF REAL GASES ON REDUCED PRESSURES AND TEMPERATURES.**<br>BUSHKOV, V. A. DOBRODEEV, V. P.<br>TEPLOENERGETIKA<br>17 ( 12 ), 85–6, 1970.<br>( FOR ENGLISH TRANSLATION SEE TPRC NO. 73875 ) |
| 73875 | **GENERALISED GRAPH OF THE DEPENDENCE OF THE SPECIFIC HEAT AT CONSTANT PRESSURE OF REAL GASES ON REDUCED PRESSURES AND TEMPERATURES.**<br>BUSHKOV, V. A. DOBRODEEV, V. P.<br>THERM. ENG., USSR<br>17 ( 12 ), 114–5, 1970.<br>( ENGLISH TRANSLATION OF TEPLOENERGETIKA, 17 ( 12 ), 85–6, 1970; FOR ORIGINAL SEE TPRC NO. 73874 ) |
| 73916 | **THERMAL CONDUCTIVITY OF A NONEQUILIBRIUM CHEMICALLY REACTING GAS.**<br>BODMAN, S. W. BRIAN, P. L. T. CHANG, T. C.<br>AICHE J.<br>17 ( 6 ), 1304–10, 1971. |
| 73917 | **DISPERSED PHASE MASS TRANSFER DURING DROP FORMATION AND COALESCENCE IN LIQUID–LIQUID EXTRACTION.**<br>SKELLAND, A. H. P. MINHAS, S. S.<br>AICHE J.<br>17 ( 6 ), 1316–24, 1971. |
| 73919 | **PROTEIN ULTRAFILTRATION: A GENERAL EXAMPLE OF BOUNDARY LAYER FILTRATION.**<br>KOZINSKI, A. A. LIGHTFOOT, E. N.<br>AICHE J.<br>18 ( 5 ), 1030–40, 1972. |
| 73920 | **INFLUENCE OF FLAME GAS THERMAL CONDUCTIVITY ON ATOM FORMATION IN FLAME SPECTROMETRY.**<br>CLAMPITT, N. C. HIEFTJE, G. M.<br>ANAL. CHEM.<br>46 ( 3 ), 382–6, 1974. |
| 73931 | **THERMAL CONDUCTIVITY OF HYDROGEN–CONTAINING GAS MIXTURES.**<br>BARNITZKE, W. MUELLER, H. D.<br>CHEM. TECH. ( LEIPZIG )<br>25 ( 11 ), 669–71, 1973. |
| 73932 | **MEASUREMENT OF VISCOSITY OF FUSED SALTS. II. VISCOSITY OF MOLTEN BINARY MIXTURES ON THE CRYOLITE BASIS.**<br>VOTAVA, I. MATIASOVSKY, K.<br>CHEM. ZVESTI<br>27 ( 5 ), 582–7, 1973. |
| 73933 | **ELECTRONIC THERMAL CONDUCTIVITY OF A COMPLETELY IONIZED PLASMA AT LOW COULOMB LOGARITHM.**<br>GAUTHIER, J. C.<br>C. R. ACAD. SCI.<br>277 B ( 21 ), 631–3, 1973. |
| 73943 | **ULTRAVIOLET REFLECTIVITY AND GEOMETRICAL ALBEDO OF TITAN.**<br>BARKER, E. S. TRAFTON, L. M.<br>ICARUS<br>20 ( 4 ), 444–54, 1973. |
| 73961 | **VISCOSITY AND NONIDEALITY OF MOLYBDENUM, TUNGSTEN, AND URANIUM HEXAFLUORIDES. DETERMINATION OF THEIR MOLECULAR PARAMETERS.**<br>MORIZOT, P. OSTORERO, J. PLURIEN, P.<br>J. CHIM. PHYS. PHYSICOCHIM. BIOL.<br>70 ( 11–12 ), 1582–6, 1973. |
| 73962 | **INTERMOLECULAR PARAMETERS AND DIMERIZATION OF CHLORINE TRIFLUORIDE.**<br>MORIZOT, P. OSTORERO, J. PLURIEN, P.<br>J. CHIM. PHYS. PHYSICOCHIM. BIOL.<br>70 ( 11–12 ), 1587–91, 1973. |
| 73964 | **OPTICAL CONSTANTS IN THE INFRARED FOR POTASSIUM SULFATE, AMMONIUM DIHYDROGEN PHOSPHATE, AND SULFURIC ACID IN WATER.**<br>QUERRY, M. R. WARING, R. C. HOLLAND, W. E.<br>EARLS, L. M. HERRMAN, M. D. NIJM, W. P.<br>HALE, G. M.<br>J. OPT. SOC. AMER.<br>64 ( 1 ), 39–46, 1974. |
| 73987 | **RELATION BETWEEN VISCOSITY OF THE DIGESTIVE FLUID AND THE DECOMPOSITION TIME OF TABLETS.**<br>HUETTENRAUCH, R. ZANGE, D. JACOB, J.<br>PHARMAZIE<br>28 ( 11–12 ), 769–72, 1973. |
| 74024 | **DEPENDENCE OF DENSITY, VISCOSITY, AND ELECTRIC CONDUCTIVITY OF AQUEOUS COPPER( II ) SULFATE SOLUTIONS ON TEMPERATURE AND CONCENTRATION.**<br>DEMICHOWICZ–PIGONIOWA, J.<br>ROCZ. CHEM.<br>47 ( 1 ), 2183–90, 1973. |
| 74072 | **A NEW CLASS OF ARTIFICIAL GRAPHITES.**<br>DEMIN, A. V. RACHKEEVA, V. I. PERKOVA, G. A.<br>SHIPKOV, N. N.<br>TSVET. MET.<br>43 ( 4 ), 61– , 1970.<br>( FOR ENGLISH TRANSLATION SEE TPRC NO. 74073 ) |
| 74073 | **A NEW CLASS OF ARTIFICIAL GRAPHITES.**<br>DEMIN, A. V. RACHKEEVA, V. I. PERKOVA, G. A.<br>SHIPKOV, N. N.<br>SOV. J. NON–FERROUS METALS<br>11 ( 4 ), 61–2, 1970.<br>( ENGLISH TRANSLATION OF TSVET. MET., 43 ( 4 ), 61– , 1970; FOR ORIGINAL SEE TPRC NO. 74072 ) |
| 74076 | **STRUCTURAL GRAPHITES WITH METALLIC BINDERS.**<br>PEREVEZENTSEV, V. P. ZOLKIN, P. I.<br>PISKUNOV, V. A. BEREZIN, I. A.<br>TSVET. METAL.<br>43 ( 10 ), 45– , 1970.<br>( FOR ENGLISH TRANSLATION SEE TPRC NO. 74077 ) |
| 74077 | **STRUCTURAL GRAPHITES WITH METALLIC BINDERS.**<br>PEREVEZENTSEV, V. P. ZOLKIN, P. I.<br>PISKUNOV, V. A. BEREZIN, I. A.<br>SOV. J. NON–FERROUS METALS<br>11 ( 10 ), 47–8, 1970.<br>( ENGLISH TRANSLATION OF TSVET. METAL., 43 ( 10 ), 45– , 1970; FOR ORIGINAL SEE TPRC NO. 74076 ) |
| 74097 | **PHASE–TRANSFORMATIONS OF COPPER – NICKEL – SULPHUR ALLOYS.**<br>BLOKHINA, L. I. CHIZHIKOV, D. M.<br>GULYANITSKAYA, Z. F.<br>IZV. AKAD. NAUK SSSR, METAL.<br>( 1 ), 82– , 1972.<br>( FOR ENGLISH TRANSLATION SEE T74098 ) |
| 74098 | **PHASE–TRANSFORMATIONS OF COPPER – NICKEL – SULPHUR ALLOYS.**<br>BLOKHINA, L. I. CHIZHIKOV, D. M.<br>GULYANITSKAYA, Z. F.<br>RUSS. MET., ( METALLY )<br>( 1 ), 60–4, 1972.<br>( ENGLISH TRANSLATION OF IZV. AKAD. NAUK SSSR, METAL., ( 1 ), 82– , 1972; FOR ORIGINAL SEE T74097 ) |
| 74103 | **THE CALCULATION OF FUGACITIES FROM VOLUMETRIC AND PHASE EQUILIBRIUM DATA BY MEANS OF THE GIBBS–DUHEM EQUATION.**<br>BARIEAU, R. E.<br>IND. ENG. CHEM. FUNDAM.<br>10 ( 3 ), 428–33, 1971. |
| 74224 | **THE THERMODYNAMIC PROPERTIES OF COMMERCIAL GRADE GASEOUS CARBON DIOXIDE.**<br>ALTUNIN, V. V. KUZNETSOV, D. O.<br>BONDARENKO, V. F.<br>THERM. ENG., USSR<br>20 ( 4 ), 88–90, 1973.<br>( ENGLISH TRANSLATION OF TEPLOENERGETIKA, 20 ( 4 ), 61–3, 1973; FOR ORIGINAL SEE T71112 ) |
| 74255 | **CONDUCTANCE OF SODIUM PERCHLORATE IN WATER–N–METHYLACETAMIDE SOLVENT SYSTEM.**<br>CASTEEL, J. F. AMIS, E. S.<br>J. CHEM. ENG. DATA<br>19 ( 2 ), 121–8, 1974. |
| 74280 | **VISCOSITY OF POTASSIUM NITRITE–POTASSIUM NITRATE AND CESIUM NITRITE–CESIUM NITRATE MELTS.**<br>GRIN'KO, L. S. PROTSENKO, P. I.<br>IZV. VYSSH. UCHEB. ZAVED., KHIM. KHIM. TEKHNOL.<br>16 ( 12 ), 1815–17, 1973. |
| 74281 | **SPECIFIC HEAT OF POTASSIUM CHLORIDE SOLUTIONS IN WATER AND A 30 AQUEOUS ISOPROPYL ALCOHOL SOLUTION AT 25.**<br>KLOPOV, V. I. KOLKER, A. M. KRESTOV, G. A.<br>IZV. VYSSH. UCHEB. ZAVED., KHIM. KHIM. TEKHNOL.<br>17 ( 1 ), 32–4, 1974. |
| 74332 | **THE PROTECTIVE VALUE OF CLOTHING.**<br>REES, W. H.<br>J. TEXTILE INST., PROC.<br>37, P132–P153, 1946. |
| 74333 | **THE TRANSMISSION OF HEAT THROUGH TEXTILE FABRICS. PART II.**<br>PEIRCE, F. T. REES, W. H.<br>J. TEXTILE INST., TRANS.<br>37, T181–T204, 1946. |
| 74338 | **INFLUENCE OF MOISTURE ON THE ABILITY OF CLOTHING MATERIAL TO CONDUCT HEAT.**<br>RUBNER, M.<br>ARCH. HYG.<br>25, 29–69, 1896. |

| TPRC Number | Bibliographic Citation |
|---|---|

74346 **TRANSPORT COEFFICIENTS OF AIR AT TEMPERATURES OF FROM 3,000 TO 25,000 K AND AT PRESSURES OF 0.1, 1, 10, AND 100 ATM.**
SOKOLOVA, I. A.
FOREIGN TECHNOLOGY DIVISION, WRIGHT PATTERSON AIR FORCE BASE, OHIO
16PP., 1974.
( ENGLISH TRANSLATION OF ZH. PRIKL. MEKH. TEKH. FIZ., ( 2 ), 80-90, 1973; FOR ORIGINAL SEE TPRC NO. 72523 )
( FTD-HT-23-907-74 )

74351 **VISCOSITY B COEFFICIENTS FOR SOME HOMOLOGOUS SERIES OF ORGANIC ELECTROLYTES IN AQUEOUS SOLUTIONS. EFFECT OF IONIC GROUPS.**
TAMAKI, J. OHARA, Y. KURACHI, H.
AKIYAMA, M. ODAKI, H.
BULL. CHEM. SOC. JAP.
47 ( 2 ), 384-8, 1974.

74356 **COMBUSTION PRODUCTS EMISSIVITY BY FE OPERATOR.**
BUETERS, K. A.
COMBUSTION
45 ( 9 ), 12-18, 1974.

74365 **VISCOSITY OF LITHIUM NITRATE-CADMIUM NITRATE, SILVER NITRATE-CADMIUM NITRATE, AND THALLIUM NITRATE-CADMIUM NITRATE MELTS.**
SMOTRAKOV, V. G. PROTSENKO, P. I.
POPOVSKAYA, N. P.
IZV. VYSSH. UCHEB. AVED., KHIM. KHIM. TEKHNOL.
17 ( 1 ), 3-6, 1974.

74376 **EFFECTS OF NEON IMPURITIES ON HEAT CAPACITY OF SOLID PARAHYDROGEN.**
MANZHELII, V. G. POPOV, V. A. CHAUSOV, G. P.
VLADIMIROVA, L. I.
J. LOW TEMP. PHYS.
14 ( 3-4 ), 397-401, 1974.

74390 **PROPERTIES OF MULTICOMPONENT MIXTURES OF MONATOMIC GASES.**
HELLEMANS, J. M. KESTIN, J. RO, S. T.
PHYSICA ( UTRECHT )
71 ( 1 ), 1-16, 1974.

74394 **THERMOPHYSICAL CHARACTERISTICS OF SUBSURFACE WATERS IN PETROLEUM AND GAS DEPOSITS OF THE AZOV-KUBAN BASIN.**
SUKHAREV, G. M. TARANUKHA, YU. K.
YAROSHENKO, A. A.
SOV. GEOL.
( 11 ), 130-6, 1973.

74430 **TRANSPORT PROPERTIES OF SELECTED ELEMENTS AND COMPOUNDS IN THE GASEOUS STATE.**
LILEY, P. E.
THERMOPHYSICAL PROPERTIES RESEARCH CENTER
83PP., 1973.
( TPRC-22, AFOSR-TR-74-1165, AD-782024 )

74450 **HEAT CAPACITY OF CEROUS MAGNESIUM NITRATE AND RELATED MATERIALS BETWEEN 0.3 AND 4 K.**
COLWELL, J. H.
J. LOW. TEMP. PHYS.
14 ( 1-2 ), 53-71, 1974.

74462 **THERMAL CONDUCTIVITY OF GASES IN A MAGNETIC FIELD. CONCENTRATION DEPENDENCE.**
HEEMSKERK, J. P. J. BULSING, G. F.
KNAAP, H. F. P.
PHYSICA ( UTRECHT )
71 ( 3 ), 515-28, 1974.

74471 **VISCOSITY OF PRODUCTS FROM THE MANUFACTURE OF THE PLASTICIZER BIS( 2-ETHYLHEXYL ) PHTHALATE.**
KRYLOV, V. S. SHITIKOVA, V. A. BRONOV, M. V.
SMIRNOV, YU. A. MAKOVSKAYA, E. B.
PLAST. MASSY
15 ( 12 ), 25-6, 1973.
( FOR ENGLISH TRANSLATION SEE TPRC NO. 77385 )

74497 **THE MEANING OF THE VISCOSITY-PRESSURE DEPENDENCE IN THE CLASSICAL AND THE ELECTRO-HYDRODYNAMICAL THEORY OF LUBRICATION.**
KUSS, E.
MINERALOELTECHNIK
18 ( 7-8 ), 48, 1973.

74519 **MICROSTRUCTURAL ASPECTS OF THERMAL EXPANSION IN GRAPHITES.**
HOLLENBERG, G. W. RUH, R.
AIP ( AMER. INST. PHYS. ), CONF. PROC.
( 17 ), 241-55. 1974.

74528 **MULTIPLE SCATTERING CALCULATION OF THE MIDDLE ULTRAVIOLET REACHING THE GROUND.**
SHETTLE, E. P. GREEN, A. E. S.
APPL. OPT.
13 ( 7 ), 1567-81, 1974.

74573 **THE PROPERTIES OF THE EQUATION OF STATE OF AIR BASED ON THE LENNARD-JONES POTENTIAL.**
ZUBAREV, V. N. KOZLOV, A. D.
KRUPINA, N. L. SPIRIDONOV, G. A.
TEPLOFIZ. VYS. TEMP.
11 ( 6 ), 1178-81, 1973.
( FOR ENGLISH TRANSLATION SEE TPRC NO. 74574 )

74574 **THE PROPERTIES OF THE EQUATION OF STATE OF AIR BASED ON THE LENNARD-JONES POTENTIAL.**
ZUBAREV, V. N. KOZLOV, A. D.
KRUPINA, N. L. SPIRIDONOV, G. A.
HIGH TEMP.
11 ( 6 ), 1053-6, 1973.
( ENGLISH TRANSLATION OF TEPLOFIZ. VYS. TEMP., 11 ( 6 ), 1178-81, 1973; FOR ORIGINAL SEE TPRC NO. 74573 )

74677 **SATURATED VAPOR PRESSURE AND SPECIFIC HEAT OF SEAWATER AT 25-350 C.**
MASHOVETS, V. P. MATVEEVA, R. P.
PUCHKOV, L. V. FEDOROV, M. K.
J. APPL. CHEM., USSR
46 ( 9 ), 2080-2, 1973.
( ENGLISH TRANSLATION OF ZH. PRIKL. KHIM., 46 ( 9 ), 1954-6, 1973; FOR ORIGINAL SEE TPRC NO. 72964 )

74678 **DENSITY, VISCOSITY, AND ELECTRICAL CONDUCTIVITY OF SEAWATER AT TEMPERATURES UP TO 300-350 C.**
MASHOVETS, V. P. PUCHKOV, L. V. SMAEV, V. N.
FEDOROV, M. K. FEDOTOV, N. V.
J. APPL. CHEM., USSR
46 ( 10 ), 1988-90, 1973.
( ENGLISH TRANSLATION OF ZH. PRIKL. KHIM., 46 ( 10 ), 1865-8, 1973; FOR ORIGINAL SEE TPRC NO. 72963 )

74679 **SPECIFIC HEATS OF SOLUTIONS OF SODIUM IODINE AND CESIUM IODIDE IN DIMETHYL SULFOXIDE AT 25.**
RYABCHENKO, O. I. MISHCHENKO, K. P.
NOVOSELOV, N. P. KLYUEVA, M. L.
J. GEN. CHEM., USSR
43 ( 9 ), 1874-6, 1973.
( ENGLISH TRANSLATION OF ZH. OBSHCH. KHIM., 43 ( 9 ), 1886-90, 1973; FOR ORIGINAL SEE TPRC NO. 73063 )

74721 **DENSITY AND VISCOSITY OF SEVERAL MOLTEN FLUORIDE MIXTURES.**
CANTOR, S.
OAK RIDGE NATL. LAB., OAK RIDGE, TENN.
33PP., 1973.
( ORNL-TM-4308 )

74736 **DYNAMICS OF FLUIDS NEAR THE CRITICAL POINT: DECAY RATE OF ORDER-PARAMETER FLUCTUATIONS.**
SWINNEY, H. L. HENRY, D. L.
PHYS. REV.
8 A ( 5 ), 2586-2617, 1973.

74745 **PHENOMENOLOGICAL HELIUM 3 QUASIPARTICLE SCATTERING AMPLITUDE IN DILUTE SOLUTIONS OF HELIUM 3 IN SUPERFLUID HELIUM 4.**
KUENHOLD, K. A. EBNER, C.
PHYS. REV.
9 A ( 6 ), 2724-32, 1974.

74757 **THERMAL CONDUCTIVITIES OF ETHENE AND ITS FLUOROCARBON ANALOGS AND OF THEIR BINARY MIXTURES WITH ARGON.**
GRAY, P. PARKINSON, C.
J. CHEM. SOC., FARADAY TRANS.
70, 1 ( 3 ), 560-71, 1974.

74759 **SPECIFIC HEAT OF HELIUM-3-HELIUM-4 MIXTURES IN VYCOR GLASS.**
EGGINGTON, M. A. MOORE, M. A.
J. LOW TEMP. PHYS.
15 ( 1-2 ), 99-110, 1974.

74763 **VISCOSITY AND DENSITY AT 20 OF THE SYSTEM POLYPHOSPHORIC ACID-POTASSIUM SULFATE.**
BUGAI, P. M. EREMENKO, R. A.
KHIM. TEKHNOL., RESPUB. MEZHVEDOM. TEMAT. NAUCH.-TEKH. SB.
( 22 ), 79-81, 1971.

74765 **STRUCTURAL VISCOSITY OF LIQUID CRYSTALS.**
GORSKII, F. K. MIKHLIN, M. E.
KUBARKO, A. I.
KINET. MEKH. KRIST.
176-80, 1973.
( EDITED BY N. N. SIROTA, NAUKA I TEKHNIKA: MINSK, USSR )

74778 **INFLUENCE OF SODIUM CHLORIDE AND LITHIUM FLUORIDE ON THE DENSITY AND VISCOSITY OF MOLTEN SODIUM HEXAFLUOROALUMINATE-BARIUM CHLORIDE SYSTEM.**
ZUCA, S. COSTIN, R. OLTEANU, M.
REV. ROUM. CHIM.
19 ( 2 ), 181-6, 1974.

74780 **VISCOSITY OF DILUTE HELIUM-3-HELIUM-4 MIXTURES AT LOW TEMPERATURES.**
KUENHOLD, K. A.
OHIO STATE UNIVERSITY, COLUMBUS, PH. D. THESIS
137PP., 1973.
( AVAIL. UNIV. MICROFILMS, ORDER NO. 73-18,919 )

| TPRC Number | Bibliographic Citation |
|---|---|
| 74787 | **CONDUCTIVITY, DENSITY, AND VISCOSITY OF ELECTROLYTES FROM DILUTE TO CONCENTRATED SOLUTIONS AT VARIOUS TEMPERATURES IN HIGH-DIELECTRIC-CONSTANT SOLVENTS.**<br>CASTEEL, J. F.<br>UNIV. ARKANSAS, FAYETTEVILLE, ARK., PHD. THESIS<br>259PP., 1973.<br>( AVAIL. UNIV. MICROFILMS, NO. 73-27,410 ) |
| 74791 | **THERMAL CONDUCTIVITY OF BINARY LIQUID MIXTURES.**<br>JAMIESON, D. T. IRVING, J. B.<br>NATIONAL ENGINEERING LABORATORY, EAST KILBRIDE, GLASGOW<br>8PP., 1974.<br>( NEL-R-567 ) |
| 74799 | **GRAPHIC CORRELATION OF EXPERIMENTAL VALUES OF THE THERMAL CONDUCTIVITY OF BINARY MONOATOMIC GAS MIXTURES.**<br>MARCHENKOV, E. I.<br>TEPLO-MASSOPERENOS VYS. TEMP.<br>121-8, 1973. |
| 74817 | **RESOLUTION OF THE VISCOSITY COEFFICIENT B IN THE JONES-DOLE EQUATION INTO IONIC COMPONENTS.**<br>KRUMGAL'Z, B. S.<br>RUSS. J. PHYS. CHEM.<br>47 ( 7 ), 956-8, 1973.<br>( ENGLISH TRANSLATION OF ZH. FIZ. KHIM., 47 ( 7 ), 1691-4, 1973; FOR ORIGINAL SEE TPRC NO. 72530 ) |
| 74820 | **ENTHALPIES AND HEAT CAPACITIES IN THE POTASSIUM HYDROXIDE-WATER SYSTEM.**<br>GINZBURG, D. M. KOCHKALDA, V. E. GUBA, N. I.<br>RUSS. J. PHYS. CHEM.<br>47 ( 8 ), 1214-5, 1973.<br>( ENGLISH TRANSLATION OF ZH. FIZ. KHIM., 47 ( 8 ), 2145-6, 1973; FOR ORIGINAL SEE TPRC NO. 72532 ) |
| 74822 | **THERMAL CONDUCTIVITY OF THE POTASSIUM CHLORIDE - LEAD( II ) CHLORIDE SYSTEM.**<br>POLYAKOV, P. V. GIL'DEBRANDT, E. M.<br>KOTEL'NIKOVA, G. A. MOZHAEVA, V. N.<br>RUSS. J. PHYS. CHEM.<br>47 ( 8 ), 1221, 1973.<br>( ENGLISH TRANSLATION OF ZH. FIZ. KHIM., 47 ( 8 ), 2155, 1973; FOR ORIGINAL SEE TPRC NO. 72533 )<br>( COMPLETE PAPER DEPOSITED AT THE ALL-UNION INST. OF SCIENTIFIC AND TECHNICAL INFORMATION( VINITI ) )<br>( NO. 6069-73 DEP. FROM 14TH MAY 1973 ) |
| 74825 | **VISCOSITY OF A MIXTURE OF ORDINARY AND HEAVY WATER AT HIGH PRESSURES AND 75-225 C.**<br>AGAEV, N. A. ABASZADE, A. KERIMOV, A. M.<br>RUSS. J. PHYS. CHEM.<br>47 ( 8 ), 1229-30, 1973.<br>( ENGLISH TRANSLATION OF ZH. FIZ. KHIM., 47 ( 8 ), 2168, 1973; FOR ORIGINAL SEE TPRC NO. 72537 )<br>( COMPLETE PAPER DEPOSITED AT THE ALL-UNION INST. OF SCIENTIFIC AND TECHNICAL INFORMATION( VINITI ) )<br>( NO. 6113-73 DEP. FROM 14TH MAY 1973 ) |
| 74827 | **FREQUENCY DEPENDENT TRANSPORT COEFFICIENTS IN BINARY MIXTURES.**<br>HAUS, J. W., JR.<br>J. CHEM. PHYS.<br>60 ( 7 ), 2638-42, 1974. |
| 74832 | **EFFECT OF PRESSURE ON DEUTERON SPIN-LATTICE RELAXATION IN SEVERAL CONCENTRATED DEUTERIUM OXIDE DIAMAGNETIC ELECTROLYTE SOLUTIONS.**<br>LEE, Y. K. CAMPBELL, J. H. JONAS, J.<br>J. CHEM. PHYS.<br>60 ( 9 ), 3537-43, 1974. |
| 74854 | **THERMAL CONDUCTIVITY OF BINARY LIQUID MIXTURES.**<br>SAKSENA, M. P. HARMINDER<br>CHEM. PHYS. LETT.<br>25 ( 3 ), 445-8, 1974. |
| 74869 | **THERMAL CONDUCTIVITY OF LACTOSE AQUEOUS SOLUTIONS.**<br>BYVAL'TSEV, YU. A. PERELYGIN, B. G.<br>POLYANSKII, K. K.<br>IZV. VYSSH UCHEB. ZAVED., PISHCH. TEKHNOL.<br>( 1 ), 70-2, 1974. |
| 74893 | **HEAT CAPACITY OF MANGANESE NICKEL CHLORIDE SYSTEM ALLOYS AT LOW TEMPERATURES.**<br>KOSTRYUKOVA, M. O. SEDOVA, L. P.<br>PIS'MA ZH. EKSP. TEOR. FIZ.<br>19 ( 5 ), 306-10, 1974.<br>( FOR ENGLISH TRANSLATION SEE TPRC NO. 76536 ) |
| 74918 | **THERMAL CONDUCTIVITY OF A MIXTURE OF MONOATOMIC AND POLYATOMIC GASES.**<br>KAMCHATOV, F. P.<br>VESTSI AKAD. NAVUK BELARUS. SSR, SER. FIZ.-ENERG. NAVUK<br>( 1 ), 97-102, 1974. |
| 74924 | **ANTIMONY RUBIDIUM SULFIDE-ANTIMONY TRISULFIDE SYSTEM.**<br>BAZAKUTSA, V. A. GNIDASH, N. I.<br>LAZAREV, V. B. SALOV, A. V. ROGACHEVA, E. I.<br>SUKHORUKOVA, L. N. VASIL'EVA, M. P.<br>BERUL, S. I.<br>ZH. NEORG. KHIM.<br>19 ( 3 ), 807-11, 1974.<br>( FOR ENGLISH TRANSLATION SEE TPRC NO. 77018 ) |
| 74929 | **VISCOSITY OF BINARY LIQUID MIXTURES INVOLVING SULFOLANE AND ALCOHOLS.**<br>JANELLI, L. RAKSHIT, A. K. SACCO, A.<br>Z. NATURFORSCH., TEIL<br>29 A ( 2 ), 355-8, 1974. |
| 74951 | **APPARENT MOLAL VOLUMES AND HEAT CAPACITIES OF UREA AND METHYL-SUBSTITUTED UREAS IN WATER AND DEUTERIUM OXIDE AT 25 DEGREES.**<br>PHILIP, P. R. PERRON, G. DESNOYERS, J. E.<br>CAN. J. CHEM.<br>52 ( 9 ), 1709-13, 1974. |
| 74952 | **TRANSLATIONAL HYDRODYNAMICS FOR BINARY MIXTURES. MEMORY FROM INTERNAL MOLECULAR RELAXATION.**<br>KAPRAL, R. DESAI, R. C.<br>CHEM. PHYS.<br>3 ( 2 ), 141-63, 1974. |
| 74962 | **VISCOSITY AND SUPERSATURATION OF SUGAR SOLUTIONS. RHEOLOGICAL PROPERTIES DURING THE COURSE OF A STRIKE.**<br>THIELE, H. LANGEN, A.<br>INT. SUGAR J.<br>76 ( 905 ), 136-40, 1974. |
| 74964 | **EFFECT OF THE MAGNETIC TREATMENT OF WATER ON SOME PROPERTIES OF ALCOHOL SOLUTIONS.**<br>MINENKO, V. I. IVANOVA, N. S.<br>KALENICHENKO, K. I.<br>IXV. VYSSH. UCHEB. ZAVED., KHIM. KHIM. TEKHNOL.<br>17 ( 2 ), 214-17, 1974. |
| 74976 | **TRANSPORT PHENOMENA IN CONCENTRATED AQUEOUS SOLUTIONS OF ALKALI METAL AND ALKALINE EARTH HALIDES.**<br>FORTES, J. M. MERCIER, M. MOLENAT, J.<br>J. CHIM. PHYS. PHYSICOCHIM. BIOL.<br>71 ( 2 ), 164-70, 1974. |
| 74999 | **VISCOSITY OF AQUEOUS SOLUTIONS OF COBALT AND NICKEL CHLORIDES AND SULFATES IN THE 20-100 RANGE.**<br>FEDOTOV, N. V. MAKSIMOVA, I. N.<br>ZH. FIZ. KHIM.<br>48 ( 3 ), 773, 1974.<br>( FOR ENGLISH TRANSLATION SEE TPRC NO. 76460 ) |
| 75005 | **EXPERIMENTAL DETERMINATION OF THE THERMAL CONDUCTIVITY OF BINARY GAS MIXTURES.**<br>SHASHKOV, A. G. KAMCHATOV, F. P.<br>J. ENG. PHYS.<br>22 ( 5 ), 586-90, 1974.<br>( ENGLISH TRANSLATION OF INZH. FIZ. ZH., 22 ( 5 ), 843-9, 1972; FOR ORIGINAL SEE TPRC NO. 67680 ) |
| 75006 | **DYNAMIC METHOD OF MEASURING THE THERMAL CONDUCTIVITY OF GASES AT HIGH TEMPERATURES.**<br>MUSTAFAEV, R. A.<br>J. ENG. PHYS.<br>22 ( 5 ), 591-7, 1974.<br>( ENGLISH TRANSLATION OF INZH. FIZ. ZH., 22 ( 5 ), 850-8, 1972; FOR ORIGINAL SEE TPRC NO. 67681 ) |
| 75016 | **THE THERMODYNAMIC PROPERTIES OF AIR.**<br>VUKALOVICH, M. P. ZUBAREV, V. N.<br>ISSLED. PO TERMODIN.<br>138-51, 1973.<br>( FOR ENGLISH TRANSLATION SEE TPRC NO. 75017 ) |
| 75017 | **THE THERMODYNAMIC PROPERTIES OF AIR.**<br>VUKALOVICH, M. P. ZUBAREV, V. N.<br>FOREIGN TECHNOLOGY DIV., WPAFB, OHIO<br>19PP., 1974.<br>( ENGLISH TRANSLATION OF ISSLED. PO TERMODIN., 138-51, 1973; FOR ORIGINAL SEE TPRC NO. 75016 )<br>( FTD-MT-24-1272-74 ) |
| 75020 | **DEPOLARIZED RAYLEIGH SCATTERING IN LIQUIDS. MOLECULAR REORIENTATION AND ORIENTATION PAIR CORRELATIONS IN A NEMATIC LIQUID CRYSTAL: MBBA.**<br>GIERKE, T. D. FLYGARE, W. H.<br>J. CHEM. PHYS.<br>61 ( 6 ), 2231-9, 1974. |
| 75021 | **DEPOLARIZED RAYLEIGH SCATTERING AND 13 C NMR STUDIES OF ANISOTROPIC MOLECULAR REORIENTATION OF AROMATIC COMPOUNDS IN SOLUTION.**<br>BAUER, D. R. ALMS, G. R. BRAUMAN, J. I.<br>PECORA, R.<br>J. CHEM. PHYS.<br>61 ( 6 ), 2255-61, 1974. |
| 75023 | **ABSORPTION SPECTRA OF MERCURY ATOMS TRAPPED IN SOLID INERT GASES.**<br>MALO, S. A.<br>J. CHEM. PHYS.<br>61 ( 6 ), 2409-11, 1974. |

| TPRC Number | Bibliographic Citation |
|---|---|

75059 INFLUENCE OF BOTTOM DEPTH AND ALBEDO ON THE DIFFUSE REFLECTANCE OF A FLAT HOMOGENEOUS OCEAN.
GORDON, H. R. BROWN, O. B.
APPL. OPT.
13 ( 9 ), 2153-9, 1974.

75063 OCEAN COLOR SPECTRUM CALCULATIONS.
MC CLUNEY, W. R.
APPL. OPT.
13 ( 10 ), 2422-9, 1974.

75069 HIERARCHY OF COMPONENTS FOR CALCULATING THE VISCOSITY OF A MIXTURE.
DUBNER, V. M. KHAIMOV, A. ZH.
TEPLOFIZ. VYS. TEMP.
12 ( 1 ), 59-63, 1974.
( FOR ENGLISH TRANSLATION SEE TPRC NO. 75070 )

75070 HIERARCHY OF COMPONENTS FOR CALCULATING THE VISCOSITY OF A MIXTURE.
DUBNER, V. M. KHAIMOV, A. ZH.
HIGH TEMP.
12 ( 1 ), 51-5, 1974.
( ENGLISH TRANSLATION OF TEPLOFIZ. VYS. TEMP., 12 ( 1 ), 59-63, 1974; FOR ORIGINAL SEE TPRC NO. 75069 )

75073 THERMAL CONDUCTIVITY OF CARBON-GRAPHITE MATERIALS WITH BARIUM OXIDE ADDITIONS.
BARYKIN, B. M. LEVINOV, B. M.
REKOV, A. I. SPIRIDONOV, E. G.
TEPLOFIZ. VYS. TEMP.
12 ( 1 ), 92-6, 1974.
( FOR ENGLISH TRANSLATION SEE TPRC NO. 75074 )

75074 THERMAL CONDUCTIVITIES OF GRAPHITIZED CARBONS CONTAINING BARIUM OXIDE.
BARYKIN, B. M. LEVINOV, B. M. REKOV, A. I.
SPIRIDONOV, E. G.
HIGH TEMP.
12 ( 1 ), 80-3, 1974.
( ENGLISH TRANSLATION OF TEPLOFIZ. VYS. TEMP., 12 ( 1 ), 92-6, 1974; FOR ORIGINAL SEE TPRC NO. 75073 )

75075 THERMODYNAMIC PROPERTIES OF THE COMBUSTION PRODUCTS OF COALS WITH ADDED POTASSIUM.
SAMUILOV, E. V. ROZHDESTVENSKII, I. B.
TSITELAURI, N. N.
TEPLOFIZ. VYS. TEMP.
12 ( 1 ), 204-6, 1974.
( FOR ENGLISH TRANSLATION SEE TPRC NO. 75076 )

75076 THERMODYNAMIC PROPERTIES OF THE COMBUSTION PRODUCTS OF COALS WITH ADDED POTASSIUM.
SAMUILOV, E. V. ROZHDESTVENSKII, I. B.
TSITELAURI, N. N.
HIGH TEMP.
12 ( 1 ), 178-80, 1974.
( ENGLISH TRANSLATION OF TEPLOFIZ. VYS. TEMP., 12 ( 1 ), 204-6, 1974; FOR ORIGINAL SEE TPRC NO. 75075 )

75142 THEORY OF TRANSIENT HOT-WIRE MEASUREMENT OF THE THERMAL CONDUCTIVITY OF ELECTROLYTIC SOLUTIONS.
BARUEL, P.
DANISH ATOMIC ENERGY COMMISSION, RISOE
75PP., 1973.
( RISO-280 )

75167 DETERMINATION OF THE THERMOPHYSICAL CHARACTERISTICS OF SUSPENSIONS.
ALTOIZ, B. A. POLISHCHUK, D. I.
FIZ. AERODISPERSNYKH SIST.
8, 91-5, 1973.

75169 THERMAL CONDUCTIVITY OF FUSED SALTS STUDIED BY THE ROTATING DISK METHOD.
SHURYGIN, P. M. LEONOV, V. V.
BUZOVKIN, V. P.
FIZ. KHIM. ELEKTROKHIM. RASPLAV. SOLEI TVERD. ELEKTROLIT.
( 1 ), 50-2, 1973.

75177 CONCENTRATION-THERMAL CONDUCTIVITY DEPENDENCE OF SALT WATER SOLUTIONS AND EFFECT OF ION SIZES ON THERMAL CONDUCTIVITY.
FATULLAEV, F. G. KERIMOV, A. M.
IZV. AKAD. NAUK AZERB. SSR, SER. FIZ.-TEKH. MAT. NAUK
( 3 ), 124-7, 1973.

75203 DENSITY AND VISCOSITY OF BINARY LIQUID SYSTEMS FORMED BY METHANOL ACETIC ACID, AND WATER.
GRANZHAN, V. A.
ZH. FIZ. KHIM.
48 ( 4 ), 1060-1, 1974.
( FOR ENGLISH TRANSLATION SEE TPRC NO. 76465 )

75204 STRUCTURE OF THE COMPLEXES OF GLYCINE AND GLYCYLGLYCINE WITH COPPER( II ) IN SOLUTION, BASED ON VISCOSITY MEASUREMENTS.
SAKELLARIDIS, P. U. KARAGEORGOPOULOS, E. K.
TSANGARIS, J. M.
Z. NATURFORSCH., TEIL
29 B ( 1-2 ), 83-5, 1974.

75220 VISCOSITY AND CONDUCTIVITY DATA. 4. PROPANE AND MIXTURES.
BONSCHER, F. S. SHIPMAN, L. M. YEN, L. C.
HYDROCARBON PROCESS.
53 ( 4 ), 169-72, 1974.

75277 EXPERIMENTAL DETERMINATION OF THE THERMAL CONDUCTIVITY OF A BINARY GAS MIXTURE CONTAINING A POLAR COMPONENT.
KAMCHATOV, F. P.
VESTSI AKAD. NAVUK BELARUS. SSR, SER. FIZ.-ENERG. NAVUK
( 4 ), 58-61, 1973.

75325 EVALUATION OF AN INFRARED OIL FILM MONITOR.
WRIGHT, D. E. WRIGHT, J. A.
WRIGHT AND WRIGHT ENVIRONMENTAL ENGINEERING, NEWTON CENTRE, MASS.
89PP., 1973.
( AD-778 814 )

75328 DENSITY AND VISCOSITY OF AQUEOUS HYDROGEN SULFIDE SOLUTIONS AT PRESSURES TO 20 ATM.
MURPHY, J. A. GAINES, G. L., JR.
J. CHEM. ENG. DATA
19 ( 4 ), 359-62, 1974.

75329 RELATIVE VISCOSITIES OF SOME AQUEOUS RARE EARTH CHLORIDE SOLUTIONS AT 25 C.
SPEDDING, F. H. WITTE, D. L. SHIERS, L. E.
RARD, J. A.
J. CHEM. ENG. DATA
19 ( 4 ), 369-73, 1974.

75339 VISCOSITIES OF LIQUID MIXTURES.
NARAYANAN, V. THAYUMANASUNDARAM, R.
LADDHA, G. S.
CHEM. IND. DEVELOP.
8 ( 5 ), 15-19, 1974.

75349 DETERMINING THE SPECIFIC HEAT OF TEA ESSENCE DRIED BY THE SUBLIMATION PROCESS.
LEBEDEV, D. P. GINZBURG, I. A.
ALEKSEEV, V. V.
INZH. FIZ. ZH.
23 ( 2 ), 284-7, 1972.
( FOR ENGLISH TRANSLATION SEE TPRC NO. 75350 )

75350 DETERMINING THE SPECIFIC HEAT OF TEA ESSENCE DRIED BY THE SUBLIMATION PROCESS.
LEBEDEV, D. P. GINZBURG, I. A.
J. ENG. PHYS.
23 ( 2 ), 1001-3, 1974.
( ENGLISH TRANSLATION OF INZH. FIZ. ZH., 23 ( 2 ), 284-7, 1972; FOR ORIGINAL SEE TPRC NO. 75349 )

75361 APPROXIMATION OF THE FUNCTIONAL DEPENDENCE OF THE HEAT CAPACITY OF PEAT COMBUSTION PRODUCTS ON SOME VARIABLE FACTORS.
OPMAN, YA. S. VENZEL, E. F.
IZV. VYSSH. UCHEB. ZAVED., ENERG.
17 ( 1 ), 145-7, 1974.

75373 INFRARED OPTICAL CONSTANTS OF AQUEOUS SOLUTIONS OF ELECTROLYTES. ACIDS AND BASES.
RHINE, P. WILLIAMS, D. HALE, G. M.
QUERRY, M. R.
J. PHYS. CHEM.
78 ( 14 ), 1405-10, 1974.

75396 RENORMALIZATION-GROUP CALCULATIONS OF DIVERGENT TRANSPORT COEFFICIENTS AT CRITICAL POINTS.
HALPERIN, B. I. HOHENBERG, P. C.
SIGGIA, E. D.
PHYS. REV. LETT.
32 ( 23 ), 1289-92, 1974.

75417 VIBRATIONAL SPECTRA OF CHROMATE ( CHROMIUM-50 ), CHROMATE ( CHROMIUM-53 ), CHROMATE ( OXYGEN-18 ), MOLYBDATE ( MOLYBDENUM-92 ), MOLYBDATE ( MOLYBDENUM-100 ), AND RHENATE ( OXYGEN-18 ) IONS. CALCULATION OF EXACT FORCE CONSTANTS OF IONS.
MUELLER, A. KOENIGER, F. WEINSTOCK, N.
SPECTROCHIM. ACTA
30 A ( 3 ), 641-9, 1974.

75430 ADIABATIC CALORIMETER FOR MEASURING SPECIFIC HEATS OF LIQUIDS IN THE TEMPERATURE RANGE FROM 0 TO 100 C.
PETROV, G. I. PUCHKOV, L. V.
J. APPL. CHEM., USSR
46 ( 10 ), 2373-5, 1973.
( ENGLISH TRANSLATION OF ZH. PRIKL. KHIM., 46 ( 10 ), 2233-6, 1973; FOR ORIGINAL SEE TPRC NO. 73064 )

75459 HEAT TRANSFER DURING UPWARD AND DOWNWARD FLOW OF GAS-LIQUID MIXTURES IN VERTICAL PIPES.
DOMANSKII, I. V. TISHIN, V. B.
SOKOLOV, V. N.
ZH. PRIKL. KHIM.
43 ( 6 ), 1392-4, 1970.
( FOR ENGLISH TRANSLATION SEE TPRC NO. 75460 )

| TPRC Number | Bibliographic Citation |
|---|---|
| 75460 | **HEAT TRANSFER DURING UPWARD AND DOWNWARD FLOW OF GAS-LIQUID MIXTURES IN VERTICAL PIPES.** DOMANSKII, I. V. TISHIN, V. B. SOKOLOV, V. N. J. APPL. CHEM., USSR 43 (6), 1398-1400, 1970. (ENGLISH TRANSLATION OF ZH. PRIKL. KHIM., 43 (6), 1392-4, 1970; FOR ORIGINAL SEE TPRC NO. 75459) |
| 75461 | **PHYSICOCHEMICAL PROPERTIES OF SOLUTIONS IN THE SYSTEM UREA-COPPER NITRATE-WATER AND OF THE COMPOUNDS COPPER NITRATE TETRA UREA AND COPPER NITRATE HEXA UREA.** PROTSENKO, P. I. KALAEVA, KH. I. PROTSENKO, G. P. BRYKOVA, N. A. ZH. PRIKL. KHIM. 43 (7), 1457-62, 1970. (FOR ENGLISH TRANSLATION SEE TPRC NO. 75462) |
| 75462 | **PHYSICOCHEMICAL PROPERTIES OF SOLUTIONS IN THE SYSTEM UREA-COPPER NITRATE-WATER AND OF THE COMPOUNDS COPPER NITRATE TETRA UREA AND COPPER NITRATE HEXA UREA.** PROTSENKO, P. I. KALAEVA, KH. I. PROTSENKO, G. P. BRYKOVA, N. A. J. APPL. CHEM., USSR 43 (7), 1469-73, 1970. (ENGLISH TRANSLATION OF ZH. PRIKL. KHIM., 43 (7), 1457-62, 1970; FOR ORIGINAL SEE TPRC NO. 75461) |
| 75483 | **STUDY OF THE PRODUCTS OF HYDROTHERMAL HARDENING OF ALUMINA-SILICA MIXTURES.** VED', E. I. LITVINOVA, Z. S. ZH. PRIKL. KHIM. 44 (5), 1163-6, 1971. (FOR ENGLISH TRANSLATION SEE TPRC NO. 75484) |
| 75484 | **STUDY OF THE PRODUCTS OF HYDROTHERMAL HARDENING OF ALUMINA-SILICA MIXTURES.** VED', E. I. LITVINOVA, Z. S. J. APPL. CHEM., USSR 44 (5), 1177-80, 1971. (ENGLISH TRANSLATION OF ZH. PRIKL. KHIM., 44 (5), 1163-6, 1971; FOR ORIGINAL SEE TPRC NO. 75483) |
| 75489 | **SOME PHYSICOCHEMICAL PROPERTIES OF AQUEOUS HYDROXYLAMINE SULFATE SOLUTIONS.** SKVORTSOV, G. A. SKVORTSOV, A. B. KLYUSHNEVA, M. N. ZH. PRIKL. KHIM. 44 (7), 1518-21, 1971. (FOR ENGLISH TRANSLATION SEE TPRC NO. 75490) |
| 75490 | **SOME PHYSICOCHEMICAL PROPERTIES OF AQUEOUS HYDROXYLAMINE SULFATE SOLUTIONS.** SKVORTSOV, G. A. SKVORTSOV, A. B. KLYUSHNEVA, M. N. J. APPL. CHEM., USSR 44 (7), 1540-2, 1971. (ENGLISH TRANSLATION OF ZH. PRIKL. KHIM., 44 (7), 1518-21, 1971; FOR ORIGINAL SEE TPRC NO. 75489) |
| 75491 | **VISCOSITY AND DENSITY OF CALCIUM, AMMONIUM, AND POTASSIUM PHOSPHATE SOLUTIONS IN PHOSPHORIC ACID.** OREKHOV, I. I. TERESHCHENKO, L. YA. CHEBOTARENKO, N. M. ZH. PRIKL. KHIM. 44 (7), 1558-62, 1971. (FOR ENGLISH TRANSLATION SEE TPRC NO. 75492) |
| 75492 | **VISCOSITY AND DENSITY OF CALCIUM, AMMONIUM, AND POTASSIUM PHOSPHATE SOLUTIONS IN PHOSPHORIC ACID.** OREKHOV, I. I. TERESHCHENKO, L. YA. CHEBOTARENKO, N. M. J. APPL. CHEM., USSR 44 (7), 1577-9, 1971. (ENGLISH TRANSLATION OF ZH. PRIKL. KHIM., 44 (7), 1558-62, 1971; FOR ORIGINAL SEE TPRC NO. 75491) |
| 75493 | **DETERMINATION OF THE HEAT CAPACITIES OF LIQUIDS OF THE SODA INDUSTRY AT 25 C BY CALCULATION COMMUNICATION II.** CHERNEN'KAYA, E. I. ZH. PRIKL. KHIM. 44 (8), 1721-5, 1971. (FOR ENGLISH TRANSLATION SEE TPRC NO. 75494) |
| 75494 | **DETERMINATION OF THE HEAT CAPACITIES OF LIQUIDS OF THE SODA INDUSTRY AT 25 C BY CALCULATION COMMUNICATION II.** CHERNEN'KAYA, E. I. J. APPL. CHEM., USSR 44 (8), 1744-7, 1971. (ENGLISH TRANSLATION OF ZH. PRIKL. KHIM., 44 (8), 1721-5, 1971; FOR ORIGINAL SEE TPRC NO. 75493) |
| 75499 | **PERCHLORIC ACID IN THE SYSTEM ACETIC ANHYDRIDE-BENZENE.** BYMENSKII, V. YA. SOROKIN, E. S. ZH. PRIKL. KHIM. 44 (8), 1908-11, 1971. (FOR ENGLISH TRANSLATION SEE TPRC NO. 75500) |
| 75500 | **PERCHLORIC ACID IN THE SYSTEM ACETIC ANHYDRIDE-BENZENE.** BYMENSKII, V. YA. SOROKIN, E. S. J. APPL. CHEM., USSR 44 (8), 1938-41, 1971. (ENGLISH TRANSLATION OF ZH. PRIKL. KHIM., 44 (8), 1908-11, 1971; FOR ORIGINAL SEE TPRC NO. 75499) |
| 75503 | **THERMODYNAMIC CHARACTERISTICS OF FORMATION OF OXALATE, PERMANGANATE AND MANGANESE IONS IN AQUEOUS SOLUTIONS AT VARIOUS TEMPERATURES.** KORNILOVA, N. V. DYMARCHUK, N. P. MISHCHENKO, K. P. ZH. PRIKL. KHIM. 44 (9), 2121-2, 1971. (FOR ENGLISH TRANSLATION SEE TPRC NO. 75504) |
| 75504 | **THERMODYNAMIC CHARACTERISTICS OF FORMATION OF OXALATE, PERMANGANATE AND MANGANESE IONS IN AQUEOUS SOLUTIONS AT VARIOUS TEMPERATURES.** KORNILOVA, N. V. DYMARCHUK, N. P. MISHCHENKO, K. P. J. APPL. CHEM., USSR 44 (9), 2160-2, 1971. (ENGLISH TRANSLATION OF ZH. PRIKL. KHIM., 44 (9), 2121-2, 1971; FOR ORIGINAL SEE TPRC NO. 75503) |
| 75513 | **PHYSICOCHEMICAL STUDY OF SYSTEMS FORMED BY METHYL ALCOHOL WITH CYCLOHEXANONE AND DIMETHYLFORMAMIDE.** GRANZHAN, V. A. KIRILLOVA, O. G. ZH. PRIKL. KHIM. 45 (1), 239-40, 1972. (FOR ENGLISH TRANSLATION SEE TPRC NO. 75514) |
| 75514 | **PHYSICOCHEMICAL STUDY OF SYSTEMS FORMED BY METHYL ALCOHOL WITH CYCLOHEXANONE AND DIMETHYLFORMAMIDE.** GRANZHAN, V. A. KIRILLOVA, O. G. J. APPL. CHEM., USSR 45 (1), 238-9, 1972. (ENGLISH TRANSLATION OF ZH. PRIKL. KHIM., 45 (1), 239-40, 1972; FOR ORIGINAL SEE TPRC NO. 75513) |
| 75521 | **CERTAIN PROPERTIES OF SYSTEMS CONTAINING AMMONIUM PHOSPHATES AND AMMONIUM NITRATE.** ZHUL'KOV, V. I. SYRKIN, L. N. KOZLOVA, T. G. SHIPYATSKAYA, R. I. SMIRNOVA, N. M. KOZINA, T. M. ZH. PRIKL. KHIM. 45 (4), 868-71, 1972. (FOR ENGLISH TRANSLATION SEE TPRC NO. 75522) |
| 75522 | **CERTAIN PROPERTIES OF SYSTEMS CONTAINING AMMONIUM PHOSPHATES AND AMMONIUM NITRATE.** ZHUL'KOV, V. I. SYRKIN, L. N. KOZLOVA, T. G. SHIPYATSKAYA, R. I. SMIRNOVA, N. M. KOZINA, T. M. J. APPL. CHEM., USSR 45 (4), 881-3, 1972. (ENGLISH TRANSLATION OF ZH. PRIKL. KHIM., 45 (4), 868-71, 1972; FOR ORIGINAL SEE TPRC NO. 75521) |
| 75523 | **ACTIVATION ENERGY FOR VISCOUS FLOW OF AQUEOUS ELECTROLYTE SOLUTIONS.** PENKINA, N. V. ZH. PRIKL. KHIM. 45 (4), 879-80, 1972. (FOR ENGLISH TRANSLATION SEE TPRC NO. 75524) |
| 75524 | **ACTIVATION ENERGY FOR VISCOUS FLOW OF AQUEOUS ELECTROLYTE SOLUTIONS.** PENKINA, N. V. J. APPL. CHEM., USSR 45 (4), 895-6, 1972. (ENGLISH TRANSLATION OF ZH. PRIKL. KHIM., 45 (4), 879-80, 1972; FOR ORIGINAL SEE TPRC NO. 75523) |
| 75525 | **NEGATIVE VISCOSITY OF AQUEOUS ELECTROLYTE SOLUTIONS.** PENKINA, N. V. KHRAMTSOVA, T. S. ZH. PRIKL. KHIM. 45 (4), 881, 1972. (FOR ENGLISH TRANSLATION SEE TPRC NO. 75526) |
| 75526 | **NEGATIVE VISCOSITY OF AQUEOUS ELECTROLYTE SOLUTIONS.** PENKINA, N. V. KHRAMTSOVA, T. S. J. APPL. CHEM., USSR 45 (4), 898, 1972. (ENGLISH TRANSLATION OF ZH. PRIKL. KHIM., 45 (4), 881, 1972; FOR ORIGINAL SEE TPRC NO. 75525) |
| 75527 | **CONCENTRATION DEPENDENCES OF VISCOSITY OF AQUEOUS ELECTROLYTE SOLUTIONS.** PENKINA, N. V. ZH. PRIKL. KHIM. 45 (4), 882, 1972. (FOR ENGLISH TRANSLATION SEE TPRC NO. 75528) |
| 75528 | **CONCENTRATION DEPENDENCES OF VISCOSITY OF AQUEOUS ELECTROLYTE SOLUTIONS.** PENKINA, N. V. J. APPL. CHEM., USSR 45 (4), 899, 1972. (ENGLISH TRANSLATION OF ZH. PRIKL. KHIM., 45 (4), 882, 1972; FOR ORIGINAL SEE TPRC NO. 75527) |

| TPRC Number | Bibliographic Citation |
|---|---|
| 75539 | **THERMODYNAMICS OF THE DISSOLUTION OF ANHYDROUS CHLORIDES OF YTTRIUM GROUP ELEMENTS YTTRIUM CHLORIDE, TERBIUM CHLORIDE, ERBIUM CHLORIDE IN WATER AT 0–100 C.** KRESTOV, G. A. KOBENIN, V. A. SEMENOVSKII, S. V. ZH. NEORG. KHIM. 18 ( 1 ), 3–, 1973. ( FOR ENGLISH TRANSLATION SEE TPRC NO. 75540 ) |
| 75540 | **THERMODYNAMICS OF THE DISSOLUTION OF ANHYDROUS CHLORIDES OF YTTRIUM GROUP ELEMENTS YTTRIUM CHLORIDE, TERBIUM CHLORIDE, ERBIUM CHLORIDE IN WATER AT 0–100 C.** KRESTOV, G. A. KOBENIN, V. A. SEMENOVSKII, S. V. RUSS. J. INORG. CHEM. 18 ( 1 ), 1–2, 1973. ( ENGLISH TRANSLATION OF ZH. NEORG. KHIM., 1, ( 1 ), 3–, 1973; FOR ORIGINAL SEE TPRC 75539 ) |
| 75541 | **THE CESIUM SULFATE–AMMONIUM SULFATE–WATER SYSTEM AT 0 C.** USHAKOV, YU. V. ZH. NEORG. KHIM. 18 ( 1 ), 273–, 1973. ( FOR ENGLISH TRANSLATION SEE TPRC NO. 75542 ) |
| 75542 | **THE CESIUM SULFATE–AMMONIUM SULFATE–WATER SYSTEM AT 0 C.** USHAKOV, YU. V. RUSS. J. INORG. CHEM. 18 ( 1 ), 146–7, 1973. ( ENGLISH TRANSLATION OF ZH. NEORG. KHIM., 18 ( 1 ), 273–, 1973; FOR ORIGINAL SEE TPRC NO. 75541 ) |
| 75555 | **THE POTASSIUM NITRATE–SODIUM NITRATE–WATER SYSTEM.** BERGMAN, A. G. SHULYAK, L. F. ZH. NEORG. KHIM. 18 ( 5 ), 1379–, 1973. ( FOR ENGLISH TRANSLATION SEE TPRC NO. 75556 ) |
| 75556 | **THE POTASSIUM NITRATE–SODIUM NITRATE–WATER SYSTEM.** BERGMAN, A. G. SHULYAK, L. F. RUSS. J. INORG. CHEM. 18 ( 5 ), 731–4, 1973. ( ENGLISH TRANSLATION OF ZH. NEORG. KHIM., 18 ( 5 ), 1379–, 1973; FOR ORIGINAL SEE TPRC NO. 75555 ) |
| 75557 | **THE ALUMINUM SULFATE–MANGANESE SULFATE–RUBIDIUM SULFATE WATER SYSTEM AT 25 C.** MAKSIMENKO, A. A. SHEVCHUK, V. G. ZH. NEORG. KHIM. 18 ( 5 ), 1401–, 1973. ( FOR ENGLISH TRANSLATION SEE TPRC NO. 75558 ) |
| 75558 | **THE ALUMINUM SULFATE–MANGANESE SULFATE–RUBIDIUM SULFATE WATER SYSTEM AT 25 C.** MAKSIMENKO, A. A. SHEVCHUK, V. G. RUSS. J. INORG. CHEM. 18 ( 5 ), 741–4, 1973. ( ENGLISH TRANSLATION OF ZH. NEORG. KHIM., 18 ( 5 ), 1401–, 1973; FOR ORIGINAL SEE TPRC NO. 75557 ) |
| 75559 | **BEHAVIOUR OF HAFNIUM ( IV ) SULPHATE IN AQUEOUS SOLUTIONS.** GODNEVA, M. M. MOTOV, D. L. OKHRIMENKO, R. F. ZH. NEORG. KHIM. 18 ( 6 ), 1450–, 1973. ( FOR ENGLISH TRANSLATION SEE TPRC NO. 75560 ) |
| 75560 | **BEHAVIOUR OF HAFNIUM ( IV ) SULPHATE IN AQUEOUS SOLUTIONS.** GODNEVA, M. M. MOTOV, D. L. OKHRIMENKO, R. F. RUSS. J. INORG. CHEM. 18 ( 6 ), 767–9, 1973. ( ENGLISH TRANSLATION OF ZH. NEORG. KHIM., 18 ( 6 ), 1450–, 1973; FOR ORIGINAL SEE TPRC NO. 75559 ) |
| 75563 | **REACTION OF PENTABORANE ( 9 ) WITH PRIMARY AMINES.** ZHIGACH, A. F. LAPTEV, V. T. PETRUNIN, A. B. NIKITIN, V. S. BEKKER, D. B. ZH. NEORG. KHIM. 18 ( 8 ), 2037–, 1973. ( FOR ENGLISH TRANSLATION SEE TPRC NO. 75564 ) |
| 75564 | **REACTION OF PENTABORANE ( 9 ) WITH PRIMARY AMINES.** ZHIGACH, A. F. LAPTEV, V. T. PETRUNIN, A. B. NIKITIN, V. S. BEKKER, D. B. RUSS. J. INORG. CHEM. 18 ( 8 ), 1080–3, 1973. ( ENGLISH TRANSLATION OF ZH. NEORG. KHIM., 18 ( 8 ), 2037–, 1973; FOR ORIGINAL SEE TPRC NO. 75463 ) |
| 75573 | **THE CHOICE OF DIFFUSION LAYER PARAMETERS DETERMINING THE KINETICS OF THE DISSOLUTION OF SALTS.** ZDANOVSKII, A. B. SERDYUK, V. V. ZH. FIZ. KHIM. 47 ( 3 ), 734–, 1973. ( FOR ENGLISH TRANSLATION SEE TPRC NO. 75574 ) |
| 75574 | **THE CHOICE OF DIFFUSION LAYER PARAMETERS DETERMING THE KINETICS OF THE DISSOLUTION OF SALTS.** ZDANOVSKII, A. B. SERDYUK, V. V. RUSS. J. PHYS. CHEM. 47 ( 3 ), 417–8, 1973. ( ENGLISH TRANSLATION OF ZH. FIZ. KHIM., 47 ( 3 ), 734–, 1973; FOR ORIGINAL SEE TPRC NO. 75573 ) |
| 75575 | **THE FORMATION OF COMPLEXES IN AQUEOUS DIETHYLAMINE SOLUTIONS.** OBRAZTSOV, V. I. KHRUSTALEVA, A. A. ZH. FIZ. KHIM. 47 ( 4 ), 926–, 1973. ( FOR ENGLISH TRANSLATION SEE TPRC NO. 75576 ) |
| 75576 | **THE FORMATION OF COMPLEXES IN AQUEOUS DIETHYLAMINE SOLUTIONS.** OBRAZTSOV, V. I. KHRUSTALEVA, A. A. RUSS. J. PHYS. CHEM. 47 ( 4 ), 525–6, 1973. ( ENGLISH TRANSLATION OF ZH. FIZ. KHIM., 47 ( 4 ), 926–, 1973; FOR ORIGINAL SEE TPRC NO. 75575 ) |
| 75591 | **BINARY LIQUID ORTHOPHOSPHORIC ACID SYSTEMS WITH PENTYL ALCOHOLS AND CYCLOHEXANOL.** KRUPATKIN, I. L. BRATUSHCHAK, A. N. ZH. FIZ. KHIM. 47 ( 6 ), 1599–, 1973. ( FOR ENGLISH TRANSLATION SEE TPRC NO. 75592 ) |
| 75592 | **BINARY LIQUID ORTHOPHOSPHORIC ACID SYSTEMS WITH PENTYL ALCOHOLS AND CYCLOHEXANOL.** KRUPATKIN, I. L. BRATUSHCHAK, A. N. RUSS. J. PHYS. CHEM. 47 ( 6 ), 907, 1973. ( ENGLISH TRANSLATION OF ZH. FIZ. KHIM., 47 ( 6 ), 1599–, 1973; FOR ORIGINAL SEE TPRC NO. 75591 ) |
| 75599 | **ELECTRON PARAMAGNETIC RELAXATION IN VISCOUS MANGANESE DIPOSITIVE SOLUTIONS. I. WATER–SUGAR SOLUTIONS OF MANGANESE NITRATE.** VISHNEVSKAYA, G. P. KARIMOVA, A. F. KOZYREV, B. M. ZH. STRUKT. KHIM. 12 ( 1 ), 40–6, 1971. ( FOR ENGLISH TRANSLATION SEE TPRC NO. 75600 ) |
| 75600 | **ELECTRON PARAMAGNETIC RELAXATION IN VISCOUS MANGANESE DIPOSITIVE SOLUTIONS. I. WATER–SUGAR SOLUTIONS OF MANGANESE NITRATE.** VISHNEVSKAYA, G. P. KARIMOVA, A. F. KOZYREV, B. M. J. STRUCT. CHEM., USSR 12 ( 1 ), 33–7, 1971. ( ENGLISH TRANSLATION OF ZH. STRUKT. KHIM., 12 ( 1 ), 40–6, 1971; FOR ORIGINAL SEE TPRC NO. 75599 ) |
| 75601 | **INVESTIGATION OF THE SELF–DIFFUSION OF MOLECULES IN WATER–NONELECTROLYTE MIXTURES BY THE SPIN ECHO METHOD.** EMEL'YANOV, M. I. NIKOLAEV, I. N. SAMIGULLIN, F. M. ZH. STRUKT. KHIM. 12 ( 1 ), 161–3, 1971. ( FOR ENGLISH TRANSLATION SEE TPRC NO. 75602 ) |
| 75602 | **INVESTIGATION OF THE SELF–DIFFUSION OF MOLECULES IN WATER–NONELECTROLYTE MIXTURES BY THE SPIN ECHO METHOD.** EMEL'YANOV, M. I. NIKOLAEV, I. N. SAMIGULLIN, F. M. J. STRUCT. CHEM., USSR 12 ( 1 ), 138–40, 1971. ( ENGLISH TRANSLATION OF ZH. STRUKT. KHIM., 12 ( 1 ), 161–3, 1971; FOR ORIGINAL SEE TPRC NO. 75601 ) |
| 75603 | **ELECTRON PARAMAGNETIC RELAXATION IN VISCOUS SOLUTIONS OF MANGANESE DIPOSITIVE. II. AQUEOUS–GLYCERIN SOLUTIONS OF MANGANESE NITRATE.** VISHNEVSKAYA, G. P. KOZYREV, B. M. KARIMOVA, A. F. ZH. STRUKT. KHIM. 12 ( 2 ), 231–6, 1971. ( FOR ENGLISH TRANSLATION SEE TPRC NO. 75604 ) |
| 75604 | **ELECTRON PARAMAGNETIC RELAXATION IN VISCOUS SOLUTIONS OF MANGANESE DIPOSITIVE. II. AQUEOUS–GLYCERIN SOLUTIONS OF MANGANESE NITRATE.** VISHNEVSKAYA, G. P. KOZYREV, B. M. KARIMOVA, A. F. J. STRUCT. CHEM., USSR 12 ( 2 ), 211–5, 1971. ( ENGLISH TRANSLATION OF ZH. STRUKT. KHIM., 12 ( 2 ), 231–6, 1971; FOR ORIGINAL SEE TPRC NO. 75603 ) |
| 75605 | **INVESTIGATION OF TRANSLATIONAL DIFFUSION OF ALCOHOL AND ACETIC ACID MOLECULES IN MIXTURES WITH CARBON TETRACHLORIDE.** EMEL'YANOV, M. I. SAMIGULLIN, F. M. ZH. STRUKT. KHIM. 12 ( 4 ), 585–90, 1971. ( FOR ENGLISH TRANSLATION SEE TPRC NO. 75606 ) |

| TPRC Number | Bibliographic Citation |
|---|---|

75606 **INVESTIGATION OF TRANSLATIONAL DIFFUSION OF ALCOHOL AND ACETIC ACID MOLECULES IN MIXTURES WITH CARBON TETRACHLORIDE.**
EMEL'YANOV, M. I. SAMIGULLIN, F. M.
J. STRUCT. CHEM., USSR
12 ( 4 ), 535-9, 1971.
( ENGLISH TRANSLATION OF ZH. STRUKT. KHIM., 12 ( 4 ), 585-90, 1971; FOR ORIGINAL SEE TPRC NO. 75605 )

75609 **STUDY OF SHORT-RANGE HYDRATION IN AQUEOUS SOLUTIONS OF CESIUM FLUORIDE.**
EMEL'YANOV, M. I. NIKIFOROV, E. A.
POPOV, G. V.
ZH. STRUKT. KHIM.
13 ( 4 ), 587-91, 1972.
( FOR ENGLISH TRANSLATION SEE TPRC NO. 75610 )

75610 **STUDY OF SHORT-RANGE HYDRATION IN AQUEOUS SOLUTIONS OF CESIUM FLUORIDE.**
EMEL'YANOV, M. I. NIKIFOROV, E. A.
POPOV, G. V.
J. STRUCT. CHEM., USSR
13 ( 4 ), 549-52, 1972.
( ENGLISH TRANSLATION OF ZH. STRUKT. KHIM., 13 ( 4 ), 587-91, 1972; FOR ORIGINAL SEE TPRC NO. 75609 )

75611 **RELATIVE PARTIAL MOLAL HEAT CAPACITIES OF WATER, METHANOL, AND ACETONE IN THE SODIUM IODIDE-WATER, SODIUM IODIDE-METHANOL, AND SODIUM IODIDE-ACETONE SYSTEMS AT DIFFERENT TEMPERATURES AND THE STRUCTURE OF THESE SOLUTIONS.**
NOVOSELOV, N. P. MISHCHENKO, K. M.
ZH. STRUKT. KHIM.
14 ( 1 ), 156-8, 1973.
( FOR ENGLISH TRANSLATION SEE TPRC NO. 75612 )

75612 **RELATIVE PARTIAL MOLAL HEAT CAPACITIES OF WATER, METHANOL, AND ACETONE IN THE SODIUM IODIDE-WATER, SODIUM IODIDE-METHANOL, AND SODIUM IODIDE-ACETONE SYSTEMS AT DIFFERENT TEMPERATURES AND THE STRUCTURE OF THESE SOLUTIONS.**
NOVOSELOV, N. P. MISHCHENKO, K. M.
J. STRUCT. CHEM., USSR
14 ( 1 ), 136-8, 1973.
( ENGLISH TRANSLATION OF ZH. STRUKT. KHIM., 14 ( 1 ), 156-8, 1973; FOR ORIGINAL SEE TPRC NO. 75611 )

75619 **ESTERS OF ALIPHATIC DIBASIC ACIDS AND 2,2-DIMETHYLALKYL ALCOHOLS.**
VOLKOVA, I. A. PROSKURYAKOV, V. A.
GROMOVA, V. V. EIDUS, YA. T.
PUZITSKII, K. V. PIROZHKOV, S. D.
ZH. PRIKL. KHIM.
45 ( 9 ), 2023-6, 1972.
( FOR ENGLISH TRANSLATION SEE TPRC NO. 75620 )

75620 **ESTERS OF ALIPHATIC DIBASIC ACIDS AND 2,2-DIMETHYLALKYL ALCOHOLS.**
VOLKOVA, I. A. PROSKURYAKOV, V. A.
GROMOVA, V. V. EIDUS, YA. T.
PUZITSKII, K. V. PIROZHKOV, S. D.
J. APPL. CHEM., USSR
45 ( 9 ), 2112-4, 1972.
( ENGLISH TRANSLATION OF ZH. PRIKL. KHIM., 45 ( 9 ), 2023-6, 1972; FOR ORIGINAL SEE TPRC NO. 75619 )

75621 **PHYSICOCHEMICAL PROPERTIES OF AQUEOUS SOLUTIONS OF SULFURIC ACID AND POTASSIUM HYDROXIDE AT LOW TEMPERATURES.**
MAKSIMOVA, I. N. OVANES'YAN, K. K.
BULLAN, V. A.
ZH. PRIKL. KHIM.
45 ( 10 ), 2173-6, 1972.
( FOR ENGLISH TRANSLATION SEE TPRC NO. 75622 )

75622 **PHYSICOCHEMICAL PROPERTIES OF AQUEOUS SOLUTIONS OF SULFURIC ACID AND POTASSIUM HYDROXIDE AT LOW TEMPERATURES.**
MAKSIMOVA, I. N. OVANES'YAN, K. K.
BULLAN, V. A.
J. APPL. CHEM., USSR
45 ( 10 ), 2280-3, 1972.
( ENGLISH TRANSLATION OF ZH. PRIKL. KHIM., 45 ( 10 ), 2173-6, 1972; FOR ORIGINAL SEE TPRC NO. 75621 )

75636 **TIME DEPENDENCE OF THERMAL CONDUCTIVITY OF CEMENT GROUT WITH ADDITIONS OF CALCIUM CHLORIDE AND SODIUM CHLORIDE.**
KLYUSOV, A. A. GORSKII, A. T. LEPNEV, E. N.
ZH. PRIKL. KHIM.
46 ( 4 ), 915-6, 1973.
( FOR ENGLISH TRANSLATION SEE TPRC NO. 75637 )

75637 **TIME DEPENDENCE OF THERMAL CONDUCTIVITY OF CEMENT GROUT WITH ADDITIONS OF CALCIUM CHLORIDE AND SODIUM CHLORIDE.**
KLYUSOV, A. A. GORSKII, A. T. LEPNEV, E. N.
J. APPL. CHEM., USSR
46 ( 4 ), 966-7, 1973.
( ENGLISH TRANSLATION OF ZH. PRIKL. KHIM., 46 ( 4 ), 915-6, 1973; FOR ORIGINAL SEE TPRC NO. 75636 )

75638 **MEASUREMENT OF DIFFUSION COEFFICIENTS IN THE SYSTEM BENZENE-N-HEPTANE WITH A POLARIZATION INTERFEROMETER USING A GATE-VALVE CELL.**
PORAI-KOSHITS, A. B. PENKINA, N. V.
OLEVSKII, V. M. POTORZHINSKAYA, S. S.
ZH. PRIKL. KHIM.
46 ( 4 ), 934-6, 1973.
( FOR ENGLISH TRANSLATION SEE TPRC NO. 75639 )

75639 **MEASUREMENT OF DIFFUSION COEFFICIENTS IN THE SYSTEM BENZENE-N-HEPTANE WITH A POLARIZATION INTERFEROMETER USING A GATE-VALVE CELL.**
PORAI-KOSHITS, A. B. PENKINA, N. V.
OLEVSKII, V. M. POTORZHINSKAYA, S. S.
J. APPL. CHEM., USSR
46 ( 4 ), 992-4, 1973.
( ENGLISH TRANSLATION OF ZH. PRIKL. KHIM., 46 ( 4 ), 934-6, 1973; FOR ORIGINAL SEE TPRC NO. 75638 )

75644 **SOME PHYSICOCHEMICAL PROPERTIES OF SOLUTIONS OF INORGANIC COMPOUNDS IN LIQUID AMMONIA.**
BAKIN, V. M. ZEL'VENSKII, YA. D.
ZH. PRIKL. KHIM.
46 ( 6 ), 1195-9, 1973.
( FOR ENGLISH TRANSLATION SEE TPRC NO. 75645 )

75645 **SOME PHYSICOCHEMICAL PROPERTIES OF SOLUTIONS OF INORGANIC COMPOUNDS IN LIQUID AMMONIA.**
BAKIN, V. M. ZEL'VENSKII, YA. D.
J. APPL. CHEM., USSR
46 ( 6 ), 1279-82, 1973.
( ENGLISH TRANSLATION OF ZH. PRIKL. KHIM., 46 ( 6 ), 1195-9, 1973; FOR ORIGINAL SEE TPRC NO. 75644 )

75646 **KINETICS OF COPPER DISSOLUTION IN SULFURIC ACID SOLUTIONS AT HIGH TEMPERATURES AND OXYGEN PRESSURES.**
NABOICHENKO, S. S.
ZH. PRIKL. KHIM.
46 ( 7 ), 1475-80, 1973.
( FOR ENGLISH TRANSLATION SEE TPRC NO. 75647 )

75647 **KINETICS OF COPPER DISSOLUTION IN SULFURIC ACID SOLUTIONS AT HIGH TEMPERATURES AND OXYGEN PRESSURES.**
NABOICHENKO, S. S.
J. APPL. CHEM., USSR
46 ( 7 ), 1575-9, 1973.
( ENGLISH TRANSLATION OF ZH. PRIKL. KHIM., 46 ( 7 ), 1475-80, 1973; FOR ORIGINAL SEE TPRC NO. 75646 )

75652 **SOLUBILITY POLYTHERM OF THE SYSTEM POLYPHOSPHORIC ACID-POTASSIUM CHLORIDE-AMMONIA-WATER.**
BUGAI, P. M. SLYN'KO, V. P.
ZH. PRIKL. KHIM.
46 ( 8 ), 1875-7, 1973.
( FOR ENGLISH TRANSLATION SEE TPRC NO. 75653 )

75653 **SOLUBILITY POLYTHERM OF THE SYSTEM POLYPHOSPHORIC ACID-POTASSIUM CHLORIDE-AMMONIA-WATER.**
BUGAI, P. M. SLYN'KO, V. P.
J. APPL. CHEM., USSR
46 ( 8 ), 2002, 1973.
( ENGLISH TRANSLATION OF ZH. PRIKL. KHIM., 46 ( 8 ), 1875-7, 1973; FOR ORIGINAL SEE TPRC NO. 75652 )

75654 **THERMOGRAPHIC AND RHEOLOGICAL STUDIES OF MELTS OF UREA PHOSPHATES AND THEIR BLENDS WITH POTASSIUM CHLORIDE.**
VOL'FKOVICH, S. I. CHEKHOVSKIKH, A. I.
MIKHALEVA, T. K.
ZH. PRIKL. KHIM.
46 ( 9 ), 1889-94, 1943.
( FOR ENGLISH TRANSLATION SEE TPRC NO. 75655 )

75655 **THERMOGRAPHIC AND RHEOLOGICAL STUDIES OF MELTS OF UREA PHOSPHATES AND THEIR BLENDS WITH POTASSIUM CHLORIDE.**
VOL'FKOVICH, S. I. CHEKHOVSKIKH, A. I.
MIKHALEVA, T. K.
J. APPL. CHEM., USSR
46 ( 9 ), 2019-23, 1973.
( ENGLISH TRANSLATION OF ZH. PRIKL. KHIM., 46 ( 9 ), 1889-94, 1973; FOR ORIGINAL SEE TPRC NO. 75654 )

75656 **ALGORITHMIZATION OF METHODS FOR DETERMINATION OF CERTAIN PHYSICOCHEMICAL PARAMETERS OF MIXED ELECTROLYTE SOLUTIONS.**
TSEITLIN, N. A.
ZH. PRIKL. KHIM.
46 ( 11 ), 2488-92, 1973.
( FOR ENGLISH TRANSLATION SEE TPRC NO. 75657 )

75657 **ALGORITHMIZATION OF METHODS FOR DETERMINATION OF CERTAIN PHYSICOCHEMICAL PARAMETERS OF MIXED ELECTROLYTE SOLUTIONS.**
TSEITLIN, N. A.
J. APPL. CHEM., USSR
46 ( 11 ), 2637-40, 1973.
( ENGLISH TRANSLATION OF ZH. PRIKL. KHIM., 46 ( 11 ), 2488-92, 1973; FOR ORIGINAL SEE TPRC NO. 75656 )

75658 **VISCOSITY OF LIQUID NITROGEN-PHOSPHORUS FERTILIZERS MADE FROM PHOSPHORIC ACID, AMMONIA, AND UREA.**
GRYNCHAROV, I. N.
ZH. PRIKL. KHIM.
46 ( 11 ), 2557-60, 1973.
( FOR ENGLISH TRANSLATION SEE TPRC NO. 75659 )

| TPRC Number | Bibliographic Citation |
|---|---|

75659 **VISCOSITY OF LIQUID NITROGEN-PHOSPHORUS FERTILIZERS MADE FROM PHOSPHORIC ACID, AMMONIA, AND UREA.**
GRYNCHAROV, I. N.
J. APPL. CHEM., USSR
46 ( 11 ), 2706–8, 1973.
( ENGLISH TRANSLATION OF ZH. PRIKL. KHIM., 46 ( 11 ), 2557–60, 1973; FOR ORIGINAL SEE TPRC NO. 75658 )

75660 **CHARACTER OF THE CHANGES OF SOLUTION PROPERTIES DURING TRANSITION FROM THE UNSATURATED TO THE SUPERSATURATED STATE.**
NIKITINA, S. D. KHAMSKII, E. V.
ZH. PRIKL. KHIM.
46 ( 11 ), 2563–4, 1973.
( FOR ENGLISH TRANSLATION SEE TPRC NO. 75661 )

75661 **CHARACTER OF THE CHANGES OF SOLUTION PROPERTIES DURING TRANSITION FROM THE UNSATURATED TO THE SUPERSATURATED STATE.**
NIKITINA, S. D. KHAMSKII, E. V.
J. APPL. CHEM., USSR
46 ( 11 ), 2711–2, 1973.
( ENGLISH TRANSLATION OF ZH. PRIKL. KHIM., 46 ( 11 ), 2563–4, 1973; FOR ORIGINAL SEE TPRC NO. 75660 )

75664 **VISCOSITIES OF LITHIUM, SODIUM, POTASSIUM, AND AMMONIUM NITRATE SOLUTIONS AT TEMPERATURES TO 275 C.**
PUCHKOV, L. V. SARGAEV, P. M.
ZH. PRIKL. KHIM.
46 ( 12 ), 2637–40, 1973.
( FOR ENGLISH TRANSLATION SEE TPRC NO. 75665 )

75665 **VISCOSITIES OF LITHIUM, SODIUM, POTASSIUM, AND AMMONIUM NITRATE SOLUTIONS AT TEMPERATURES TO 275 C.**
PUCHKOV, L. V. SARGAEV, P. M.
J. APPL. CHEM., USSR
46 ( 12 ), 2793–5, 1973.
( ENGLISH TRANSLATION OF ZH. PRIKL. KHIM., 46 ( 12 ), 2637–40, 1973; FOR ORIGINAL SEE TPRC NO. 75664 )

75670 **PHYSICOCHEMICAL PROPERTIES OF SOLUTIONS IN THE SYSTEM UREA–COPPER SULFATE–WATER, SATURATED AT 25 C.**
KARCHEVSKII, A. V. GANZ, S. N.
BRAGINSKAYA, R. I. SELEZNEVA, E. P.
ZH. PRIKL. KHIM.
47 ( 1 ), 92–5, 1974.
( FOR ENGLISH TRANSLATION SEE TPRC NO. 75671 )

75671 **PHYSICOCHEMICAL PROPERTIES OF SOLUTIONS IN THE SYSTEM UREA–COPPER SULFATE–WATER, SATURATED AT 25 C.**
KARCHEVSKII, A. V. GANZ, S. N.
BRAGINSKAYA, R. I. SELEZNEVA, E. P.
J. APPL. CHEM., USSR
47 ( 1 ), 91–4, 1974.
( ENGLISH TRANSLATION OF ZH. PRIKL. KHIM., 47 ( 1 ), 92–5, 1974; FOR ORIGINAL SEE TPRC NO. 75670 )

75672 **VISCOSITIES OF LITHIUM, SODIUM, AND POTASSIUM SULFATE SOLUTIONS AT TEMPERATURES UP TO 275 C.**
PUCHKOV, L. V. SARGAEV, P. M.
ZH. PRIKL. KHIM.
47 ( 1 ), 96–8, 1974.
( FOR ENGLISH TRANSLATION SEE TPRC NO. 75673 )

75673 **VISCOSITIES OF LITHIUM, SODIUM, AND POTASSIUM SULFATE SOLUTIONS AT TEMPERATURES UP TO 275 C.**
PUCHKOV, L. V. SARGAEV, P. M.
J. APPL. CHEM., USSR
47 ( 1 ), 95–7, 1974.
( ENGLISH TRANSLATION OF ZH. PRIKL. KHIM., 47 ( 1 ), 96–8, 1974; FOR ORIGINAL SEE TPRC NO. 75672 )

75674 **INFLUENCE OF TEMPERATURE ON DENSITY AND VISCOSITY OF CERTAIN SALT SOLUTIONS.**
KONONENKO, A. F. SASHEVSKAYA, Z. G.
ZH. PRIKL. KHIM.
47 ( 1 ), 211–2, 1974.
( FOR ENGLISH TRANSLATION SEE TPRC NO. 75675 )

75675 **INFLUENCE OF TEMPERATURE ON DENSITY AND VISCOSITY OF CERTAIN SALT SOLUTIONS.**
KONONENKO, A. F. SASHEVSKAYA, Z. G.
J. APPL. CHEM., USSR
47 ( 1 ), 208–9, 1974.
( ENGLISH TRANSLATION OF ZH. PRIKL. KHIM., 47 ( 1 ), 211–2, 1974; FOR ORIGINAL SEE TPRC NO. 75674 )

75678 **DEPENDENCE OF VISCOSITY OF ELECTROLYTE SOLUTIONS ON TEMPERATURE AND CONCENTRATION.**
PUCHKOV, L. V. SARGAEV, P. M.
ZH. PRIKL. KHIM.
47 ( 2 ), 280–5, 1974.
( FOR ENGLISH TRANSLATION SEE TPRC NO. 75679 )

75679 **DEPENDENCE OF VISCOSITY OF ELECTROLYTE SOLUTIONS ON TEMPERATURE AND CONCENTRATION.**
PUCHKOV, L. V. SARGAEV, P. M.
J. APPL. CHEM., USSR
47 ( 2 ), 283–6, 1974.
( ENGLISH TRANSLATION OF ZH. PRIKL. KHIM., 47 ( 2 ), 280–5, 1974; FOR ORIGINAL SEE TPRC NO. 75678 )

75680 **APPLICATION OF MATHEMATICAL EXPERIMENTAL DESIGN TO STUDY OF THE INFLUENCE OF TEMPERATURE ON VISCOSITY OF SODIUM IODIDE SOLUTIONS IN ISOBUTYL ALCOHOL.**
KRUNCHAK, E. G. KRUMGAL'Z, B. S.
UVAROVA, N. A.
ZH. PRIKL. KHIM.
47 ( 2 ), 285–8, 1974.
( FOR ENGLISH TRANSLATION SEE TPRC NO. 75681 )

75681 **APPLICATION OF MATHEMATICAL EXPERIMENTAL DESIGN TO STUDY OF THE INFLUENCE OF TEMPERATURE ON VISCOSITY OF SODIUM IODIDE SOLUTIONS IN ISOBUTYL ALCOHOL.**
KRUNCHAK, E. G. KRUMGAL'Z, B. S.
UVAROVA, N. A.
J. APPL. CHEM., USSR
47 ( 2 ), 287–9, 1974.
( ENGLISH TRANSLATION OF ZH. PRIKL. KHIM., 47 ( 2 ), 285–8, 1974; FOR ORIGINAL SEE TPRC NO. 75680 )

75682 **HYDRODYNAMIC CHARACTERISTICS OF VARIOUS SYSTEMS IN PULSED EXTRACTION COLUMNS WITH KRIMZ PLATES.**
KARPACHEVA, S. M. ZAKHAROV, E. I.
ZH. PRIKL. KHIM.
47 ( 2 ), 339–43, 1974.
( FOR ENGLISH TRANSLATION SEE TPRC NO. 75683 )

75683 **HYDRODYNAMIC CHARACTERISTICS OF VARIOUS SYSTEMS IN PULSED EXTRACTION COLUMNS WITH KRIMZ PLATES.**
KARPACHEVA, S. M. ZAKHAROV, E. I.
J. APPL. CHEM., USSR
47 ( 2 ), 339–43, 1974.
( ENGLISH TRANSLATION OF ZH. PRIKL. KHIM., 47 ( 2 ), 339–43, 1974; FOR ORIGINAL SEE TPRC NO. 75682 )

75684 **PHYSICAL PROPERTIES OF SOLUTIONS OF NITROBENZANILIDES IN N–METHYLPYRROLIDONE AND N,N–DIMETHYLFORMAMIDE.**
MYASNIKOVA, L. F. BUSHINSKII, V. I.
VAISMAN, I. L. NOVOKHATKA, D. A.
ZH. PRIKL. KHIM.
47 ( 2 ), 464–6, 1974.
( FOR ENGLISH TRANSLATION SEE TPRC NO. 75685 )

75685 **PHYSICAL PROPERTIES OF SOLUTIONS OF NITROBENZANILIDES IN N–METHYLPYRROLIDONE AND N,N–DIMETHYLFORMAMIDE.**
MYASNIKOVA, L. F. BUSHINSKII, V. I.
VAISMAN, I. L. NOVOKHATKA, D. A.
J. APPL. CHEM., USSR
47 ( 2 ), 465–7, 1974.
( ENGLISH TRANSLATION OF ZH. PRIKL. KHIM., 47 ( 2 ), 464–6, 1974; FOR ORIGINAL SEE TPRC NO. 75684 )

75740 **THE TOTAL EMISSIVITIES OF LUMINOUS AND NON–LUMINOUS FLAMES.**
TAYLOR, P. B. FOSTER, P. J.
INT. J. HEAT MASS TRANSFER
17 ( 12 ), 1591–1605, 1974.

75751 **TRANSPORT PROPERTIES OF COMBUSTION PRODUCTS FROM COALS WITH ADDED POTASSIUM IV.**
SAMUILOV, E. V. ROZHDESTVENSKII, I. B.
TSITELAURI, N. N. OLEVINSKII, K. K.
TEPLOFIZ. VYS. TEMP.
12 ( 2 ), 468–70, 1974.
( FOR ENGLISH TRANSLATION SEE TPRC NO. 75752 )

75752 **TRANSPORT PROPERTIES OF COMBUSTION PRODUCTS FROM COALS WITH ADDED POTASSIUM IV.**
SAMUILOV, E. V. ROZHDESTVENSKII, I. B.
TSITELAURI, N. N. OLEVINSKII, K. K.
HIGH TEMP.
12 ( 2 ), 400–2, 1974.
( ENGLISH TRANSLATION OF TEPLOFIZ. VYS. TEMP., 12 ( 2 ), 468–70, 1974; FOR ORIGINAL SEE TPRC NO. 75751 )

75755 **THERMAL CONDUCTIVITY OF POTASSIUM CHLORIDE ALLOYS.**
ANTHONY, P. J. WURST, J. C.
J. AMER. CERAM. SOC.
57 ( 11 ), 504, 1974.

75781 **THERMODYNAMIC AND TRANSPORT PROPERTIES OF SODIUM BENZOATE AND HYDROXY BENZOATES IN WATER AT 25 C.**
DESNOYERS, J. E. PAGE, R. PERRON, G.
FORTIER, J.–L. LEDUC, P.–A. PLATFORD, R. F.
CAN. J. CHEM.
51 ( 13 ), 2129–37, 1973.

75782 **HIGH TEMPERATURE AQUEOUS ELECTROLYTE CONCENTRATION CELLS AND THE IONIZATION OF LIQUID WATER TO 200 C.**
MAC DONALD, D. D. BUTLER, P. OWEN, D.
CAN. J. CHEM.
51 ( 15 ), 2590–5, 1973.

75783 **EFFECT OF SOLVENTS ON THE QUANTUM EFFICIENCY OF RUBRENE ELECTROGENERATED CHEMILUMINESCENCE.**
PIGHIN, A.
CAN. J. CHEM.
51 ( 20 ), 3467–72, 1973.

75784 **VIBRATIONAL SPECTRA AND STRUCTURES OF THE SOLID PHASES OF GERMANE.**
THE, N. D. GAGNON, J. M. BELZILE, R.
CABANA, A.
CAN. J. CHEM.
52 ( 2 ), 327–35, 1974.

| TPRC Number | Bibliographic Citation |
|---|---|

75790 EXPERIMENTAL HEAT CAPACITIES OF NON-IDEAL BINARY GASEOUS MIXTURES FROM PROPANE, ISOBUTANE, CHLORODIFLUORO METHANE, AND CHLORO-PENTA FLUOROETHANE.
BIER, K. BUSSER, J. ERNST, G.
J. CHEM. THERMODYN.
5 ( 1 ), 83-96, 1973.

75791 THERMODYNAMIC AND OTHER PROPERTIES OF METHANOL + ACETONE, CARBON DISULPHIDE + ACETONE, CARBON DISULPHIDE + METHANOL, AND CARBON DISULPHIDE + METHANOL + ACETONE.
CAMPBELL, A. N. KARTZMARK, E. M.
J. CHEM. THERMODYN.
5 ( 2 ), 163-72, 1973.

75800 THERMOCHEMISTRY OF SOLUTIONS OF BIOCHEMICAL MODEL COMPOUNDS. 2. ALKOXYETHANOLS AND 1,2-DIALKOXYETHANES IN WATER.
KUSANO, K. SUURKUUSK, J. WADSO, I.
J. CHEM. THERMODYN.
5 ( 5 ), 757-67, 1973.

75810 EXCESS FUNCTIONS, EXCESS VOLUMES AND EXCESS HEAT CAPACITIES OF ISOOCTANE + BENZENE AND + TOLUENE.
RAJAGOPAL, E. SUBRAHMANYAM, S. V.
J. CHEM. THERMODYN.
6 ( 9 ), 873-6, 1974.

75815 THE TIN CHLORIDE-PHOSPHORYL CHLORIDE AND TIN CHLORIDE-SELENINYL CHLORIDE SYSTEMS.
NISEL'SON, L. A. TRET'YAKOVA, K. V.
TORBINA, E. N. LEBEDEV, V. G. ELLERT, G. V.
ZH. NEORG. KHIM.
18 ( 10 ), 2822- , 1973.
( FOR ENGLISH TRANSLATION SEE TPRC NO. 75816 )

75816 THE TIN CHLORIDE-PHOSPHORYL CHLORIDE AND TIN CHLORIDE-SELENIYL CHLORIDE SYSTEMS.
NISEL'SON, L. A. TRET'YAKOVA, K. V.
TORBINA, E. N. LEBEDEV, V. G. ELLERT, G. V.
RUSS. J. INORG. CHEM.
18 ( 10 ), 1500-3, 1973.
( ENGLISH TRANSLATION OF ZH. NEORG. KHIM., 18 ( 10 ), 2822- , 1973; FOR ORIGINAL SEE TPRC NO. 75815 )

75819 REACTION OF SULPHURIC ACID WITH THIONYL CHLORIDE.
MAKITRA, R. G. PYRIG, YA. N. DIDYCH, M. N.
ZH. NEORG. KHIM.
18 ( 11 ), 2898- 1973.
( FOR ENGLISH TRANSLATION SEE TPRC NO. 75820 )

75820 REACTION OF SULPHURIC ACID WITH THIONYL CHLORIDE.
MAKITRA, R. G. PYRIG, YA. N. DIDYCH, M. N.
RUSS. J. INORG. CHEM.
18 ( 11 ), 1541-3, 1973.
( ENGLISH TRANSLATION OF ZH. NEORG. KHIM., 18 ( 11 ), 2898- , 1973; FOR ORIGINAL SEE TPRC NO. 75819 )

75826 PROPERTIES OF SOLUTIONS OF THE BERYLLIUM NITRATE-NICKEL NITRATE-WATER AND CALACIUM NITRATE-NICKEL NITRATE-WATER SYSTEMS AT 25 C.
ANDREEVA, T. A. POZHARSKAYA, S. S.
GOLOVANOVA, T. G.
ZH. NEORG. KHIM.
18 ( 11 ), 3078- , 1973.
( FOR ENGLISH TRANSLATION SEE TPRC NO. 75827 )

75827 PROPERTIES OF SOLUTIONS OF THE BERYLLIUM NITRATE-NICKEL NITRATE-WATER AND CALCIUM NITRATE-NICKEL NITRATE-WATER SYSTEMS AT 25 C.
ANDREEVA, T. A. POZHARSKAYA, S. S.
GOLOVANOVA, T. G.
RUSS. J. INORG. CHEM.
18 ( 11 ), 1638-40, 1973.
( ENGLISH TRANSLATION OF ZH. NEORG. KHIM., 18 ( 11 ), 3078- , 1973; FOR ORIGINAL SEE TPRC NO. 75826 )

75828 THE ALUMINUM OXIDE-IRON OXIDE-NITROGEN OXIDE-WATER SYSTEM AT ELEVATED TEMPERATURES.
BOGACHEVA, L. M. ISMATOV, KH. R.
KARIMOV, R. Z.
ZH. NEORG. KHIM.
18 ( 11 ), 3086- , 1973.
( FOR ENGLISH TRANSLATION SEE TPRC NO. 75829 )

75829 THE ALUMINUM OXIDE-IRON OXIDE-NITROGEN OXIDE-WATER SYSTEM AT ELEVATED TEMPERATURES.
BOGACHEVA, L. M. ISMATOV, KH. R.
KARIMOV, R. Z.
RUSS. J. INORG. CHEM.
18 ( 11 ), 1642-5, 1973.
( ENGLISH TRANSLATION OF ZH. NEORG. KHIM., 18 ( 11 ), 3086- , 1973; FOR ORIGINAL SEE TPRC NO. 75828 )

75830 THE POTASSIUM CARBONATE-POTASSIUM NITRATE-WATER SYSTEM AT 0, 25, AND 40 C.
BOGOYAVLENSKII, P. S. GASHPAR, E. D.
ZH. NEORG. KHIM.
18 ( 11 ), 3125- , 1973.
( FOR ENGLISH TRANSLATION SEE TPRC NO. 75831 )

75831 THE POTASSIUM CARBONATE-POTASSIUM NITRATE-WATER SYSTEM AT 0, 25, AND 40 C.
BOGOYAVLENSKII, P. S. GASHPAR, E. D.
RUSS. J. INORG. CHEM.
18 ( 11 ), 1662-3, 1973.
( ENGLISH TRANSLATION OF ZH. NEORG. KHIM., 18 ( 11 ), 3125- , 1973; FOR ORIGINAL SEE TPRC NO. 75830 )

75833 THE PROPERTIES OF AQUEOUS STRONTIUM NITRATE-NICKEL NITRATE AND BARIUM NITRATE-NICKEL NITRATE SOLUTIONS AT 25 C.
ANDREEVA, T. A. POZHARSKAYA, S. S.
ZH. NEORG. KHIM.
18 ( 12 ), 3323- , 1973.
( FOR ENGLISH TRANSLATION SEE TPRC NO. 75834 )

75834 THE PROPERTIES OF AQUEOUS STRONTIUM NITRATE-NICKEL NITRATE AND BARIUM NITRATE-NICKEL NITRATE SOLUTIONS AT 25 C.
ANDREEVA, T. A. POZHARSKAYA, S. S.
RUSS. J. INORG. CHEM.
18 ( 12 ), 1769-72, 1973.
( ENGLISH TRANSLATION OF ZH. NEORG. KHIM., 18 ( 12 ), 3323- , 1973; FOR ORIGINAL SEE TPRC NO. 75833 )

75837 INVESTIGATION OF THE SULPHURIC ACID-WATER EQUILIBRIUM IN CONCENTRATED SULPHURIC ACID SOLUTIONS BY THE DISTORTED TOTAL INTERNAL REFLECTION METHOD.
MAIOROV, V. D. LIBROVICH, N. B.
ZH. FIZ. KHIM.
47 ( 9 ), 2298- , 1973.
( FOR ENGLISH TRANSLATION SEE TPRC NO. 75838 )

75838 INVESTIGATION OF THE SULPHURIC ACID-WATER EQUILIBRIUM IN CONCENTRATED SULPHURIC ACID SOLUTIONS BY THE DISTORTED TOTAL INTERNAL REFLECTION METHOD.
MAIOROV, V. D. LIBROVICH, N. B.
RUSS. J. PHYS. CHEM.
47 ( 9 ), 1295-7, 1973.
( ENGLISH TRANSLATION OF ZH. FIZ. KHIM., 47 ( 9 ), 2298- , 1973; FOR ORIGINAL SEE TPRC NO. 75837 )

75840 THERMAL CONDUCTIVITY OF MIXTURES OF CARBOXYLIC ACIDS.
USMANOV, I. U. SALIKHOV, A. S.
ZH. FIZ. KHIM.
47 ( 10 ), 2532- , 1973.
( FOR ENGLISH TRANSLATION SEE TPRC NO. 75841 )

75841 THERMAL CONDUCTIVITY OF MIXTURES OF CARBOXYLIC ACIDS.
USMANOV, I. U. SALIKHOV, A. S.
RUSS. J. PHYS. CHEM.
47 ( 10 ), 1426-7, 1973.
( ENGLISH TRANSLATION OF ZH. FIZ. KHIM., 47 ( 10 ), 2532- , 1973; FOR ORIGINAL SEE TPRC NO. 75840 )

75844 THE PROPERTIES OF CONCENTRATED ELECTROLYTE SOLUTIONS. III. VISCOSITY OF AQUEOUS MAGNESIUM CHLORIDE, NITRATE, AND SULPHATE SOLUTIONS.
ROMAN'KOV, YU. I. KOMAROVA, T. A.
RUSS. J. PHYS. CHEM.
47 ( 10 ), 1528-9, 1973.
( ENGLISH TRANSLATION OF ZH. FIZ. KHIM., 47 ( 10 ), 2712- , 1973; FOR ORIGINAL SEE TPRC NO. 73061 )

75855 SPECIFIC HEATS AND SPECIFIC VOLUMES OF ISOPIESTIC AQUEOUS SOLUTIONS OF BERYLLIUM-SUBGROUP METAL CHLORIDES.
VASILEV, Y. A. FEDYAINOV, N. V.
KURENKOV, V. V.
RUSS. J. PHYS. CHEM.
47 ( 11 ), 1570-3, 1973.
( ENGLISH TRANSLATION OF ZH. FIZ. KHIM., 47 ( 11 ), 2799- , 1973; FOR ORIGINAL SEE TPRC NO. 73009 )

75861 CONSTRUCTION OF AN ADIABATIC CALORIMETER. HEAT CAPACITY OF THE N-HEXANE + N-HEXADECANE MIXTURES.
DIAZ PENA, M. RENUNCIO, J. A. R.
AN. QUIM.
70 ( 2 ), 113-20, 1974.

75862 REFLECTIVITY STUDIES OF LASER-PRODUCED PLASMAS OF VARIABLE DENSITY.
GRUHL, T. PUELL, H. KAISER, W.
APPL. PHYS. LETT.
25 ( 3 ), 135-7, 1974.

75867 ESTIMATING THE VISCOSITIES OF AQUEOUS ELECTROLYTIC SOLUTIONS.
BRESLAU, B. R. WELSH, P. B. MILLER, I. F.
CHEM. ENG. ( N. Y. )
81 ( 8 ), 112-13, 1974.

75924 CONCENTRATION DEPENDENCE OF THE WIDTH OF RAMAN EFFECT LINES IN SOLUTIONS WITH INCREASED VISCOSITY.
SIDOROV, N. K. STAL'MAKHOVA, L. S.
NOVIKOVA, G. L.
OPT. SPEKTROSK.
37 ( 1 ), 91-4, 1974.
( FOR ENGLISH TRANSLATION SEE T76441 )

75958 HEAT CAPACITY OF AQUEOUS SOLUTIONS OF GROUP III METAL PERCHLORATES.
BABAKULOV, N. LATYSHEVA, V. A.
ZH. FIZ. KHIM.
48 ( 4 ), 1012-14, 1974.
( FOR ENGLISH TRANSLATION SEE TPRC NO. 76461 )

***TPRC Number*** ***Bibliographic Citation***

75961 **HEAT CAPACITY OF AQUEOUS SOLUTIONS OF GROUP II ELEMENT PERCHLORATES.**
LATYSHEVA, V. A. ANDREEVA, I. N.
ZH. FIZ. KHIM.
48 ( 4 ), 1059, 1974.
( FOR ENGLISH TRANSLATION SEE T76464 )

75983 **ULTRASONIC ATTENUATION IN 1-1 ELECTROLYTES IN 50 PERCENT GLYCOL-WATER MIXTURE.**
KOR, S. K. PANDEY, S. K.
ACUSTICA
30 ( 5 ), 286-7, 1974.

76016 **THERMOPHYSICAL PROPERTIES OF AIR.**
VUKALOVICH, M. P. ZUBAREV, V. N.
ALEKSANDROV, A. A. KOZLOV, A. D.
HEAT TRANSFER SOV. RES.
6 ( 5 ), 152-66, 1974.

76115 **THERMAL CONDUCTIVITY OF MOLTEN MIXTURES OF URANIUM TETRACHLORIDE WITH SODIUM AND POTASSIUM CHLORIDES.**
BYSTRAI, G. P. DESYATNIK, V. N.
ZLOKAZOV, V. A.
AT. ENERG.
36 ( 6 ), 517-18, 1974.
( FOR ENGLISH TRANSLATION SEE T77220 )

76146 **MACROSCOPIC VISCOSITY AND THE DIELECTRIC RELAXATION OF ALIPHATIC CHAIN COMPOUNDS.**
CROSSLEY, J.
J. CHEM. PHYS.
61 ( 3 ), 866-71, 1974.

76169 **REFLECTION FROM CARBON DIOXIDE LASER PRODUCED PLASMA.**
DYER, P. E. JAMES, D. J. RAMSDEN, S. A.
SHIPPER, M. A.
PHYS. LETT.
48 A ( 4 ), 311-12, 1974.

76185 **TECHNIQUE FOR COMPUTER CALCULATION OF THE ISOBARIC HEAT CAPACITY OF MULTICOMPONENT LIQUID MIXTURES OF LIGHT HYDROCARBONS.**
SOLDATENKO, YU. A. VASHCHENKO, D. M.
MINEHENKO, YU. B.
TEPLOFIZ. SVOISTVA UGLEVODORODOV, IKH SMESEI, NEFTEI
( 1 ), 89-98, 1973.

76211 **OPTICAL CONSTANTS OF SULFURIC ACID; APPLICATION TO THE CLOUDS OF VENUS.**
PALMER, K. F. WILLIAMS, D.
APPL. OPT.
14 ( 1 ), 208-19, 1975.

76224 **NEW COMPUTATIONAL PROCEDURE FOR DETERMINING THE APPARENT THERMAL DIFFUSIVITY OF A SOLID BODY APPROXIMATED WITH AN INFINITIVE SLAB.**
HAYAKAWA, K.-I. BAKAL, A.
JOURNAL OF FOOD SCIENCE
38 ( 4 ), 623-9, 1973.

76260 **THERMODYNAMIC PROPERTIES OF ALKALI HALIDES. II. ENTHALPIES OF DILUTION AND HEAT CAPACITIES IN WATER AT 25 DEGREES.**
FORTIER, J. L. LEDUC, P. A. DESNOYERS, J. E.
J. SOLUTION CHEM.
3 ( 4 ), 323-49, 1974.

76272 **ENGINEERING PROPERTY DATA ON ROCKET PROPELLANTS.**
CONSTANTINE, M. T. YOUEL, K. J. HON, J. F.
UNTERBERG, W. RODRIGUEZ, S. E. LECCE, J. V.
MELVOLD, R. W. QUAGLINO, J.
AIR FORCE ROCKET PROPULSION LAB., EDWARDS, CALIF.
264PP., 1968.
( AD-390 864, AFRPL-TR-68-100 )

76330 **K XI DEPENDENCE OF HIGH-FREQUENCY SHEAR VISCOSITY IN THE CRITICAL REGION.**
CHU, B. LIN, F. L.
J. CHEM. PHYS.
61 ( 2 ), 739-40, 1974.
( AD-786 339, AROD-8014.25-C )

76333 **INVESTIGATION OF THE AMMONIUM PERCHLORATE-CALCIUM PERCHLORATE-WATER, AND CALCIUM CHROMATE-CALCIUM PERCHLORATE-WATER TERNARY SYSTEMS AT 40 C.**
LEPESHKOV, I. N. VASIL'EVA, S. I.
ZH. KHIM.
74-81, 1970.
( FOR ENGLISH TRANSLATION SEE TPRC NO. 76334 )

76334 **INVESTIGATION OF THE AMMONIUM PERCHLORATE-CALCIUM PERCHLORATE-WATER, AND CALCIUM CHROMATE-CALCIUM PERCHLORATE-WATER TERNARY SYSTEMS AT 40 C.**
LEPESHKOV, I. N. VASIL'EVA, S. I.
ARMY FOREIGN SCIENCE AND TECHNOLOGY CENTER, CHARLOTTESVILLE, VIRGINIA
9PP., 1973.
( ENGLISH TRANSLATION OF ZH. KHIM., 74-81, 1970; FOR ORIGINAL SEE TPRC NO. 76333 )
( AD-784 360, FSTC-HT-23-1302-73 )

76337 **ABSOLUTE VISCOSITY OF AIR DOWN TO CRYOGENIC TEMPERATURES AND UP TO HIGH PRESSURES.**
LATTO, B. SAUNDERS, M. W.
J. MECH. ENG. SCI.
15 ( 4 ), 266-70, 1973.

***TPRC Number*** ***Bibliographic Citation***

76343 **VISCOSITY AND PHASE-SEPARATION CURVE OF PRESSURIZED HELIUM-3 - HELIUM-4 SOLUTIONS NEAR THE TRICRITICAL POINT.**
LAI, C.-M.
STATE UNIVERSITY OF NEW YORK, STONY BROOK, N. Y.,
PH.D. THESIS
185PP., 1973.
( UM 74-14004, AVAIL. UNIV. MICROFILMS )

76376 **GAS PROPERTIES COMPUTATIONAL PROCEDURE SUITABLE FOR ELECTRONIC CALCULATORS.**
ANDREWS, J. R. BIBLARZ, O.
NAVAL POSTGRADUATE SCHOOL, MONTEREY, CALIF.
88PP., 1974.
( AD-783 527, NPS-57ZI74071A )

76377 **PYROLYTIC GRAPHITE COATED THROAT INSERTS.**
PAYNE, W. F.
AIR FORCE ROCKET PROPULSION LAB., EDWARDS AFB, CALIF.
56PP., 1974.
( AD-783 849, AFRPL-TR-74-42, AIAA-PAPER-73-1260 )

76378 **TEMPERATURE REQUIREMENTS FOR THE MAINTENANCE OF THERMAL BALANCE IN HIGH PRESSURE HELIUM OXYGEN ENVIRONMENTS.**
FLYNN, E. T. VOROSMARTI, J. MODELL, H. I.
NAVY EXPERIMENTAL DIVING UNIT, WASHINGTON, D.C.
49PP., 1974.
( AD-784 983, NEDU-RR-21-73 )

76384 **INFRARED REFLECTANCE MEASUREMENTS OF MISSOURI WATER FOR WATER QUALITY APPLICATIONS.**
WARING, R. C. QUERRY, M. R.
DEPT. OF PHYSICS, MISSOURI UNIV., KANSAS CITY
88PP., 1973.
( PB-225 420/9, N74-18081 )

76389 **KINETIC THEORY OF MODERATELY DENSE GAS MIXTURES. II. TRANSPORT COEFFICIENTS FOR BINARY MIXTURES OF HARD SPHERES.**
ROBLES DOMINGUEZ, J. A. GARCIA-COLIN, L. S.
PHYSICA ( UTRECHT )
73 ( 3 ), 510-26, 1974.

76406 **AMMONIA ABSORPTION RELEVANT TO THE ALBEDO OF JUPITER. II - INTERPRETATION.**
TOMASKO, M. G.
ASTROPHYSICAL JOURNAL
187, PT. 1, 641-50, 1974.

76407 **A NEW UPPER LIMIT FOR AN ATMOSPHERE OF CARBON DIOXIDE, CARBON OXIDE ON MERCURY.**
FINK, U. LARSON, H. P. POPPEN, R. F.
ASTROPHYSICAL JOURNAL
187, PT. 1, 407-15, 1974.

76427 **A MODEL OF HEAT TRANSFER IN GAS FLUIDIZED BEDS.**
KUBIE, J. BROUGHTON, J.
INT. J. HEAT MASS TRANSFER
18 ( 2 ), 289-99, 1975.

76441 **CONCENTRATION DEPENDENCE OF RAMAN LINEWIDTH IN HIGH-VISCOSITY SOLUTIONS.**
SIDOROV, N. K. STALMAKHOVA, L. S.
NOVIKOVA, G. L.
OPT. SPECTROS., USSR
37 ( 1 ), 48-9, 1974.
( ENGLISH TRANSLATION OF OPT. SPEKTROSK., 37 ( 1 ), 91-4, 1974; FOR ORIGINAL SEE TPRC NO. 75924 )

76460 **THE VISCOSITIES OF AQUEOUS SOLUTIONS OF COBALT AND NICKEL CHLORIDES IN THE RANGE 20-100 DEGREE C.**
FEDOTOV, N. V. MAKSIMOVA, I. N.
RUSS. J. PHYS. CHEM.
48 ( 3 ), 449, 1974.
( ENGLISH TRANSLATION OF ZH. FIZ. KHIM., 48 ( 3 ), 773, 1974; FOR ORIGINAL SEE TPRC NO. 74999 )
( COMPLETE ARTICLE DEPOSITED AT THE ALL-UNION INST. SCIENTIFIC TECHNICAL INFORMATION( VINITI ) )
( NO. 7279-73 DEP. 1973 )

76461 **HEAT CAPACITIES OF AQUEOUS SOLUTIONS OF GROUP III METAL PERCHLORATES.**
BABAKULOV, N. LATYSHEVA, V. A.
RUSS. J. PHYS. CHEM.
48 ( 4 ), 587-9, 1974.
( ENGLISH TRANSLATION OF ZH. FIZ. KHIM., 48 ( 4 ), 1012-4, 1974; FOR ORIGINAL SEE TPRC NO. 75958 )

76464 **HEAT CAPACITY OF AQUEOUS SOLUTIONS OF GROUP II PERCHLORATES.**
LATYSHEVA, V. A. ANDREEVA, I. N.
RUSS. J. PHYS. CHEM.
48 ( 4 ), 618, 1974.
( ENGLISH TRANSLATION OF ZH. FIZ. KHIM., 48 ( 4 ), 1059, 1974; FOR ORIGINAL SEE TPRC NO. 75961 )
( COMPLETE ARTICLE DEPOSITED AT THE ALL-UNION INST. SCIENTIFIC TECHNICAL INFORMATION( VINITI ) )
( NO. 7677-73 DEP. 1973 )

*TPRC Number* | *Bibliographic Citation*

76465 **DENSITY AND VISCOSITY OF BINARY LIQUID SYSTEMS FORMED BY METHANOL, ACETIC ACID, AND WATER.**
GRANZHAN, V. A.
RUSS. J. PHYS. CHEM.
48 ( 4 ), 619–20, 1974.
( ENGLISH TRANSLATION OF ZH. FIZ. KHIM., 48 ( 4 ), 1060–1, 1974; FOR ORIGINAL SEE TPRC NO. 75203 )
( COMPLETE ARTICLE DEPOSITED AT THE ALL-UNION INST. SCIENTIFIC TECHNICAL INFORMATION( VINITI ) )
( NO. 7670–73 DEP. 1973 )

76466 **EXPERIMENTAL INVESTIGATION OF THERMODYNAMIC PROPERTIES OF BINARY GAS MIXTURES ENTHALPY. ISOBARIC SPECIFIC HEAT.**
ALTUNIN, V. V. KUZNETSOV, D. O.
BONDARENKO, V. F.
TEPLOFIZ. VYS. TEMP.
12 ( 3 ), 513–8, 1974.
( FOR ENGLISH TRANSLATION SEE TPRC NO. 76467 )

76467 **EXPERIMENTAL INVESTIGATION OF THERMODYNAMIC PROPERTIES OF BINARY GAS MIXTURES ENTHALPY. ISOBARIC SPECIFIC HEAT.**
ALTUNIN, V. V. KUZNETSOV, D. O.
BONDARENKO, V. F.
HIGH TEMP.
12 ( 3 ), 440–4, 1974.
( ENGLISH TRANSLATION OF TEPLOFIZ. VYS. TEMP., 12 ( 3 ), 513–8, 1974; FOR ORIGINAL SEE TPRC NO. 76466 )

76524 **A LASER LIGHT SCATTERING STUDY OF TRANSPORT AND CRITICAL PHENOMENA.**
GUELARI, E.
CALIFORNIA INSTITUTE OF TECHNOLOGY, PASADENA, PH.D. THESIS
282PP., 1973.
( UM 73–26463, AVAIL. UNIV. MICROFILM, N74–23077 )

76530 **VISCOSITIES OF BINARY MIXTURES OF CARBON TETRACHLORIDE WITH METHANOL AND ISOPROPYL ALCOHOL.**
KATTI, P. K. PRAKASH, O.
J. CHEM. ENG. DATA
11 ( 1 ),46–7, 1966.

76536 **HEAT CAPACITY OF MANGANESE NICKEL CHLORIDE SYSTEM ALLOYS AT LOW TEMPERATURES.**
KOSTRYUKOVA, M. O. SEDOVA, L. P.
JETP LETTERS
19 ( 5 ), 177–8, 1974.
( ENGLISH TRANSLATION OF PIS'MA ZH. EKSP. TEOR. FIZ., 19 ( 5 ), 306–10, 1974; FOR ORIGINAL SEE TPRC NO. 74893 )

76568 **COMPUTED RELATIONSHIPS BETWEEN THE INHERENT AND APPARENT OPTICAL PROPERTIES OF A FLAT HOMOGENEOUS OCEAN.**
GORDON, H. R. BROWN, O. B. JACOBS, M. M.
APPL. OPT.
14 ( 2 ), 417–27, 1975.

76631 **COLLISION INTEGRALS OF ELECTRONICALLY EXCITED STATES AND TRANSPORT COEFFICIENTS OF THERMAL PLASMAS.**
CAPITELLI, M. LAMANNA, U.
J. PLASMA PHYS.
12, 71–9, 1974.

76685 **VISCOSITY OF DIMETHYL SULFOXIDE SOLUTIONS OF AMMONIUM BROMIDE, AMMONIUM CHLORIDE, POTASSIUM THIOCYANATE, AND CESIUM IODIDE.**
BLIKHRA, R. L. PARMAR, M. L.
AUST. J. CHEM.
27 ( 7 ), 1407–11, 1974.

76691 **THERMAL CONDUCTIVITY OF BINARY LIQUID MIXTURES.**
SAKSENA, M. P. HARMINDER
CHEM. PHYS. LETT.
27 ( 3 ), 448–50, 1974.

76793 **SPONTANEOUS MAGNETIZATION OF ELECTRONIC THERMAL CONDUCTIVITY IN A LASER PLASMA.**
BOL'SHOV, L. A. DREIZIN, YU. A.
DYKHNE, A. M.
ZH. EKSP. TEOR. FIZ. PIS'MA RED.
19 ( 5 ), 288–91, 1974.
( FOR ENGLISH TRANSLATION SEE TPRC NO. 76794 )

76794 **SPONTANEOUS MAGNETIZATION OF ELECTRONIC THERMAL CONDUCTIVITY IN A LASER PLASMA.**
BOL'SHOV, L. A. DREIZIN, YU. A.
DYKHNE, A. M.
JETP LETT.
19 ( 5 ), 168–70, 1974.
( ENGLISH TRANSLATION OF ZH. EKSP. TEOR. FIZ. PIS'MA RED., 19 ( 5 ), 288–91, 1974; FOR ORIGINAL SEE TPRC NO. 76793 )

76830 **SHEAR VISCOSITY OF LIQUID MIXTURES IN THE HARD-SPHERE APPROXIMATION.**
MC DONALD, I. R.
PHYSICA
65 ( 3 ), 630–3, 1973.

76831 **TEMPERATURE, PRESSURE, AND CONCENTRATION DEPENDENCE OF THE THERMAL CONDUCTIVITY OF VERY DILUTE SOLUTIONS OF HELIUM IN SUPERFLUID HELIUM.**
ROSENBAUM, R. L. LANDAU, J. ECKSTEIN, Y.
J. LOW TEMP. PHYS.
16 ( 1–2 ), 131–43, 1974.

76851 **A NOTE ON THE CALCULATION OF VISCOSITIES FOR MULTICOMPONENT GAS MIXTURES.**
SAXENA, S. C. TANZMAN, A.
HIGH TEMP. SCI.
6, 215–20, 1974.

76857 **DEVELOPMENT OF SPECIAL INFRARED TRANSMITTING GLASSES.**
KING, B. W. DUCKWORTH, W. H.
BATTELLE MEMORIAL INSTITUTE, COLUMBUS, OHIO
54PP., 1956.
( AD–112 833 )

76925 **EMISSIVITY OF A CARBON DIOXIDE PLASMA IN THE VISIBLE REGION.**
KISELEVSKII, L. I. SOLOV'YANCHIK, D. A.
FESHCHENKO, Z. P.
J. APPL. SPECTROSC., USSR
16 ( 2 ), 173–5, 1972.
( ENGLISH TRANSLATION OF ZH. PRIKL. SPEKTROSK., 16 ( 2 ), 234–8, 1972; FOR ORIGINAL SEE TPRC NO. 67143 )

76987 **SOME PHYSICAL PROPERTIES OF ACETIC ANHYDRIDE PLUS WATER SOLUTIONS.**
DUTCHAK, YA. I.
UKR. FIZ. ZH.
12 ( 3 ), 454–6, 1967.
( FOR ENGLISH TRANSLATION SEE TPRC NO. 76988 )

76988 **SOME PHYSICAL PROPERTIES OF ACETIC ANHYDRIDE PLUS WATER SOLUTIONS.**
DUTCHAK, YA. I.
ISRAEL PROGRAM FOR SCIENTIFIC TRANSLATIONS, LTD.
3PP., 1967.
( ENGLISH TRANSLATION OF UKR. FIZ. ZH., 12 ( 3 ), 454–6, 1967; FOR ORIGINAL SEE TPRC NO. 76987 )
( TT–67–51421/3, N69–26768 )

76997 **RADIATION BASE HEATING FROM SOLID PROPELLANT LAUNCH VEHICLE EXHAUST PLUMES.**
PEARCE, B. E.
JOINT PROPULSION SPECIALIST CONF., AIAA, 8TH
12PP., 1972.
( AIAA–PAPER–72–1168 )

77018 **THE ANTIMONY RUBIDIUM SULFIDE - ANTIMONY TRISULFIDE SYSTEM.**
BAZAKUTSA, V. A. GNIDASH, N. I.
LAZAREV, V. B. SALOV, A. V. ROGACHEVA, E. I.
SUKHORUKOVA, L. N. VASIL'EVA, M. P.
BERUL', S. I.
RUSS. J. INORG. CHEM.
19 ( 3 ), 438–41, 1974.
( ENGLISH TRANSLATION OF ZH. NEORG. KHIM., 19 ( 3 ), 807–11, 1974; FOR ORIGINAL SEE TPRC NO. 74924 )

77027 **DIFFUSE REFLECTION SPECTRA AS A METHOD OF INVESTIGATION AND CONTROL IN THE PRODUCTION OF ZINC - CADMIUM SULFIDE PHOSPHORS.**
KRONGAUZ, V. G. PIVNEVA, S. P.
PAN'KOVA, L. T.
J. APPL. SPECTROSC., USSR
17 ( 1 ), 944–6, 1972.
( ENGLISH TRANSLATION OF ZH. PRIKL. SPEKTROSK., 17 ( 1 ), 146–9, 1972; FOR ORIGINAL SEE TPRC NO. 68627 )

77039 **RADIOMETRIC OCEAN COLOR SURVEYS THROUGH A SCATTERING ATMOSPHERE.**
CURRAN, R. J. HOVIS, W. A.
GODDARD SPACE FLIGHT CENTER, GREENBELT, MD.
9–1–9–12, 1972.
( N72–29310, NASA–TM–X–68564 )

77118 **THEORETICAL DESCRIPTION OF AC ARCS IN MERCURY AND ARGON.**
LOWKE, J. J. ZOLLWEG, R. J.
LIEBERMANN, R. W.
J. APPL. PHYS.
46 ( 2 ), 650–60, 1975.

77220 **THERMAL CONDUCTIVITY OF MOLTEN MIXTURES OF URANIUM TETRACHLORIDE WITH SODIUM AND POTASSIUM CHLORIDES.**
BYSTRAI, G. P. DESYATNIK, V. N.
ZLOKAZOV, V. A.
SOV. J. AT. ENERGY
36 ( 6 ), 654–5, 1974.
( ENGLISH TRANSLATION OF AT. ENERG., 36 ( 6 ), 517–8, 1974; FOR ORIGINAL SEE TPRC NO. 76115 )

77228 **HEAT AND MASS TRANSPORT IN GASES AND THEIR MIXTURES.**
SHASHKOV, A. G. ABRAMENKO, T. N.
IVASHKEVICH, E. V.
INZH. FIZ. ZH.
20 ( 1 ), 166–77, 1971.
( FOR ENGLISH TRANSLATION SEE TPRC NO. 77229 )

| TPRC Number | Bibliographic Citation |
|---|---|
| 77229 | **HEAT AND MASS TRANSPORT IN GASES AND THEIR MIXTURES.**<br>SHASHKOV, A. G. ABRAMENKO, T. N.<br>IVASHKEVICH, E. V.<br>J. ENG. PHYS., USSR<br>20 ( 1 ), 125–34, 1971.<br>( ENGLISH TRANSLATION OF INZH. FIZ. ZH., 20 ( 1 ),<br>166–77, 1971; FOR ORIGINAL SEE TPRC NO. 77228 ) |
| 77249 | **REFRACTORY MATERIALS SUITABLE FOR USE IN GUIDED MISSILE PROPULSION SYSTEMS.**<br>MICCIOLI, B. R. JULIEN, H. P.<br>RESEARCH AND DEVELOPMENT DIVISION, THE CARBORUNDUM COMPANY<br>19PP., 1965.<br>( AD–358 137 ) |
| 77262 | **REFRACTORY MATERIALS SUITABLE FOR USE IN GUIDED MISSILE PROPULSION SYSTEMS.**<br>MICCIOLI, B. R. NIESSE, J. E.<br>PRENTICE, J. A. VERDE, E. T.<br>RESEARCH AND DEVELOPMENT DIVISION, THE CARBORUNDUM COMPANY<br>27PP., 1963.<br>( AD–342 005 ) |
| 77281 | **REFRACTORY MATERIALS SUITABLE FOR USE IN GUIDED MISSILE PROPULSION SYSTEMS.**<br>MICCIOLI, B. PRENTICE, J. A.<br>RESEARCH AND DEVELOPMENT DIVISION, THE CARBORUNDUM COMPANY<br>23PP., 1962.<br>( AD–329 631 ) |
| 77305 | **INVESTIGATION OF MAGNETOHYDRODYNAMIC POWER GENERATION. VOLUME IV. MATERIALS.**<br>MARKOWSKI, S. J. SEAWARD, E. T.<br>EMANUELSON, R. C. HARRISON, W. J. BACON, J.<br>PRATT AND WHITNEY AIRCRAFT, DIV. OF UNITED AIRCRAFT CORP., EAST HARTFORD, CONNECTICUT<br>115PP., 1962.<br>( AD–411 915, RADC–TR–62–464–VOL IV, PWA–2125 ) |
| 77354 | **THERMAL EXPANSION OF BERYLLIUM OXIDE - SILICON CARBIDE COMPOSITES.**<br>GARDNER, W. J.<br>BALLASTIC SYSTEMS AND SPACE SYSTEMS DIVISION,<br>LOS ANGELES AIR FORCE STATION, LOS ANGELES<br>14PP., 1964.<br>( AD–608 653, SSD–TDR–64–187 ) |
| 77377 | **FABRICATION AND PROPERTIES OF SILICON COMPOUNDS. TASK I. FABRICATION, MICROSTRUCTURE AND SELECTED PROPERTIES OF SILICON NITRIDE-ALUMINUM OXIDE COMPOSITIONS. TASE II. OXIDATION OF SILICON CARBIDE HOT PRESSED WITH ALUMINUM OXIDE.**<br>LANGE, F. F. SINGHAL, S. C.<br>WESTINGHOUSE RESEARCH LABS., PITTSBURGH, PA.<br>82PP., 1974.<br>( AD–786 750 ) |
| 77385 | **THE VISCOSITY OF MATERIALS INVOLVED IN THE PRODUCTION OF THE PLACTICISER DI( 2-ETHYLHEXYL ) PHTHALATE.**<br>KRYLOV, V. S. SHITIKOVA, V. A.<br>BRONOV, M. V. SMIRNOV, YU. A.<br>MAKOVSKAYA, E. B.<br>SOV. PLAST.<br>( 12 ), 23, 1973.<br>( ENGLISH TRANSLATION OF PLAST. MASSY, 15 ( 12 ),<br>25–6, 1973; FOR ORIGINAL SEE TPRC NO. 74471 ) |
| 77413 | **DIFFUSIVITIES AND VISCOSITIES OF SOME BINARY LIQUID NONELECTROLYTES AT 25 DEGREES.**<br>GHAI, R. K. DULLIEN, F. A. L.<br>J. PHYS. CHEM.<br>78 ( 22 ), 2283–91, 1974. |
| 77458 | **CORRELATION FUNCTION METHOD FOR THE EVALUATION OF PLASMA TRANSPORT COEFFICIENTS. SHEAR VISCOSITY COEFFICIENT.**<br>BARTIS, J. T. OPPENHEIM, I.<br>PHYS. REV.<br>10 A ( 4 ), 1263–71, 1974. |
| 77500 | **CORRELATION OF OPTICAL PARAMETERS WITH THE ELECTRONIC BAND STRUCTURE OF TITANIUM CARBONITRIDE SOLID SOLUTIONS.**<br>BOEHM, G. GORETZKI, H.<br>J. LESS COMMON METALS<br>27 ( 3 ), 311–23, 1972. |
| 77523 | **ANISOTROPIC SOLAR REFLECTANCE OVER WHITE SAND, SNOW AND STRATUS CLOUDS.**<br>SALOMONSON, V. V. MARLATT, W. E.<br>J. APPL. METEOROL.<br>7, 475–83, 1968. |
| 77561 | **ADVANCED CERAMIC SYSTEMS FOR ROCKET NOZZLE APPLICATIONS.**<br>JUN, C. K. SHAFFER, P. T. B.<br>THE CARBORUNDUM COMP., NIAGARA FALLS, NEW YORK<br>21PP., 1970.<br>( AD–874 298 ) |
| 77563 | **ADVANCED CERAMIC SYSTEMS FOR ROCKET NOZZLE APPLICATIONS.**<br>JUN, C. K. SHAFFER, P. T. B.<br>THE CARBORUNDUM COMP., NIAGARA FALLS, NEW YORK<br>64PP., 1970.<br>( AD–877 211 ) |
| 77573 | **RESEARCH AND DEVELOPMENT OF REFRACTORY OXIDATION RESISTANT DIBORIDES: PART II, VOLUME V: THERMAL, PHYSICAL, ELECTRICAL AND OPTICAL PROPERTIES.**<br>CLOUGHERTY, E. V. WILKES, K. E. TYE, R. P.<br>MAN LABS, INC., CAMBRIDGE, MASS.<br>90PP., 1969.<br>( AD–865 321, AFML–TR–6,–190–PT. 2–VOL. 5 ) |
| 77706 | **INFRARED EMISSIVITY, SHORT-WAVE ALBEDO, AND THE MICROPHYSICS OF STRATIFORM WATER CLOUDS.**<br>PALTRIDGE, G. W.<br>J. GEOPHYS. RES. ( USA )<br>79 ( 27 ), 4053–8, 1974. |
| 77761 | **CRITICAL VISCOSITY AND DIFFUSION IN THE BINARY-LIQUID PHASE TRANSITION.**<br>FERRELL, R. A. PERL, R.<br>DEPT. OF PHYSICS AND ASTRONOMY, MARYLAND UNIV., COLLEGE PARK, MD.<br>13PP., 1972.<br>( AD–747 733, TR–72–103 ) |
| 77762 | **DECOUPLED MODE THEORY OF CRITICAL VISCOSITY AND DIFFUSION IN THE PHASE TRANSITION.**<br>FERRELL, R. A. PERL, R.<br>DEPT. OF PHYSICS AND ASTRONOMY, MARYLAND UNIV., COLLEGE PARK, MD.<br>40PP., 1972.<br>( AD–747 735, TR–72–106 ) |
| 77784 | **OCEAN COLOR ANALYSIS.**<br>DUNTLEY, S. Q. AUSTIN, R. W. WILSON, W. H.<br>EDGERTON, C. F. MORAN, S. E.<br>SCRIPPS INSTITUTION OF OCEANOGRAPHY, SAN DIEGO, CALIF.<br>72PP., 1974.<br>( AD–783 101, SIO–REF–74–10 ) |
| 77928 | **CODEPOSITED PYROLITIC GRAPHITE/SILICON CARBIDE NOZZLE LINERS FOR ADVANCED ICBM SYSTEMS. VOL. II. COATING CHARACTERIZATION.**<br>HUGHES, M. C. SINGLETON, R. H.<br>AIR FORCE ROCKET PROPULSION LAB., AIR FORCE SYSTEMS COMMAND, EDWARDS, CALIF.<br>476PP., 1974.<br>( AD–922 533L, AFRPL–TR–74–15–VOL–2 ) |
| 77930 | **HIGH THERMAL CONDUCTIVITY MICROWAVE SUBSTRATES.**<br>POPE, B. J. HORTON, M. D. BOWMAN, L. S.<br>HALL, H. T. ADANIYA, H.<br>U. S. ARMY ELECTRONICS COMMAND, FORT MONMOUTH, N. J.<br>29PP., 1973.<br>( AD–918 531L, ECOM–0289–1–73 ) |
| 77936 | **RESEARCH IN OPTICAL MATERIALS AND STRUCTURES FOR HIGH POWER LASERS.**<br>HORRIGAN, F. A. DEUTSCH, T. F.<br>RAYTHEON RESEARCH DIV., WALTHAM, MASS.<br>73PP., 1970.<br>( AD–878 519L ) |
| 77946 | **A STUDY ON THE SINTERED SILICON NITRIDE - MAGNESIUM OXIDE SYSTEM.**<br>OYAMA, Y. KAMIGAITO, O.<br>YOGYO KYOKAI SHI<br>81 ( 7 ), 34–7, 1973.<br>( FOR ENGLISH TRANSLATION SEE TPRC NO. 77947 ) |
| 77947 | **A STUDY ON THE SINTERED SILICON NITRIDE - MAGNESIUM OXIDE SYSTEM.**<br>OYAMA, Y. KAMIGAITO, O.<br>FOREIGN TECHNOLOGY DIV., WRIGHT PATTERSON AFB, OHIO<br>15PP., 1975.<br>( ENGLISH TRANSLATION OF YOGYO KYOKAI SHI, 81 ( 7 ),<br>34–7, 1973; FOR ORIGINAL SEE TPRC NO. 77946 )<br>( AD–922 220L, FTD–HC–23–2219–74 ) |
| 78042 | **ADVANCED PROPELLANTS INVESTIGATION FOR PREPACKAGED LIQUID ENGINES.**<br>TANNENBAUM, S.<br>REACTION MOTORS DIVISION, THIOKOL CHEMICAL CORP.<br>48PP., 1964.<br>( AD–357 057L, RMD–5046–Q2 ) |
| 78046 | **ANALYTICAL STUDIES OF AIRCRAFT AND MISSILES POWERED BY AIR-BREATHING PROPULSION SYSTEMS: VOLUME II-THEORETICAL COMBUSTION CHARACTERISTICS.**<br>ROBACK, R. COON, W.<br>UNITED AIRCRAFT CORP., RESEARCH LAB., EAST HARTFORD, CONN.<br>84PP., 1965.<br>( AD–360 101L ) |

| TPRC Number | Bibliographic Citation |
|---|---|
| 78292 | **HALIDE MATERIALS PROCESSING FOR HIGH-POWER, INFRARED LASER WINDOWS.**<br>HARRISON, W. B. BEAUDOIN, T. L.<br>HENDRICKSON, G. O. KOEHLER, L. E.<br>STARLING, J. E.<br>AIR FORCE CAMBRIDGE RESEARCH LABS., L. G. HANSCOM FIELD, BEDFROD, MASS.<br>391–411, 1972.<br>( AD–914 501L, AFCRL–SR–162–VOL–2 ) |
| 78311 | **THE SENSITIVITY, PERFORMANCE, AND MATERIAL PROPERTIES OF SOME HIGH EXPLOSIVE FORMULATIONS.**<br>COSTAIN, T. S. MOTTO, R. V.<br>PICATINNY ARSENAL, FELTMAN RESEARCH LAB., DOVER, N. J.<br>73PP., 1973.<br>( AD–915 795L, PA–TR–4587 ) |
| 78345 | **ADVANCED CERAMIC SYSTEMS FOR ROCKET NOZZLE APPLICATIONS.**<br>SPECK, D. A. MICCIOLI, B. R.<br>THE CARBORUNDUM COMPANY<br>42PP., 1968.<br>( AD–835 351L ) |
| 78380 | **MOLAR HEAT CAPACITIES OF BINARY SYSTEMS OF CYCLOHEXANE, CARBON TETRACHLORIDE, SILICON TETRACHLORIDE AND TIN TETRACHLORIDE.**<br>WILHELM, E. SACKMANN, H. ZETTLER, M.<br>BER BUNSENGES. PHYS. CHEM. ( GERMANY )<br>78 ( 8 ), 795–804, 1974. |
| 78428 | **SYNTHESES AND USES OF AQUOAMMONOSULFURIC COMPOUNDS. IV. RECOVERY OF AMMONIA GAS BY SULFONYL DIAMIDE.**<br>TSURUGA, N. UCHIDA, S.<br>NIPPON KAGAKU KAISHI<br>( 10 ), 1888–92, 1974. |
| 78662 | **THERMODYNAMIC AND THERMOPHYSICAL PROPERTIES OF COMBUSTION PRODUCTS. VOLUME 8 – FUELS BASED ON FLUORIDE OXIDIZERS.**<br>ALEMASOV, V. E. DREGALIN, A. F.<br>TISHIN, A. P. KHUDYAKOV, V. A.<br>KOSTIN, V. N.<br>IZDATEL'STVO VINITI<br>716PP., 1974. |
| 78687 | **EMISSIVITIES AND FORWARD SCATTERING OF NATURAL AND MAN–MADE MATERIAL AT THREE MILLIMETER WAVELENGTH.**<br>SCHANDA, E. HOFER, R.<br>INTERN. SYMP. REMOTE SENSING ENVIRONMENT, 9TH, ANN ARBOR, MICH.<br>1585–92, 1974. |
| 78814 | **EMISSIVITY OF CLOUDS IN THE 8–12 MICRON BAND.**<br>BELICH, R. B. KALACHINSKII, S. F.<br>ADVANCES IN SATELLITE METEROLOGY<br>95–101, 1974. |
| 78825 | **REFLECTANCE OF A SCATTERING MEDIUM CONTAINING POLYDISPERSE NONSPHERICAL PARTICLES.**<br>HOWARD, F. H. NOVOTNY, J. L.<br>HEAT TRANSFER, PROC. INT. HEAT TRANSFER CONF., 5TH<br>1, 26–30, 1974. |
| 78855 | **RADIATIVE TRANSFER IN HOMOGENEOUS NONGRAY GASES WITH NONISOTROPIC PARTICLE SCATTERING.**<br>DOMOTO, G. A. WANG, W. C.<br>TRANS. ASME<br>96 C ( 3 ), 385–90, 1974. |
| 79055 | **GAS PHASE CHEMICAL EQUILIBRIUM IN DINITROGEN TRIOXIDE AND DINITROGEN TETROXIDE.**<br>CHAO, J. WILHOIT, R. C. ZWLINSKI, B. J.<br>THERMOCHIM. ACTA<br>10 ( 4 ), 359–71, 1974. |
| 79079 | **REALIZATION OF A ROTATING DISK ELECTRODE USABLE AT HIGH TEMPERATURES. APPLICATION TO THE DETERMINATION OF THE DIFFUSION COEFFICIENT OF THE NICKEL( II ) SPECIES IN A MOLTEN LIGHIUM CHLORIDE–POTASSIUM CHLORIDE EUTECTIC AT 450 DEGREES.**<br>BOUTEILLON, J. BARBIER, M. J.<br>J. ELECTROANAL. CHEM. INTEFACIAL ELECTROCHEM.<br>56 ( 3 ), 399–407, 1974. |
| 83558 | **RADIANT HEAT TRANSFER FROM LUMINOUS FLAME.**<br>SATO, T. MATSUMOTO, R.<br>INTERN. HEAT TRANSFER CONF.<br>804–11, 1961. |
| 83560 | **STUDY ON THE EMISSIVITY OF A GAS WHICH CONTAINS PARTICLES.**<br>YOKOBORI, S.<br>INTERN. HEAT TRANSFER CONF.<br>843–55, 1961. |
| 88635 | **RAPID DETERMINATION OF THE THERMOPHYSICAL CHARACTERISTICS OF LIQUIDS IN THE 30–95 RANGE.**<br>TERESHIN, I. V. KOZLOV, V. A.<br>FILIPPOV, S. N.<br>ZAVOD. LAB.<br>43 ( 2 ), 187–9, 1977.<br>( FOR ENGLISH TRANSLATION SEE T93697 ) |
| 89755 | **TRANSPORT COEFFICIENTS OF HIGH TEMPERATURE NITROGEN – HYDROGEN MIXTURES.**<br>CAPITELLI, M. GORSE, C. FAUCHAIS, P.<br>J. PHYS. ( PARIS )<br>38 ( 6 ), 653–7, 1977. |
| 91157 | **TRANSPORT COEFFICIENTS OF ARGON – HYDROGEN HIGH TEMPERATURE MIXTURES.**<br>CAPITELLI, M. GORSE, C. FAUCHAIS, P.<br>J. CHIM. PHYS.<br>73 ( 7–8 ), 755–9, 1976. |
| 91174 | **LOW TEMPERATURE PROPERTIES AND USES OF SALT AND BRINE.**<br>KAUFMANN, D. W.<br>SODIUM CHLORIDE<br>( CHAPT. 23 ), 547–68, 1960. |
| 93596 | **SOLAR ABSORPTION PROPERTIES OF A HIGH TEMPERATURE DIRECT–ABSORBING HEAT TRANSFER FLUID.**<br>DROTNING, W. D.<br>SYMP. THERMOPHYSICAL PROPERTIES, 7TH.<br>304–11PP., 1977.<br>( N78–15577, SAND–76–9104, CONF–770537–7 ) |
| 93601 | **DETERMINATION OF THE ENTHALPY AND SPECIFIC HEAT OF A ( UR1NIUM, ZIRCONIUM ) CARBIDE – GRAPHITE REACTOR FUEL M1TERIAL TO 2000 K.**<br>STEPHENS, H. P.<br>SYMP. THERMOPHYSICAL PROPERTIES, 7TH.<br>351–7, 1977. |
| 93604 | **THERMODYNAMIC STATE, SPECIFIC HEAT, AND ENTHALPY FUNCTION OF SATURATED VAPOR OVER URANIUM DIOXIDE BETWEEN 3000 K 1ND 5000 K.**<br>KAROW, H. U.<br>SYMP. THERMOPHYSICAL PROPERTIES, 7TH.<br>373–8PP., 1977. |
| 93611 | **EQUILIBRIUM AND TRANSPORT ESTIMATES FOR HYDROGEN AND HELIUM VIA LENNARD–JONES FLUID.**<br>ASHURST, W. T.<br>SYMP. THERMOPHYSICAL PROPERTIES, 7TH.<br>431–41, 1977. |
| 93615 | **AN ADIABATIC NON–FLOW CALORIMETER APPARATUS TO MEASURE THE THERMODYNAMIC PROPERTIES OT THE LIQUID AND GASEOUS PHASES OF MIXTURE SYSTEMS.**<br>NOWAK, E. S. CHAN, J. S.<br>SYMP. THERMOPHYSICAL PROPERTIES, 7TH.<br>531–7, 1977. |
| 93616 | **A MIXTURE SATURATION THEORY AND EXPERIMENTAL MEASUREMENTS OF THE THERMAL PROPERTIES OF A TWO COMPONENT–TWO PHASE SYSTEM.**<br>NOWAK, E. S. CHAN, J. S.<br>SYMP. THERMOPHYSICAL PROPERTIES, 7TH.<br>538–45, 1977. |
| 93622 | **TRANSPORT PROPERTIES OF MIXTURES.**<br>KESTIN, J.<br>SYMP. THERMOPHYSICAL PROPERTIES, 7TH.<br>628–37, 1977. |
| 93626 | **PREDICTION OF THE THERMAL CONDUCTIVITY OF FLUID MIXTURES.**<br>HANLEY, H. J. M.<br>SYMP. THERMOPHYSICAL PROPERTIES, 7TH<br>668–73, 1977. |
| 93630 | **VISCOSITY OF R12, R13, R13BL, R22, AND R500 IN THE VAPOUR PHASE.**<br>LATTO, B.<br>SYMP. THERMOPHYSICAL PROPERTIES, 7TH<br>705–11, 1977. |
| 93634 | **CORRELATIONS FOR THE PRANDTL NUMBER OF SATURATED LIQUIDS, SATURATED VAPORS AND DILUTE GASES.**<br>JOHNSON, M. W. LILEY, P. E.<br>SYMP. THERMOPHYSICAL PROPERTIES, 7TH<br>739–43, 1977. |
| 93697 | **QUICK METHODS OF DETERMINING THERMOPHYSICAL PROPERTIES OF LIQUIDS OVER THE 30–95 C TEMPERATURE RANGE.**<br>TERESHKIN, I. V. KOZLOV, V. A.<br>FILIPPOV, S. N.<br>IND. LAB., USSR<br>43 ( 2 ), 228–30, 1977.<br>( ENGLISH TRANSLATION OF ZAVOD. LAB., 43 ( 2 ), 187–9, 1977; FOR ORIGINAL SEE T88635 ) |
| 94260 | **PHYSICAL PROPERTIES OF SODIUM CHLORIDE IN CRYSTAL, LIQUID, GAS, AND AQUEOUS SOLUTION STATES.**<br>KAUFMAN, D. W.<br>SODIUM CHLORIDE<br>( CHAPT. 25 ), 587–625, 1960. |

# PERSONAL AUTHOR INDEX

# Part D
# AUTHOR INDEX

PART D

# USE OF AUTHOR INDEX

The *Author Index* is given in two parts, namely, personal authors and coauthors followed by corporate authors. The numbers shown along with the name of each author refer to the TPRC serial numbers of the references cited in the *Bibliography* (Part C).

For simplification only the author's last name and his initials are used. Naturally, in the case of some of the more popular last names having the same initials, this practice is likely to lead to the unfortunate result of improperly crediting papers to similarly named authors. For this the Editors express their sincere apology. In such instances, the user of this *Author Index* should inspect the titles of the documents cited in the *Bibliography* to identify those technical papers which may have been improperly credited.

The term "corporate author" applies to industrial organizations, government agencies, colleges, universities, etc. In alphabetizing corporate names, the main listing is made under the key word of the parent company or organization's name, and names of divisions, laboratories, departments, etc., are cross-referenced under the key words of their names.

In the case of universities, the word "university" always appears second in abbreviated form. In identification of corporate names, information is listed in the following sequence: name of organization, city, state, division, laboratory, etc.

# PERSONAL AUTHOR INDEX

# CORPORATE AUTHOR INDEX

## CONDENSED MATERIALS INDEX TO SIX-VOLUME RETRIEVAL GUIDE SUPPLEMENT II*



Volume 1: Elements and Inorganic Compounds
Volume 2: Organic Compounds and Polymeric Materials
Volume 3: Alloys, Intermetallic Compounds, and Cermets
Volume 4: Oxide Mixtures and Minerals
Volume 5: Mixtures and Solutions
Volume 6: Coatings, Systems, and Composites

Earlier Publications of the Retrieval Guide:

Basic Edition (Plenum, 1967)
Supplement I, 1964–1970 (IFI/Plenum, 1973)

*Material classes such as woods, pharmaceuticals, foods, natural products, etc., or references concerning major compendia are not reported in this six-volume Supplement II. Special bibliographic searches on these may be requested from CINDAS directly.